The first edition of this textbook was widely praised. In this new edition the authors have taken the opportunity to bring the work completely up to date by the addition of new material on mesoporous materials, fullerenes, molecular magnets, organic conductors and high-temperature superconductors. All of the chapters have been revised with new additional sections.

In recent years, solid state chemistry has emerged as a very important element of mainstream chemistry and modern materials science. Students, teachers and researchers of condensed matter science and materials science need to understand the chemistry of solids because this plays a crucial role in determining the properties of materials. An understanding of solid state chemistry is also essential in materials design. For example, innovative ways of preparing unusual solids chemically have been discovered, often giving rise to metastable materials which could not otherwise be made. Many fascinating relationships between the structure and properties of solids that help in the design of materials have been discovered by chemists. The authors survey the highlights of this vital area of chemistry by citing the most recent examples, case studies and references. Not only does the book bring out the essence of the subject, it will also help the reader understand the subject in all its breadth and depth.

This text requires only an understanding of basic physics, chemistry and crystallography. It will be of value to advanced students and researchers studying solid state chemistry as a text and reference work.

New Directions in Solid State Chemistry
Second Edition

New Directions in Solid State Chemistry

Second Edition

C. N. R. Rao, F.R.S.,

and

J. Gopalakrishnan

Solid State and Structural Chemistry Unit, Indian Institute of Science
Bangalore-560012, India

CAMBRIDGE
UNIVERSITY PRESS

PUBLISHED BY THE PRESS SYNDICATE OF THE UNIVERSITY OF CAMBRIDGE
The Pitt Building, Trumpington Street, Cambridge CB2 1RP, United Kingdom

CAMBRIDGE UNIVERSITY PRESS
The Edingburgh Building, Cambridge CB2 2RU, United Kingdom
40 West 20th Street, New York, NY10011-4211, USA
10 Stamford Road, Oakleigh, Melbourne 3166, Australia

First published 1986
First paperback edition 1989
Second edition 1997

Printed in Great Britain at the University Press, Cambridge

Typeset in Times 10/13 pt

*A catalogue record for this book is available from
the British Library*

Library of Congress cataloguing in publication data available

ISBN 0 521 49559 8 hardback
ISBN 0 521 49907 0 paperback

Contents

Preface to the second edition

The first edition of this book published in 1986 was well received by the chemistry and materials science communities and this resulted in the paperback edition published in 1989. We are most gratified by this warm reception to the book which has been found useful by students and teachers as well as practising solid state chemists and materials scientists. Since we first wrote the book, there have been many new developments in the various aspects of solid state chemistry covering synthesis, structure elucidation, properties, phenomena and reactivity. The discovery of high-temperature superconductivity in the cuprates created a great sensation and gave a boost to the study of solid state chemistry. Many new types of materials such as the fullerenes and carbon nanotubes have been discovered. We have now revised the book taking into account the new developments so that it reflects the present status of the subject adequately and points to new directions.

In this edition, we have incorporated new material in all the chapters and updated references to the literature. New sections dealing with porous solids, fullerenes and related materials, metal nitrides, metal tellurides, molecular magnets and other organic materials have been added. Under preparative strategies, we have included new types of synthesis reported in the literature, specially those based on soft chemistry routes. We have a new section covering typical results from empirical theory and electron spectroscopy. There is a major section dealing with high-temperature oxide superconductors. We hope that this edition of the book will prove to be a useful text and reference work for all those interested in solid state chemistry and materials science.

C. N. R. Rao
J. Gopalakrishnan
Bangalore
1997

Preface to the first edition

Although solid state science is an area of intense research activity pursued by physicists and materials scientists, the contributions of chemists to this area have a distinct identity. The great skill of chemists in developing novel methods for the synthesis of complex materials, and their understanding of the intricacies of structure and bonding, make their contributions to solid state science unique. At the present time, solid state chemistry is mainly concerned with the development of new methods of synthesis, new ways of identifying and characterizing materials and of describing their structure and above all, with new strategies for tailor-making materials with desired and controllable properties be they electronic, magnetic, dielectric, optical, adsorptive or catalytic. It is heartening that solid state chemistry is increasingly coming to be recognized as an emerging area of chemical science.

In this monograph, we have attempted to present the highlights of modern solid state chemistry and indicate the new directions in a concise manner. In doing so, we have not described the varied principles, properties and techniques that embody this subject at length, but have concerned ourselves with the more important task of bringing out the flavour of the subject to show how it works. We believe that the material covered is up to date, taking the reader to the very frontiers of the subject. We have been careful to include some introductory material for each aspect in order to enable students and beginners to benefit from the book. Instead of dividing the book into the traditional chapters (dealing with crystal chemistry, properties of solids, reactivity and so on) we have tried to present the subject in a style that would reflect the way the subject is growing today. Because of this approach, the lengths of the different chapters have inevitably become somewhat variable.

We hope that the book will be found useful by practitioners of solid state science, especially chemists interested in the study of condensed matter. While the book can certainly be used as a supplementary text in a broad course on solid state science, it could form the basis of a well-planned course in solid state chemistry. We shall be more than rewarded if the book is found useful by students, teachers and practitioners of solid state chemistry.

We have cited important material from the very recent literature including some of the latest references, but in dealing with some of the new concepts we had to be all too brief in order to limit the size of the book. It is possible that we have not included some references by error of judgement or oversight for which we would like to be excused.

Much of the book was planned and written when one of us (CNRR) was Jawaharlal

Nehru Visiting Professor at the University of Cambridge. His thanks are due to Professor J. M. Thomas, FRS, and other colleagues of the Department of Physical Chemistry and to the members of King's College, Cambridge, for their kindness and hospitality.

The authors thank Professor Robert Cahn for his valuable suggestions and encouragement. Their thanks are also due to the University Grants Commission, New Delhi, for support of this work.

C. N. R. Rao
J. Gopalakrishnan
Bangalore
June 1985

1   Structure of solids: old and new facets

1.1   Introduction

Solid state chemistry deals with a variety of solids, inorganic as well as organic; the solids can be crystalline or noncrystalline. A sound knowledge of the structure of solids as well as of the nature of bonding is essential for an appreciation of solid state chemistry since properties of solids are, by and large, determined by the structure. Crystal chemistry of inorganic solids has been reviewed widely in the literature (see, for example, Adams, 1974; Rao, 1974; Wells, 1984), but there has been effort of late to explore new ways of looking at inorganic structures and to understand their stabilities. In this chapter, we shall briefly review the highlights of inorganic crystal chemistry after summarizing some of the basic information related to crystals and the different types of bonding found in them. We shall also discuss polytypism, organic crystal structures and related topics before finally presenting the models employed to understand the structures of noncrystalline or amorphous solids.

1.2   Description of crystals

A regular crystal consists of an infinite array of constituent units in three dimensions. Since the nature of the constituent unit does not affect translational periodicity, the periodicity is generally represented by replacing the repeating unit by a point, the resulting array of such points in space being called a lattice. In a *space lattice*, the translation vectors a, b and c in the three crystallographic directions define a parallelepiped called a *primitive cell*. A primitive cell or a suitable combination thereof, chosen as the repeating unit of the lattice, is called the *unit cell*. In order to define a crystallographic unit cell in three dimensions, we require the three translation vectors and the interaxial angles α, β and γ. We can define seven *crystal systems* based on the six lattice parameters (Table 1.1). There are fourteen independent ways of arranging points in space (in three dimensions) giving rise to the 14 *Bravais lattices* (Fig. 1.1) listed in Table 1.1. In two dimensions, five plane lattices are possible, whereas in one dimension there is only one way of arranging points, in the line lattice.

Based on extensive studies of the symmetry in crystals, it is found that crystals possess one or more of the ten basic symmetry elements (five proper rotation axes 1, 2, 3, 4, 6 and five inversion or improper axes, $\bar{1}$ = centre of inversion i, $\bar{2}$ = mirror plane m, $\bar{3}$, $\bar{4}$ and $\bar{6}$). A set of symmetry elements intersecting at a common point within a crystal is called the *point group*. The 10 basic symmetry elements along with their 22 possible combinations constitute the 32 *crystal classes*. There are two additional symmetry

Table 1.1. *Crystal systems and Bravais lattices*

System	Unit cell specification	Essential symmetry[a]	Bravais lattice[b]
Cubic	$a = b = c$ $\alpha = \beta = \gamma = 90°$	Four 3s	P, I, F
Tetragonal	$a = b \neq c$ $\alpha = \beta = \gamma = 90°$	One 4 or $\bar{4}$	P, I
Orthorhombic	$a \neq b \neq c$ $\alpha = \beta = \gamma = 90°$	Three 2s mutually perpendicular or one 2 intersecting with two *ms*	P, I, C, F
Rhombohedral	$a = b = c$ $\alpha = \beta = \gamma \neq 90°$	One 3	R (P)
Hexagonal	$a = b \neq c$ $\alpha = \beta = 90°$ $\gamma = 120°$	One 6	P
Monoclinic	$a \neq b \neq c$ $\alpha = \gamma = 90° \neq \beta$	One 2 or one *m*	P C
Triclinic	$a \neq b \neq c$ $\alpha \neq \beta \neq \gamma$	None	P

[a] 3, 4 etc. are the rotation axes; $\bar{3}$, $\bar{4}$ etc. are the inversion axes; *m* is mirror plane.
[b] P = primitive lattice containing lattice points at the corners of the unit cell; F = face-centred lattice; I = body-centred lattice.

elements, *screw axis* and *glide plane*, which arise in crystals but have no counterpart in molecular symmetry. A combination of these elements involving translation with the point group symmetry is called a *space group*. For a triclinic system, for example, there are only two possible space groups P1 and P$\bar{1}$ while for a monoclinic system there are 13 possible space groups arising from the two possible Bravais lattices P and C and three point groups 2, *m* and 2/*m*. In all, there are 230 possible space groups.

In discussing the symmetry of point groups and space groups, we assigned to every point in the lattice a set of spatial coordinates (x, y, z) and studied the effect of various symmetry operations on the points. Shubnikov suggested that, in addition to the three spatial coordinates, a fourth coordinate, *s*, may be added to each point. This coordinate *s* is allowed to take one of two possible values at a time; *s* may refer to the spin of a particle, in which case the two allowed values are spin-up and spin-down. In abstract terms, they may correspond to two colours, black and white. A new symmetry operation, called operation of antisymmetry *R*, is introduced to reverse the value of *s* from black to white or from spin-up to spin-down. If this operation is considered in conjunction with all the other symmetry operations of point groups and space groups, the total number of possible crystallographic point groups and space groups swells to

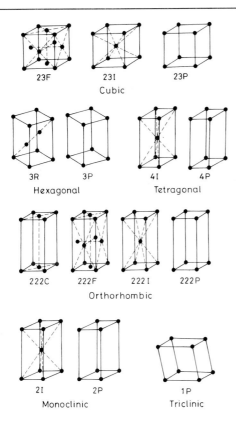

Figure 1.1   The fourteen Bravais lattices. The numbers 2, 3 etc. designate the symmetry axes and the letters P, F, I and C designate the lattice type.

122 and 1651 respectively. These enlarged numbers of groups are often referred to as *magnetic groups* or *coloured groups* and are of use in describing the symmetry of magnetically ordered crystalline solids (Cracknell, 1975).

1.3   Bonding in crystals

A classification of crystals based on bonding is useful in understanding structure–property relations in solids. Five types of solids are readily defined on bonding considerations: ionic, covalent, metallic and molecular (van der Waals) and hydrogen-bonded. In Table 1.2, the important characteristics of the five types of solids are presented. In real situations, however, solids may exhibit features of more than one type of bonding.

1.3.1   *Ionic crystals*

Ionic crystals are formed between highly electropositive and highly electronegative elements when electron transfer from the former to the latter occurs, resulting in oppositely charged ions with closed shell electronic configuration. X-ray diffraction studies of ionic crystals reveal an essentially spherical charge distribution around the

Table 1.2. *Types of solids*

Type	Units present	Characteristics	Examples	Approximate cohesive energy, kJ mol^{-1}
Ionic	Positive and negative ions	Brittle, insulating and fairly high melting	NaCl LiF	795 1010
Covalent	Atoms (bonded to one another)	Hard, high melting and nonconducting (when pure)	Diamond SiC	715 1010
Metallic	Positive ions embedded in a collection of electron 'gas'	High conductivity	Na Fe	110 395
van der Waals (Molecular)	Molecules or atoms	Soft, low melting, volatile and insulating	Argon CH_4	7.6 10
Hydrogen-bonded	Molecules held together by hydrogen bonds	Low melting insulators	H_2O (ice) HF	50 30

ions. Alkali and alkaline earth metal halides are typical examples of ionic solids. The spherical charge distribution around the ions in these solids results in symmetric structures, with the ions of one charge surrounding those of the opposite charge and the ions of the same charge remaining as far apart as possible.

The cohesive energy of ionic crystals is mainly due to electrostatic interaction and can be calculated on the basis of a point-charge model. Following Born, the cohesive energy (U) of a crystal containing oppositely charged ions with charges Z_1 and Z_2 is written as the sum of two terms, one due to attraction and the other due to repulsion:

$$U = -\frac{AZ_1Z_2e^2}{R} + B \exp\left(-\frac{R}{\rho}\right) \tag{1.1}$$

Here, the *Madelung constant*, A, accounts for the fact that we are dealing with a crystal and not just a pair of ions, and A is characteristic of the geometric arrangement of ions in crystals (namely, the crystal structure); B is the repulsion constant, ρ the repulsion exponent and R the distance between the two oppositely charged ions. The repulsion term in equation (1.1) accounts for the stability of ionic crystals without collapsing and arises from the fact that ions with closed electron shells resist overlap of their electron clouds with neighbouring ions. The constants B and ρ are respectively a measure of the strength and the range of the repulsive interaction.

For a uni–univalent crystal like NaCl ($Z_1 = Z_2 = 1$), equation (1.1) becomes

$$U = -\frac{Ae^2}{R} + B \exp\left(-\frac{R}{\rho}\right) \tag{1.2}$$

B can be evaluated from the fact that the potential energy U is a minimum at the equilibrium separation, R_e, between a pair of oppositely charged ions,

$$\left(\frac{dU}{dR}\right)_{R=R_e} = 0 = \frac{Ae^2}{R_e^2} - \frac{B}{\rho}\exp\left(-\frac{R_e}{\rho}\right) \tag{1.3}$$

Accordingly,

$$U = -\frac{Ae^2}{R_e}\left(1 - \frac{\rho}{R_e}\right) \tag{1.4}$$

Equation (1.4) is an expression for the lattice energy of an ionic solid like NaCl, first derived by Born and Mayer. The equation can be used directly for the calculation of cohesive energy of ionic solids provided we know A and ρ.

In the early work on cohesion of ionic solids, the repulsion energy was assumed to vary as an inverse power of the distance between the ions, (B/R^n). Both n and ρ are obtained from the compressibility of solids, the compressibility at zero kelvin being

Table 1.3. *Cohesive energy parameters of some alkali halides with NaCl structure*

Crystal	$R_e(\text{Å})$	$B(10^{-8}$ erg$)$	$\rho(\text{Å})$	$U(\text{kJ mol}^{-1})$
NaCl	2.820	1.05	0.321	795
NaBr	2.989	1.33	0.328	757
KCl	3.147	2.05	0.326	724
KBr	3.298	2.30	0.336	695
RbCl	3.291	3.19	0.323	695
RbBr	3.445	3.03	0.338	673

given by

$$\frac{1}{K_0} = V_0 \left(\frac{d^2 U}{dV^2}\right)_{V=V_e} \tag{1.5}$$

where V_e is the volume of the crystal corresponding to R_e. Values of B and ρ for various alkali halides are given in Table 1.3. Pauling (1960) has suggested the following n values for various closed shell electronic configurations: He, 5; Ne, 7; Ar, 9; Kr, 10; Xe, 12. The value of n for an alkali halide is taken as the average between the values for the two ions. For example, n for NaCl according to this method is 8.

The Madelung constant, A, is a function of crystal structure and can, therefore, be computed from the geometrical arrangement of ions in the crystal. Thus the Madelung constant for the NaCl structure can be written as a summation series,

$$A = \frac{6}{1} - \frac{12}{\sqrt{2}} + \frac{8}{\sqrt{3}} - \frac{6}{2} + \cdots$$

which converges to a value of 1.74756. The series is, however, only conditionally convergent because the summation can be stopped at any finite point. The value obtained at a finite point is characteristic of a finite crystal. The value of A quoted above for the NaCl structure is obtained for a nearly infinite crystal. Cohesive energies of alkali halides calculated using equation (1.4) are given in Table 1.3. These values refer to a static crystal and do not include the contribution from van der Waals forces and the correction for zero-point energy. These are, however, minor terms, accounting for a small percentage of the total lattice energy. A modified expression for the lattice energy of a solid MX incorporating all the four terms reads as

$$U_c = -\frac{Ae^2}{R_e} + B \exp\left(-\frac{R_e}{\rho}\right) - \left(\frac{C}{R_e^6} + \frac{D}{R_e^8}\right) + \frac{9}{4} h\nu_{max} \tag{1.6}$$

where the terms on the right-hand side represent respectively the Madelung energy, the repulsion energy, the van der Waals contribution and the zero-point correction, ν_{max} being the highest frequency of the lattice vibrational mode. In the van der Waals term, the R_e^{-6} and R_e^{-8} terms represent dipole–dipole and dipole–quadrupole interactions respectively. Cohesive energies of ionic solids have been extensively reviewed in the literature (Tosi, 1964). Experimental lattice energies of ionic solids are obtained from thermodynamic data using the Born–Haber cycle. The agreement between the experimental and theoretical values of lattice energies is good, thereby lending support to the ionic model for alkali halides and such solids. The ionic model is, however, a poor approximation for crystals containing large anions and small cations, where the covalent contribution to bonding becomes significant. In addition, cohesive energy calculations cannot be used *a priori* to predict the structure of an ionic compound for the reason that the method makes use of experimental interatomic distances in conjunction with formal ionic charges. We shall discuss some limitations of the ionic model later when discussing inorganic crystal structures.

The electrostatic (Madelung) part of the lattice energy (MAPLE) has been employed to define *Madelung potentials* of ions in crystals (Hoppe, 1975). MAPLE of an ionic solid is regarded as a sum of contributions of cations and anions; the Madelung constant, A, of a crystal would then be the sum of partial Madelung constants of cation and anion subarrays. Thus,

$$A(\text{NaCl}) = A_c(\text{Na}^+) + A_a(\text{Cl}^-)$$
$$1.747 \times 56\cdots = 1 \times 0.873 \times 78\cdots + 1 \times 0.873 \times 78\cdots$$
$$A(\text{CaF}_2) = A_c(\text{Ca}^{2+}) + 2A_a(\text{F}^-)$$
$$5.038 \times 78\cdots = 3.276 \times 12\cdots + 2 \times 0.881 \times 33\cdots$$

where A_c and A_a are the partial Madelung constants of cation and anion subarrays respectively. Madelung potentials are nearly independent of the crystal structure and MAPLE values for the different polymorphic modifications of a given composition differ only by about 1%. Calculations based on the ionic model indeed predict wrong structures in certain instances; for example, LiF was predicted to have the wurtzite structure instead of the rocksalt structure.

1.3.2 Covalent crystals

When atoms in a crystal have similar electronegativities, bond formation occurs through the sharing of valence electrons, each atom contributing one electron to the bond. The force of attraction between two atoms of similar electronegativity arises from the overlap of atomic orbitals and the consequent net lowering of energy. The resulting linkage, the covalent bond, is the classical electron-pair bond. Typical covalent solids are formed by elements of Group IV in the periodic table such as carbon, silicon and germanium. These elements crystallize in the diamond structure, where each atom is bonded to four others through covalent bonds. Atoms in covalent solids tend to achieve closed shell electronic configuration by electron-sharing with neighbours. For example, in diamond, carbon atoms (with $2s^2 2p^2$ valence electronic configuration) form four equivalent tetrahedrally disposed sp^3 hybridized orbitals, each filled with one electron. These hybrid orbitals overlap with similar ones on neighbouring carbon atoms forming four covalent bonds; the valence shell around each atom is thus complete with eight electrons. Unlike an ionic bond, a covalent bond has distinct directional character.

A large number of binary AB compounds formed by elements of groups IIIA and VA or IIA and VIA (the so-called III–V and II–VI compounds) also crystallize in diamond-like structures. Among the I–VII compounds, copper (I) halides and AgI crystallize in this structure. Unlike in diamond, the bonds in such binary compounds are not entirely covalent because of the difference in electronegativity between the constituent atoms. This can be understood in terms of the fractional *ionic character* or *ionicity* of bonds in these crystals.

Pauling (1960) expressed the ionicity of a covalent bond A–B in terms of the difference between the electronegativities of the two atoms:

$$f_i = 1 - \exp\left[-(X_A - X_B)^2/4\right] \tag{1.7}$$

Here f_i is the fractional ionic character of the bond while X_A and X_B are the Pauling electronegativities of the elements A and B. It is seen that $f_i = 0$ when $X_A = X_B$ and $f_i = 1$ when $X_A - X_B \ll 1$. In crystals, this expression is modified to account for the fact that an atom forms more than one bond and the number of bonds formed is not always equal to its formal valence. For $A^N B^{8-N}$ crystals, the expression for ionicity reads as

$$f_i^c = 1 - (N/M) + (Nf_i/M) \tag{1.8}$$

where M is the number of nearest neighbours. Pauling's scale of ionicity explains the structures of many of the binary compounds. Mooser & Pearson (1959) have similarly shown that a plot of the average principal quantum number against the difference in electronegativity clearly separates the differently coordinated structures of binary compounds.

Phillips (1973a) has proposed a new scale of ionicity for AB-type crystals based on the assumption that the average band gap, E_g, of these crystals consists of both covalent and ionic contributions as expressed by $E_g^2 = E_h^2 + E_i^2$, where E_h and E_i are the homopolar and ionic parts. The *ionicity parameter* is given by

$$f_i^c = \frac{E_i^2}{E_g^2} \tag{1.9}$$

The parameter is obtained by relating the static dielectric constant to E_g and taking E_h in such crystals to be proportional to $a^{2.5}$ where a is the lattice constant. Phillips' parameters for a few crystals are listed in Table 1.4. Phillips has shown that all crystals with a f_i^c below the critical value of 0.785 possess the tetrahedral diamond (or wurtzite) structure; when $f_i^c \geqslant 0.785$, six-fold coordination (rocksalt structure) is favoured. Pauling's ionicity scale also makes such structural predictions, but Phillips' scale is more universal. Accordingly, MgS ($f_i^c = 0.786$) shows a borderline behaviour. Cohesive energies of tetrahedrally coordinated semiconductors have been calculated making use

Table 1.4. *Phillips & Pauling ionicity parameters for some $A^N B^{8-N}$ crystals*

Crystal[a]	E_h (eV)	E_i (eV)	f_i^c (Phillips)	f_i^c (Pauling)
Si(D)	4.77	0	0	0
BN(Z)	13.10	7.71	0.256	0.42
GaAs(Z)	4.32	2.90	0.310	0.26
ZnO(W)	7.33	9.30	0.616	0.86
CuBr(Z, W)	4.10	6.90	0.735	0.80
MgS(W, R)	3.71	7.10	0.786	0.67, 0.78
NaCl(R)	3.10	11.8	0.935	0.94
RbI(R)	1.60	7.1	0.951	0.92

[a] Crystal structures are given in parentheses:
D = diamond; Z = zinc blende; W = wurtzite; R = rocksalt.

of the ionicity parameters. It should be noted that the Born model of ionic solids would not be applicable to such partly ionic solids.

An approach based on *orbital radii* of atoms effectively rationalizes the structures of 565 AB solids (Zunger, 1981). The orbital radii derived from hard-core pseudopotentials provide a measure of the effective size of atomic cores as felt by the valence electrons. Linear combinations of orbital radii, which correspond to the Phillips' structural indices E_h and E_i, have been used as coordinates in constructing structure maps for AB solids.

1.3.3 Metallic crystals
Metals are significantly different from nonmetallic solids both in structure and physical properties. They usually crystallize in close-packed structures with large coordination numbers (12 to 8). Characteristically, metals show high thermal and electrical conductivities, indicative of the presence of electrons that are relatively free to move through the crystal. It is clearly not possible to account for these features in terms of localized bonds of the covalent type between near neighbours. Pauling (1960) described a metal as a structure in which one-electron bonds and electron-pair bonds resonate among the various pairs of atoms. Brewer (1967, 1984) has pointed out the value of the Engel correlation (relating structures to electronic configurations) in understanding metal and alloy structures. According to Brewer, the importance of acid–base interactions in metallic systems is that electrons are not free to occupy all parts of physical space; they are primarily restricted to orbital volumes. Electrons in metal systems therefore occupy the allowed space and interact with two or more nuclei.

A theory for the metallic state proposed by Drude at the turn of this century explained many characteristic features of metals. In this model, called the *free-electron theory*, all the atoms in a metallic crystal are assumed to take part collectively in bonding, each atom providing a certain number of (valence) electrons to the bond. These 'free electrons' belong to the crystal as a whole. The crystal is considered to be held together by the electrostatic interaction between the ion cores and the free electrons. Within the crystal, the free electrons move under a constant potential, while at the boundaries there is a large potential difference which prevents escape of electrons. The free electrons inside the potential-well do not, however, obey the Maxwell–Boltzmann statistics; these electrons do not contribute significantly to either the specific heat or paramagnetism of metals.

Sommerfeld modified the Drude theory by introducing the laws of quantum mechanics. According to quantum mechanics, electrons are associated with a wave character, the wavelength λ being given by $\lambda = h/p$ where p is the momentum, mv. It is convenient to introduce a parameter, k, called the *wave vector*, to specify free electrons in metals; the magnitude of the wave vector is given by

$$k = \frac{2\pi}{\lambda} = \frac{2\pi m v}{h}$$

(1.10)

The energy of a free electron in a metal is essentially the kinetic energy due to its motion, given by

$$E = \tfrac{1}{2}mv^2 = \frac{h^2 k^2}{2m} = \frac{h^2 k^2}{8\pi^2 m}$$

(1.11)

The relationship between E and k would, therefore, be parabolic. It was soon realized that the free-electron parabola does not describe the real situation. Electrons in crystals experience a periodic potential due to the presence of nuclei and not a constant potential. In the present-day picture, one talks of energy bands in solids which are separated by forbidden gaps (see also Chapter 6). One can understand the formation of energy bands separated by forbidden gaps in terms of the 'tight-binding approximation'. The approach is essentially the same as in molecular orbital theory used for discrete molecules. In metals the highest occupied energy band is partly filled, so electrons are readily excited to unoccupied levels within the band under the stimulus of electrical or thermal energy, giving rise to high conductivity.

1.3.4   van der Waals crystals (molecular crystals)

Noble gases like argon and krypton as well as molecules which possess no permanent dipole moment such as CO_2, methane and benzene condense to form crystals with small cohesive energies. The fact that atoms and molecules in which all the chemical binding forces are apparently saturated can be condensed into the crystalline state shows that there are certain weak residual forces of attraction besides the usual ionic, covalent or metallic type. The relative weakness of this force is evident from the low melting points, small cohesive energies and low heats of melting of such solids. The origin of these forces is in the momentary fluctuations of electron distribution, arising from the induced dipole–dipole interaction between atoms or molecules. If the charge distribution, say, in a noble gas atom were rigid, there would be no fluctuation of charge density and hence no interaction would occur. However, even the charge cloud of a noble gas atom is deformable to some extent, creating a momentary nonspherical distribution that induces a dipole moment on neighbouring atoms. It is this induced moment that causes an attractive interaction between atoms. The energy of the induced dipole–dipole interaction varies as R^{-6}. Such attractive forces are referred to as van der Waals or dispersive forces. At a very close distance of approach, repulsive forces set in because of the overlap of charge clouds of the different atoms. The total van der Waals potential, therefore, consists of both dispersive and repulsive terms as given by

$$U_r = -\frac{A}{R^6} + \frac{B}{R^{12}} \qquad (1.12)$$

Equation (1.12) represents the well-known *Lennard–Jones potential*.

It is possible to evaluate cohesive energies, packing characteristics and other properties of molecular crystals quantitatively on the basis of nonbonded interactions between atoms (Kitaigorodsky, 1973). The analytical expressions are of the form

$$U_{ij} = -\frac{A_{ij}}{r_{ij}^6} + B_{ij}\exp(-C_{ij}r_{ij})$$

$$U_{ij} = -\frac{A_{ij}}{r_{ij}^6} + \frac{B_{ij}}{r_{ij}^n} \qquad (1.13)$$

where the constants A,B,C and n are obtained by fitting the equation to experimental data, such as crystal structures of hydrocarbons. For example, Williams (1967) has evaluated these parameters by carefully choosing a large number of observational equations of aliphatic and aromatic hydrocarbons. Equation (1.13) only takes into account dispersive and repulsive interactions, electrostatic interactions in molecular crystals being insignificant. Coulombic interactions can be included by taking the point charges, q_e, on atoms (Williams & Starr, 1977) to be as follows:

$$U_{ij} = -\frac{A_{ij}}{r_{ij}^6} + B_{ij}\exp(-C_{ij}r_{ij}) + \frac{q_iq_je^2}{r_{ij}} \qquad (1.14)$$

The lattice energy of a crystal of known structure (atomic positions) is thus calculated by compiling all possible distances between pairs of atoms in different molecules. The method of atom–atom potentials has been employed to investigate phenomena pertaining to static as well as dynamic lattices and the subject has been reviewed by Kitaigorodsky (1973) as well as by Ramdas & Thomas (1980). Typical of the problems that have been investigated by this method are: defects and planar faults, phase transitions and molecular rotation in crystals.

1.3.5   *Hydrogen-bonded crystals*

There are many instances where hydrogen bonds formed between two electronegative atoms like oxygen or fluorine determine the stability and properties of crystals. When a hydrogen atom is bonded to a strongly electronegative atom, the electron pair constituting the bond is so unevenly displaced that the hydrogen atom has a considerable positive charge and is able to attract the negative end of an atom with a lone pair of electrons in another molecule. The hydrogen atom is usually disposed unsymmetrically between two such electronegative atoms in hydrogen-bonded crystals. Compared to covalent and ionic bonds, hydrogen bonds are weak, having energies in the range 12–40 kJ mol^{-1}. The hydrogen bond brings about cohesion of crystals of substances such as ice and proteins.

Ice is the prototype of a hydrogen-bonded crystal. Even in the liquid state, water is exceptional in its physical properties owing to hydrogen bonding. In the crystalline state, the structure of H_2O is anomalous. Whereas many such hydrides crystallize in close-packed structures just like molecular crystals, ice has a hexagonal structure with a fairly short (2.76 Å) intermolecular O–O distance. Each oxygen in the structure is surrounded tetrahedrally by four other oxygen atoms. If we ignore the presence of hydrogen atoms, the structure of ice is exactly like that of wurtzite, the oxygens of water being present at the positions of zinc and sulphur. Hydrogen bonding plays a major role in determining properties of crystals of acid salts like KH_2PO_4.

1.4   Inorganic structures

Structures of metals as well as of many of the nonmolecular solids with nondirectional bonding are readily described in geometrical terms by assuming that the atoms are hard spheres, packing themselves as closely as possible in order to economize on space and minimize energy. The *closest packing arrangements* of cubic (ABCABC) and hexagonal (ABAB) symmetry (CCP and HCP respectively) both have 74.1% packing efficiency. Both the close-packed arrangements possess two types of voids (or interstices): tetrahedral, with a coordination number of four, and octahedral, with a coordination number of six. In an assembly of N spheres of the same size, there will be $2N$ tetrahedral and N octahedral voids. The body-centred cubic (BCC) arrangement of spheres has a packing efficiency of 68.1%.

1.4.1   *Metallic elements*

Most metals crystallize in one or more of the three crystal structures, CCP (A1), BCC (A2) and HCP (A3), shown in Fig. 1.2. Many of them are polymorphic and show transitions from one structure to another with change in temperature or pressure:

Ti(HCP) $\xrightarrow{1155K}$ BCC

Fe(BCC) $\xrightarrow{1182K}$ CCP $\xrightarrow{1676\ K}$ BCC

Cs(BCC) $\xrightarrow{41\ kbar}$ CCP

High-temperature modifications generally have lower coordination numbers, just as low-pressure modifications do. Structures of certain metals such as Zn, Cd and Hg do not conform to this pattern. Mercury has a rhombohedral structure in which each atom is six-coordinated following the $(8 - N)$ rule indicating the presence of the covalent

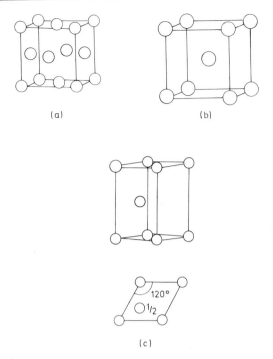

Figure 1.2   Structures of metals: (a) A1; (b) A2; (c) A3. In (c), projection of the hexagonal cell in the xy-plane is shown.

component in bonding. Zinc and cadmium crystallize in a distorted HCP structure with a larger spacing between the individual layers than the ideal case (c/a ratio > 1.633), indicating that the bonding among atoms within a layer is stronger than that between layers. Atomic radii for metallic elements in 12-coordination have been defined by Goldschmidt, and the relation between bond lengths and the nature of bonding in metals has been discussed by Pauling (1960).

1.4.2   Nonmetallic and semimetallic elements

Among the nonmetallic elements, noble gases crystallize in close-packed structures (similar to metals), van der Waals forces being responsible for cohesion. Nonmetallic solids of group V (N, P, As, Sb), group VI (O, S, Se, Te) and group VII (halogens) crystallize in molecular forms in which the intramolecular bonding is covalent while the intermolecular bonding is of van der Waals type. Among these, nitrogen and oxygen form molecular solids composed of multiple bonded diatomic molecules. In the case of other elements, an atom from the Nth group of the periodic table has $(8 - N)$ nearest neighbours, since there are N electrons in the valence shell which can form $(8 - N)$ covalent bonds. Chlorine, bromine and iodine crystallize in an orthorhombic structure, in which diatomic molecules are arranged in a tubular fashion along the c-axis. Among the chalcogens, sulphur, selenium and tellurium form clusters involving puckered rings

(S_6, S_8 etc.) or zig-zag chains with two neighbours. The finite rings or infinite chains are held together by van der Waals forces. The fifth group elements other than nitrogen crystallize in structures containing three near neighbours. Phosphorus exists in several polymorphs; white phosphorus contains unstable P_4 units and a form of red phosphorus (Hittorf's phosphorus) crystallizes in a tubular structure in which each P atom is surrounded by three others pyramidally. Arsenic, antimony and bismuth crystallize in isomorphous layer structures with each atom in a layer covalently bonded to three others, forming an extended puckered sheet; the whole structure is obtained by superposition of such layers giving a rhombohedral arrangement. Besides having three neighbours in its own layer, each atom has three neighbours in adjacent sheets at a larger distance. The difference in distance between the two sets of neighbours decreases as we go from arsenic to bismuth, indicating an approach to the metallic bond in bismuth.

Elements of the fourth group (C, Si, Ge, Sn) form four covalent bonds (sp^3 hybridization), giving rise to completely covalent bonded structures. Diamond, silicon, germanium and grey tin belong to this type in which each atom is bonded to four others and the crystal as a whole is a giant molecule. The unit cell of diamond (Fig. 1.3) consists of an FCC array of carbon in which additional carbon atoms are present at alternate tetrahedral sites (A4). Graphite crystallizes in a hexagonal layer structure (Fig. 1.3) in which each carbon is bonded covalently to three others (sp^2 hybridization). Adjacent sheets are mutually displaced so that the hexagonal rings of carbon atoms in a layer do not lie exactly one over the other; if two such layers are designated A and B, the crystal structure consists of the ABAB ... sequence in hexagonal graphite; in the rare, rhombohedral modification of graphite, the layer sequence is ABCABC .... In the graphite structure, the C—C distance is 1.42 Å within the layer and 3.35 Å between the layers. Apart from carbon, the graphite structure is not found in other elements of the fourth group. Tin is dimorphous, the α (grey) form (stable below 291 K) having the diamond structure and the β (white) form having a tetragonal structure. Lead crystallizes in a cubic close-packed structure typical of metals.

1.4.3   *Inorganic compounds*

Nonmolecular inorganic solids containing two elements, one metallic and the other nonmetallic, form binary compounds of compositions AB, A_2B_3, AB_2, AB_3 and so on. Bonding in such compounds is ionic if the electronegativity difference between A and B is sufficiently large. Crystal structures of these compounds are mainly determined by the nature of bonding and the relative sizes of A and B. When the bonding is predominantly ionic, as in fluorides and oxides, the structure can be described in geometric terms on the basis of close-packing of ions. Since in most such compounds anions are much larger in size than cations, the structures can be thought of as consisting of a closest-packed array (HCP or CCP) of anions, the interstices (voids) of which are occupied by the cations. Where the cations are larger than anions, the closest-packed arrangement would be that of the cations, with the anions occupying the voids. Crystal structures of binary ionic compounds are, therefore, determined by the relative sizes of the voids and the ions forming the closest packing arrangement. It

Table 1.5. *Radius ratios for various cation coordination geometries in ionic compounds*

Ratio r_c/r_a	Coordination number (CN) and geometry
0–0.155	2; Linear
0.155–0.225	3; Triangular
0.225–0.414	4; Tetrahedral
0.414–0.732	6; Octahedral
0.732–1.000	8; Cubic
>1.000	12

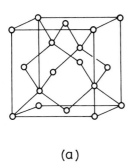   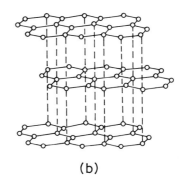

(a)   (b)

Figure 1.3   Structures of (a) diamond and (b) graphite.

can be shown that if R is the radius of a large sphere forming the closest packing, a smaller sphere of radius $r = 0.414\,R$ can be snugly fitted into an octahedral void, and a still smaller sphere of radius $r = 0.225\,R$ into a tetrahedral void. If we imagine the larger spheres as anions and the smaller ones as cations, the sizes of voids define the limiting *radius ratios*, r_c/r_a, for octahedral and tetrahedral coordination of cations by anions. Table 1.5 gives the range of radius ratios for various coordination geometries in binary ionic compounds.

The ionic model, according to Table 1.5, predicts transitions in AB-type solids from 8:8 to 6:6 and from 6:6 to 4:4 coordinated structures at specific values of the radius ratio. The actual structures of alkali halides, however, differ from the predicted ones, as shown in Fig. 1.4. Such failures are due to limitations of the simple ionic model (O'Keeffe, 1977). The ionic model assumes that the anions are considerably larger in size than the cations. The radii of oxide and fluoride ions (1.40 and 1.33 Å respectively) given by Shannon & Prewitt (1969) are indeed larger than the radii of most cations. *Ionic radii* listed in the literature are obtained by apportioning bond lengths on the basis of certain properties of free ions and therefore correspond to free ions and not to ions in crystals. Anions in crystals are subject to a positive Madelung potential and this has the effect of contracting the charge cloud; similarly, cations are subject to a negative potential which has an opposite effect. When these effects are taken into consideration,

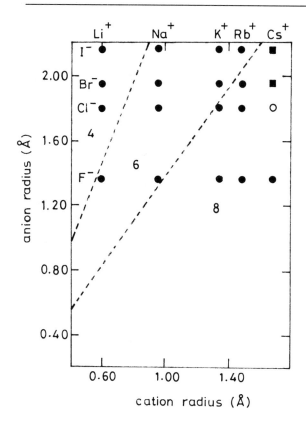

Figure 1.4   Actual structures of alkali halides compared with the predictions of the ionic model. The figure is divided into three regions by lines corresponding to $r_c/r_a = 0.414$ and 0.732. The regions correspond to four-, six- and eight-coordinated structures. Filled circles denote six-coordinated structures, open circle, six- or eight-coordinated structure and filled squares, eight-coordinated structures. Note that the predictions of the ionic model do not entirely correspond with the actual structures of alkali halides (Following Phillips, 1973b).

the differences between the anion and cation radii turn out to be not so significant. Bond lengths in relation to bond strengths therefore seem to be more meaningful than ionic radii; in the absence of an unambiguous and satisfactory procedure to divide bond lengths into cation and anion parts, tables of ionic radii could at best be regarded as tables of bond lengths for certain rational bond strengths. While the experimental electron density has a minimum at a point between the cation and the anion, it is a shallow minimum. If we use the minimum point for determining the cation and anion radii, we end up with charges on the ions that are significantly smaller than the formal charges. If the division is made keeping integral charges on the ions, the radii of anions turn out to be smaller and those of cations larger than the conventional values.

    Tables of ionic radii available in the literature (Shannon & Prewitt, 1969; Shannon, 1976) are derived from bond-length data of oxides and fluorides, because they are the most ionic of the solids. These values are, however, not of general applicability; when used for solids such as nitrides and sulphides, experimental bond lengths are not correctly reproduced. A new set of radii applicable to sulphides has been proposed (Shannon, 1981). Ionic radii values derived from oxides and fluorides do not appear to be useful even for oxyfluorides. For instance, the observed La—O and La—F bond lengths in LaOF are 2.42 and 2.60 Å respectively, while the corresponding values calculated using the Shannon–Prewitt radii are 2.54 and 2.47 Å. According to Pauling's concept of electrostatic bond strength and the electrostatic valence rule, the sum of bond strengths at an anion must be equal to the anion valence and this would predict the La—F bond strength to be $\frac{1}{4}$ and La—O bond strength $\frac{1}{2}$ (O'Keeffe, 1981). This implies a situation where La is coordinated by twelve fluoride ions or by six oxide ions. When we use lanthanum radii appropriate to these coordinations, the bond lengths of LaOF are predicted somewhat reasonably. These considerations reveal that ionic radius is a nebulous concept eluding quantitative treatment and lacking in general applicability. On the contrary, Pauling's concept of electrostatic bond strength and subsequent developments of bond strength–bond length relationships (see, for example, Brown, 1981) seem to be of greater applicability.

In the ionic model, the anion array is considered to approximate to closest packing of equal-sized spheres. In real crystals, however, anion packing densities are much smaller than the ideal value of 0.74. In α-Al_2O_3, for example, possessing the HCP array of anions, the anion packing density is only 0.595. Although the anions are topologically arrayed as if they are in closest packing, they are actually not in contact with one another. O'Keeffe (1977) and O'Keeffe & Hyde (1981) have coined the term *eutaxy* to describe such a situation. According to O'Keeffe, ionic crystal structures are best described as maximum volume (minimum electrostatic energy) structures for fixed cation–anion distances. He has illustrated the idea by making use of several common ionic structures. We shall cite the example of ZnO (wurtzite structure). The wurtzite structure is hexagonal, space group $P6_3mc$ with Zn in 2(a): $0, 0, u_1; \frac{1}{3}, \frac{2}{3}, \frac{1}{2} + u_1$, oxygen in 2(a): $0, 0, u_2$ etc. Two parameters $\gamma = c/a$ and $u = u_1 - u_2$ specify the structure completely. The four tetrahedral Zn—O distances are equal when $\gamma^2 = 4/(12u - 3)$ and therefore, volume $V = \sqrt{27}l^3(u - \frac{1}{4})/2u^3$ where l is the cation–anion distance; V given by this equation is maximum when $u = 3/8$ and $\gamma = 1.633$. For ZnO, $u = 0.383$ and $\gamma = 1.6022$ and for BeO, $u = 0.378$ and $\gamma = 1.602$. For several other ionic solids of this structure, u and γ are close to the ideal values predicted by maximum volume considerations.

The cation coordination number is determined by the r_c/r_a ratio, which implies that the larger the cation radius, the larger is its coordination number. Many examples can be given to show that this assumption is wrong. In $MgAl_2O_4$ (spinel), it is the larger Mg^{2+} that has tetrahedral coordination and the smaller Al^{3+} octahedral coordination.

Bond valence–bond length relations and their applications to solids have been discussed by several authors (see for example, Ziolkowski, 1985; Sävborg, 1985; Domenges et al., 1985). The relations have been extended to equibond enthalpy and bond energy. Of particular use in understanding ionic solids is the concept of *bond valence* (Brown, 1977; Brown & Altermatt, 1985; Pauling, 1929). Bond valence, s, is related to the bond length, r, by

$$s = (r/r_o)^{-N} \qquad\qquad\qquad (1.15)$$
$$s = \exp\left[(r_o - r)/B\right] \qquad\qquad\qquad (1.16)$$

Here, r_o, N and B are constants for a given pair of atoms. A significant feature of bond valence in crystals is that the sum around a cation or an anion, i, to its coordinating ions of opposite charge, j, is a constant.

$$V_i = \sum_j s_{ij} \qquad\qquad\qquad (1.17)$$

Equation 1.16 is more useful since B is generally around 0.37 for a large number of systems, leaving only r_o which can be expressed in terms of additive parameters for the cation and the anion. Equation (1.17) cannot alone fix the topology of the crystal. There is another rule which is useful in this regard. According to this rule, the individual valences around each centre will tend to be as nearly equal as possible. Or, the sum of bond valences around any closed loop in the structure will be equal to zero, if we take the bond valence from anion to cation as positive, and that from cation to anion as negative. Bond valence sums provide a way to distinguish oxide species from hydroxide species in a crystal; the latter being associated with lower values. They also indicate the relative strengths of contacts of different types. For example, six-coordinated potassium will form bonds with a bond valence of around $\frac{1}{6}$. A four-coordinated oxygen forms bonds with a bond valence of $\frac{1}{2}$ while a four-coordinated nitrogen forms bonds with a bond valence of $\frac{3}{4}$. Burdett (1988) has discussed the usefulness of the bond valence rule and suggests how it can be used in conjunction with the orbital picture. He derives the expression of the type $\sum (r_e/r_o)^{-N} = \text{constant}$, where r_e is the equilibrium distance.

We discussed ionic radii earlier. The so-called *pseudopotential radii* have also been useful in crystal chemistry (St. John & Block, 1974; Chelikowsky & Phillips, 1978; Zunger, 1980). A valence electron is attracted by the positively charged nucleus which is screened by the other electrons, the screening increasing with increasing distance. The electron also experiences a repulsion which increases with decreasing distance from the nucleus. The sum of the two gives the pseudopotential, $V(r)$. The distance at which $V(r) = 0$ is the pseudopotential radius. There are different pseudopotentials for s, p and d electrons and accordingly different radii (for different values of the l quantum number). Burdett *et al.* (1981, 1982) have drawn structure maps for AB octets using pseudopotential radii and have shown how inverse and normal spinels are delineated by using pseudopotential radii. Cohen (1981) has calculated energy–volume relations in GaAs and such compounds using pseudopotential radii and has predicted stable structures.

AB structures.   The five principal structures of AB type compounds are rocksalt (B1), CsCl (B2), zinc blende (B3), wurtzite (B4) and NiAs (B8) and these are shown in Fig. 1.5. In the first four of the structures, the cation and anion sublattices are entirely equivalent and the coordination geometry around the cation and anion is the same. The *rocksalt (NaCl) structure* is exhibited by a large number of AB type compounds. The structure (Fig. 1.5) may be thought of as consisting of two interpenetrating

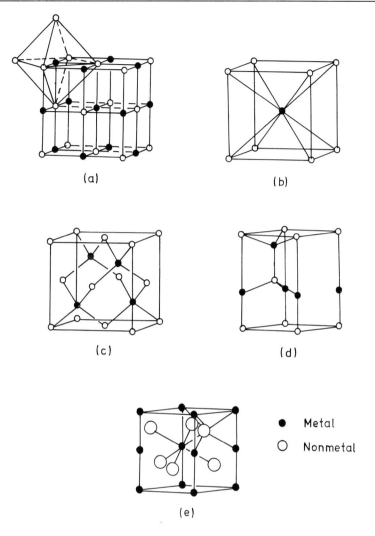

Figure 1.5   Structures of AB compounds: (a) rocksalt; (b) CsCl; (c) zinc blende; (d) wurtzite and (e) NiAs.

FCC arrays of Na^+ and Cl^- ions so as to give octahedral coordination around each ion by the other (6:6 coordination); there are four NaCl formula units per cell. Another way to describe the structure would be to consider it as resulting from cubic close packing of chloride ions ($r_{Cl^-} = 1.81$ Å) with the sodium ions ($r_{Na^+} = 0.95$ Å) occupying octahedral interstices. The structure may also be viewed as cation-centred $NaCl_6$ octahedra extending in three dimensions sharing all its edges. Typical examples of AB compounds crystallizing in this structure are: (a) alkali halides excepting CsCl, CsBr and CsI; (b) NH_4Cl, NH_4Br and NH_4I above 457 411 and 255 K; (c) alkaline earth metal oxides and chalcogenides excepting beryllium salts and MgTe; (d) divalent 3d-metal oxides excepting CuO and (e) oxides and chalcogenides of divalent lanthanides and actinides.

Although this structure is commonly believed to be adopted by ionic compounds, some of the less ionic materials like nitrides, phosphides and arsenides of trivalent lanthanides as well as carbides and nitrides of titanium and vanadium group elements also crystallize in this structure. Several III–V compounds such as GaP and InP and II–VI compounds like CdS and CdSe undergo transformations to this structure at high pressures. Besides the foregoing examples of binary compounds, several ternary compounds containing two different metal ions also crystallize in NaCl-related structures. For example, in $AA'B_2$ type compounds, the two different cations present may be statistically distributed as in γ-LiAlO$_2$, α-NaTlO$_2$ and NaLaS$_2$ or ordered in alternate (111) planes along the body diagonal as in LiNiO$_2$ and AgBiS$_2$; other types of cation ordering are also possible. In AMO$_2$-type oxides (M = Fe, Ni) (A = Ag, Cu) possessing the delafossite structure, the linear $(AO_2)^{3-}$ groups are packed closely parallel to the c-axis and M^{3+} ions occupy the octahedral sites.

The NaCl structure is also found in compounds like TiO, VO and NbO, possessing a high percentage of cation and anion vacancies. Ternary oxides of the type Mg$_6$Mn^{4+}O$_8$ crystallize in this structure with $\frac{1}{8}$ of the cation sites vacant. Solid solutions such as Li$_{(1-x)}$Mg$_{x/2}$Cl ($0 \leqslant x \leqslant 1$) crystallize in the rocksalt structure; stoichiometric MgCl$_2$ may indeed be considered as having a defect rocksalt structure with 50% of ordered cation vacancies.

The CsCl structure (Fig. 1.5) consisting of interpenetrating primitive cubic arrays of Cs$^+$ and Cl$^-$ ions with 8 : 8 coordination is exhibited by many solids: (a) CsCl, CsBr and CsI; (b) NH$_4$Cl, NH$_4$Br and NH$_4$I in their low-temperature modification; (c) thallous halides; (d) alloys of β-brass type, CuZn, AuZn; (e) CuCN, CuSH and TlCN in their high-temperature form and (f) alkali halides (with the exception of lithium salts) at high pressures.

The zinc blende (ZnS) structure is closely related to the diamond structure (Fig. 1.5). If one half of the carbon atoms in the diamond structure forming the FCC array are replaced by zinc and the other half by sulphur, the zinc blende structure with 4 : 4 coordination is obtained. Typical examples of AB compounds crystallizing in zinc blende structure are: copper (I) halides; γ-AgI; beryllium chalcogenides (excepting BeO); zinc, cadmium and mercury chalcogenides (excepting the oxides); many III–V compounds like BN, BP, BAs, AlP, AlAs, AlSb, GaP, GaAs, GaSb and SiC. Derivative structures of zinc blende of the following types are known: (a) chalcopyrite, CuFeS$_2$, with ordering of monovalent and trivalent cations at the zinc sites; (b) structures with a statistical distribution of different cations as found in CuSi$_2$P$_3$ and Cu$_2$GeS$_3$ and (c) Cu$_3$SbS$_4$, famatinite obtained by substitution of Cu$^+$ and Sb^{5+} ions in place of Fe^{3+} in chalcopyrite. The zinc blende structure can also sustain cation and anion vacancies; for example, Ga$_2$S$_3$ crystallizes in this structure wherein $\frac{1}{3}$ of cation sites are vacant. Likewise, Ag$_2$HgI$_4$ has a zinc-blende-derived structure with $\frac{1}{4}$ vacant cation sites.

The wurtzite structure is closely related to the zinc blende structure, having the same 4 : 4 tetrahedral coordination arising from a hexagonal close packing of anions in which half the tetrahedral sites are occupied by cations. Examples of AB compounds crystallizing in this structure are CuCl, CuBr, CuI, β-AgI, BeO, ZnO, ZnS, MnS, MnSe,

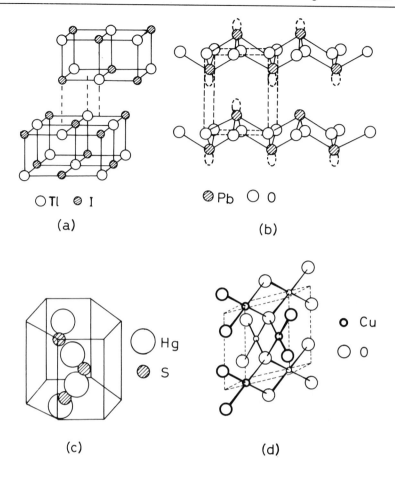

O Tl   ⊘ I

(a)

⊘ Pb   O O

(b)

○ Hg

⊘ S

(c)

o Cu

O O

(d)

Figure 1.6   Structures of AB Compounds: (a) TlI; (b) red PbO; (c) HgS and (d) CuO.

MnTe, AlN, GaN, InN, SiC and NH_4F. The wurtzite structure is, in a sense, intermediate between NaCl and ZnS structures.

In the hexagonal NiAs structure, nickel is octahedrally coordinated by arsenic atoms but arsenic is surrounded by six nickel atoms at the corners of a trigonal prism (Fig. 1.5). Furthermore, each nickel has two other nickel atoms as near neighbours along the c-axis. The structure may be thought of as resulting from a hexagonal close packing of arsenic in which all the octahedral sites are occupied by metal atoms; the $NiAs_6$ octahedra share common faces along the c-axis, bringing metal atoms close together. As compared to the NaCl structure, the NiAs structure has a smaller Madelung constant (1.748 and 1.693 respectively). In AB compounds of NiAs structure, the A—B bond is fairly covalent and there is metal–metal bonding between A atoms. Compounds with the NiAs structure are generally metallic and occur over a wide composition range. Typical examples of compounds with this structure are the sulphides, selenides, tellurides, phosphides, arsenides, antimonides and bismuthides of transition metals.

Among the less common AB structures, mention may be made of the following: PbO, SnS, TlI, HgS and PtS (Fig. 1.6). Among these, PbO, SnS and TlI illustrate the effect of the cation inert pair of electrons on the structure. Although the inert pair is formally considered to be a pair of electrons occupying the 5s or the 6s-orbital, there appears to be a pronounced directional character associated with these electrons that influences the structure adopted. PbO is dimorphous; the red variety (litharge) has a tetragonal layer structure in which lead occupies the apex of a tetragonal pyramid and oxygen has a tetrahedral coordination of metal atoms. Each lead carries an inert pair that points in between two metals in the adjacent layers. The blue–black modification of SnO is isomorphous with red PbO. Yellow PbO (massicot) has an orthorhombic chain structure consisting of zigzag chains of —Pb—O—Pb— parallel to the b-axis. HgS (cinnabar) consists of spiral chains that wind along the c-axis with each mercury having a linear coordination of two sulphur atoms. The occurrence of a linear coordination in this and other d^{10} systems appears to be related to the formation of strongly directional linear sp or sd hybrid orbitals because of the close proximity of ns, np and (n-1)d orbital energies. PtS, PtO, PdO and PdS crystallize in tetragonal structures in which the metal has a square planar (dsp^2) coordination and the anion has a tetrahedral coordination. CuO crystallizes in a distorted PtS structure.

AB_2 structures. Fluorides and oxides of the formula AB_2, which are distinctly ionic, crystallize in structures determined by size considerations. As in the case of AB structures, it is the coordination geometry of anions around the cation that determines the structural arrangement. The coordination may be 8-, 6-, or 4-fold, fixing the corresponding anion coordination numbers to 4, 3 or 2. We thus have the following structures for ionic AB_2 compounds: fluorite (8:4), rutile (6:3) and silica (4:2) and these structures are shown in Fig. 1.7.

The fluorite structure (C1) consists of a cubic close-packed array of cations in which all the tetrahedral sites are occupied by anions (Fig. 1.7). The relation to the zinc blende structure is obvious. Anions are tetrahedrally coordinated by cations; cations occur in cubic coordination. There is also a close relationship to the CsCl structure, with the anions forming a primitive cubic structure in which alternate cubes are occupied by cations. Fluorite structure generally occurs in compounds with radius ratio greater than 0.732: (a) fluorides of large divalent cations like Ca, Sr, Ba, Cd, Pb, Hg; ternary fluorides of mono- and trivalent cations of approximately equal size as, for example, $NaYF_4$, $KLaF_4$, Na_2ThF_6 and K_2UF_6; (b) dioxides of large quadrivalent cations like Zr, Hf, Ce, Pr, Tb, Po, Th, Pa, U, Pu and Am; oxyfluorides of trivalent lanthanides and (c) a number of intermetallic and interstitial compounds like $AuAl_2$, $SiMg_2$, $GeMg_2$, $CoSi_2$, Be_2C as well as the dihydrides of many lanthanides. The fluorite structure is also versatile in accommodating nonstoichiometry involving both anion deficiency and anion excess. Several nonstoichiometric oxides and oxyfluorides of lanthanides and actinides crystallize in fluorite-related structures. The antifluorite structure adopted by many oxides, sulphides, selenides and tellurides of the alkali metals (e.g. Li_2O, Na_2S,

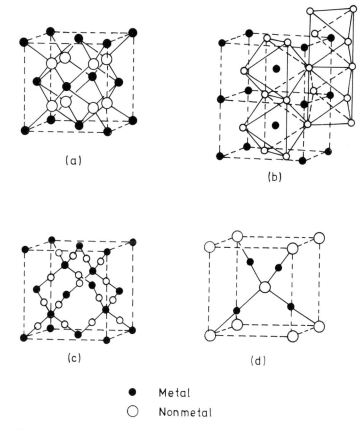

●   Metal

○   Nonmetal

Figure 1.7   Structures of AB_2 compounds: (a) fluorite; (b) rutile; (c) cristobalite and (d) cuprite.

etc.) is derived from the fluorite structure by interchange of the positions of the anions and cations (4:8 coordination).

The rutile (TiO_2) structure (C4) is a tetragonal structure stable for ionic compounds with $r_c/r_a < 0.73$. It consists of an infinite array of TiO_6 octahedra linked through opposite edges along the c-axis (Fig. 1.7). Divalent metal fluorides (Mg, Cr, Mn, Fe, Co, Ni, Cu, Zn) and quadrivalent metal oxides (Ti, V, Cr, Mn, Nb, Ta, Mo, W, Ru, Os, Ir, Sn, Pb, Te) crystallize in this structure. As with the fluorite structure, it is possible to substitute in this structure stoichiometrically equivalent amounts of appropriate cations to give ternary compounds; thus $AlSbO_4$, $GaSbO_4$, $CrNbO_4$, $FeNbO_4$, $RhVO_4$ and so on have the rutile structure. This structure is also the basis for the *Magnéli phases* of the general formula M_nO_{2n-1} found in reduced titanium and vanadium oxides.

SiO_2 occurs in the following three forms:

Low quartz (C8)   Tridymite (C10)   Cristobalite (C9)
(below 1143 K)   (1143–1743 K)   (Above 1743 K)

The various forms of silica generally consist of SiO_4 tetrahedra sharing all the corners. The ideal structure of cristobalite (Fig. 1.7) may be described as an arrangement of silicon atoms occupying the position of the carbons in the diamond structure with the oxygens inserted midway between each pair of silicon atoms. In the resulting structure, silicon has a tetrahedral coordination and oxygen a linear coordination (4:2). BeF_2 also possesses this structure. In tridymite, too, we have 4:2 coordination but it is related to wurtzite in exactly the same way as cristobalite is related to zinc blende. In quartz, the SiO_4 tetrahedra are arranged in hexagonal spirals; depending on the direction of the spiral, quartz exhibits l or d optical activity. $AlPO_4$ analogues of all the SiO_2 forms are known, the quartz analogue berlinite crystallizing with a chiral space group.

The cuprite (Cu_2O) structure consists of a body-centred cubic array of oxygen; the copper atoms occupy centres of four of the eight cubelets into which the BCC cell may be divided (Fig. 1.7). In this structure, copper has a linear coordination and oxygen tetrahedral coordination (2:4). This structure is unique among inorganic materials in that it consists of two identical interpenetrating frameworks which are not directly linked to each other.

In AB_2 compounds where B is a halogen other than fluorine or a chalcogen (S, Se, Te), the A—B bond is considerably covalent and a different set of structures is encountered. Some of the important structures are shown in Fig. 1.8.

The pyrite (FeS_2) structure (C2) consists of S_2^{2-} molecular ions (Fig. 1.8). The structure is closely related to NaCl from which it may be derived by replacing Na by Fe and Cl by S_2^{2-}, with the centre of the S_2^{2-} ion occupying the chloride position. Each iron is octahedrally coordinated but each sulphur is tetrahedrally surrounded (one S and three Fe). Several transition metal chalcogenides crystallize in this structure.

The CdI_2 and $CdCl_2$ structures (C6 and C19) are closely related. In the CdI_2 structure, the iodine atoms form a hexagonal close-packed layer sequence in which the cadmium atoms occupy alternate interlayer octahedral sites (Fig. 1.8). The layer sequence runs as AcB AcB ... where A and B denote iodine layers and c denotes cadmium atoms. $CdCl_2$ also has a similar layer structure but the anion layers are in cubic close-packing instead of hexagonal close-packing; the cations occupy alternate interlayer octahedral sites as in CdI_2. Several metal dihalides and dichalcogenides crystallize in these layer structures with weak (van der Waals) interlayer bonding. Usually the more ionic of them prefer the $CdCl_2$ structure (Madelung constants for the $CdCl_2$ and CdI_2 structures are respectively 4.489 and 4.383). $MgCl_2$, $MnCl_2$, $FeCl_2$, $NiCl_2$, NbS_2 and TaS_2 are some examples of solids possessing the $CdCl_2$ structure. Di-iodides of divalent metals and dichalcogenides of tetravalent metals crystallize generally in the CdI_2 structure. MoS_2 of CdI_2 structure has been prepared recently (Wypych & Schöllhorn, 1992).

The molybdenite (MoS_2) structure (C7) is a layer structure closely related to $CdCl_2$ and CdI_2. Each layer of MoS_2 consists of a sheet of metal atoms sandwiched between two sheets of nonmetal atoms (Fig. 1.8). These layers are stacked one upon the other along the hexagonal c-axis, the sequence of sheets of atoms being AbA BaB where the capital letters denote anions and the lower case letters cations. In this structure,

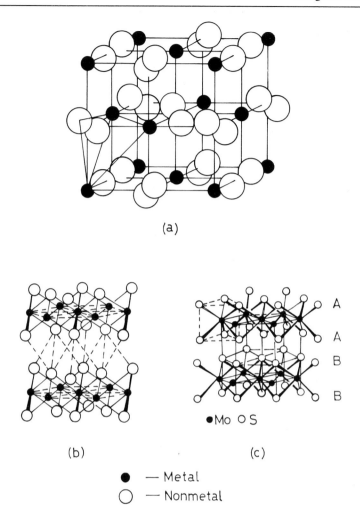

(a)

(b)   •Mo O S   (c)

● — Metal

○ — Nonmetal

Figure 1.8   Structures of AB_2 compounds: (a) pyrite; (b) CdI_2 and (c) MoS_2.

molybdenum has a trigonal prismatic coordination of anions. A d^4sp hybridization of metal orbitals is believed to give rise to this coordination. MoS_2, $MoSe_2$, $MoTe_2$, WS_2, WSe_2, α-$NbSe_2$ and α-$TaSe_2$ are examples of compounds crystallizing in this structure. The HgI_2 structure (C13) is also a layer structure involving tetrahedral coordination of mercury and a HCP arrangement of iodine atoms.

A_2B_3 *structures.*   Crystal structures adopted by A_2B_3 compounds fall into two major categories depending on whether the A—B bond is predominantly ionic or covalent. Ionic A_2B_3 compounds typified by the sesquioxides of d-transition metals and lanthanides crystallize mainly in two structures: Al_2O_3 (corundum) and C-rare earth oxide. The corundum structure (D5₁ type) adopted typically by Ti_2O_3, V_2O_3, Cr_2O_3 and α-Fe_2O_3 consists of a hexagonal close-packed array of anions in which the cations occupy two-thirds of the octahedral interstices; the cations have octahedral coordination and the oxygens nearly tetrahedral coordination (6:4). The ilmenite structure is derived from the corundum structure by replacing the cations in alternate layers with Fe and Ti. Sesquioxides of iron and aluminium also exist in the metastable γ-form which has a cation-deficient cubic spinel structure. The C-rare earth oxide structure commonly adopted by sesquioxides of lanthanides and actinides as well as by Mn_2O_3 is derived from the fluorite structure by removal of one-fourth of the anions so as to confer a six-fold coordination to the metal atoms, but the coordination geometry is not exactly octahedral. Rare earth oxides also crystallize in two other structures, A (hexagonal) and B (monoclinic) forms, wherein the lanthanide ion is seven-coordinated. Among the sesquioxides, Bi_2O_3 is unique, crystallizing in at least four polymorphic modifications, α (monoclinic), β (tetragonal), γ (BCC) and δ (FCC), of which the δ-phase, having a defect fluorite structure, exhibits high ionic conductivity.

Among the covalent A_2B_3 structures, the Bi_2Te_3 (C33-type) structure consists of a close packing of chalcogen atoms in which bismuth occurs in octahedral sites. Sb_2S_3 and Bi_2S_3, on the other hand, crystallize in a chain structure.

AB$_3$ structures.   The important AB_3 structure which forms the basis for the structures of several metal oxides, fluorides and oxyfluorides is that of ReO_3 (DO_9-type). This structure (Fig. 1.9(a)) is cubic, comprising corner-shared ReO_6 octahedra with linear Re—O—Re links. WO_3 possesses a distorted ReO_3 structure and many complex niobium oxides contain ReO_3 type blocks along the major axis of a tetragonal structure. Other compounds crystallizing in the ReO_3 structure are NbF_3, TaF_3, $TiOF_2$ and $MoOF_2$. Many transition metal trifluorides also consist of a corner-shared AF_6 octahedral network, but the A—F—A angles are not linear. In the ideal ReO_3 structure, the anions occupy three-quarters of the positions of a CCP arrangement. When the anions are arranged in HCP, the angle becomes 132°. Typical examples of the latter are PdF_3 and RhF_3. Other transition metal fluorides, AF_3(A = Ti, V, Cr, Fe, Co, Mo and Ru), possess structures intermediate between those of ReO_3 and PdF_3, where the A—F—A angle is around 150°. The structures of $Sc(OH)_3$ and $In(OH)_3$ are related to ReO_3. Pnictides of certain transition metals, AB_3(A = Co, Ni, Rh, Pd, Ir; B = P, As, Sb), adopt the *skutterudite* ($CoAs_3$) structure, which is related in a simple way to the ReO_3 structure; when the nonmetal atoms of the ReO_3 structure are displaced from the centre of the edges so as to form square-planar B_4 groups, the skutterudite structure is the result. Anti-ReO_3 structures are also known: e.g. Cu_3N, Li_3N, which shows fast lithium ion conduction (Huggins & Rabenau, 1978), adopts a hexagonal structure consisting of alternating Li and Li_2N layers. Contrary to the common belief that a dilute statistical occupation of conducting ions is necessary for fast ion conduction, the lithium ion sites responsible for ionic conduction are almost completely occupied in Li_3N.

Lanthanide and actinide trifluorides adopt a different structure called the tysonite structure (DO_6). In this structure, the metal atom has the unusual eleven (5 + 6) coordination. Covalent AB_3 halides crystallize in layer structure (DO_5) typified by $CrCl_3$ and BiI_3; these structures are related respectively to $CdCl_2$ and CdI_2.

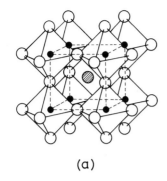

(a)

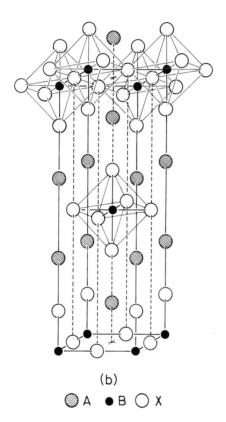

(b)

◍ A   ● B   ○ X

Figure 1.9 (a) The perovskite structure. Without the large A atom at the body centre position, the structure becomes that of cubic ReO_3; (b) The K_2NiF_4 structure consisting of rocksalt (KF) and perovskite ($KNiF_3$) layers. The NiF_6 octahedra share equatorial corners restricting the Ni–F–Ni interaction to the xy-plane.

Perovskites.   Perovskites of the general formula ABX_3 may be regarded as derived from the ReO_3 structure as shown in Fig. 1.9(a). The BX_3 framework in the perovskite is similar to that in the ReO_3 structure consisting of corner-shared BX_6 octahedra. The large A cation occupies the body-centre, 12-coordinate position. In an ideal cubic perovskite structure where the atoms are just touching one another, the B—X distance is equal to $a/2$ and the A—X distance is $\sqrt{2}(a/2)$ where a is the cube unit cell length and the following relation between radii of ions holds: $(R_A + R_X) = \sqrt{2}(R_B + R_X)$. Goldschmidt found that the perovskite structure is still retained in ABX_3 compounds even when this relation is not exactly obeyed and defined a *tolerance factor*, t, as

$$t = \frac{R_A + R_X}{\sqrt{2}(R_B + R_X)} \tag{1.15}$$

For the ideal perovskite structure, t is unity. The perovskite structure is, however, found for lower values of t ($\sim 0.75 < t \leqslant 1.0$) also. In such cases, the structure distorts to tetragonal, rhombohedral or orthorhombic symmetry. This distortion arises from the smaller size of the A ion which causes a tilting of the BX_6 octahedra in order to optimize A—X bonding. Katz & Ward (1964) have given an alternative description of the perovskite structure in terms of close-packing of A and O ions. In this model of the ABO_3 perovskite structure, close-packed AO_3 layers are stacked one over the other with the B cations occupying octahedral holes surrounded by oxygen.

A large number of oxides and fluorides crystallize in perovskite-related structures. The important ferroelectric oxides $BaTiO_3$, $NaNbO_3$, $KNbO_3$, $NaTaO_3$ and $KTaO_3$ have distorted perovskite structures at ordinary temperatures and become cubic at high temperatures. The perovskite structure can tolerate vacancies at the A or X sites giving rise to nonstoichiometric compositions, $A_{(1-x)}BX_3$ and ABX_{3-x}. Typical examples are the tungsten bronzes, A_xWO_3, and the brownmillerite, $CaFeO_{2.5}$. Perovskite-type oxides rarely show anion excess nonstoichiometry. In $LaMnO_{3+x}$, the apparent anion excess arises from La and Mn vacancies.

Besides perovskites of the type $A^{II}B^{IV}O_3$ and $A^IB^VO_3$, some of which are ferroelectric, there are the orthorhombic $GdFeO_3$ type and the rhombohedral $LaAlO_3$ type perovskites. There are also numerous complex stoichiometric oxides with two or more metal atoms at the octahedral B-sites. Many such oxides show ordering of B-site atoms leading to multiple perovskite cells. There are two major categories of such oxides, $A_2BB'O_6$ and $A_3BB_2'O_9$, where B and B' possess different oxidation states. An interesting example of B-site ordering giving rise to a layered perovskite superstructure is that of La_2CuSnO_6 (Anderson *et al.*, 1993).

A number of complex oxides of the general formula A_2BO_4 crystallize in the K_2NiF_4 structure, which is closely related to the perovskite structure. The tetragonal structure of K_2NiF_4 (Fig. 1.9(b)) can be regarded as consisting of perovskite slabs of one unit cell thick, which are stacked one over the other along the c-direction. The adjacent slabs are displaced relative to one another by $\frac{1}{2}\frac{1}{2}\frac{1}{2}$, such that the c-axis of the tetragonal

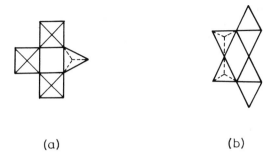

(a)   (b)

Figure 1.10   (a) Replacement of a BO_6 octahedron by a PO_4 tetrahedron in an octahedral B_4O_{20} unit and (b) accommodation of a diphosphate group.

structure is roughly equal to three times the cell edge of the cubic perovskite. The structure is two-dimensional in the sense that only the equatorial anions of the NiF_6 octahedra are linked through corners. Oxides of the K_2NiF_4 structure have been investigated extensively (Ganguly & Rao, 1984) and the 'two-dimensional' magnetic and electrical properties are quite interesting. Tolerance factors for the K_2NiF_4 structure have been defined just as for perovskites.

Oxide bronzes.   These are ternary compounds of the general formula, $A_xB_yO_z$, where B is a transition metal, B_yO_z its highest valency oxide and A is generally an electropositive metal like the alkalis and $0 < x \leqslant 1$. The oxide bronzes may be regarded as a solution of metal A in the matrix of the host oxide B_yO_z. Well-known in this class of solids are the tungsten bronzes, A_xWO_3. They display interesting properties like bright colours, metallic lustre and high electrical conductivity similar to a metal. Besides tungsten, oxide bronzes have also been obtained with metals like V, Nb, Mo and Re. Less electropositive metals such as the lanthanides, Cu, Zn, Cd, In, Tl, Sn and Pb have also been incorporated as A metals in the bronzes. Structurally, cubic tungsten bronzes are the simplest, being regarded as perovskites whose A sites are incompletely occupied. Depending on the nature of the A ion and the value of x, tungsten bronzes may have hexagonal, tetragonal or cubic structures. Insulating oxides of the type A_xBO_3 which are isotypic with bronzes are referred to as *bronzoids*.

The *phosphate tungsten bronzes* form a large family wherein a monophosphate group, PO_4, or a diphosphate group, P_2O_7 is accommodated in the octahedral ReO_3-type framework. These oxides result from the replacement of one BO_6 octahedron by one PO_4 or P_2O_7 group in the octahedral B_4O_{20} structural unit (Fig. 1.10). The first family, called the diphosphate tungsten bronzes with hexagonal tunnels ($DPTB_H$), have a monoclinic structure. They form a series of phases with the general formula $A_x(P_2O_4)_2(WO_3)_{2m}$ with A = K, Rb, Tl, Ba, $x \approx 1$ and m being an integer. The host lattice (Fig. 1.11(a)) is built up of corner-sharing WO_6 octahedra and P_2O_7 groups and the structure can be described as the stacking of ReO_3-type slices connected through single layers of diphosphate groups. At the junction between ReO_3-type slices, WO_6 octahedra and P_2O_7 groups form six-sided tunnels running along b, where the A

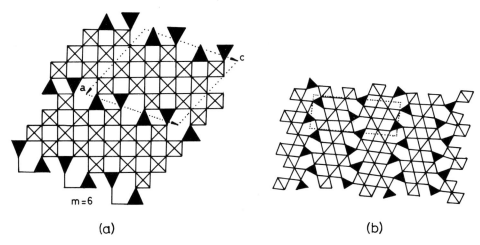

Figure 1.11   (a) Host lattice of the diphosphate tungsten bronzes $A_xP_2O_4$ $(WO_3)_{2m}$ with hexagonal tunnels (an even m-member, $m = 6$). The WO_6 octahedra (crossed squares) and the P_2O_7 groups (black triangles) form six-sided tunnels where the A cations (not represented) are present. (b) Host lattice of the monophosphate bronzes, $(PO_2)_4(WO_3)_{2m}$, with pentagonal tunnels. The WO_6 octahedra (lozenges) form ReO_3-type layers interconnected with PO_4 tetrahedra (black triangles). (The even-m member ($m = 4$) $P_4W_8O_{32}$ consisting of strings of four WO_6 octahedra forming a fish-bone array) [Following Raveau (1986)].

cations are located. The ReO_3-type slices are of variable width (characterized by the number of octahedra, m) and form strings between two diphosphate planes parallel to [101]. The even-m members differ from the odd-m members by a relative translation of the diphosphate planes. Monophosphate tungsten bronzes with pentagonal tunnels $(MPTB_P)$ represent the second family of this series. These oxides have an orthorhombic structure and conform to the general formula $(PO_2)_4(WO_3)_{2m}$. Their host lattice (Fig. 1.11(b)) is composed of corner-sharing WO_6 octahedra and PO_4 tetrahedra; each PO_4 tetrahedron sharing the four corners with WO_6 octahedra. This mixed framework is characterized by strong anisotropy. It can also be described as the stacking of ReO_3-type slices with a variable width, connected to each other through single layers of PO_4 tetrahedra parallel to (001). At the junction between two ReO_3-type slices, the PO_4 tetrahedra and WO_6 octahedra form pentagonal tunnels which are empty. The geometry of these tunnels is characterized by 120° and 90° angles. The members of this family differ in the thickness of the ReO_3-type slices and by the parity of the number m of octahedra forming the octahedral string between two successive tetrahedral planes. The third family is that of monophosphate tungsten bronzes with hexagonal tunnels $(MPTB_H)$, closely related to $MPTB_P$s.

Spinels.   Several ternary oxides of AB_2O_4 composition have the structure of the mineral spinel, $MgAl_2O_4$, the unit cell containing eight formula units $(Mg_8Al_{16}O_{32})$. The structure (Fig. 1.12) consists of a cubic close-packed array of oxide ions in which the magnesium ions occupy $\frac{1}{8}$ of the tetrahedral (A) sites and the

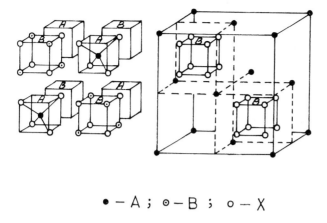

$$\bullet - A \; ; \; \circ\!\!\!- B \; ; \; \circ - X$$

Figure 1.12   The spinel structure. The characteristic A and B octants containing the tetrahedral and octahedral site cations in the structure are shown on the left. The FCC lattice of A site ions in a unit cell of spinel is shown on the right.

aluminium ions occupy half the octahedral (B) sites. Such a distribution of ions represented by $(A)_t[B_2]_oO_4$ is called the 'normal' spinel structure of which $MgAl_2O_4$ is the prototype. An important variation is the 'inverse' spinel structure, $(B)_t[AB]_oO_4$; in this case, half of the B ions are in tetrahedral sites and the other half along with A ions occupy octahedral sites. In between the two extremes, intermediate cation distributions such as complete randomization of cations over all the available 24 cation sites are also known. Well-known examples of oxide spinels are the so-called 2:3 spinels, $A^{2+}B_2^{3+}O_4$, where A = Mg, Mn, Fe, Co, Ni, Cu and Zn and B = V, Cr, Mn, Fe, Co, Rh and so on. Of these, spinel ferrites, AFe_2O_4, which have the inverse structure, constitute the technically important ferrimagnetic materials. Magnetite, Fe_3O_4, has the inverse spinel structure with $Fe_t^{3+}[Fe^{2+}Fe^{3+}]_oO_4$ distribution.

Olivines and the double-hexagonal (DH) form of $LiFeSnO_4$ possess an anion close-packing closely related to that in spinels. Existence of nonstoichiometric spinels such as δ-Fe_2O_3 or $Zn_2Mo_3O_8$ shows that defects can be introduced at the metallic sites, maintaining the anionic close-packing. Hexagonal ferrites possess spinel layers which alternate with close-packed layers where barium partially replaces the oxygen. Structures of these ferrites are also related to that of DH–$LiFeSnO_4$. The *hexagonal ferrites* belong to the AO–Fe_2O_3–MeO system with A = Ba, Pb, Sr and Me = Mg, Fe, Co, Ni, Zn. Many coupled substitutions in these oxides, such as the replacement of bivalent Me^{II} by a couple, Li^+/Fe^{3+} or Li^+/Sn^{4+}, are possible without changing the structure. Among these oxides, the barium compounds are the most numerous, with more than 60 different oxides known up to the present time. The structure of the hexagonal ferrites ($a \approx 6$ Å) results from the stacking of close-packed oxygen slabs denoted as O_4 and the A substituted anionic layers, AO_3. Close-packing of these anionic layers with the B cations located in their cavities gives rise to four types of structural units called S, R, T and Q, defined by the cationic planes perpendicular to the c axis. The general formula of these ferrites is given by, $(B_6O_8)_{n_1}^S (B_6BaO_{11})_{n_2}^R$

$(B_8Ba_2O_{14})_{n_3}^T (B_7Ba_2O_{14})_{n_4}^Q$ where n_1, n_2, n_3 and n_4 are integers. The stacking sequences are denoted by the letters S,R,T and Q. The theoretical number of the possible sequences in these ferrites is infinite and many of these combinations have been observed.

Pyrochlores.   Ternary oxides of the composition $A_2B_2O_7$ crystallize frequently in this structure. It may be regarded as derived from the fluorite structure from which $\frac{1}{8}$ of anions are removed systematically so as to confer near-octahedral coordination around B ions; A cations retain the cubic coordination as in fluorite. The anions have two types of coordination environment. Six anions have (2A + 2B) tetrahedral coordination while the seventh anion has (A + 3B) tetrahedral neighbours. Well-known examples of the structure are the 3:4 compounds $Ln_2B_2O_7$ where Ln = lanthanide ion and B = tetravalent Ti, Zr, Ru and Sn ions. Among the 2:5 pyrochlores, $Cd_2Nb_2O_7$ is an important ferroelectric material. Pyrochlore structure also occurs for certain anion-deficient compositions like $Pb_2B_2O_{7-x}(x < 1.0)$ where B = Ru, Re or Ir, as well as $TlNbO_3$ and $TlTaO_3$.

Octahedral tunnel structures.   Metal–oxygen octahedra, BO_6, form host lattices characterized by large tunnels where cations are located. Thus, perovskites form a large family which are described by four-sided tunnels. There are a large number of tunnel structures which can be classified according to the size and the shape of the tunnels. Tunnel structures with angles of 90° or 60°–120° occur commonly. We shall briefly examine the general features of tunnel structures. (Note that the bronzes we discussed earlier are tunnel structures).

Na$_x$TiO$_2$ (which is a bronze) exhibits tunnels, with a square section (angle of 90°) similar to those observed in perovskites. The host lattice, $[TiO_2]_\infty$, is similar to that of a perovskite, but differs in that the TiO_6 octahedra share their edges instead of their corners along two directions (Fig. 1.13), whereas along the third direction, they share corners as in a perovskite. The $[TiO_2]_\infty$ framework results from the ReO_3 type framework by shearing the structure in order to form edge-sharing octahedra in place of corner-sharing octahedra (see Chapter 5 for a discussion of shear structures). The sodium content that can be incorporated in the tunnels is not well-defined and the tunnels are never fully occupied. The $[TiO_2]_\infty$ framework is distorted and Na^+ ions have only four near neighbours in the perovskite edge. Ti exhibits mixed valence, Ti(III)/Ti(IV), and the material has metallic conductivity. $AlNbO_4$ which is isotypic has all its tunnels empty and does not exhibit electron delocalization since Nb is in the 5+ oxidation state. (It is a bronzoid).

Perovskite tunnels can get associated by their faces to form rectangular tunnels. This is found in $A_2Ti_6O_{13}$ (A = Na, K, Rb) which has a monoclinic cell, with the host lattice, $[Ti_6O_{13}]_\infty$, consisting of structural units of 2 × 3 edge-sharing octahedra (Fig. 1.14(a)). In the (010) plane, the units share their corners (Fig. 1.14(b)), resulting in rectangular tunnels which consist of three face-sharing perovskite tunnels (called 3P). In these tunnels, two perovskite cages out of three are occupied by the A cations which are displaced towards each other with respect to the ideal positions. A '4P' structure (where

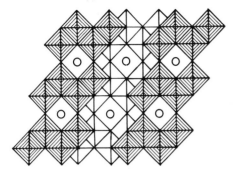

Figure 1.13 Projection of the structure of the bronze Na_xTiO_2 onto the (110) plane. The TiO_6 octahedra (crossed squares) share edges forming square tunnels where the sodium ions (open circles) are located [Following Wadsley (1964)].

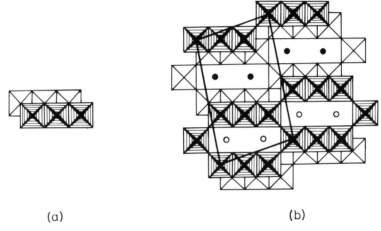

(a)   (b)

Figure 1.14 Structure of the $K_2Ti_6O_{13}$-type oxides showing (a) units of 2×3 edge-sharing octahedra and (b) the projection onto (010) showing the rectangular tunnels [Following Andersson & Wadsley (1962)].

the host lattice is $K_2Ti_8O_{17}$) consisting of units of 2×4 edge-sharing octahedra is known. The tetragonal hollandite structure (A_xMnO_2) can be derived from the rutile structure by edge-sharing of the octahedra in the B_4O_{20} units. This results in large square tunnels which are generally less than half occupied, x being close to 0.25 in the mineral. The tunnels can be empty or partially occupied by water molecules as in α-MnO_2. Some of the oxyhydroxides (e.g. FeOOH) adopt this structure. The structures of ramsdellite and psilomelane are related to those of hollandite.

The structure of the hexagonal tungsten bronze (HTB),A_xWO_3 (A = K, Rb, Tl, Cs, In) has tunnels with 120° angles. The host lattice $[WO_3]_\infty$ (Fig. 1.15) consists of corner-sharing WO_6 octahedra; the latter share all their corners in the three directions of the hexagonal cells resulting in the formation of hexagonal tunnels running along c. The A ions are located in these tunnels. The large size of these tunnels shows that such a

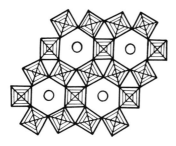

Figure 1.15   The hexagonal tungsten bronze (HTB), $K_{0.33}WO_3$ (projection onto the (001) plane). Notice the small triangular tunnels surrounding the big hexagonal tunnels. [Following Magnéli (1953)].

structure is readily obtained under normal pressure, for large A cations (with a size equal to or higher than that of potassium). For smaller A cations, the structure can be stabilized by application of high pressures. For $x = \frac{1}{3}$, the tunnels are fully occupied, whereas for $x < \frac{1}{3}$, the A cations are distributed statistically in the tunnels. Many tungsten, molybdenum and tantalum oxide systems form hexagonal tunnels. Diaspore, α-AlO(OH) has small hexagonal tunnels and its structure is derived from that of ramsdellite.

Oxides with pentagonal tunnels generally involve structures with both 90° and 120° angles. Representative of these oxides, is the large family of tetragonal tungsten bronzes (TTB), A_xWO_3 with A = K, In, Na, Ba, Pb. The corresponding tetragonal bronzes of molybdenum K_xMoO_3 are synthesized under high pressure. The tetragonal cell of these bronzes has a $[WO_3]_\infty$ host lattice but with corner-sharing WO_6 octahedra, giving rise to three kinds of tunnels: perovskite with a square section, triangular and pentagonal (Fig. 1.16). The first two types of tunnels were encountered earlier. The pentagonal tunnels are not generally characterized by a regular section, but exhibit one 90° angle, and four 120° angles. This geometry results from the fact that the host lattice consists of the perovskite structural units B_4O_{20} (90° angles) and HTB units (60° angles). In the TTBs, the perovskite and pentagonal tunnels can be partially filled by A cations (maximum $x = 0.6$). Several tantalates and niobates of divalent metals adopt the TTB structure.

Lamellar structures.   In these structures, octahedra form layers which are electrically neutral or charged and cohesion of the layers is through van der Waals interaction, hydrogen bonding or ionic bonding. Two cases must be considered in this category of oxides. When the octahedra share only their corners, the layers are sufficiently thick to avoid distortion. When thin layers are formed, the octahedra share both edges and corners.

Oxides of the formula $A_nB_nO_{3n+2}$ form a large family in which A is an alkaline earth and/or a lanthanide ion and B is titanium and/or niobium. Different members of this series are found in $A_2Nb_2O_7$–$CaTiO_3$ (A = Ca, Sr) and $Ln_2Ti_2O_7$–$CaTiO_3$ (Ln = La, Nd) systems. The ideal structure is orthorhombic, but monoclinic distortion often

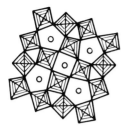

Figure 1.16   Structure of the tetragonal tungsten bronze (TTB), K_xWO_3 (projection onto the (001) plane). [Following Magnéli (1949)].

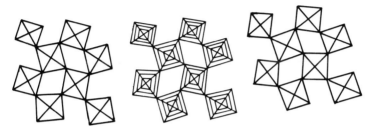

Figure 1.17   Layered perovskites, $A_nB_nO_{3n+2}$ ($n = 4$). [After Nanot et al., (1974)].

occurs. Thus, $Ca_2Nb_2O_7$ ($n = 4$), is found in both orthorhombic and monoclinic forms, differing with respect to the distortions and tilting of the NbO_6 octahedra. The other members of the family generally exhibit an orthorhombic cell with the a and c parameters close to those of $Ca_2Nb_2O_7$, but with a b parameter varying with the thickness of the octahedral layer. Structures of these oxides (Fig. 1.17) consist of the anionic layers, $[B_nO_{3n+2}]^{n'-}$ ($2n < n' < 3n$). The octahedral layers have a distorted perovskite configuration, leading to diamond-shaped tunnels (running along a), where the A cations are located. The integral number n corresponds to the number of BO_6 octahedra which determines the thickness of the perovskite slab. Non-integral n values lead to the formation of intergrowths between members of integral n (see Chapter 5 for a discussion of intergrowths).

In order to build thin octahedral layers, it is necessary to have a rigid chain of octahedra along one dimension. This implies that the octahedra should share their edges in order to prevent a tilting around the mean direction of the chain. Two ways of forming such chains can be considered. In the first, each octahedron shares two adjacent edges with its neighbours forming infinite chains of edge-sharing octahedra. These chains, called double ReO_3-type chains (DRC), share the octahedral edges (Fig. 1.18). The second way is simple, the octahedron sharing two opposite edges with its neighbours forming infinite chains of edge-sharing octahedra. These are the rutile chains (Fig. 1.18). It is easy to visualize how thin layers result from the association of such identical chains of octahedra.

Structures of many of the lamellar oxides result from the association of double ReO_3-type chains. Isolated chains have been observed in $NaMoO_3F$. Association of

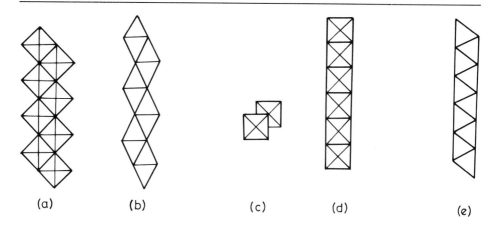

Figure 1.18   Double ReO_3-type chains (DRC) resulting from the association of two ReO_3-type chains through the edges.

such chains through octahedral corners leads to the formation of two ReO_3-type layers sharing their edges (which are in fact the (001) crystallographic shear planes encountered in many of the B_nO_{3n-1} oxides (see Chapter 5). The structure of MoO_3 involves such layers held together by van der Waals forces.

When two adjacent DRCs share one edge per four octahedra, we get oxides of the type $ThTi_2O_6$ and $A_xV_xMo_{1-x}O_3$ (A = alkali metal). When two successive DRCs share three edges per unit of four octahedra, one obtains a structure characterized by identical layers built up of infinite double ribbons of edge-sharing octahedra (Fig. 1.19(a)), found in several families of oxides. Four different families differing only in the relative positions of the octahedral layers are known: (i) $A_xTi_{2-y}B_yO_4$ (A = Rb, Tl, Cs, B = Mn, Sc, Al, Mg, Ni, Zn, Fe), exemplified by $Rb_xB_2O_4$ (Fig. 1.19(b)), (ii) $K_xTi_{2-x}B_yO_4$ (B = Mg, Ni, Cu, Fe, Zn), shown as $K_xB_2O_4$ (Fig. 1.19(c)), (iii) γ-FeOOH (Fig. 1.18(d)) and (iv) β-NaMnO$_2$ and $LiMnO_2$. From an examination of Fig. 1.19, we see that the structure of $K_xB_2O_4$ can be deduced from that of the $Rb_xB_2O_4$ by a translation of one layer out of two by half an octahedral edge (along the length of the octahedral ribbons), changing the pseudo-cubic coordination of the A ion (rubidium) into the prismatic trigonal coordination (potassium). The γ-FeOOH structure can be similarly derived from the $Rb_xB_2O_4$ structure by a simple translation of one layer out of two, by half the height of an octahedron in the direction perpendicular to the plane of the projection. The structures of β-NaMnO$_2$ and $LiMnO_2$ (which are close-packed) are closely related to the three other structures.

There are many other ways double ReO_3-type chains can be associated. Layered titantes, titanoniobates and tantalates of the general formula $A_nB_{2n}O_{4n+2}$ which exhibit ion exchange properties constitute an interesting mode of associating DRCs. These structures can also be described as built up of structural units of $2 \times n$ edge-sharing octahedra forming infinite ribbons which are n octahedra wide and

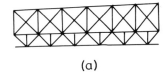

(a)

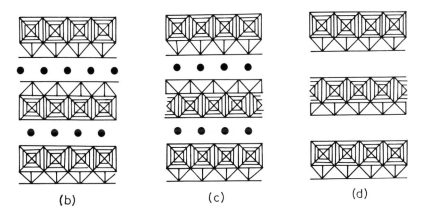

(b)   (c)   (d)

Figure 1.19 $A_xB_2O_4$: (a) infinite layer of octahedra built up from edge-sharing DRCs, (b) $Rb_xB_2O_4$, (c) $K_xB_2O_4$ and (d) -FeOOH [Following Raveau (1987)].

connected to each other through the corners of the octahedra. Three oxides corresponding to the $n = 3$ member of this series are: $Na_2Ti_3O_7$, $CsTi_2NbO_7$ and $A_3Ti_5BO_{14}$ (A = K, Rb Tl; B = Ta, Nb). These structures have identical $[B_3O_7]_\infty$ layers which result from the association of three DRCs through the edges and two DRCs through the corners alternately. Structures of these three compounds differ in the relative positions of the layers.

There are several families of layered oxides whose structures are built up of DRCs where some octahedra are missing. Typical examples are $K_4Nb_6O_{17}\cdot xH_2O$ and $K_{0.33}MoO_3$.

Mixed framework structures containing octahedra and tetrahedra.   Silicates, germanates and phosphates commonly form such mixed framework structures. These are the garnets, nasicons, benitoites, wadeites and milarites. The garnets constitute a large family of minerals $A_3B_2Si_3O_{12}$. $Y_3Fe_5O_{12}$ belongs to this family. Nasicons are silicophosphates of the formula $Na_{1+x}Zr_2P_{3-x}Si_xO_{12}$. Besides the phosphate tungsten bronzes discussed earlier, there are many phosphates where octahedral frameworks are accommodated.

We have not discussed the structures of intermetallic compounds. Girgis (1983) has reviewed the systematics in the structures of intermetallic compounds, classifying them into valence compounds, electron compounds and size-factor compounds.

1.5   Silicates and aluminosilicates

The basic building unit of all silicates is the SiO_4 tetrahedron. The low coordination number and the directional property of Si—O bonds owing to partial covalency render silicate structures less close-packed than other oxide structures. The complexity of silicate structures is due to the possibility of linking the SiO_4 tetrahedra in many ways and also because of the possibility of substitution of Si^{4+} by ions like Al^{3+}. Depending on the number of corners (0, 1, 2, 3 or 4) of the SiO_4 tetrahedron shared with other tetrahedra, various kinds of silicates (islands, single or double chains, rings, sheets, or three-dimensional networks) are obtained. When two tetrahedra are brought close together by sharing a corner, an edge or a face, the stability of the resulting unit decreases in the ratio 1.0:0.58:0.33, owing to electrostatic repulsion of the cations. In the silicate structure, only corners are shared. A variety of cations, mostly of alkali, alkaline earth or transition metals, is found in these silicates as charge-balancing entities. The Al^{3+} ion with a radius ratio of 0.36 with respect to an oxide ion can exist in four or six coordination; substitution of Al^{3+} for Si^{4+} in silicates has the effect of increasing the negative charge of the anion framework which is compensated by additional positive charge. These factors are responsible for the innumerable silicate structures (Fig. 1.20).

In orthosilicates, the SiO_4 tetrahedra do not share corners with one another. The isolated tetrahedra are joined by cations such as Mg^{2+}, Fe^{2+} or Al^{3+} (which occupy octahedral sites) or by Zr^{4+} (which is 8-coordinated). For example, forsterite (Mg_2SiO_4) can be considered as a packing of SiO_4 tetrahedra and MgO_6 octahedra. Forsterite and fayalite (Fe_2SiO_4) form a complete solid solution series of *olivines*. In *garnets* $(A_3^{2+} B_2^{3+} (SiO_4)_3)$, extensive ionic substitution with A^{2+} = Mg, Fe, Mn or Ca (CN = 8) and B^{3+} = Cr, Al or Fe (CN = 6) is possible. The Si:O ratio can become larger than 4 if the cation requires oxygen to form a complex ion, as in $(Al_2O)^{4+}$ of kyanite (Al_2SiO_5) and its polymorphs, sillimanite and andalusite. Island structures are also obtained when two SiO_4 tetrahedra share a common oxygen, giving rise to a $(Si_2O_7)^{6-}$ unit as in pyrosilicates.

When two oxygens of each SiO_4 tetrahedra are shared with others, ring anions as well as linear chain anions are formed. Three SiO_4 tetrahedra can form a ring anion $(Si_3O_9)^{6-}$ as in benitoite $(BaTiSi_3O_9)$; six SiO_4 tetrahedra can form a hexagonal ring anion $(Si_6O_{18})^{12-}$ which occurs in beryl $(Be_3Al_2Si_6O_{18})$. Linear single chain anions $(SiO_3^{2-})_n$ are found in enstatite $(MgSiO_3)$ and diopside $(MgCaSi_2O_6)$; these are called *pyroxenes*. When alternate SiO_4 tetrahedra share two and three corners, $2\frac{1}{2}$ oxygens are shared on the average, resulting in the double chain anions $(Si_4O_{11}^{6-})_n$ present in *amphiboles*. A typical amphibole is tremolite, $Ca_2Mg_5(Si_4O_{11})(OH)_2$. Amphiboles are similar to pyroxenes, except for the presence of OH^- and F^- ions in the former. Amphiboles and pyroxenes exhibit directional characteristics in their properties as in asbestos because of the presence of chain silicate anions. Although the term asbestos was originally used to denote fibrous amphiboles such as tremolite containing Si_4O_{11} double chains, commercial asbestos mainly consists of chrysotile which is a layer silicate, $Mg_3Si_2O_5(OH)_4$.

When three corners of each SiO_4 tetrahedron are shared, a two-dimensional sheet

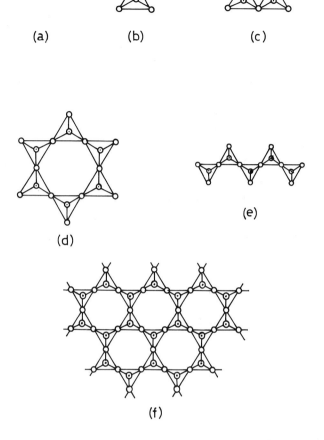

(a)   (b)   (c)

(d)

(e)

(f)

Figure 1.20   Anions in silicate structures: (a) SiO_4^{4-}; (b) $Si_2O_7^{6-}$; (c) $Si_3O_9^{6-}$; (d) $Si_6O_{18}^{12-}$; (e) $(SiO_3^{2-})_n$ chain and (f) $(Si_2O_5^{2-})_n$ layer.

structure with the formula $(Si_2O_5^{2-})_n$ is formed. *Clay minerals* belong to such a family of phyllosilicates containing continuous tetrahedral sheets of composition $T_2O_5 (T =$ Si, Al, Fe, etc.); the tetrahedral sheets are linked in the unit structure to octahedral sheets and to groups of coordinated (or individual) cations (Brindley & Brown, 1980). The unit formed by linking one octahedral sheet to one tetrahedral sheet is referred to as a 1:1 layer and the exposed surface of the octahedral sheet consists of OH groups; a similar linkage can occur on the other side of an octahedral sheet to form a 2:1 layer, both surfaces of such a layer consisting of the hexagonal mesh of basal oxygens (Fig. 1.21). There are structures where the layers are not neutral and charge balance is maintained by interlayer cations. Clay minerals can be classified on the basis of the layer type (1:1 or 2:1), charge on the layers and the material between the layers.

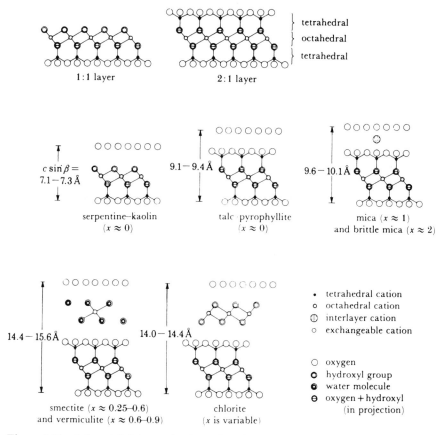

Figure 1.21   1:1 and 2:1 layers in clay minerals and structures of typical members. (After Brown, 1984.)

Structures of typical groups are shown in Fig. 1.21. Further subdivision is possible on the basis of the nature of the octahedral sheets (di- or tri-octahedral) and stacking sequences.

Kaolinite is obtained by a stacking of $(Si_2O_5^{2-})_n$ and $Al_2(OH)_4^{2-}$ to give $Al_2Si_2O_5(OH)_4$; here, two-thirds of the octahedral voids in a close-packed layer of OH^- ions are ocupied by Al^{3+} ions, giving the $Al_2(OH)_4$ di-octahedral unit. Pyrophyllite, $[Al_2Si_4O_{10}(OH)_2]_2$ is obtained by sandwiching a sheet of gibbsite, $Al(OH)_3$, between two silicate sheets. If the di-octahedral gibbsite is replaced by tri-octahedral brucite, $Mg(OH)_2$, we obtain talc, $[Mg_3Si_4O_{10}(OH)_2]_2$. If part of the Si^{4+} ions are replaced by Al^{3+} ions and alkali ions are introduced to maintain charge neutrality, we obtain micas, a typical example being muscovite, $[KAl_3Si_3O_{10}(OH)_2]_2$. In Fig. 1.21, structures of typical clay minerals are shown.

If all the four corners of the SiO_4 tetrahedra are shared, three-dimensional networks result. The different forms of silica (quartz, tridymite and cristobalite) discussed earlier in this section are typical examples. Feldspars are generated when part of the Si^{4+} is

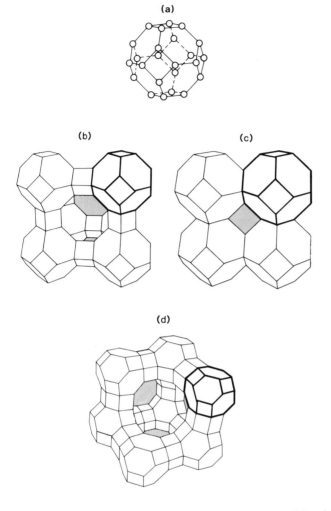

Figure 1.22   (a) Sodalite cage or β-cage; (b) zeolite A; (c) sodalite and (d) faujasite (zeolite X and Y).

replaced by Al^{3+} and charge balance achieved by locating large cations in interstitial positions. Examples are albite, $NaAlSi_3O_8$, and orthoclase, $KAlSi_3O_8$. If the channels are large with a relatively open framework occupied by Na^+, K^+ or Ca^{2+}, zeolites and ultramarines are obtained. A brief introduction to *zeolite* structures is in order in view of their great importance as sorbents and catalysts.

1.5.1   Porous solids

Zeolites are intrinsically microporous aluminosilicates of the general formula $M_{x/n}$ $[(AlO_2)_x(SiO_2)_y]\cdot mH_2O$ and may be considered as open structures of silica in which aluminium has been substituted in a fraction $x/(x + y)$ of the tetrahedral sites. The net negative charge of the aluminosilicate framework is neutralized by exchangeable cations, M, of valence n. The void space, which can be greater than 50% of the volume, is occupied by m molecules of water in the unit cell. The truncated octahedron (cubo-octahedron) shown in Fig. 1.22(a) is the building block of zeolites; this is also called a β-cage or sodalite cage. Tetrahedral atoms denoted as o in Fig. 1.22(a) are present at the corners of polygons with oxygen atoms (not shown in the figure) approximately half-way between them. The porosity consists of one-, two- or three-dimensional networks of interconnected channels and cavities of molecular dimensions, depending on the structure of the zeolite. Accordingly, zeolite A is formed by linking sodalite cages through double four-membered rings while sodalite is formed by direct face-sharing of four-membered rings in neighbouring sodalite cages; faujasite (zeolite X and Y) is formed by linking the sodalite cages through double six-membered rings (see Fig. 1.22). Thirty-seven zeolite minerals and a large number of synthetic ones are described in the literature (Meier & Olson, 1978). Much of the advance in zeolite chemistry is due to Barrer (1978) who suggested the generic term porotectosilicate to describe siliceous crystalline materials including both zeolites and solids that resemble them in crystal structures but do not possess ion exchange properties.

The unit cell stoichiometry of natural faujasite is $Na_{56}[(AlO_2)_{56}(SiO_2)_{136}]\cdot 250H_2O$ and that of Linde A (synthetic) is $\{Na_{12}[(AlO_2)_{12}(SiO_2)_{12}]\cdot 27H_2O\}_8$. Elements such as Be, Mg, B, Ga and P can take the place of Si and Al in the zeolitic framework. Zeolites can be entirely dealuminated using $SiCl_4$ vapour to give rise to zeolite silicas; 'ultrastabilization' of zeolites is effected by heat treatment of NH_4^+-exchanged zeolites where NH_4^+ have replaced Na^+ ions. Specially noteworthy is the new class of highly siliceous zeolites with an optimal pore diameter of 5.5 Å; the ZSM-5 (Zeolite Socony Mobil) is one such zeolite with fantastic catalytic properties (e.g. methanol–gasoline conversion). ZSM-5 has the formula $H_x[(AlO_2)_x(SiO_2)_{96-x}]\cdot 16H_2O$ and has a pentasil unit as its building block; the pentasil units form chains which in turn are interlinked to form the three-dimensional framework with 10-membered ring openings of a diameter of ~ 5.5 Å as shown in Fig. 1.23. The Si, Al ordering within the tetrahedral framework is difficult to determine by conventional structural methods, but one of the guidelines has been that two Al atoms cannot be adjacent on tetrahedral sites (*Lowenstein's rule*).

The *chabazitic group* of zeolites which includes offretite, erionite and levyne appears at first to consist of unrelated members with unusual compositions; however, if we recognize that the tetrahedral sites can be occupied by AlO_4 or SiO_4 tetrahedra, we readily see the systematics in the stacking sequence in this family just as in polytypes (see section 1.8). For example, offretite, erionite and levyne with the compositions $(Na_2, Ca\cdots)_2Al_4Si_{14}O_{36}$, $(Na_2, Ca\cdots)_{4.5}Al_9Si_{27}O_{72}$ and $Ca_9Al_{18}Si_{36}O_{108}$ respectively (and repeating stack distances of 7.6, 15.1 and 23 Å) have the stacking sequences AAB, AABAAC and AABCCABBC.

Besides the conventional zeolites, several novel zeolite analogues such as the ALPOs (aluminophosphates), MeALPOs (divalent-metal (Me) substituted aluminophosphates), SAPOs (silicon substituted aluminophosphates) and so on have been synthesized (Davis & Lobo, 1992). Wilson *et al.* (1982) first reported the synthesis of microporous ALPOs. ALPO synthesis differs from zeolite synthesis in that it involves acidic or mildly basic conditions and no alkali metal ions. Some members in the ALPO

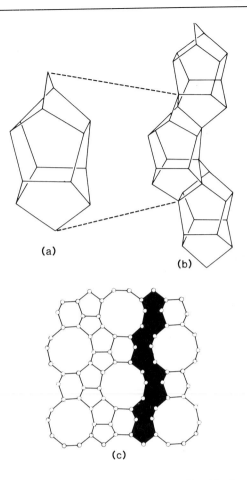

Figure 1.23. (a) Pentasil unit (building block of ZSM-5 and ZSM-11); (b) chain of pentasil units; (c) chains are interlinked to form the framework with 5.5 Å openings (tetrahedral sites are shown by circles).

family have structures identical to those of zeolites (for example, ALPO-20 is isostructural with sodalite, ALPO-17 has the erionite structure); others, such as VPI-5, have no zeolite analogue. VPI-5 is noteworthy in that it contains 18-ring hexagonal channels with a free pore diameter of ~ 13 Å (Davis et al., 1988, 1989). A gallophosphate called cloverite has a 30 Å supercage which can be accessed through six clover-shaped pores (Estermann et al., 1991). DAF-1 is a member of the MeALPO family that exhibits strong Bronsted acidity (Wright et al., 1993). The material having the framework composition $Mg_{0.22}Al_{0.78}PO_4$ contains an unique double-barrelled pore structure. Zeolitic phosphates containing exclusively divalent cations are also known (Harvey & Meier, 1989; Gier & Stucky, 1991). For instance, beryllium and zinc phosphate analogues of sodalite, rho, gismondine, edingtonite and pollucite have been synthesized. Specially interesting is the cobalt (II) phosphate, $CoPO_4 0.5H_2en$

(en = ethylenediamine), named DAF-2, where Co(II) is exclusively in tetrahedral coordination (Chen *et al.*, 1994).

$[C_5H_5NH]^+$ $[CoGa_2P_3O_{12}]^-$ is a microporous phosphate that contains tetrahedrally coordinated Co(II) (Chippindale & Walton, 1994). Use of different templating agents in hydrothermal synthesis (see section 3.2.3) has yielded a rich diversity of vanadium phosphates with open-framework or layered structures (Zhang *et al.*, 1995; Soghomonian *et al.*, 1995). Ethylenediamine template has yielded two open-framework phosphates, abbreviated as V_9P_8-en and V_7P_8-en, containing 8- and 16-membered rings formed by VO_5 pyramids and PO_4 tetrahedra. On the other hand, use of piperazine and 1,4-diazabicyclo[2,2,2]octane as templating agents results in two-dimensional V–P–O solids containing alternating inorganic and organic layers. Holscher *et al.* (1994) have synthesized a unique microporous solid, $[H_3N(CH_2)_6NH_3]$ $[W_{18}P_2O_{62}]\cdot3H_2O$, built up from Dawson anions, $[W_{18}P_2O_{62}]^{8-}$ and protonated 1,6-diamino-hexane units. Discrete $[W_{18}P_2O_{62}]^{8-}$ units are cross-linked *via* hydrogen bonds with the $(H_3N(CH_2)_6NH_3)^{2+}$ units, the network being unique in the sense that there are no iono-covalent bonds between the hetero-polymetallate ions. The role of templates in the formation of microporous solids has been examined by computer modelling techniques (Lewis *et al.*, 1995). Two complementary criteria seem to be important: a favourable nonbonding interaction between the template and the framework and the efficient packing of the template molecules within the framework.

A major development in this field is the synthesis of silica-based *mesoporous solids* with adjustable pore size (Kresge *et al.*, 1992; Beck *et al.*, 1992). A typical example is MCM-41 that possesses a hexagonal arrangement of uniform mesopores whose dimensions can be engineered in the range of ~ 15 to $\gtrsim 100$ Å. The synthesis of these materials has been achieved by an entirely new strategy that makes use of a self-assembled molecular array, instead of an isolated molecule or ion as the templating agent. Typically, quaternary ammonium ions, $C_nH_{2n+1}(CH_3)_3N^+$, which are known to form micelles in aqueous solution, are used as templates in the synthesis. Formation of mesoporous solids of this kind seems to involve a 'liquid crystal templating' mechanism wherein the inorganic silicate/aluminate ions build 'walls' around organic template assemblies which are already ordered in the gel. Significantly, while the arrangement of pores is very regular in these materials, considerable disorder exists in the inorganic 'walls' that form the porous structure. The alkyl chain length n in $C_nH_{2n+1}(CH_3)_3N^+$ determines the pore size; when $n = 12, 14$ and 16, the pore sizes are 30, 34 and 38 Å respectively. Stucky and coworkers (Huo *et al.*, 1994) have generalized this approach to enable the synthesis of periodic mesoporous materials of a variety of metal oxides including tungsten, antimony, lead, aluminium, iron and zinc. The strategy is to make use of cationic surfactants as templates for anionic inorganic species and anionic surfactants for cationic inorganic species.

While most of the microporous solids exclusively consist of tetrahedrally coordinated metal atoms, progress has also been made in the synthesis of materials containing octahedrally coordinated metal atoms. Typical examples of this category include reduced molybdenum phosphates containing octahedral- frameworks, such as $(Me_4N)_{1.3}(H_3O)_{0.7}[Mo_4O_8(PO_4)_2]\cdot2H_2O$ and $(NH_4)[Mo_2P_2O_{10}]\cdot H_2O$. (Haushalter

& Mundi, 1992). Interestingly, the latter is isostructural with the mineral leucophosphite, $K[Fe_2(PO_4)_2(OH)] \cdot 2H_2O$. Cacoxenite, $AlFe_{24}O_6(OH)_{12}(PO_4)_{17} \cdot 17H_2O$, is another mineral with an octahedral–tetrahedral framework containing large pores of 14.2 Å diameter. Microporous materials consisting of only metal–oxygen octahedra are also known, typical examples being hexagonal molybdenum–vanadium oxides, $A_{0.13}V_{0.13}Mo_{0.87}O_3$ (Feist & Davies, 1991) and their tungsten analogues (Gopalakrishnan *et al.*, 1994), as well as the 3×3 octahedral manganese molecular sieve that is analogous to the mineral todorokite (Shen *et al.*, 1993).

Nonoxide microporous materials containing polychalcogenide open-frameworks have been synthesized, an example being $(Ph_4P)[M(Se_6)_2]$ (M = Ga, In, Tl) (Dhingra & Kanatzidis, 1992). Microporous solids composed of organic molecules 'organic zeolites', have also been realized (Moore & Lee, 1994). The strategy here is to prevent closest packing of organic molecules in the solid state by choosing multiple functional groups that engage in strong and directional bonding (hydrogen bonding or dative bonds) symmetrically disposed on a rigid skeleton. Hexaphenylacetylene is an example of an organic zeolite crystallizing in a porous structure with pores of 8 Å diameter running parallel to the *c*-axis.

1.6   Nonbonded interactions in ionic crystals

Values of ionic radii used at present are based on the notion of anion–anion contact. Anion–anion contact does not, however, exist in real crystals (eutaxy), as pointed out earlier, and the possibility of cation–cation contact in crystals has been ignored until recently. The existence of a nonbonded radius for cations, which limits cation–cation approach in crystals, has been demonstrated by O'Keeffe & Hyde (1981, 1982) by consideration of crystal structures of SiO_2 and other B9 type solids. The idea of nonbonded radii for atoms themselves was originally suggested by Bartell (1968) to explain certain molecular geometries; for example, the nonbonded Si $\cdots$ Si distance in the disilyl ether, amine and methane is constant (~ 3.06 Å) irrespective of the bridging atom. Bragg had earlier noticed the constancy of Si $\cdots$ Si distances in singly bridged chain silicates, but its significance was not realized. It can now be appreciated that a conspicuous feature of single-bonded silicate structures is the large angle ($\sim 145°$) at the oxygen atom bridging two SiO_4 tetrahedra. In the high and low forms of cristobalite, Si—O—Si angle is $\sim 145°$, although the angle in principle can vary from 180° to 109.5°. An early interpretation of the large Si—O—Si angles invoked d_π–p_π bonding between silicon and oxygen, but the fact that BeF_2 also adopts the cristobalite structure with a similar angle rules out this explanation. The real reason seems to be the nonbonded interaction between silicon atoms; Si—O—Si and other similar singly bridged T—X—T (T = tetrahedrally coordinated cation and X = anion) angles in 'tetrahedral' structures are determined by nonbonded Si $\cdots$ Si(T $\cdots$ T) contacts. From an examination of 141 silicate structures, O'Keeffe & Hyde (1979) find that Si $\cdots$ Si contact distances are around 3.06 Å in a majority of them, yielding a nonbonded radius of 1.53 Å for silicon. The case for nonbonded radii of cations in crystals becomes much more compelling since similar nonbonded contacts seem to exist in structures containing other angular species such as P—O—P, Ge—O—Ge and Al—O—Si. In mixed cases, the nonbonded distances are the sum of the individual nonbonded radii.

A natural question that arises is whether there is any correlation between Si—O—Si (T—X—T) angles (θ) and Si—O (T—X) bond lengths, l. Geometrically, the Si $\cdots$ Si(T $\cdots$ T) distance, d, and l are related by the expression, $l = (d/2) \operatorname{cosec}(\theta/2)$. A correlation between l and θ should exist when cations are in contact (d is constant). When θ is large, no Si $\cdots$ Si contact exists and therefore there should be no correlation between d and l. A correlation would exist between l and d when $\theta \leqslant 145°$, thus indicating that 'tetrahedral' structures are determined by T—X and T $\cdots$ T interactions rather than X $\cdots$ X interactions.

Nonbonded radii enable us to rationalize aspects of crystal chemistry that cannot be understood in terms of the ionic model. For instance, it is difficult to understand why silicon is four-coordinated in most oxides. This is unlikely to be because of the radius ratio of the classical ionic model; Si^{4+} is too small to have four oxygens around it. Six-coordinated silicon is known in certain cases (e.g. SiP_2O_7); silicon is octahedrally coordinated in fluorides, although the ionic radii of O^{2-} and F^- are similar. The reason seems to be that if the coordination number of silicon is more than four, that of oxygen must be more than two; a three coordination of oxygen would require a Si—O—Si angle of 120°. With a normal Si—O bond length of 1.6 Å, $\theta \leqslant 120°$ requires $d(Si \cdots Si) \leqslant 2.77$ Å, which is much smaller than the normal $d(Si \cdots Si)$ of 3.06 Å. In SiP_2O_7, however, the presence of higher-valence phosphorus raises the anion/cation ratio and thus retains oxygen in two coordination. Structures of this type seem to be therefore determined by the cation–cation nonbonded repulsion and not by the cation-to-anion radius ratio or anion–anion repulsion. Anion-centred polyhedra, on the other hand, could be more important in deciding the stability of phases than the conventional cation-centred polyhedra. Thus, the different forms of SiO_2 transform to stishovite (rutile type) at high pressure, not because SiO_4 tetrahedra are converted to SiO_6 octahedra at high pressures but because pressure is necessary to convert OSi_2 to OSi_3 units. In TiO_2, the presence of two long and four short Ti—O distances is due to nonbonded interaction between the oxide ions.

O'Keeffe & Hyde (1978, 1979) discuss several applications of nonbonded radii in crystal chemistry. It is possible that the ratio of nonbonded radius to T—X bond length, R/l, is more significant than the r_c/r_a ionic radius ratio in deciding structures. Values of R and l in 'tetrahedral' oxygen compounds together with the R/l ratios are given in Table 1.6. In one-angle situations, the critical ratios for anion coordination are: $R/l \leqslant \sin 180°/2 = 1.0$ for two coordination; $R/l \leqslant \sin 180°/3 = 0.866$ for three coordination; $R/l \leqslant \sin 109°28'/2 = 0.816$ for four coordination; $R/l \leqslant \sin 90°/2 = 0.707$ for six coordination and $R/l \leqslant 1/\sqrt{3} = 0.577$ for eight coordination. When the R/l ratio is unfavourable in a solid, alternative structures become competitive. In diamond, R/l is 0.812 and it is therefore not the stable modification; three-coordinated graphite is more stable. Another mode of accommodating unfavourable R/l ratios is to replace one-angle connection by two-angle connection between polyhedra (edge-sharing of cation-centred polyhedra in place of corner-sharing).

Table 1.6. *Nonbonded radius R and bond length to oxygen l in tetrahedral compounds and their ratio*[a]

Atom	R(Å)	l(Å)	R/l
Li	1.50	1.97	0.76
Be	1.35	1.65	0.82
B	1.26	1.49	0.85
O	1.12	—	—
Na	1.68	2.37	0.71
Mg	1.66	1.95	0.85
Al	1.62	1.77	0.92
Si	1.53	1.64	0.93
P	1.46	1.55	0.94
S	1.45	1.50	0.97
Zn	1.65	1.98	0.83
Ge	1.58	1.77	0.89

[a] From O'Keeffe & Hyde (1981).

1.7 New ways of looking at the structures of inorganic solids

Description of crystal structures of solids so as to bring out interrelationships among different families is an important aspect of crystal chemistry. The commonly used description is based on linkage of cation-centred coordination polyhedra through corners, edges and/or faces: The ReO_3 structure is described as ReO_6 octahedra connected through all corners; NaCl structure as $NaCl_6$ octahedra sharing all edges and so on. The approach has been extended by Simon (1981, 1982) to describe structures of condensed metal cluster compounds. Two important octahedral metal clusters M_6X_8 and M_6X_{12} (Fig. 1.24) are recognized as fundamental structure-building units from which the structures of more complex metal cluster solids can be derived. Condensation of the M_6X_8 clusters through the opposite corners of the M_6 octahedron results in a one-dimensional column structure typified by Ti_5Te_4 (Fig. 1.25(a)). Similarly, linking of M_6 octahedra of M_6X_8 by *trans*-edges produces the chain structure of Gd_2Cl_3 (Fig. 1.25(b)); linking of opposite faces of the M_6 octahedra of M_6X_8 results in one-dimensional *Chevrel* phases, KMo_3S_3 and $TlFe_3Te_3$ (Fig. 1.25(c)). The cluster compound $NaMo_4O_6$ is built up of Mo_6O_{12} clusters by *trans*-edge linking of Mo_6 octahedra. The structure consists of chains of Mo_4O_6 running parallel to the tetragonal c-axis and alkali metal atoms are present in between the chains (Fig. 1.26). The structure of NbO itself can be regarded as made up of Nb_6O_{12} clusters linked three-dimensionally through the corners of Nb_6 octahedra, $Nb_{6/2}O_{12/4}$.

Although the connected coordination polyhedron approach is useful in crystal chemistry, it has certain implicit rigidity which makes it difficult to extend to less regular and more complex structures. Several alternative approaches to the description

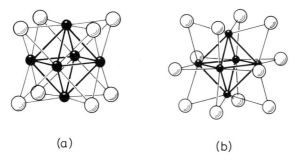

(a)   (b)

Figure 1.24   The structures of (a) M_6X_8 and (b) M_6X_{12} clusters.

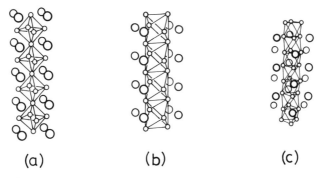

(a)   (b)   (c)

Figure 1.25   The structures of (a) Ti_5Te_4; (b) Gd_2Cl_3 and (c) KMo_3S_3 formed by condensation of octahedral metal clusters. (After Simon, 1981.)

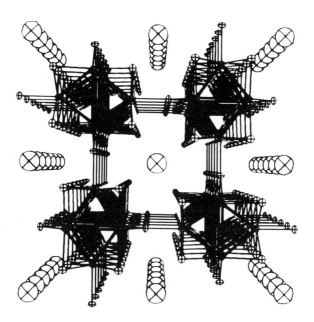

Figure 1.26   The structure of $NaMo_4O_6$. (After McCarley, 1982.)

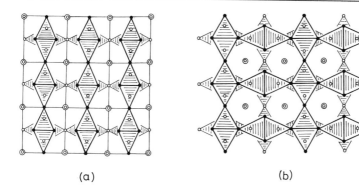

Figure 1.27 (a) The structure of Cr_3Si. Filled and open circles are Cr on $\frac{1}{2}$ and 0 respectively. Double circles, two Cr on $\frac{1}{4}$ and $\frac{3}{4}$. Si on 0 or $\frac{1}{2}$ where tetraedersterns share corners; (b) The structure of Zr_4Al_3 projected down [110]. Open and filled circles are at 0 and $\frac{1}{2}$. Double circles (aluminium) at $\frac{1}{4}$ and $\frac{3}{4}$. The aluminium atoms are also situated at the tetraedersterns sharing corners. The rest are Zr. (After Andersson, 1978.)

of crystal structures of nonmolecular solids have emerged in recent years. One of them is to regard complex structures as built up of simple structural units which are related by common crystallographic operations such as translation, rotation and reflection as well as intergrowth (Hyde *et al.*, 1974; Hyde, 1979; Andersson, 1983). When these operations are repeated at regular intervals, large superstructures are generated from simple prototype sructures, but when the operations are carried out irregularly, it provides a means of describing 'defective' structures. By employing the translation operation, *crystallographic shear structures* (e.g. $W_{20}O_{58}$) can be shown to be derived from the parent ReO_3 and rutile structures (Magnéli, 1978) and we shall discuss this later in Chapter 5. The translation operation can be used to derive structures of alloys (for example, $CuAl_2$ and Zr_4Al_3 structures from β-W (or Cr_3Si) structure (Douxing *et al.*, 1983)). In Fig. 1.27 we show the structure of Cr_3Si consisting of strings of Cr atoms with the Si atoms in the interstices (the basic unit of the structure is the tetraederstern formed by five irregular tetrahedra) along with the structure of Zr_4Al_3. Structures of transition-metal borides are also related by the translation operation.

By employing the rotation operation, structures of tetragonal tungsten bronze as well as Mo_5O_{14} can be generated from the ReO_3 framework, the operation converting eight square tunnels around each column into four pentagonal and four triangular tunnels (Hyde *et al.*, 1974) as shown in Fig. 1.28. The structure of W_5Si_3 is related to Cr_3Si by means of such a rotation as shown by Andersson (1983). Hyde *et al.* (1980) and Parthe (1981) have shown how the reflection (*twinning*) operation is applicable in describing the structures of a large number of solids such as metal alloys, metal carbides, borides, oxides, sulphides and silicates. In a *twinned crystal*, the twin plane constitutes a mirror plane which relates the twin individuals. The reflection operation resulting from twinning can be readily understood in the case of close-packed arrays of atoms. For example, reflection twinning in a HCP array of atoms gives rise to a new type of interstitial sites (trigonal prisms) at the twin plane, $(11\bar{2}2)$ as shown in Fig. 1.29; if

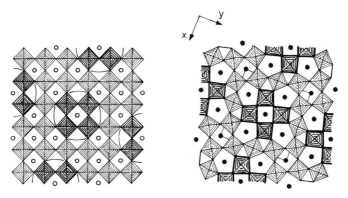

Figure 1.28   Conversion of ReO_3 framework (left) to the TTB framework (right) by ordered rotation operation. (After Hyde, 1979.)

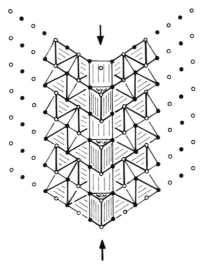

Figure 1.29   An HCP array of atoms projected on ($1\bar{1}00$) showing reflection twinning on ($11\bar{2}2$) plane. (After Hyde *et al.*, 1980.)

these sites are occupied by a different set of atoms, the twin plane becomes a composition plane where the two twin individuals meet. Repeated twinning operations can generate new families of superstructures. Fe_3C (cementite) is an example of a structure based on HCP twinning; structures of Fe_5C_2, Pd_5B_2 and several rare earth alloys are related to Fe_3C. Such twinned structures are seen in $LuFeO_3$, A-site deficient perovskites, $CaTa_2O_6$ and $LaNb_3O_9$. The concept of twinning can be extended to 'multiling'; the silicate, paulingite, is shown to result by a multiling operation on gismondine (Andersson & Fälth, 1983).

Other ideas found useful in describing complex crystal structures are *planar nets* and *rod packing*. The idea of planar nets, which has been explored at length by O'Keeffe & Hyde (1980), is to regard complex crystal structures as stackings of two-dimensional

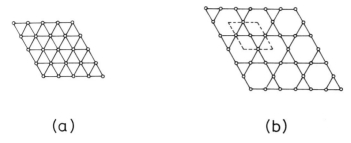

Figure 1.30   (a) 3^6 net and (b) 3.6.3.6 Kagomé net. (After O'Keeffe & Hyde, 1980.)

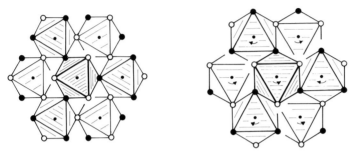

Figure 1.31   Transformation of octahedral network of ReO_3(left) to that of PdF_3 (right). Arrows represent the sense of octahedral rotations for the reverse transformation. (After O'Keeffe & Hyde, 1980.)

planar nets of atoms. The nets need not be close-packed (eutactic) arrangements. These workers have systematically derived all the two-dimensional periodic nets, giving examples of their occurrence in compounds of established structure, and have discussed transformations between nets with examples. One of the simplest planer nets is 3^6; the notation indicates that there are six triangles meeting at any point in the net. Another related net is the so-called Kagomé net, 3.6.3.6 (Fig. 1.30). The 3^6 net can be transformed into the 3.6.3.6 net by rotation of triangular groups of atoms. The {111} anion layers of the ReO_3 structure correspond to the Kagomé net. In this structure, the stacking of the nets is as in cubic close-packing, ABC .... Such a sequence of Kagomé nets can be transformed into hexagonal close-packing, ABAB .... or primitive hexagonal packing, AAA ... , or BBB ... , or a mixture of the two by a simple transformation of the 3.6.3.6 net into the 3^6 net. Such a CCP ↔ HCP transformation, which retains corner linking of the octahedra, would exactly correspond to the transformation of the ReO_3 structure to the PdF_3 structure as was first pointed out by Jack & Guttmann (1951). The transformation is elegantly described as rotation of alternate octahedra by $\pm 30°$ about a trigonal axis (Fig. 1.31). The topotactic transformation of hexagonal $LiNbO_3$ to cubic $HNbO_3$ (Rice & Jackel, 1982) suggests that the mechanism is indeed operative in real chemical systems.

Certain crystal structures can be considered as built up of closely packed cylindrical rods of polyhedra or strings of atoms as shown in Fig. 1.32 (O'Keeffe & Andersson, 1977). The hexagonal honeycomb-like packing of cylindrical rods corresponds to the

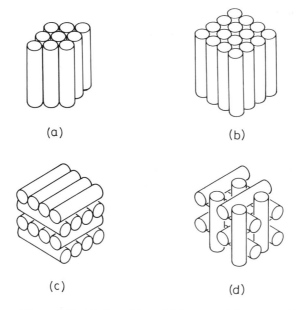

(a)   (b)

(c)   (d)

Figure 1.32   Rod packings: (a) hexagonal (honeycomb) packing; (b) tetragonal packing; (c) tetragonal layer packing and (d) primitive cubic rod packing. (After O'Keeffe & Andersson, 1977).

densest packing, with a density 0.9069. In the square-packing of rods giving a tetragonal cell, the packing density is 0.7854; an example of this packing is found in the structure of $CuCl_2 \cdot 2H_2O$, which consists of strings of copper atoms along [001] with approximately square groups normal to the strings. A two-layer rod packing (Fig. 1.32(c)) in which the rods are not parallel gives the same packing density as in square-packing; an example of a structure which corresponds to this packing is PtO. The rod packing shown in Fig. 1.32(d), possessing primitive cubic symmetry, has a packing density 0.589; structures of β-W(A15) and Cr_3Si correspond to this type of packing.

1.8   Polytypism

Certain solids crystallize in more than one modification, which are chemically and crystallographically similar but differ in one of the unit cell dimensions (usually the c-parameter). This phenomenon, which may be regarded as polymorphism in one-dimension, is called polytypism (Verma & Trigunayat, 1974; Rao & Rao, 1978; Krishna, 1983). Polytypism is generally exhibited by solids having close-packed and layered structures where the primary coordination around an atom is satisfied in more than one way, as in cubic versus hexagonal close-packing. Consequently, polytypes of a substance have the same primary coordination environment. Polytypes are composed of structural units that are identical in two dimensions (layers) and these units are stacked one over the other. Because of the structural identity of the constituent layers, polytypes of a given solid have the same unit cell dimensions in the layer plane but a different dimension in the direction normal to the layers. Another consequence of the identity of the layers is that the variable unit cell dimensions of all polytypes of a particular substance are integral multiples of a basic repeat distance which corresponds to the thickness of a single layer.

Although polytypism appears to be structurally similar to polymorphism, in terms of thermodynamics they are distinct phenomena with marked differences. Whereas polymorphic changes are in many ways analogous to changes of state from solid to liquid or liquid to vapour, describable in terms of the Gibbs' phase rule and the Clausius–Clapeyron equation (see Chapter 4), polytypism defies such a thermo-dynamic description. Different polytypes of a substance are frequently formed under similar thermodynamic conditions, often coexisting in the same preparation. Polytypes are known having giant unit cell dimensions (of a few thousand Å) along the unique axis, where the long-range order extends over large distances.

Polytypism was first discovered in SiC by Baumhauer in 1912. SiC exists in two polymorphic forms: the α form having a hexagonal structure and the β form with a cubic structure. It is the α form that exhibits a large number (about 130) of polytypes. The hexagonal c-parameter in these polytypes varies from ~ 5 Å in the case of the two-layer (2H) polytype, to ~ 15 Å in the case of the most common six-layer (6H) polytype and to $\sim 12\,000$ Å in the 4680-layer polytype. Polytypism is now known to occur in a variety of inorganic solids such as ZnS, CdI_2, PbI_2, layered transition metal chalcogenides, layered silicates and several ABX_3 perovskites. In CdI_2, each I—Cd—I sandwich is 3.42 Å in thickness; two such sandwiches constitute the simplest (2H) polytype. The c-parameter of CdI_2 polytypes vary from 6.84 Å in the case of 2H to ~ 410 Å in the case of 120 R. PbI_2 crystallizing in the same structure as CdI_2 also exists in several polytypes (about 40). Layered transition-metal dichalcogenides MX_2 (M = Nb, Ta, Mo; X = S, Se) crystallize in hexagonal or rhombohedral structures consisting of X—M—X sandwiches. The coordination of the M atom is either octahedral (AbC), as in the CdI_2 structure, or trigonal prismatic (AbA), as in the MoS_2 structure. The X—M—X sandwiches stack in different ways, giving purely octahedral (1T), purely trigonal prismatic (2H, 3R, $4H_c$) or mixed coordination ($4H_b$, 6R) polytypes; some of these are shown in Fig. 1.33. The octahedral form (1T) is usually stable at high temperatures and may be retained at room temperature by quenching from above 1100 K in an atmosphere of the chalcogen.

Polytypism in layered silicates has been well characterized. There are as many as 19 polytypes known among micas. The repeat unit in muscovite mica, $[KAl_2(OH)_2(Si_3Al)O_{10}]$, for instance, consists of a sheet of octahedrally coordinated aluminium ions sandwiched between two identical sheets of (Si, Al)O_4 tetrahedra, the large K^+ ions being located in interlayer positions. Surface oxygens of the tetrahedra in the individual layers possess pseudohexagonal symmetry, which permits individual composite layers to be oriented in one of six possible ways. The fact that all the orientations are equivalent appears to be responsible for the occurrence of polytypism in micas.

Several ABO_3 oxides, where A is a large cation like Ba and B is a small cation of the d-transition series, are known to exhibit polytypism. These oxides consist of close-packed, ordered, AO_3 layers which are stacked one over the other and the B cations

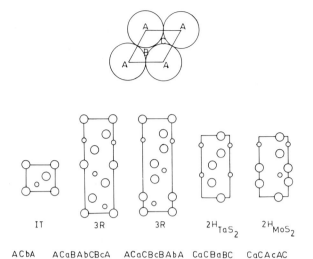

Figure 1.33. Polytypes of transition-metal dichalcogenides ($11\bar{2}0$ sections): *large* circles, chalcogen; *small* circles, metal.

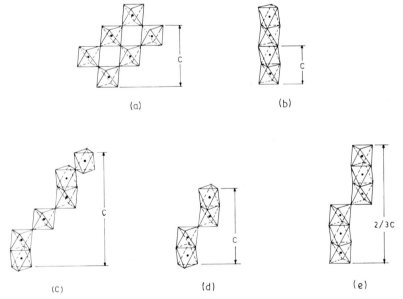

Figure 1.34   Linking of BO_6 octahedra in ABO_3 perovskite polytypes: (a) 3C; (b) 2H; (c) 6H; (d) 4H and (e) 9R.

occupy all the interlayer oxygen octahedra. The stacking of an AO_3 layer in the structure may be cubic (c) or hexagonal (h) with respect to its two adjacent layers, depending on whether it is in the middle of ABA or ABC sequence. If the stacking is entirely cubic, the B-cation octahedra share only corners in three dimensions to form the perovskite (3C) structure (Fig. 1.34). If the stacking is all-hexagonal, the B-cation octahedra share opposite faces forming chains along the c-axis as in $BaNiO_3(2H)$. In between the two extremes, there can be several polytypic structures consisting of mixed cubic and hexagonal stacking of AO_3 layers; for example, the 6H and 4H polytypes have the stacking sequences cch cch and chch respectively. Typical ABO_3 oxides showing polytypism are $BaCrO_3$, $BaMnO_3$, and $BaRuO_3$.

One of the interesting features of polytypes is the presence of different stacking sequences, especially in the large-period polytypes. Such coexistence of different polytypes would appear to be a contradiction of the Gibbs' phase rule. Energy differences between polytypic forms of a solid are small and interpolytypic transitions would therefore be associated with small changes in enthalpy, unlike intrapolytypic transitions. In spite of the extensive studies of polytypes of a variety of solids reported in the literature, the phenomenon is far from being well understood. Several explanations have been advanced to account for polytypism (Krishna, 1983; Rao & Rao, 1978). The basic question, however, remains without a completely satisfactory answer. What are the forces responsible for long-range ordering over large distances that give rise to the specific stacking sequences and giant unit cells in polytypes?

It has been found convenient to draw an analogy between one-dimensional spin-half Ising chains and polytypes. On the basis of such an analogy, formation of different kinds of polytypes has been explained (Ramasesha & Rao, 1977; Uppal *et al.*, 1980). In order to explain the existence of large numbers of polytypes of a given substance, the competing interaction model or the anisotropic next nearest neighbour Ising (ANNI) model first described by Elliott (1961) has been employed with success (Ramasesha, 1984). The study of Ramasesha & Rao (1977), where they employ an infinite-range interaction term (besides a short-range term), suggests that elastic forces may be responsible for the long-range order in polytypes.

1.9   Organic crystal structures

Organic solids have received much attention in the last 10 to 15 years especially because of possible technological applications. Typically important aspects of these solids are superconductivity (of quasi one-dimensional materials), photoconducting properties in relation to commercial photocopying processes and photochemical transformations in the solid state. In organic solids formed by nonpolar molecules, cohesion in the solid state is mainly due to van der Waals forces. Because of the relatively weak nature of the cohesive forces, organic crystals as a class are soft and low melting. Nonpolar aliphatic hydrocarbons tend to crystallize in approximately close-packed structures because of the nondirectional character of van der Waals forces. Methane above 22 K, for example, crystallizes in a cubic close-packed structure where the molecules exhibit considerable rotation. The intermolecular C—C distance is 4.1 Å, similar to the van der Waals bonds present in krypton (3.82 Å) and xenon (4.0 Å). Such close-packed structures are not found in molecular crystals of polar molecules.

With nonpolar aromatic molecules such as benzene, naphthalene and anthracene, cohesion is almost entirely due to van der Waals interaction, but the cohesive energies are significantly larger than in aliphatic nonpolar molecules. The larger cohesive energies arise from the greater polarizability of the π-clouds. Arrangement of molecules in these aromatic solids is dictated by the planar geometry of the molecules. Benzene, for example, crystallizes in an orthorhombic structure with the corner and face centre positions of the unit cell occupied by molecules. Naphthalene and anthracene crystallize in monoclinic structures where the molecules are so arranged that their long axis lies nearly parallel to the z-direction. The a and b parameters of the unit cells are the same for both the crystals, but the c-axis is longer in anthracene to the extent necessary to accommodate the larger anthracene molecule. In all these three aromatic crystals, the nonbonded C—C distance is $\sim 3.7\,\text{Å}$. Robertson (1951) classified crystalline aromatic solids into three categories. In the A-type, of which benzene, naphthalene and anthracene are typical examples, adjacent molecules overlap only slightly. In the B-type crystals there is considerable overlap between neighbouring molecules. The B-type is further subdivided into B_1 and B_2 types; in the B_1-type, the overlapping molecules are arranged in infinite stacks along one of the crystallographic axes, an example being 9-substituted anthracenes. In the B_2 category, overlapping molecules exist as pairs, each pair occupying a lattice point as in pyrene crystals.

Although it has long been recognized that crystal structures of nonpolar molecules are more or less distorted variants of closest-packed arrangements, it was only in 1973 that a systematic enunciation of the principles of organic crystal structures was put forward (Kitaigorodsky, 1973). Kitaigorodsky's basic idea is simple and stems from the fact that organic molecules are not exactly spherical in shape and hence their packing in crystals cannot be ideal HCP or CCP. If we were to approximate an organic molecule as a sphere, parts of the molecule would protrude from the imaginary sphere in certain directions while there would be hollows in the sphere in certain other directions. Kitaigorodsky points out that protrusions of one molecule will pack into the hollows of the other (the key-to-lock principle). Although the principle sounds simple, it has far-reaching consequences as far as the crystallography of organic solids is concerned. Application of this principle has shown, for example, that, of the 230 space groups, only 13 are most probable for molecular crystals composed of odd-shaped molecules. The principle is illustrated by the structures of halogens and benzene (Fig. 1.35). At low temperatures chlorine, bromine and iodine have the same orthorhombic structure. The structure consists of chains of diatomic molecules, related to one another by a two-fold screw axis, forming layers. These layers are stacked one over the other such that protrusions of one layer fit into the hollows of the other. Similarly, in the orthorhombic structure of benzene, molecules in the xy-plane are stacked along the z-direction in such a manner that they do not come one over the other in the adjacent layers. Kitaigorodsky's approach has been useful in calculating lattice energies and molecular conformations as mentioned earlier (see section 1.3). Mason & Rae (1968) have computed the unit cell constants of benzene crystal by this method within 3% error. The computed lattice energy of $52.3\,\text{kJ mol}^{-1}$ at 270 K compares favourably with the latent heat of sublimation.

The arrangement of molecules in many organic crystals is determined by intermolecular hydrogen bonding, molecules being disposed in crystals so as to facilitate hydrogen bond formation. Methanol, for example, crystallizes in an orthorhombic structure containing extended chains (along the c-axis) of CH_3OH molecules linked

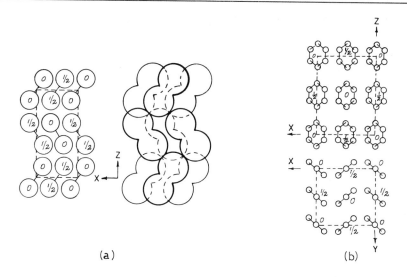

Figure 1.35   Structures of (a) iodine and (b) benzene crystals.

through hydrogen bonds, each oxygen being involved in two hydrogen bonds. These chains are held together laterally by van der Waals forces. Oxalic acid is another typical example of an organic crystal where cohesion is primarily through hydrogen bonding. In both the α- and β-forms, individual molecules have a *trans*-planar configuration and form extended chains.

1.9.1   *Fullerenes and related materials*

There has been considerable interest in understanding the nature of carbon in the universe for some time. It has been known for sometime that some of the features in the absorption spectra of the interstellar region arise from large molecules of carbon. In an attempt to generate these molecules in the laboratory, Kroto *et al.* (1985) studied the mass spectra of carbon vapours resulting from the ablation of graphite crystals with high-energy lasers. To their surprise, they found mass peaks due to various carbon species in the vapour, the most abundant amongst them being a cluster containing 60 carbon atoms. The cluster of 60 carbon atoms has remarkable stability and has a shape similar to a soccer ball involving a framework with an icosahedral hollow cage (Fig. 1.36). Such a cage structure has six- and five-membered rings (20 and 12 respectively). Since geodesic domes of a similar shape had earlier been designed by the famous American architect Buckminster Fuller, the C_{60} molecule was named *buckminsterfullerene*. Popularly, this molecule has been called buckyball. This class of cage molecules is called *fullerenes*.

Along with C_{60}, the other prominent carbon cage molecule found in graphite vapours is C_{70}, containing 70 atoms of carbon. The C_{70} molecule has the shape of a rugby ball (Fig. 1.36). This entire family of carbon molecules is referred to as fullerenes. Other interesting species (e.g. C_{76}, C_{78}, C_{84}) have also been identified. One of the large

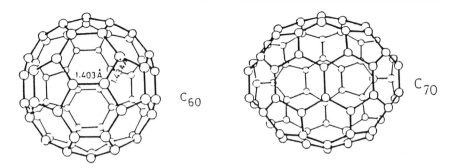

Figure 1.36   Structures of C_{60} and C_{70}.

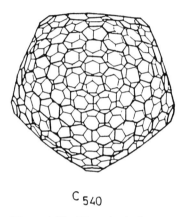

C_{540}

Figure 1.37   Hypothetical structure of C_{540}.

fullerenes one can visualise is C_{540} made up of 540 carbon atoms (Fig. 1.37). The fullerenes clearly constitute a new allotrope of carbon, different from graphite and diamond.

In spite of its stability, it was not possible to study C_{60} in the laboratory until 1990, since there was no way to obtain sufficient quantities of the material. Theoreticians predicted some properties based on various types of calculations, but they could not be verified. The situation however changed when Krätschmer et al. (1990) found a simple laboratory method to prepare macroscopic quantities of C_{60} and C_{70}. The method involves vapourizing graphite by passing a high current through two graphite electrodes in an atmosphere of helium. The soot collected from the arc is digested with an organic solvent such as toluene. The solid obtained after removing the solvent from the toluene solution contains C_{60} mixed with a smaller proportion of C_{70}. C_{60} and C_{70} can be separated by column chromatography. C_{60} when dissolved in organic liquids gives a magenta coloured solution while the solution of C_{70} is wine red.

With the availability of a method to produce fullerenes in the laboratory, the topic has become the rage of the day. It has created great excitement in the scientific world comparable only to that of high-temperature superconductors in early 1987. C_{60} and C_{70} have been characterized in terms of the crystal structure, UV-visible, NMR, infrared and Raman spectra as well as high-resolution electron microscopy. At room temperature, the ^{13}C NMR spectrum of C_{60} gives a single line showing the equivalent of all the carbons. At room temperature, C_{60} has a face-centered cubic structure. The X-ray study confirms the high symmetry of C_{60} and shows an intercluster separation of 10 Å in the crystal. The icosahedral hollow cage has a diameter of 7 Å (Fig. 1.36). The icosahedral symmetry, I_h, of C_{60} is the highest possible point group symmetry. Since the crystals at room temperature contain spherical molecules which can randomly orient themselves, there is considerable orientational disorder. On cooling to 250K, the molecules order themselves, giving rise to a more crystalline phase. C_{70} also shows orientational disorder at room temperature.

In diamond, the distance between two carbon atoms is that of a C—C single bond (1.54 Å) while in graphite, it is smaller (1.39 Å). In buckminsterfullerene, C_{60}, there are two distinct C—C bond lengths of around 1.40 Å and 1.43 Å (Fig. 1.36). These distances are between those in diamond and graphite. Diamond is a perfect insulator while graphite is a conductor; C_{60} is an insulator with a band gap of 2.2 eV. The compressibility of C_{60} is comparable to that of diamond. C_{60} is a reactive molecule and can undergo several types of reactions. It has been reduced to give $C_{60}H_{18}$ and $C_{60}H_{36}$, methylated to get $C_{60}(CH_3)_6$ and brominated to $C_{60}Br_{24}$. A problem that one faces while functionalising C_{60} is the large number of isomers that are possible in these derivatives. For example, $C_{60}H_2$ can have 59 isomers.

One of the aspects that has been of interest is the incorporation of an external atom in the spheroidal cavity. A variety of metal atoms can, in principle, be trapped in this cavity. Some of the studies have claimed that it is possible to push atoms such as lanthanum, iron and helium inside the spheroidal cavity of C_{60} and other fullerenes. Substitution of the carbon in C_{60} by boron and nitrogen has been attempted. Interestingly, nitrogen not only substitutes for carbon in the cage but also adds on to C_{60} and C_{70}.

Excitement in fullerene research has been enhanced by the discovery of superconducting fullerene compounds as well as ferromagnetic derivatives. A variety of problems related to C_{60} and other fullerenes will undoubtedly occupy the interest of chemists, physicists and materials scientists in the next few years. In addition to high-temperature superconductivity, molecular ferromagnets, molecular traps and molecular bearings are of interest. C_{60} offers immense possibilities in synthetic organic chemistry. Use of C_{60} in catalysis and organo-metallic chemistry and also as part of polymeric structures promises to be of interest. Properties of fullerenes other than C_{60} and C_{70} are equally fascinating; thus, C_{76} has a chiral D_2 symmetry. The higher fullerenes are yet to be prepared in large quantities. Recently, other cage compounds (not containing carbon) such as $Na_{96}In_{97}Z_2$ (Z = Ni, Pd, Pt) containing fullerene type cages in the structure have been reported (Sevov & Corbett, 1993; Nesper, 1994).

In addition to fullerenes, many other carbon structures formed of six-, five- and seven-membered rings (predominantly six) are likely to be discovered. Special mention must be made of *carbon nanotubes* formed in the direct-current arc evaporation of graphite (Iijima, 1991). The nanotubes are essentially made up of graphite sheets and have an inner core of around 1 nm with a variable number of graphite sheaths (Fig.

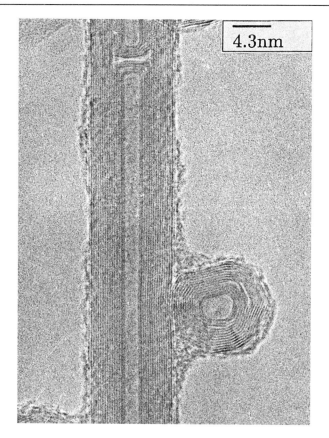

4.3nm

Figure 1.38   Electron micrograph of a carbon nanotube and an onion.

1.38). The tubes are closed at either end with the closed tip having a 5-membered ring (Fig. 1.39). The tubes can be opened by oxidation at the tip. Oxidation is carried out by heating in oxygen or in nitric acid (Seshadri *et al.*, 1994, Tsang *et al.*, 1994). Single-layer nanotubes (Fig. 1.39) without graphitic sheaths can also be obtained by using appropriate catalysts. Metals and metal oxides have been incorporated into the opened tubes (Seshadri *et al.*, 1994, Ajayan *et al.*, 1995, Lago *et al.*, 1995). Besides the nanotubes, *carbon onions* (Ugarte, 1992) consisting of closed concentric graphitic shells (giant fullerenes) have been discovered (Fig. 1.38). Onions can also be stuffed with metals. Fullerenes and related carbon structures have been dealt with extensively in special issues of journals (*Accts Chem. Res*, 1992; *Indian J. Chem*, 1992; *J. Phys. Chem. Solids* 1992 and 1994; *MRS Bulletin*, 1994). Solid State properties of these materials have been reviewed (Rao & Seshadri, 1994, Rao *et al.*, 1995).

Nested fullerene-type structures with other inorganic layered materials have been investigated recently. Tenne *et al.* (1992) have shown that MoS_2 is formed in a nested (onion-like) structure when MoO_{3-x} is reduced by H_2S in the gas phase at 1070–1220 K (Feldman *et al.*, 1995). Similar results have been obtained with WS_2. The idea seems to be more general, being applicable to other inorganic layered structures, as

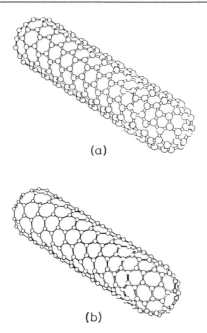

(a)

(b)

Figure 1.39   Schematic diagram of single-layer carbon nanotube: (a) a nanotube
with a sawtooth configuration; (b) a nanotube with a zig-zag configuration.

revealed by the synthesis of $(VO)_6(O_3POSiMe_3)_8Cl$ which is structurally related to
layered α-VOPO$_4$, the same way the fullerenes are to graphite (Thorn *et al.*, 1995).

1.9.2   *Charge-transfer (CT) complexes*

This class of organic solids formed between two different organic molecules, one acting
as electron donor (D) and the other electron acceptor (A), show interesting electronic
properties in the solid state (Soos & Klein, 1976). The D and A molecules in these solids
are arranged in infinite stacks; the stacking may be mixed, ... DADA ... , as in the
CT-complex between tetramethyl-*p*-phenylene diamine (TMPD) and tetracyanoben-
zene (TCNB), or segregated regular stacks, ... DDDD ... and ... AAAA ... , as, for
example, in the 5,10-dihydro-5,10-dimethylphenazine(M$_2$P)-tetracyano-*p*-quino-
dimethane (TCNQ) complex (Fig. 1.40). These examples of CT complexes belong to the
category of 1:1 complexes where there is a complete transfer of an electron from D to A
forming radical ion salts. There are several complexes where the ratio of donor to
acceptor is different, 1:2, 2:3 etc. In many of them, the CT interaction is weak, forming
only neutral molecular complexes. The tetrathiafulvalene (TTF)–TCNQ family of
complexes has excited the interest of chemists and physicists (Torrance, 1979), because
of their unusually high metal-like conductivity down to 60 K. The structure of this
complex consists of segregated regular stacks of donors and acceptors and the cohesive
energy of the solid has considerable contribution from coulomb interaction, which
depends on the extent of charge transfer. The CT salt formed between hexa-

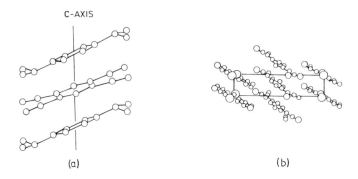

C-AXIS

(a)   (b)

Figure 1.40   Structures of charge-transfer (CT) complexes: (a) TMPD–TCNB and
(b) M_2P–TCNQ.

methylenetetraselenafulvalene (HMTSF) and TCNQ shows metallic conductivity
down to very low temperatures (see p. 63 for the structures of D and A molecules).

1.10   Inclusion compounds and clathrates

These are solid compounds with a host structure which accommodates certain guest
species (Atwood *et al.*, 1984). The host is usually a crystal with a large open space within
which the guest species can be accommodated. The guest species can be atoms, ions or
molecules. Depending on whether the guest species are found in tunnels, interlayer
spacings or cages of the host, inclusion compounds can be classified into one-, two- or
three-dimensional types. Typical examples of inclusion compounds are the *gas
hydrates*, *β-quinol clathrates* and inclusion compounds of urea and thiourea. When
water is solidified in the presence of gases like halogens and noble gases, the volatile
component is trapped in a cage-like structure formed by water molecules. Such gas
hydrates occur in two forms, type I and type II. A typical example of the type I gas
hydrate is $Cl_2 \cdot 7\frac{1}{4} H_2O$, first isolated by Sir Humphry Davy. The structure is cubic
($a \simeq 12$ Å), consisting of a hydrogen-bonded network of 46 water molecules which
encloses two small and six large cavities. The larger cavities are fully occupied and the
smaller ones partly occupied by chlorine molecules. In the type II structure, the cubic
unit cell is ~ 17 Å dimension with 136 water molecules and 16 small and 8 large
cavities. In quinol clathrates, molecules like N_2, HCl and HCOOH are trapped in the
available space of the hydrogen-bonded host network; the guest molecules rotate
relatively freely in the structure. In the inclusion compounds of urea and thiourea, guest
molecules are accommodated in the channels of the host structure. The inclusion
compounds of urea with guest molecules such as C_{10} to C_{50} *n*-hydrocarbons, straight
chain alcohols, acids and esters crystallize in a hexagonal structure with $a \sim 8.2$ and
$c \sim 11$ Å. There are six urea molecules per unit cell forming an interpenetrating helical
spiral. Each oxygen is hydrogen-bonded to four nitrogen atoms and each nitrogen to
two oxygen atoms. The structure provides circular channels of ~ 5 Å diameter parallel
to the *c*-axis which are occupied by the guest molecules. It is interesting that urea itself
crystallizes in a different (tetragonal) structure and transforms to the hexagonal
structure when inclusion occurs. Furthermore, the hexagonal structure collapses on

NC   CN

(a)

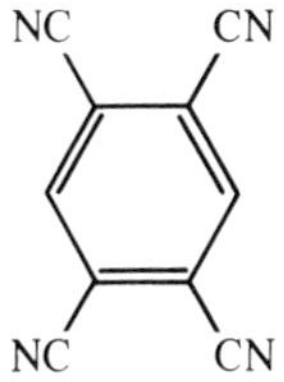

NC   CN

TCNB

(b)

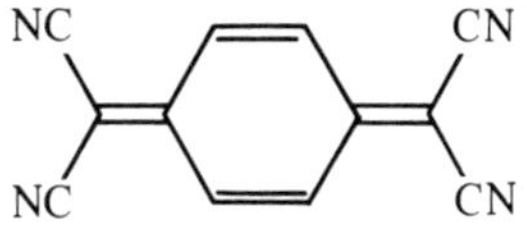

NC   CN

NC   CN

TCNQ

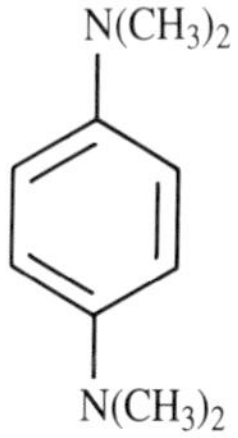

N(CH3)2

N(CH3)2

TMPD

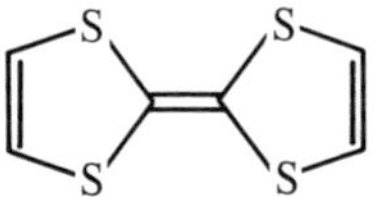

S   S

S   S

TTF

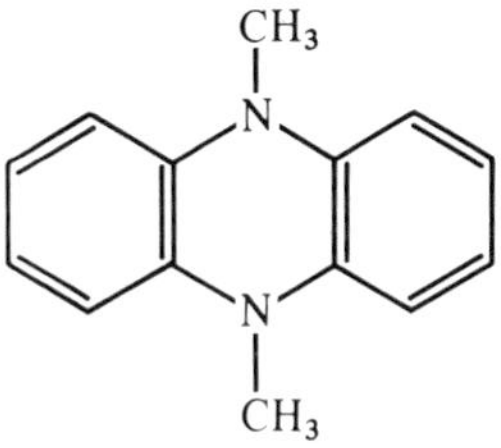

CH3

N

N

CH3

M₂P

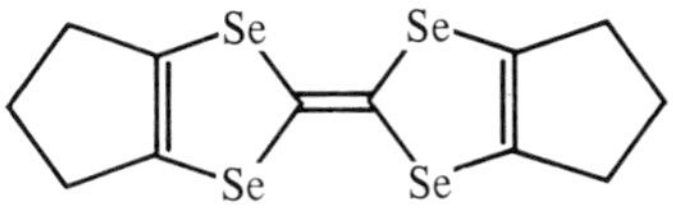

Se   Se

Se   Se

HMTSF

removal of the guest molecules. Inclusion compounds of thiourea are similar, but the larger channel diameter accommodates molecules of larger diameter such as branched chain hydrocarbons.

It has been recognized for some time that SiO_2 and H_2O show a similar tendency to form tetrahedrally linked framework structures (Kamb, 1965). The mineral *melanophlogite* is isostructural with cubic gas hydrate I, while the synthetic *clathrasil, dodecasil-3C* is isostructural with cubic gas hydrate II (Gies et al., 1981, 1982). A section of dodecasil-3C is shown in Fig. 1.41.

Besides the molecular clathrates where the bonding between the host and the guest is weak (van der Waals type), certain metallic clathrates are known where the host–guest interaction is stronger. Silver oxide clathrate salts, $[Ag_7O_8]^+X^-(X = NO_3, ClO_4, BF_4)$ belong to this category. These compounds, prepared by electrolytic oxidation of acidic silver salt solutions are shiny black metal-like crystals; they crystallize in a fairly complex cage structure consisting of Ag_6O_8 polyhedra, each one accommodating one X^- ion. Four such polyhedra are fused together in one unit cell with one silver in between. As such, there are two types of silver in the structure with an average

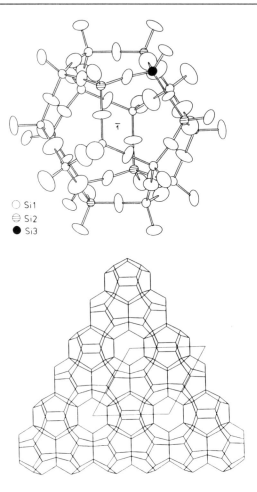

○ Si1
⊖ Si2
● Si3

Figure 1.41   The pentagonododecahedron and a section of a layer of linked pentagonododecahedra of dodecasil-3C; the unit cell is also indicated. (After Gies et al., 1982.)

oxidation state of 17/7. The mixed oxidation state of silver together with the unique structures renders these substances metallic and Pauli paramagnetic.

Another class of three-dimensional inclusion compounds is that of metal borides consisting of boron clusters; alkaline earth and lanthanide metal hexaborides of the formula MB_6 are typical examples of this class of compounds. The structure of hexaborides is similar to CsCl, the B_6 clusters occupying corners of a cube and the large M atom sitting at the body-centre position. Physical properties of hexaborides can be understood on the basis of a bonding model proposed by Longuet-Higgins & Roberts (1968). There are six boron atoms per unit cell of the MB_6 crystal which require 20 electrons (6 to form covalent bonds in B_6 clusters and 14 to inter-connect B_6 clusters) to fill the valence-band completely; 18 of these are provided by six borons and two additional electrons must come from the cations. Divalent metal hexaborides like CaB_6

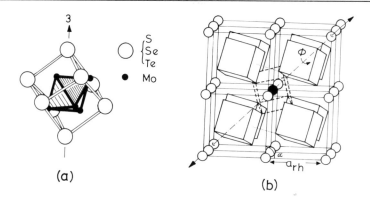

Figure 1.42   Structure of $M_xMo_6X_8$ Chevrel phases: (a) the Mo_6X_8 building unit and (b) the stacking of eight Mo_6X_8 units in the rhombohedral structure. The cube-shaped cavity at the centre is occupied by the large M atom. (After Yvon, 1979.)

(fomulated as $Ca^{2+}B_6^{2-}$) are insulators or semiconductors, while those of trivalent metals like the lanthanide hexaborides are metallic because an electron from the cation occupies the lowest nonbonding metal band.

Ternary molybdenum chalcogenides (*Chevrel phases*) of the formula $M_xMo_6X_8$ (M = Pb, rare earth, etc.; X = S, Se, Te and $x \sim 1$) are an important class of inclusion compounds, exhibiting remarkable physical properties (for a comprehensive review, see, for example, Yvon, 1979). The structure (Fig. 1.42) is rhombohedral consisting of eight Mo_6X_8 units, which enclose a number of cavities. The largest one is a cube-shaped cavity at the centre formed by eight chalcogen atoms belonging to different Mo_6X_8 units. Smaller cavities are present close to the middle of the rhombohedral unit cell vectors. Large M atoms such as Pb occupy the cubical cavity, while smaller atoms like Cu occupy the other cavities as well. The interesting electronic properties of Chevrel phases, which include high field, high T_c superconductivity and coexistence of superconductivity and magnetic order, arise from electron transfer from the guest M atoms to the Mo_6X_8 host.

We shall be discussing other types of inclusion compounds formed by layered graphite and TaS_2 as well as other varieties of hosts in Chapter 8, which deals with intercalation chemistry.

1.11   Noncrystalline or amorphous solids

Although much of solid state chemistry deals with crystalline materials, there has been a tremendous increase in interest in noncrystalline (or amorphous) solids in the last ten to fifteen years, probably because of the universality with which almost all materials can be transformed into the amorphous state as well as the diverse applications of these solids. While the term amorphous or noncrystalline solid is a general one, the term glass is used only with reference to amorphous solids prepared by slow- or fast-cooling of melts. Today we have metallic glasses, organic glasses and polymeric glasses. Glasses are characterized by the so-called glass transition which we shall discuss in Chapter 4.

At the glass transition temperature, the glass melts or an undercooled liquid freezes (Zallen, 1983; Elliott *et al.*, 1986).

In crystal chemistry, the existence of long-range order forms the basis of all discussion; crystal symmetry, Brillouin zones and associated aspects of crystals are of paramount importance in solid state chemistry because of the presence of long-range order. In the noncrystalline state there is no long-range order, but short-range order is very much in evidence. Although the absence of long-range order makes it difficult to obtain exact structural information from diffraction and other experiments, we are able to say much about the structure of noncrystalline solids. Some of the recent techniques such as EXAFS and MASNMR are able to provide precise information on the structural features of these solids (see Chapter 2 for details), but much has been gained in our understanding of these solids by the structural models of the amorphous state.

One of the early models to describe the amorphous state was by Zachariasen (1932), who proposed the continuous random network model for covalent inorganic glasses. We are now able to distinguish three types of continuous random models:

> *Continuous random network* (applicable to covalent glasses)
> *Random close packing* (applicable to metallic glasses)
> *Random coil model* (polymeric glasses).

All these models involve a description of the amorphous state in terms of statistical distributions. These models have been discussed widely in the recent literature.

In the continuous random network model (RNM) of inorganic covalent glasses, the structure was supposed to be formed by polyhedra such as SiO_4 (in the case of SiO_2) connected together in such a way that long-range order is destroyed by allowing freedom of rotation of neighbouring polyhedra about the connecting atom (the T—O—T angle is allowed to take up a variety of values). We should note that the structure is not truly random in the statistical sense as implied by the term random network. Well-defined local order exists in glasses by virtue of the existence of the polyhedra, and if the polyhedra connect together to form, say, rings, then the distribution of dihedral angles for neighbouring units is not uniform (i.e. random), but instead certain dihedral angles are preferred. In Fig. 1.43 we show a typical random network of an A_2B_3 glass. RNM has been used for IV–V–VI chalcogenides as well as elemental glasses of Si and Ge. RNM of glasses can be topologically of one, two or three dimensions as exemplified by Se, As_2Se_3 and SiO_2 respectively.

Polk (1971) demonstrated that it was possible to build an extended continuous random network (consisting of 440 atoms), based on an invariant coordination number of four, for tetrahedrally bonded amorphous semiconductors such as Si and Ge. Even when all bond lengths were kept close to the equilibrium value, no serious bond-length strain was produced if a modest spread in angles was permitted. Significantly, 5-, 6- and 7-membered rings were present in such a continuous network, although there are only 6-membered rings in the crystalline semiconductors; furthermore, the density of the resulting noncrystalline structure differed from that of the diamond-like structure by about one per cent. The conclusion is that, in simple covalently bonded glasses (Si, SiO_2

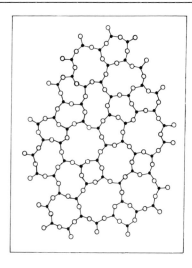

Figure 1.43   Zachariasen's random network model for A_2B_3 glass.

etc.), the short-range order is close to that present in the corresponding crystals. Even in the ionic glass formed by BeF_2, the 4:2 coordination is retained.

The *intermediate-range order*, beyond the short-range local order as characterized by the polyhedral units in covalent glasses, is of considerable interest. There is increasing evidence for the existence of fairly well-defined clusters of atoms, containing 10 or more atoms, in the structure of certain covalent glasses (e.g. Ge–Se alloys); vibrational spectroscopy as well as diffraction experiments indicate their presence.

Glassy metals and alloys can be structurally considered as made up of a *random packing of hard* (RCP) *spheres*. The notion of dense random packing figured in the classic work of Bernal (1965) on the structure of monoatomic liquids. Bernal's work on liquids involved ball bearings packed into rubber bladders, kneaded, set in black paint with the positions analyzed by hand and eye. Later variants of this approach have entailed computer simulations (Fig. 1.44). All these approaches have shown that it is possible to generate a randomly packed model which is sufficiently dense (63.7% occupancy of available space compared with 74.05% in cubic or hexagonal close-packing of spheres) to reproduce experiment. Such a model involves no crystalline regions provided the spheres are not bounded by a regular surface. It is well known that, in the closest packing of spheres (as in crystalline metals such as Ag, Au and Pt), there are holes which are tetrahedral and octahedral in shape. Bernal noted that, in the dense random packing of hard spheres model, there are holes of five distinct types; in addition to the tetrahedral and octahedral holes, there are three others. These holes are polyhedra with equal triangular facets (i.e. deltahedral objects) which are small enough not to admit another sphere of the same size and are referred to as Bernal's canonical holes. A large number of structures possessing nearly identical nearest-neighbour distances can be formed by packing the five deltahedra together. Finney (1970) has constructed a dense randomly packed structure, composed of 7994 'atoms' with a

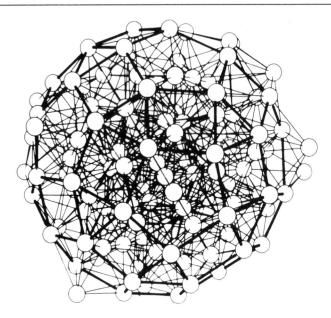

Figure 1.44   Cluster obtained spontaneously on cooling a simulated
Lennard–Jones drop of 129 atoms in a molecular dynamics study. The heavy lines
indicate elements of pentagonal symmetry. (After Hoare & Barker, 1976).

packing efficiency of 44 per cent. The (total) radial distribution function for this
structure compares well with that determined experimentally for amorphous $Ni_{76}P_{24}$,
although the agreement for the individual partial pair correlation functions is not as
satisfactory.

The RCP model accounts well for the structure of amorphous metal alloys such as
Zr–Cu. However, it is unsatisfactory for the so-called '*metglas*' alloys formed from
approximately 80% transition metal and 20% metalloid (B, Si, P, etc.), this being the
eutectic composition. In these materials, chemical ordering seems to exist and
crystalline motifs such as trigonal prismatic units are suggested to be present. Gaskell
(1979) has proposed that the structure of the glassy alloys may also consist of such
well-ordered local units. In this way, metalloid–metalloid nearest-neighbour avoidance
is achieved (consistent with diffraction evidence), which is not the case for the RCP
model, in which it is supposed that the small metalloid atoms, such as P, reside in the
canonical polyhedral holes in the structure, and, as such, there is nothing to stop them
being nearest neighbours. The extent of chemical ordering in amorphous materials is of
great interest. The means by which amorphous materials can accommodate gross
nonstoichiometry is also noteworthy. There being no well-defined architectural unit in
the amorphous state, continuous variation in composition is possible. This introduces
a special situation, unrealizable in crystalline solids, whereby there is local saturation of
valence requirements in covalently bonded systems or creation of structural groupings
in ionic systems, which achieve electrical neutrality without the need for compensating
defects. The random close-packing model can be used for ionic glasses in a restricted
way.

Polymeric solids such as polystyrene are most often noncrystalline. The random coil model would be most appropriate to describe such solids. In many polymers, both crystalline and amorphous regions are present; in such materials, well-defined coiled regions are embedded in a randomly coiled matrix.

A characteristic of the amorphous state is the presence of both compositional and density fluctuations; associated with these are the fluctuations in the nature of bonding. Most multi-component oxide glasses consist of charged, ionically bonded oxygens as well as two-coordinated, covalently bonded oxygens, their relative proportions being variable. In the As_2Se_3–As_2Te_3 system, both ionic and covalent bonds are present between similar sets of atoms. Networks like Ge_xSe_{1-x} may contain rigid regions and floppy regions made up of bridging Se_n fragments.

Chemical reactions and structural rearrangements in amorphous solids are features governed by the chemistry of these materials. For example, the occurrence of photochromism (which is commercially exploited) involves charge-transfer and reionization of silver halide particles, often in the presence of sensitizer ions such as Cu^+. The oxygen coordination of germanium varies as a function of the modifier oxide in germania glass, varying from four to six and to four again in modifier oxide rich regions. A continuous variation of the concentration of tetrahedral and planar (3-coordinated) borons is achieved in borate glasses by modifier oxides. It is known that both over- and under-coordinated charged chalcogen (or pnicogen) defect centres are present in all chalcogenide (and pnictide) glasses and they play a crucial role in transport and related properties. These charged pairs result from a favourable reaction of bonds which initially suffer homopolar scission.

1.12   Quasicrystals

Since the discovery of Shechtman *et al.* (1984) that the rapidly quenched Al-14% Mn alloy is quasicrystalline with icosahedral symmetry, there has been intense activity to explore such alloys to find out whether there would be a conceptual revolution in crystallography. Pauling (1985) proposed that the icosahedral symmetry results from multiple twinning of cubic crystals, but many features of these unique alloys, especially their electron diffraction patterns, cannot be understood on the basis of twinning alone.

The unusual feature of quasicrystals is the presence of five-fold symmetry forbidden by conventional crystallography. Alloys with other forbidden symmetries have since been discovered (e.g., decagonal phases). These materials have long-range quasi-periodic order (following known quasiperiodic equations) and long-range orientational order disallowed by crystallographic symmetry. They are not like the usual crystals and at the same time they are not random on the scale of glasses or liquids. They are however metastable and can be incorporated in a metastable phase diagram. They can be transformed to the crystalline state. These materials have been reviewed widely in the literature (Kelton, 1993; Ranganathan & Chattopadhyay, 1991).

1.13   Models and graphics

The importance of models in visualizing, understanding and illustrating structures of solids need not be overemphasized. Besides the well-known ball and stick and

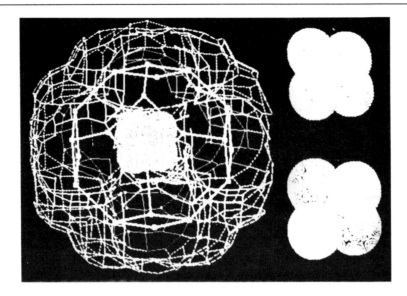

Figure 1.45 Na_4^{3+} cluster in the sodalite cage of zeolite Y. On the right K_4^{3+} and Rb_4^{3+} are shown. (Courtesy of S. Ramdas & J.M. Thomas, University of Cambridge.)

space-filling models, one often builds models specially suited to a particular structural problem. Models built of tiny flexible tubes with atom centres (of different colours) of different bond angles and vertices (tetrahedral, octahedral, etc.) are specially useful. It is not only useful but essential in some areas of solid state chemistry to build models. For example, oxide layers built of cardboard octahedra have been most effective in visualizing shear and block structures. The cages and channels in zeolites and clays are best understood by making use of stereo models of Dreiding type. Random networks in amorphous materials are best studied with the help of models.

In the last few years, computer graphics with colour display are being more commonly used not only to visualize complex structures better, but also to examine unusual structural features, defects and transformations as well as reactions. In Fig. 1.45, we show the presence of a Na_4^{3+} cluster within the sodalite cage of zeolite Y as depicted by computer graphics; the cluster fits well within the cavity bounded by the van der Waals surface (net) of the framework atoms. The immense power of computer graphics has been exploited widely in recent years. Structural transitions in solids and sorbate dynamics in zeolites are typical areas where computer simulation and graphics have been used (Ramdas *et al.*, 1984; Rao *et al.*, 1992).

Let us briefly examine modelling of solids, taking silicates as a case study. In order to model silicate structures, it is necessary to develop accurate potentials for the various silicate forms. Such potentials should be able to predict the known structures of zeolites and related materials accurately. Attempts have been made by a number of workers in this direction. The SiO_4 tetrahedron is the basic building block in silicates. While in orthosilicates, tetrahedra are connected to octahedra, in metasilicates the tetrahedra form chains (e.g., diopside, wollastonite, rhodonite). Tectosilicates consist of tetrahedra linked to form three-dimensional networks referred to as frameworks. Examples of this class are the zeolites and the various forms of silica such as quartz, tridymite and cristobalite. There are nonframework cations such as Na, K, Ca, etc. which balance the charge on the framework caused by the substitution of Si by Al.

A number of techniques have been employed to model the framework structure of silica and zeolites (Catlow & Cormack, 1987). Early attempts at calculating the lattice energy of a silicate assumed only electrostatic interactions. These calculations were of limited use since the short-range interactions had been ignored. The short-range terms are generally modelled in terms of the Buckingham potential,

$$\phi(r) = -A/r^6 + B \exp(-r/\rho)$$

The repulsive term is modelled in terms of the Born–Mayer function. These central force pair-potentials can model orthosilicates and metasilicates satisfactorily, but fail in the case of tectosilicates (Parker et al., 1984). This is because the SiO_4 units cannot retain the tetrahedral geometry in the absence of explicit covalent effects. By incorporating an additional term, $E_\theta = K_b(\theta - \theta_o)^2$, where θ is the O–Si–O angle, θ_o is the tetrahedral angle and K_b is an harmonic force constant, pentasil zeolites and SrA zeolite could be modelled accurately (Hope, 1985). Polarization of the oxide ion and of the extra-framework cations, are not included in such a treatment. Later workers have taken polarization effects into account. Such potentials have been employed in modelling several zeolites and silicates. Studies have been carried out to predict the position of extra-framework cations in fixed framework zeolites. Sanders & Catlow (1983) have investigated the distribution of potassium ions in faujasite for $Si/Al = 1.4$. The results show among other things that the S_1 and S_1' sites have equal occupancy for a range of Si/Al ratios, in agreement with the experimental results. There have also been studies on other aspects such as Sr^{2+} ordering in SrA zeolite. Crystal morphology, surface energies and other features have been obtained by simulation techniques by Catlow and coworkers. Catlow et al. (1994) have recently reviewed computer modelling applications in materials chemistry.

1.14   Solitons

Many phenomena such as dislocations, electronic structures of polyacetylenes and other solids, Josephson junctions, spin dynamics and charge density waves in low-dimensional solids, fast ion conduction and phase transitions are being explained by invoking the concept of *solitons*. Solitons are exact analytical solutions of non-linear wave equations corresponding to bell-shaped or step-like changes in the variable (Ogurtani, 1983). They can move through a material with constant amplitude and velocity or remain stationary; when two of them collide they are unmodified. The soliton concept has been employed in solid state chemistry to explain diverse phenomena.

References

Adams, D. M. (1974) *Inorganic Solids, An Introduction to Concepts in Solid-State Structural Chemistry*, Wiley, London.

Ajayan, P. M., Stephan, O., Redlich, Ph. & Colliex, C. (1995) *Nature*, **375**, 564.

Anderson, M. T., Greenwood, K. B., Taylor, G. A. & Poeppelmeier, K. R. (1993) *Prog. Solid State Chem.* **22**, 197–233.

Andersson, S. (1978) *J. Solid State Chem.* **23**, 191.

Andersson, S. (1983) *Angew. Chem. Int. Ed. (Engl.)* **22**, 69.

Andersson, S. & Wadsley, A. D. (1962) *Acta Crystallogr.* **15**, 94.

Andersson, S. & Fälth, L. (1983) *J. Solid State Chem.* **46**, 265.

Atwood, J. L., Davies J. E. D. & MacNicol, D. D. (eds) (1984) *Inclusion Compounds*, Vols 1 and 2, Academic Press, London.

Barrer, R. M. (1978) *Zeolites and Clay Minerals as Sorbents and Molecular Sieves*, Academic Press, New York.

Bartell, L. S. (1968) *J. Chem. Ed.* **45**, 754.

Beck, J. S., Vartuli, J. C., Roth, W. J., Lenowicz, M. E., Kresge, C. T., Schmitt, K. D., Chu, C. T-W., Olson, D. H., Sheppard, E. W., McCullen, S. B., Higgins, J. B. & Schlenker, J. L. (1992) *J. Amer. Chem. Soc.* **114**, 10834.

Bernal, J. D. (1965) in *Liquids: Structure, Properties, Solid Interactions* (ed. Hughel, T. J.) Elsevier, Amsterdam.

Brewer, L. (1967) *Acta Met.* **15**, 553.

Brewer, L. (1984) *J. Chem. Ed.* **61**, 101.

Brindley, G. W. & Brown, G. (eds) (1980) *Crystal Structure of Clay Minerals and their X-ray Diffraction*, Mineralogical Society, London.

Brown, G. (1984) *Phil. Trans. Roy. Soc. London* **A311**, 221.

Brown, I. D. (1977) *Acta Crystallogr.* **B33**, 1305.

Brown, I. D. (1981) in *Structure and Bonding in Crystals*, Vol. II (eds O'Keeffe, M. & Navrotsky, A.), Academic Press, New York.

Brown, I. D. & Altermatt, D. (1985) *Acta Crystallogr.* **B41**, 244.

Burdett, J. K. (1988) *Chem. Rev.* **88**, 3.

Burdett, J. K., Price, S. L. & Price, G. D. (1981) *Solid State Commun.* **40**, 923.

Burdett, J. K., Price, G. D. & Price, S. L. (1982) *J. Amer. Chem. Soc.* **104**, 92.

Catlow, C. R. A. & Cormack, A. N. (1987) *Int. Rev. Phys. Chem.* **6**, 227.

Catlow, C. R. A., Bell, R. G. & Gale, J. D. (1994) *J. Mater. Chem.* **4**, 781.

Chelikowsky, J. R. & Phillips, J. C. (1978) *Phys. Rev.* **B17**, 2453.

Chen, J., Jones, R. H., Natarajan, S., Hursthouse, M. B. & Thomas, J. M. (1994) *Angew. Chem. Int. Ed. (Engl.)* **33**, 639.

Chippindale, A. M. & Walton, R. I. (1994). *JCS Chem. Commun* 2453.

Cohen, M. L. (1981) in *Structure and Bonding in Crystals* (eds O'Keeffe, M. & Navrotsky, A.) Academic Press, New York.

Cracknell, A. P. (1975) *Magnetism in Crystalline Materials*, Pergamon Press, Oxford.

Davis, M. E., Saldarriaga, C., Garces, J. M. & Crowder, C. (1988) *Nature* **331**, 698.

Davis, M. E., Montes, C., Hathaway, P. E., Arhancet, J. P., Hasha, D. L. & Garces, J. M. (1989) *J. Amer. Chem. Soc.* **111**, 3919.

Davis, M. E. & Lobo, R. F. (1992) *Chem. Mater.* **4**, 756.

Dhingra, S. & Kanatzidis, M. G. (1992) *Science* **258**, 1769.

Domenges, B., McGuire, N. K. & O'Keeffe, M. (1985) *J. Solid State Chem.* **56**, 94.

Douxing, L., Stenberg, L. & Andersson, S. (1983) *J. Solid State Chem.* **48**, 368.

Elliott, R. J. (1961) *Phys. Rev.* **124**, 346.

Elliott, S. R., Rao, C. N. R. & Thomas, J. M. (1986) *Angew. Chem. Int. Ed. (Engl.)* **25**, 31.

Estermann, M., McCusker, L. B., Baerlocher, C., Merrouche, A. & Kessler, H. (1991) *Nature* **352**, 320.

Feist, T. P. & Davies, P. K. (1991) *Chem. Mater.* **3**, 1011.

Feldman, Y., Wasserman, E., Srolovitz, D. J. & Tenne, R. (1995) *Science* **267**, 222.

Finney, J. L. (1970) *Proc. Roy. Soc. London* **A319**, 479.

Ganguly, P. & Rao, C. N. R. (1984) *J. Solid State Chem.* **53**, 193.

Gaskell, P. H. (1979) *J. Non-Cryst. Solids* **32**, 207.

Gier, T. E. & Stucky, G. D. (1991) *Nature* **349**, 508.

Gies, H. & Liebau, F. (1981) *Acta Crystallogr.* **A37**, C187.

Gies, H., Liebau, F. & Gerke, H. (1982) *Angew. Chem.* **94**, 241.

Girgis, K. (1983) in *Physical Metallurgy* (3rd Edn) (eds Cahn, R. W. & Hassen, P.) North-Holland, Amsterdam.

Gopalakrishnan, J., Bhuvanesh, N. S. P. & Raju, A. R. (1994) *Chem. Mater.* **6**, 373.

Harvey, G. & Meier, W. M. (1989) in *Zeolites: Facts, Figures and Future* (eds Jacobs, P. A. & van Santen, R. A.) Elsevier, Amsterdam; pp. 411–20.

Haushalter, R. C. & Mundi, L. A. (1992) *Chem. Mater.* **4**, 31.

Hoare, M. R. & Barker, J. (1976) in *The Structure of Non-Crystalline Materials* (ed. Gaskell, P. H.) Taylor & Francis, London.

Hoffmann, R. & Zheng, C. (1985) *J. Phys. Chem.* **89**, 4175.

Holsche, M., Englert, U., Zibrowins, B. & Holderich, W. F. (1994) *Angew. Chem. Int-Ed (Engl)* **33**, 2491.

Hope, A. J. T. (1985) Ph. D. Thesis, University of London.

Hoppe, R. (1975) in *Crystal Structure and Chemical Bonding in Inorganic Chemistry* (eds Rooymans, C. J. M. & Rabenau, A.) North-Holland, Amsterdam.

Huggins, R. A. & Rabenau, S. (1978) *Mater. Res. Bull.* **13**, 1315.

Hughbanks, T. & Hoffmann, R. (1983) *J. Amer. Chem. Soc.* **105**, 1150.

Huo, Q., Margolese, D. I., Clesla, U., Feng, P., Gier, T. E., Sieger, P., Leon, R., Petroff, P. M., Schüth, F. & Stucky, G. D. (1994) *Nature* **368**, 317.

Hyde, B. G. (1979) in *Modulated Structures – 1979*, AIP Conference Proceedings (eds. Cowley, J. M., Cohen, J. B., Salamon, M. B. & Wuensch, B. J.) American Institute of Physics, New York.

Hyde, B. G., Bagshaw, A. N., Andersson, S. & O'Keeffe, M. (1974) *Ann. Rev. Mater. Sci.* **4**, 43.

Hyde, B. G., Andersson, S., Bakker, M., Plug, C. M. & O'Keeffe, M. (1980) *Prog. Solid State Chem.* **12**, 273.

Iijima, S. (1991) *Nature* **356**, 56.

Jack, K. H. & Guttman, V. (1951) *Acta Crystallogr.* **4**, 246.

Kamb, B. (1965) *Science* **148**, 232.

Katz, L. & Ward, R. (1964) *Inorg. Chem.* **3**, 205.

Kelton, K. F. (1993) *Intnl. Mater. Revs.* **38**, 105.

Kitaigorodsky, A. I. (1973) *Molecular Crystals and Molecules*, Academic Press, New York.

Krätschmer, W., Lamb, L. D., Fostiropoulous, K. & Huffman, D. R. (1990) *Nature* **347**, 354.

Kresge, C. T., Leonowicz, M. E., Roth, W. J., Vartuli, J. C. & Beck, J. S. (1992) *Nature* **359**, 710.

Krishna, P. (ed.) (1983) *Crystal Growth and Characterization of Polytype Structures*, Pergamon Press, Oxford.

Kroto, H. W., Heath, J. R., O'Brien, S. C., Curl, R. F. & Smalley, R. E. (1985) *Nature* **318**, 162.

Kroto, H. W., Allaf, A. W. & Balm, S. P. (1991) *Chem. Rev.* **91**, 1213.

Lago, R. M., Tsang, S. C., Lu, K. L., Chen, Y. K. & Green, M. L. H. (1995) *JCS Chem. Commun.* 1355.

Lewis, D. W., Freeman, C. M. & Catlow, C. R. A. (1995) *J. Phys Chem.* **99**, 11194.

Longuet-Higgins, H. C. & Roberts, M. de V. (1968) *Proc. Roy Soc. London* **A224**, 336.

McCarley, R. E. (1982) *Phil. Trans. Roy. Soc. London*, **A308**, 141.

Magneli, A. (1949) *Arkiv Kemi* **1**, 213.

Magneli, A. (1953) *Acta Chem. Scand.* **7**, 315.

Magneli, A. (1978) *Pure & Applied Chem.* **54**, 1261.

Mason, R. & Rae, A. I. M. (1968) *Proc. Roy. Soc. London* **A304**, 501.

Meier, W. M. & Olson, D. H. (1978) *Atlas of Zeolite Structure Types*, Int. Zeolite Assoc., Polycrystal Book Serv., Pittsburgh, Pennsylvania.

Moore, J. S. & Lee, S. (1994) *Chemistry & Industry*, 556.

Mooser, E. & Pearson, W. B. (1959) *Acta Crystallogr.* **12**, 1015.

Nanot, M., Queyroux, F., Gilles, J-C., Carpy, A. & Galy, J. (1974) *J. Solid State Chem.* **11**, 272.

Nesper, R. (1994) *Angew. Chem. Int. Ed. Engl.* **33**, 843.

Ogurtani, T. O. (1983) *Ann. Rev. Mat. Sci.* **13**, 67.

O'Keeffe, M. (1977) *Acta Crystallogr.* **A33**, 924.

O'Keeffe, M. (1981) in *Structure and Bonding in Crystals*, Vol. I (eds O'Keeffe, M. & Navrotsky, A.) Academic Press, New York.

O'Keeffe, M. & Andersson, S. (1977) *Acta Crystallogr.* **A33**, 914.

O'Keeffe, M. & Hyde, B. G. (1978) *Acta Crystallogr.* **34**, 3519.

O'Keeffe, M. & Hyde, B. G. (1979) *Trans. Amer. Crystallogr. Assoc.* **15**, 65.

O'Keeffe, M. & Hyde, B. G. (1980) *Phil. Trans. Roy. Soc. London* **A295**, 553.

O'Keeffe, M. & Hyde, B. G. (1981) in *Structure and Bonding in Crystals*, Vol. I (eds O'Keeffe, M. & Navrotsky, A.) Academic Press, New York.

O'Keeffe, M. & Hyde, B. G. (1982) *J. Solid State Chem.* **44**, 24.

Parker, S. C., Catlow, C. R. A. & Cormack, A. N. (1984) *Acta Crystallogr* **B40**, 200.

Parthe, E. (1981) in *Structure and Bonding in Crystals*, Vol. II (eds O'Keeffe, M. & Navrotsky, A.) Academic Press, New York.

Pauling, L. (1929) *J. Amer. Chem. Soc.* **51**, 1010.

Pauling, L. (1960) *The Nature of the Chemical Bond,* (3rd edn) Cornell University Press, New York.

Pauling, L. (1985) *Nature,* **317**, 512.

Phillips, J. C. (1973a) *Bonds and Bands in Semiconductors,* Academic Press, New York.

Phillips, J. C. (1973b) in *Treatise on Solid State Chemistry,* Vol. 1 (ed. Hannay, N. B.) Plenum Press, New York.

Polk, D. E. (1971) *J. Non-Cryst. Solids* **5**, 365.

Ramasesha, S. (1984) *Pramana* **23**, 745.

Ramasesha, S. & Rao, C. N. R. (1977) *Phil. Mag.* **34**, 827.

Ramdas, S. & Thomas, J. M. (1980) in *Chemical Physics of Solids and their Surfaces, Specialist Periodical Report,* Vol. 7 (eds Roberts, M. W. & Thomas, J. M.) The Chemical Society, London.

Ramdas, S., Thomas, J. M., Betteridge, P. W., Cheetham, A. K. & Davies, E. K. (1984) *Angew. Chem. Int. Ed. Engl.* 23.

Ranganathan, S. & Chattopadhyay, K. (1991) *Ann. Rev. Mater. Sci.* **21**, 437.

Rao, C. N. R. (ed.) (1974) *Solid State Chemistry,* Marcel Dekker, New York.

Rao, C. N. R. & Rao, K. J. (1978) *Phase Transitions in Solids,* McGraw-Hill, New York.

Rao, C. N. R., Rao, K. J., Ramasesha, S., Sarma, D. D. & Yashonath, S. (1992) *Annual Reports.* (C) Royal Soc. Chem., **7**, 179.

Rao, C. N. R. & Seshadri, R. (1994) *MRS Bulletin* **19** (11), 28.

Rao, C. N. R., Seshadri, R., Sen, R. & Govindaraj, A. (1995) *Mat. Sci. Eng.* **R15**, 1.

Raveau, B. (1986) *Proc. Ind. Nat. Sci. Acad.* **52A**, 67.

Raveau, B. (1987) *Reviews of Inorg. Chem.* **9**, 37.

Rice, C. E. & Jackel, J. L. (1982) *J. Solid State Chem.* **41**, 308.

Robertson, J. M. (1951) *Proc. Roy. Soc. London* **A207**, 101.

Sanders, M. J. & Catlow, C. R. A. (1983) *Proc. 6th Intnl. Zeolite Conference* (eds. Orsen, D. & Bisio, S.) Butterworths, London.

Seshadri, R., Govindaraj, A., Aiyer, H. N., Sen, R., Subbana, G. N., Raju, A. R. & Rao, C. N. R. (1994) *Curr. Sci.,* **66**, 839.

Sevov, S. C. & Corbett, J. D. (1993) *Science* **262**, 880.

Shannon, R. D. (1976) *Acta Crystallogr.* **A32**, 751.

Shannon, R. D. (1981) in *Structure and Bonding in Crystals,* Vol. II (eds. O'Keeffe, M. & Navrotsky, A.) Academic Press, New York.

Shannon, R. D. & Prewitt, C. T. (1969) *Acta Crystallogr.* **B25**, 925.

Shectman, D., Blech, I., Gratias, D. & Cahn, J. W. (1984) *Phys. Rev. Lett.* **53**, 1951.

Shen, Y. F., Zerger, R. P., DeGuzman, R. N., Suib, S. L., McCurdy, L., Potter, D. I. & O'Young, C. L. (1993) *Science* **260**, 511.

Simon, A. (1981) *Angew Chem. Int. Ed. (Engl.)* **20**, 1.

Simon, A. (1982) *Ann. Chim. Fr.* **7**, 539.

Soghomonian, V., Chen, Q., Zhang, Y., Haushalter, R. C., O'Connor, C. J., Tao, C. & Zubieta, J. (1995) *Inorg. Chem.* **34**, 3509.

Soos, Z. G. & Klein, D. J. (1976) in *Treatise on Solid State Chemistry,* Vol. 3 (ed. Hannay, N. B.) Plenum Press, New York.

St. John, J. & Block, A. N. (1974) *Phys. Rev. Lett.,* **14**, 107.

Sävborg, O. (1985) *J. Solid State Chem.* **57**, 154.

Tenne, R., Margulis, L., Genut, M. & Hodes, G. (1992) *Nature*, **360**, 444.

Thorn, D. L., Harlow, R. L. & Herron, N. (1995) *Inorg. Chem.* **14**, 2629.

Torrance, J. B. (1979) *Acc. Chem. Res.* **12**, 79.

Tosi, M. P. (1964) *Solid State Phys.* Vol. 16 (eds Seitz, F. & Turnbull, D.) Academic Press, New York.

Tsang, S. C., Chen, Y. K., Harris, P. J. F. & Green, M. L. H. (1994) *Nature* **372**, 159.

Ugarte, D. (1992) *Nature* **359**, 707.

Uppal, M. K., Ramasesha, S. & Rao, C. N. R. (1980) *Acta Crystallogr.* **A36**, 356.

Verma, A. R. & Trigunayat, G. C. (1974) in *Solid State Chemistry* (ed Rao, C. N. R.) Marcel Dekker, New York.

Wadsley, A. D. (1964) in *Nonstoichiometric Compounds* (ed. Mandelcorn, L.) Academic Press, New York.

Wells, A. F. (1984) *Structural Inorganic Chemistry*, (5th edn) Oxford University Press.

Williams, D. E. (1967) *J. Chem. Phys.* **47**, 4680.

Williams, D. E. & Starr, R. (1977) *Comput. Chem.* **1**, 173.

Wilson, S. T., Lok, B. M., Messina, C. A., Cannon, T. R. & Flanigen, E. M. (1982) *J. Amer. Chem. Soc.* **104**, 1146.

Wright, P. A., Jones, R. H., Natarajan, S., Bell, R. G., Chen, J., Hursthouse, M. B. & Thomas, J. M. (1993) *J. Chem. Soc. Chem. Commun.* 633.

Wypych, F. & Schöllhorn, R. (1992) *J. Chem. Soc. Chem. Commun.* 1386.

Yvon, K. (1979) in *Current Topics in Materials Science*, Vol. 3 (ed. Kaldis, E.) North-Holland, Amsterdam.

Zachariasen, W. H. (1932) *J. Amer. Chem. Soc.* **54**, 3841.

Zallen, R. (1983) *The Physics of Amorphous Solids*, John Wiley, New York.

Zhang, Y., Clearfield, A. & Haushalter, R. C. (1995) *Chem. Mater* **7**, 1221.

Ziolkowski, J. (1985) *J. Solid State Chem.* **57**, 269.

Zunger, A. (1980) *Phys. Rev.* **B22**, 5839.

Zunger, A. (1981) in *Structure and Bonding in Crystals*, Vol. I (eds O'Keeffe, M. & Navrotsky, A.) Academic Press, New York.

2 New and improved methods of characterization

2.1 Introduction

Characterization is an essential part of all investigations in solid state chemistry and materials science. The various aspects of characterization are: (i) chemical composition and compositional homogeneity of the specimen, (ii) impurities that may affect the properties, (iii) structure, revealing the crystallinity or otherwise of the specimen, crystal system, unit cell and where possible (or necessary) atomic coordinates, bonding and ultrastructure and (iv) the nature and concentration of imperfections (defects) influencing properties. While it may not be possible to achieve complete characterization of a given solid by a single investigator or at any given time, yet without a minimum level of characterization no investigation can be initiated or completed. The scope of 'characterization' is so vast that nearly all aspects of solid state chemistry can be included under its domain. According to the US Materials Advisory Board Committee, 'characterization describes those features of the composition and structure (including defects) of a material that are significant for a particular preparation, study of properties or use, and suffice for reproduction of the material'. The subject of characterization has been reviewed sufficiently in the literature (Cheetham & Day, 1987; Meinke, 1973; Newnham & Roy, 1973; Honig & Rao, 1981; West, 1985) and we shall therefore recount the main essentials and highlight some of the recent developments. The advances made in the last few years in characterization techniques, especially in structure elucidation, have been truly remarkable and have opened new vistas in solid state chemistry.

Since the early part of this century, X-ray diffraction has played a central role in identifying and characterizing solids. Our ideas on the nature of bonding and the working criteria for distinguishing short-range and long-range order of crystalline from amorphous substances are largely derived from X-ray diffraction. While X-ray diffraction still remains a most useful tool to obtain structural information averaged over a large number of unit cells, it has several limitations. New and more powerful methods therefore became necessary and several of these are now available. Efforts will undoubtedly continue to develop newer and better techniques for the characterization of solids.

2.2 Structural characterization

The various techniques available for structural characterization may be grouped under the following categories: optical methods, diffraction methods, electron microscopic methods, spectroscopic methods, and use of physical properties as characterization tools (see Table 2.1). Optical methods involve examination of the substance under an optical (polarizing) microscope, measurement of refractive indices, surface topography and so on. These studies provide valuable information on the morphology, crystal symmetry, physical defects, magnetic and electric domains, etc. Detailed evaluation of the optical indicatrix and refractive index measurement can tell us whether crystals are optically isotropic (cubic and amorphous solids), uniaxial (tetragonal, hexagonal and rhombohedral) or biaxial (orthorhombic, monoclinic and triclinic). Optical studies of etched surfaces provide information on internal surfaces and on line defects. Emerging dislocations show etch pits on the surface and these provide a standard method for determination of dislocation densities in metals. The symmetries of *etch pits* give additional information on crystal symmetry and crystal orientation since they can be classified into ten two-dimensional point groups. Ferroelectric domains etch at different rates according to the orientation of the polarization vector with respect to the surface. Ferromagnetic domains can be viewed by decorating the surface. Information about particle shape and size distribution in a polycrystalline aggregate can also be obtained from optical studies. Spectroscopic methods provide various types of information on solids: local symmetry, point defects, chemical environment of atoms, bonding, phase transition, etc.

A complete description of the structure of a solid requires the determination of the crystal system, space group, unit cell dimensions, atomic coordinates and finally the actual electron density distribution around the atoms. In order to provide such information, one uses diffraction involving X-rays, electrons or neutrons. The diffraction condition is described by the well-known Bragg's law, $n\lambda = 2d \sin \theta$, where λ is the wavelength of the radiation or particle-wave, d the spacing between atomic planes in a crystal and θ the Bragg angle. The relation between the diffraction pattern and the structure of the crystal is understood in terms of the reciprocal lattice. Amongst the different diffraction techniques, X-ray diffraction has been most commonly used for routine characterization as well as for detailed structural elucidation, but solid state chemists are making increasing use of electron diffraction and neutron diffraction to obtain information not provided by X-ray diffraction.

2.2.1   X-ray diffraction

The three standard methods of X-ray diffraction are: (a) the Laue method, involving a stationary single crystal and 'white' X-rays, (b) the rotating crystal method involving a single crystal rotated in a beam of monochromatic X-rays and (c) the powder method involving a polycrystalline sample rotated in a beam of monochromatic X-rays. The Laue method can be used to determine crystal symmetry and orientation, but not for structure determination. The rotating crystal method can be used to determine the single crystalline nature and unit cell parameters of a sample. The main difficulty with the rotating crystal method is overlapping of reflections due to different hkl values. This problem is overcome in the Weissenberg method by moving the film parallel to the axis of rotation as the crystal rotates. Furthermore, a screen with a narrow aperture between the crystal and the film allows reflections only from a single-layer line

Table 2.1. *Structural characterization of solids*

	Structural detail	Experimental techniques
1. (i)	Morphology, phase identification, amorphous or crystalline nature, crystal system, space group, position of atoms	X-ray diffraction, optical microscopy, electron diffraction, electron microscopy, neutron diffraction, EXAFS and XANES
(ii)	Ultra-structure	High-resolution electron microscopy, STM and AFM
2.	Site symmetry	Spectroscopic techniques (IR, Raman, UV-visible, NMR, ESR)
3.	Perfection of crystals	Microscopy and X-ray topography
4.	Thermal amplitudes of vibration	X-ray diffraction, neutron diffraction, Mössbauer spectroscopy
5.	Phase transitions	Differential thermal analysis (DTA) or differential scanning calorimetry (DSC), light scattering and various spectroscopic techniques, diffraction methods (especially X-ray diffraction), measurement of thermal expansion and any other property changing with the transition
6.	Bonding and electronic structure	Electron spectroscopies[a], electron density by X-ray diffraction and spin density by neutron diffraction
7.	Defects and ordering	Thermogravimetry, transmission electron microscopy, neutron diffraction, ESR and optical spectroscopy
8.	Magnetic ordering and spin configuration	Magnetic susceptibility, neutron diffraction, Mössbauer spectroscopy
9.	Chemical environment of nuclei	Various spectroscopic and related techniques especially HRNMR spectroscopy of solids, Mössbauer spectroscopy, EXAFS and XANES, electron spectroscopies[a]
10.	Oxidation states of metals	Redox titrations, XPS, EELS, AES and X-ray absorption spectroscopy

[a] including X-ray and UV photoelectron spectroscopy (XPS and UPS), and Auger electron spectroscopy (AES) and electron energy loss spectroscopy (EELS).

LAMBDA 1.2019 A,

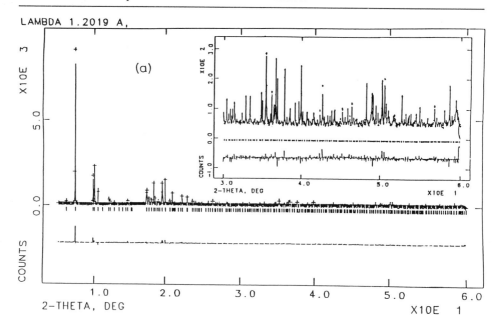

Figure 2.1   X-ray diffraction profile of silicon ferrierite. Difference Profile is also shown. Data from Brookhaven Laboratory.
(Courtesy A.K. Cheetham)

(reflections for which one of the hkl indices is constant) to be recorded on the moving film. Movement of the film avoids overlapping of reflections. A more advanced version of this, the precession on 'retigraph' method, allows a layer of the reciprocal lattice to be recorded in undistorted form. Powder diffraction patterns readily give crystal structure data. X-ray powder diffractometers are commonly used for identification of crystal structures, analysis, etc.; for accurate determination of cell dimensions, the Guinier focussing camera is often employed.

In order to obtain detailed structure, a knowledge of diffraction intensities is essential, the intensities being related to the structure factor. Computer-controlled single-crystal X-ray diffractometers with structure (software) packages have made structure elucidation a routine matter. The availability of synchrotron X-radiation of continuously variable wavelength has made X-ray diffraction a still more powerful structural tool for the study of solids. A technique of great utility to solid state chemists is the *Rietveld treatment* of powder X-ray diffraction profiles (Rietveld, 1969; Manohar, 1983). Automated structure packages for the determination of unknown structures by this method are now commercially available (see section 2.2.3). In Fig. 2.1, we show a typical set of profile data.

Besides structure, X-ray diffraction gives a host of other valuable information. For example, powder diffraction, especially with counter diffractometers, has been widely used for phase identification, quantitative analysis of a mixture of phases, particle size analysis, characterization of physical imperfections (the last two being obtainable from line broadening) and *in situ* studies of reactions. In the case of amorphous solids, X-ray diffraction data provide radial distribution functions which give the number of atoms per unit volume at any distance r from the reference atom. The radial distribution function (RDF) of an amorphous solid, obtained by Fourier transformation of the scattered intensity data, exhibits peaks at distances corresponding to the average interatomic separation of the first shell, second shell and so on, superimposed on a background curve. Although different shells can contribute to the higher lying peaks, the area under the first (and the second) peak gives the coordination number directly. Structural interpretation of the RDF of an amorphous material is difficult because the RDF is an average quantity describing the spatially averaged structural arrangement about a given atom; it is also a chemical average over all the types of atom in the sample. The difficulties encountered in describing liquid structure are also present in the case of amorphous solids. The RDF itself is not very sensitive to variations in the model.

Variable-temperature X-ray diffraction studies of crystalline substances are useful in the study of phase transitions, thermal expansion and thermal vibrational amplitudes of atoms in solids. Similarly, diffraction studies at high pressures are employed to examine pressure-induced phase transitions. Time-resolved X-ray diffraction studies (Clark & Miller, 1990) will be of great value for examining reactions and other transformations.

2.2.2 *Electron diffraction*

Electron beams differ from X-rays in two respects which render electron diffraction a valuable technique for structural studies of solids. These are (i) the small wavelength of electron beams and (ii) the charge carried by them. The smaller wavelength leads to smaller Bragg angles in electron diffraction. The radius of the Ewald sphere, $1/\lambda$, is therefore much larger for electron diffraction than X-ray diffraction. This makes it possible to record extensive sections of the reciprocal lattice directly with a small stationary crystal. Because of the charge, interaction of electrons with atoms is about 10^3 times stronger than that of X-rays and this makes it possible to record electron diffraction patterns almost instantaneously; however, the strong interaction places stringent conditions on sample thickness. Electron diffraction patterns are readily obtained with commercial electron microscopes. Employing electron diffraction one can profitably investigate defect ordering, superstructures, fine particle samples and so on. Accordingly, ordering of cation and anion vacancies (15% of each) in TiO or of anion vacancies in $CaFeO_{2.5}$ and $CaMnO_{2.5}$ is best examined by electron diffraction; such ordering gives rise to superstructure spots in the diffraction patterns. For example, La_2NiO_4 exhibits superlattice spots corresponding to a $\sqrt{2} \times a$ type of unit cell (Fig. 2.2), not detected by X-ray diffraction; in La_4LiCoO_8, Li^+ and Co^{3+} ions are shown to be ordered in two dimensions. Fig. 2.2 also shows characteristic five-fold symmetry in a quasicrystal.

Electrons with energies in the range 10–200 eV have wavelengths of the same order as interatomic spacings in crystals. These electrons have very little penetrating power. Electron diffraction carried out with low-energy electrons (LEED) gives patterns characteristic of the atomic arrangement of surface layers. Such diffraction patterns can be interpreted in terms of two-dimensional lattices (Somorjai, 1981).

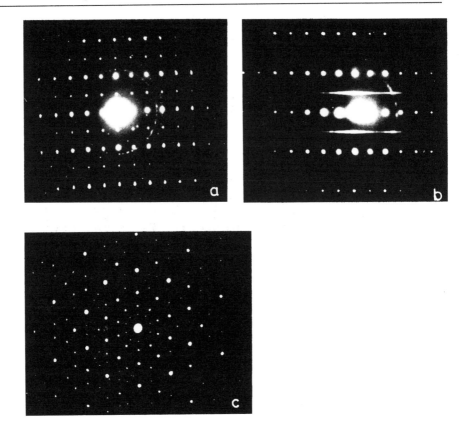

Figure 2.2   Electron-diffraction patterns of (a) La_2NiO_4 and (b) La_4LiCoO_8 showing superlattice spots. In (b) streaking is seen due to some disorder. (After Ganguly & Rao, 1984.) (c) shows the pattern of Al-14%Mn quasicrystalline alloy (Courtesy of S. Ranganathan of this Institute).

2.2.3   *Neutron diffraction and related techniques*

Thermal neutrons with a velocity of about $4000 \, ms^{-1}$ associated with a wavelength of $\sim 1.0 \, \text{Å}$ are used in neutron diffraction experiments. Whereas X-rays are scattered primarily by the extranuclear electrons in atoms, neutrons are scattered mainly by the atomic nucleus. Since the neutron-scattering amplitude does not show a smooth dependence on the atomic number of the atoms, neutron diffraction is particularly useful in locating light atoms (especially hydrogen) in crystals. Additional scattering of neutrons can arise owing to the magnetic moment of neutrons (due to the interaction of the neutron magnetic moment with the permanent magnetic moments of atoms or ions in a crystal). In the absence of an external magnetic field, the magnetic moments of atoms in a paramagnetic crystal are arranged at random, so that the magnetic scattering of neutrons by such a crystal is also random. It only contributes a diffuse background to the diffraction pattern. In magnetically ordered materials (e.g. ferromagnets and antiferromagnets), however, the magnetic moments are regularly aligned. Neutron diffraction provides an experimental means whereby the different

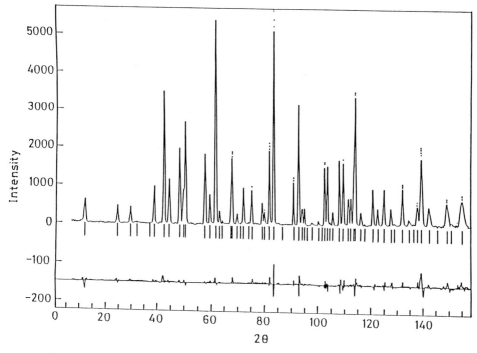

Figure 2.3   Neutron-diffraction profile analysis of $Tl_{0.5}Pb_{0.5}Sr_2CuO_5$. Difference profile and reflection positions are shown. *Data from ILL, Grenoble* (From Kovatcheva *et al.*, 1991).

magnetic structures can be determined. In addition to the two scattering effects that are elastic, neutrons can also undergo inelastic scattering by crystals. This involves energy exchange between the lattice and neutrons. Inelastic neutron scattering by crystals is used in the study of quantized vibrational modes (phonons) and dynamics in solids (Satyamurthy *et al.*, 1981; Lovesey & Springer, 1977).

Because neutron beams are much weaker in intensity than X-rays, neutron diffraction requires large single crystals (10–$100\,mm^3$ in volume as compared to the $\sim 0.1\,mm^3$ crystal volume used in X-ray diffraction work). However, it is possible to obtain useful structural data by analysis of neutron-diffraction profiles from polycrystalline materials (Cheetham & Taylor, 1977).

In the profile analysis method developed by Rietveld (1969), the structural parameters are fitted to the overall profile of the powder pattern instead of the traditional approach of fitting the intensities of individual peaks (see Figure 2.3). The pattern is assumed to be the sum of a number of Gaussian-shaped Bragg reflections. The method involves calculation of an intensity function Y_i at every value of the Bragg angle θ and comparison with the corresponding observed values. The function Y_i, which is a measure of the intensity contribution to the profile at the point 2θ, can be written as $Y_i = W_{i,k} \cdot F_k^2$, where $W_{i,k}$ gives the intensity contribution of a Bragg peak at position $2\theta_k$ to the profile at the point $2\theta_i$ and F_k is the structure factor appropriate to the Bragg peak k. Profile analysis is particularly suitable in neutron-diffraction work because the peaks are accurately described by a Gaussian function. Structures of a large number of oxides, hydrides, zeolites etc., (Cheetham, 1986) have been refined using powder neutron diffraction data and the accuracy achieved is often around 10% ($R = 0.1$) which is as good as the accuracy obtained in single crystal X-ray work (see Fig. 2.3). Several modifications of the original Reitveld programme are now available and these include allowance for anisotropic temperature factors and flexible molecular constants. The availability of intense pulsed spallation neutron sources has made this technique still more effective (Lovesey & Sterling, 1984). The power of the powder method today is illustrated by the solution of the structure of $La_3Ti_5Al_{15}O_{37}$ with 60 atoms in the asymmetric unit by a combined use of X-ray and neutron-diffraction profiles (Morris *et al.*, 1994).

With single crystals, neutron-diffraction studies provide a wealth of structural information. It is widely used to locate light atoms like hydrogen, to determine charge distribution and for surface studies. Spin ordering in magnetic solids is readily investigated by this technique. Information about covalency can be obtained from magnetic form factors. Many of the surface studies carried out hitherto have been on oriented graphite crystals. Neutron diffraction has been used to investigate amorphous solids such as glasses and liquids as well as spin glasses. Partial structure factors have been obtained in these studies by isotopic substitution. X-rays (from a synchrotron) can also provide such information if anomalous scattering near an absorption edge is made use of.

Neutron scattering provides valuable information on dynamics of processes occurring in solid state. Thus, quasielastic scattering gives us information on molecular motion, hydrogen diffusion, tunnelling states and critical scattering (central mode) as well as about spin diffusion. Very high resolution is obtained in the back-scattering geometry and it is then possible to separate elastic and quasielastic components. An advantage of neutron scattering over light scattering is that it gives $S(Q, \omega)$ whereas light scattering gives $S(O, \omega)$; the scattering function $S(Q, \omega)$ describes the dynamic properties of the target system where Q is the change in the neutron wave vector and $\hbar\omega$ the change in energy accompanying the scattering process. The vibrational density of states is readily obtained with incoherent inelastic scattering. The information is similar to that obtained by IR and Raman spectroscopy, but there are no selection rules; powder samples can be used for these studies. Coherent inelastic scattering studies (with single crystals) are most useful to investigate phonons and soft modes as well as magnons. With recent neutron sources, an energy resolution of a few microelectron volts and a time resolution extended to 10^{-6}–10^{-8} s are possible.

Many interesting examples of neutron studies on problems related to solid surfaces can be cited. A variety of materials has been investigated by neutron techniques to obtain (nuclear or magnetic) structural information. We shall cite a few examples from the literature. Neutron diffraction and scattering have been employed to investigate the structure and vibrational modes as well as tunnelling transitions of methane adsorbed on graphite (Bomchil *et al.*, 1980). Tunnelling of hydrogen in alkali metal–intercalated graphite has been investigated (Beanfils *et al.*, 1981). *In situ* inelastic neutron-scattering studies (Vasudevan *et al.*, 1982) have revealed the existence of two sites at which

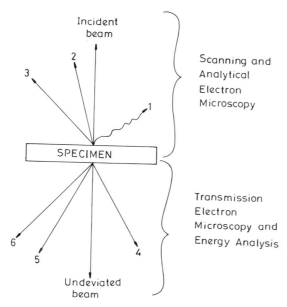

Figure 2.4   Different types of interactions of electrons with a solid: 1, X-ray or optical photons; 2, back-scattered electrons; 3, secondary electrons; 4, coherent elastic scattering; 5, inelastic scattering; 6, incoherent elastic scattering.

hydrogen may be sorbed on MoS_2. The structure of NH_3-intercalated TaS_2 and the translational diffusion of NH_3 have been studied; making use of these results along with those from NMR, it is shown that the ammonia molecule is arranged with the nitrogen lone-pair approximately parallel to the sheets. The motion of H_2O molecules in aluminosilicate minerals has similarly been examined (Olejnik *et al.*, 1970). *Neutron spin echo* method is specially useful to study magnetic phase transitions.

2.2.4   *Electron microscopy*

The electron microscope has emerged to become the most versatile tool to study the ultrastructure of materials, to identify new or known phases and simultaneously to yield information on composition. Usually electron diffraction and imaging are employed together in transmission electron microscopy (TEM). Images of isolated atoms are well within the capability of modern instruments. Recent advances in scanning instruments (which record signals due to back-scattered electrons or emitted secondary electrons) enable us to detect extremely small quantities of materials and also to map the topography of surfaces of a wide range of materials (minerals, metals, catalysts, polymers and so on). These capabilities are further enhanced in the scanning transmission electron microscope (STEM). In Fig. 2.4, different types of interaction of an electron beam with a solid sample are shown schematically.

The resolving power of a microscope is related, among other things, to the wavelength of the radiation used. With ordinary white light of effective wavelength 550 nm, the resolving power is around 200 nm. Beyond this, there seems no way of

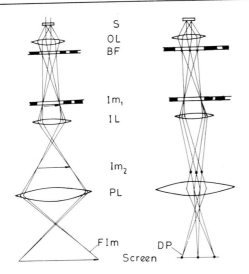

Figure 2.5 Operational principles of a transmission electron microscope. S, specimen; OL, objective lens; BF, back focal plane; Im$_1$, intermediate image; IL, intermediate lens; Im$_2$, second intermediate image; PL, projector lens; FIm, final image; DP, diffraction pattern.

improving the resolving power of an optical microscope. Although electromagnetic radiation of wavelengths much shorter than those of visible light exists, e.g. X-rays, there is no suitable lens for focussing such rays. With the recognition that moving electrons possess small wavelengths and with the discovery that electron beams can be focussed by magnetic as well as electrostatic fields, microscopes of much higher resolving power which make use of electron beams have come into existence from around the 1930s. Since typical wavelengths of electron beams are about 10^5 times smaller than those of visible radiation, an electron microscope is capable of imaging atomic structures, if wavelength is the only factor determining the resolving power.

The mechanism of image formation in TEM involves both elastic and inelastic scattering. The image contrast in the electron microscope arises from (i) the amplitude contrast (the phases of transmitted and diffracted beams do not recombine) and (ii) the phase contrast (transmitted and diffracted beams recombine preserving their amplitudes and phases). For crystalline specimens, there is an additional contrast arising from a strong preferential Bragg scattering in certain well-defined directions. This is called diffraction contrast, which is a special case of amplitude contrast. A schematic diagram that illustrates the mode of image formation is shown in Fig. 2.5 along with the diffraction mode. By a flick of a switch, which activates the intermediate lens, one can obtain the image or the diffraction pattern from the same region of the specimen on the screen or on the photographic plate. Various factors that decide whether the experimental image corresponds to the actual structure are specimen thickness, the angular range of diffracted beams passing through the objective aperture, the extent of defocus (employed to obtain proper contrast) and so on. The minimum resolvable

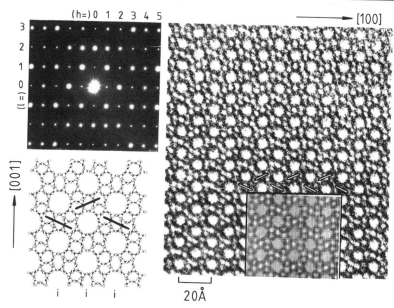

Figure 2.6   High-resolution electron microscopic (HREM) image of
ZSM-5/silicalite. The computer-simulated image is shown as an inset. The
corresponding electron diffraction pattern and structural drawing are also given.
Notice how the large channels (~ 6.5 Å diameter) as well as the small ones are
revealed in the image. The zig-zag path of inversion symmetry lines joining the
interchannel apertures is indicated by the bars drawn. (Courtesy of G.R. Millward
& J.M. Thomas. University of Cambridge.)

distance is given by $d \approx 0.6S^{1/4}\lambda^{3/4}$ where S is the spherical aberration coefficient of the
objective lens and λ the wavelength of electrons. It is important to remember that in
producing the image, the electron microscope carries out Fourier transformations: the
crystal in real space is transformed to a diffraction pattern in reciprocal space and the
latter is then transformed to the image in real space. In Fig. 2.6, a lattice image of a
ZSM-5 type material is shown to illustrate the power of the technique.

   TEM generally produces a projection image where all depths of thin specimens are in
focus at the same time, making it difficult to interpret surface features of specimens.
Several important improvements have emerged recently to obviate this difficulty. With
appropriate modifications, an electron microscope can be converted into a veritable
laboratory with high/low-temperature stages, facility to operate in different atmospheres and so on. High-voltage microscopes operating at a million volts or more are
now being employed to study samples under very high resolution as well as under
realistic conditions. The Cambridge University high-resolution electron microscope
operating at 500 kV does, in fact, give nearly atomic resolution in favourable instances,
as will be shown later.

   In scanning electron microscopy (SEM), a finely focussed electron beam probe
moves from one point on the specimen to the next to form a raster pattern, just as in
television imaging. The intensity of scattered or secondary electrons is continuously

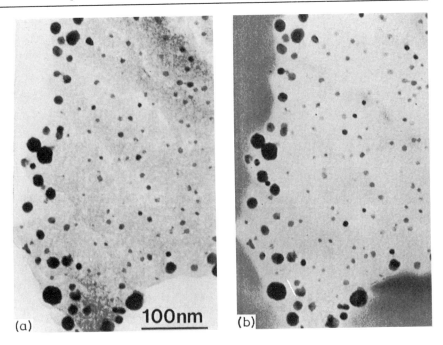

Figure 2.7   STEM images of 1 wt% Pd on charcoal. Mean particle size is around 100 Å. (After Treacy *et al.*, 1978.)

measured and displayed on a TV monitor to form a raster image. With modern high-intensity electron sources (LaB_6), a surface resolution of 20–50 Å is readily attained. The resolution is limited by electron penetration underneath the surface and also by surface cleanliness. Surface scanning electron microscopy produces topographic images which are readily interpreted. Magnetic domains can also be examined in the reflective mode in a suitably modified scanning microscope.

In *scanning transmission electron microscopy*, one combines the transmission of electrons through the specimen with scanning and analytical facilities. The information obtained about the specimen involves an analysis of the perturbations suffered by the electrons in their collisions with the atoms/molecules of the specimen. In this technique, a scanned image can be produced with a 5 Å electron beam and the image resolution is similar to that obtained with a high-resolution transmission electron microscope. Furthermore, several scanning signals are collected simultaneously, allowing enhancement of image contrast. The technique is used to study a variety of minerals and materials, especially catalysts. In Fig. 2.7, micrographs of catalyst particles are shown to illustrate the power of the technique; particles as small as 10 Å can be seen in this manner.

Analytical electron microscopy is the most sophisticated tool available for microstructural analysis today. In this method, we can obtain both the high-resolution structure and elemental composition of a specimen. This is probably the best technique to obtain local elemental composition of small regions of heterogeneous solids. When a high-energy electron beam is incident on a specimen, we get elastically and inelastically scattered (transmitted) electrons, characteristic emitted X-rays, back-scattered electrons and secondary (emitted) electrons (Fig. 2.4). All these can be collected at the same time in the microscope from a small specimen area in order to carry out analysis of the structure and composition. Elastically scattered electrons give images and diffraction patterns. Electrons which have lost energy give information on composition as does the emitted X-ray spectrum. Data from both spectra are generally collected in a microcomputer, processed and appropriately displayed. In modern instruments, the operating vacuum has been improved greatly (10^{-9} torr) in order to maintain high-intensity electron sources and to minimize contamination. In Fig. 2.8 an example of X-ray emission analysis on the electron microscope is shown in the case of a crystal of nominal composition $Bi_{0.2}WO_3$; the actual composition was found to be $Bi_{0.07}WO_3$ by X-ray emission analysis and was also verified by X-ray photoelectron spectroscopy. The presence of Bi atoms in strips of hexagonal tungsten bronze periodically occurring in WO_3 could be established only with the aid of electron microscopy combined with X-ray emission analysis. Another example of electron microscopic analysis is provided by lizardite, which is a serpentine where the misfit between $Mg(O,OH)_6$ octahedral layers and SiO_4 tetrahedra is removed by aluminium substitution; that such is the case could be established by X-ray emission analysis. Other examples of X-ray emission analysis in the electron microscope are described by Jefferson (1982). Electron energy loss spectroscopy (EELS) can be used for analysis and also for obtaining information on the oxidation states of metals and electronic structures of oxides and other materials. The use of EELS carried out in an electron microscope in qualitative and quantitative analysis of ultramicroquantities has been reviewed (Egerton, 1982). The L absorption edges of transition metals and the K absorption edge of oxygen in transition-metal oxides (see Fig. 2.9) reveal considerable structural information of significance (Rao *et al.*, 1984). The plasmon bands of metals in EELS are similarly of diagnostic value (Sparrow *et al.*, 1983).

A present-day state-of-the-art analytical microscope readily produces images with a point-to-point resolution 3 to 4.5 Å (with magnification in the range 300 000 X–800 000 X); a dedicated high-resolution microscope (200 kV) instrument can even give a resolution of 2.5 Å. Under optimum conditions, we can get diffraction patterns from crystals smaller than 5 Å in diameter, but readily with crystals of 100 Å diameter or more. The crystal thickness has to be small for the best analytical results (500 Å thickness or less being preferred). When the specimen is in the free state or surrounded by a weakly scattering matrix, 10^{-18} g of a material can be easily detected by this technique; even smaller quantities (several orders of magnitude less) can be detected under optimum conditions. When a matrix is present, the sensitivity depends on the atomic number(s) of the elements composing the matrix.

The most important aspect of electron microscopy in solid state chemistry lies in its ability to elucidate problems that are beyond the capability of X-ray or neutron crystallography. High-resolution electron microscopic (HREM) images show local structures of crystals in remarkable detail; in Fig. 2.10 we show the HREM image of $Bi_{0.1}WO_3$ obtained with the Cambridge University 500 kV microscope to show the improved resolution compared with the image obtained with a 200 kV microscope

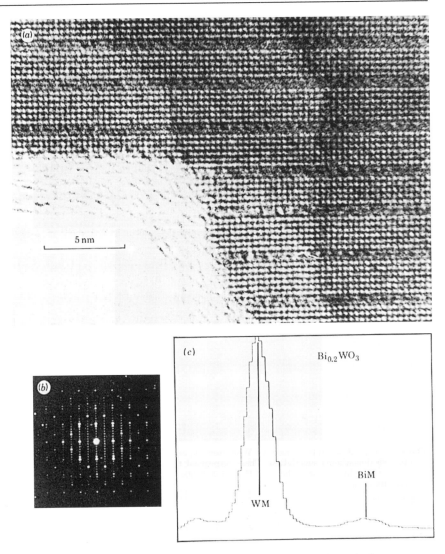

Figure 2.8   HREM image (recorded at 200 kV) of a sample of nominal composition $Bi_{0.2}WO_3$. The electron diffraction pattern and X-ray emission profile are also shown. The stripes are due to Bi atoms present in hexagonal tunnels. (After Ramanan et al., 1984.)

(Fig. 2.8). Ultramicrostructures of imperfections are readily seen in the images. Selected area diffraction along with imaging can often reveal atomic arrangements in regions containing only a few atoms. Since the scattering factors for electrons are a thousand times those for X-rays, structural information can be obtained with very small specimens (~ 5 Å diameter). HREM is also very useful in identifying new phases, some often occurring in one or two rows or layers as imperfections or as intergrowths in other phases (see Fig. 2.11).

Once the experimental image is obtained using a thin specimen (< 100 Å), the

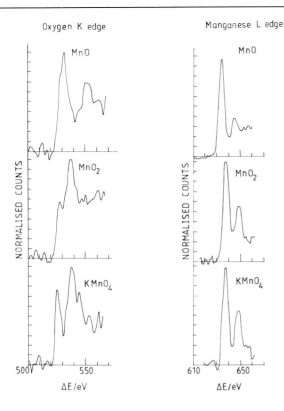

Oxygen K edge

Manganese L edge

Figure 2.9   Mn(L) edge and O(K) edge in the EELS of manganese oxides measured in an electron microscope. (Rao *et al.*, 1984.)

interpretation is checked with the aid of computed images obtained by taking into account multiple scattering of electrons, lens characteristics and specimen thickness. The so-called multislice method (Goodman & Moodie, 1974) is generally employed for image simulations and the images calculated for a series of values of crystal thickness and defocus. Applications of electron microscopy for investigating various types of phenomena, structures, defects and reactions of importance to solid state chemistry have been reviewed adequately (Allpress & Sanders, 1973; Anderson, 1978; Kihlborg, 1979; Beer *et al.*, 1981; Thomas, 1982) and we shall be dealing with some of them in the appropriate sections.

2.2.5   *X-ray absorption spectroscopy (EXAFS and XANES)*

Traditionally, X-ray absorption edge measurements have been used to determine oxidation states of metals in complex materials. The extended X-ray absorption fine structure (EXAFS), on the other hand, provides structural information such as bond distances and coordination numbers even with powdered samples, crystalline or amorphous, the fine structure essentially resulting from short-range order around the absorbing atom. The technique is also useful for studying solid surfaces (SEXAFS). The observation of fine structure beyond the K-absorption edges of materials dates back to

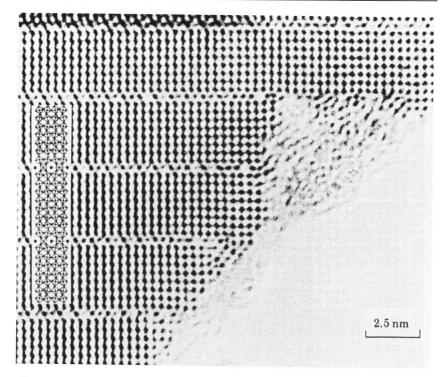

Figure 2.10   HREM image (recorded at 500 kV) of a sample of nominal composition $Bi_{0.1}WO_3$. The structural drawing is shown. Bi atoms in the hexagonal tunnels can be seen. (After Ramanan *et al.*, 1984.)

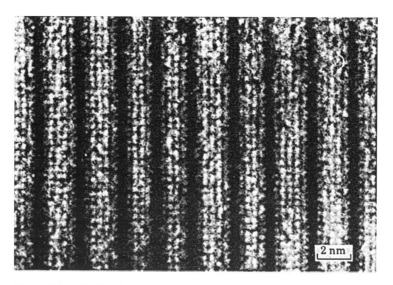

Figure 2.11   Lattice image of a nominal $n = 8$ member of the oxide family $(Bi_2O_2)^{2+}(A_{n-1}B_nO_{3n+1})^{2-}$, showing random occurrence of $n = 4$ and $n = 5$ members. (After Hutchison *et al.*, 1977.)

the early 1920s when it was called Kronig structure, but it is only since 1965 that much advance has been made in this area. It is the sensitivity of EXAFS to local structure that renders it a unique structural tool (Stern *et al.*, 1978; Sandstrom & Lytle, 1979; Teo, 1980; Parthasarathy *et al.*, 1982). The absorption edge itself is a characteristic feature of an element and EXAFS is a consequence of (low-energy) electron diffraction that occurs locally. In a sense, EXAFS can be considered to be a LEED experiment 'conducted with an electron gun and a phase-sensitive detector located within the sample itself' (Sandstrom & Lytle, 1979). It is evident that a study of the EXAFS of different elements in a given material results in the complete definition of the local structure. This is tantamount to a knowledge of the (short-range) partial structure factor for each component element.

It was noted earlier that EXAFS is a result of two fundamental processes: (a) K- (or L-) absorption of an X-ray photon which is the photoelectric effect; and (b) an effective diffraction of the electron so emitted. In the case of an isolated absorbing atom (absorber) one sees only the characteristic rise in absorbance, μ ($= \ln I_0/I$), at the energy corresponding to the edge and an exponential decrease thereafter. When the absorber is surrounded by other atoms, μ exhibits undulations sometimes up to 2000 eV beyond the edge. Undulations starting 30 eV beyond the edge constitute the EXAFS. As an example, the EXAFS of a bimetallic catalyst is shown in Fig. 2.12.

Photoelectrons are emitted, following the absorption of the X-ray photon, by the

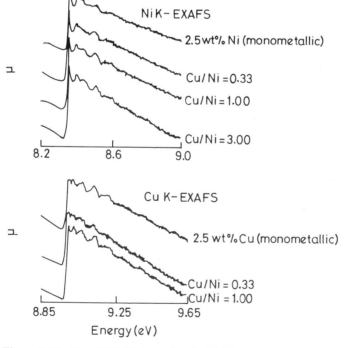

Figure 2.12   Raw EXAFS data of a Cu-Ni/Al$_2$O$_3$ bimetallic catalyst (Courtesy G.U. Kulkarni & C.N.R Rao).

absorber atoms. The outgoing (spherical) photoelectron wave emitted by the absorber is reflected by the surrounding 'scatterers'. The phase of the reflected wave is dependent on photoelectron wavelength, the atomic number of the scatterer and the distance between absorber and scatterer atoms. The reflected (phase-shifted) wave undergoes self-interference with the outgoing wave at the emitter site. That this self-interference is responsible for the EXAFS may be seen as follows. The probability, W, of X-ray absorption (i.e. photoemission) is governed by

$$W = 2\pi^2 e^2 (\omega c^2 m)^{-1} |M_{fs}|^2 \rho(E_f) \tag{2.1}$$

where $M_{fs} = \langle f | \mathbf{p}.\varepsilon | s \rangle$, e is the charge of the electron, $2\pi\omega$ is the frequency of the X-ray photon, c is the velocity of light, m is the mass of the electron and $\mathbf{p}$ its momentum vector, ε is the electric field vector of the radiation and $\rho(E_f)$ is the density of final states. $|s\rangle$ is the K-shell wave-function and $\langle f|$ is the wave-function of the final state to which the electron is excited. The selection rule requires that $\langle f|$ have p-character.

For $E \gtrsim 30\mathrm{eV}$, the final state is essentially a continuum state. It is at once obvious from equation (2.1) that the only term which can cause the fine structure is $|M_{fs}|$ since the other term, $\rho(E_f)$, is a monotonically varying function. The $|s\rangle$ state in $|M_{fs}|$ corresponds to a very low potential energy and hence remains unaffected; thus it is only the $\langle f|$ state that can cause the undulations. The final state of a photoelectron which corresponds to a spherical wave propagating away from the absorber is shown schematically in Fig. 2.13 (Wong, 1980) along with the back-scattered wave, which, we may recall, is generally phase-shifted with respect to the outgoing wave. The reflected wave is further phase-shifted at the absorber atom, and interference between this wave and the outgoing one at the K-shell (which is a tenth or a fiftieth of the atomic radius) causes alterations in $\langle f|$.

Stern and coworkers have treated the effect of such self-interference on the final state with the inclusion of relevant phase-shifts and have shown that the EXAFS, $\chi(\mathbf{k})$, can be written as

$$\chi(\mathbf{k}) = \sum_j A_j(\mathbf{k}) \sin[2kR_j + \xi_j(\mathbf{k})] \tag{2.2}$$

where $k = |\mathbf{k}|$, summation is over j coordination shells and R_j is the distance from the absorber to the jth shell while $\xi_j(\mathbf{k})$ is the total phase-shift due to the jth shell ($\eta_j(\mathbf{k})$) and the absorber ion potential ($\delta(\mathbf{k})$). $\xi_j(\mathbf{k})$ can be fitted to a second-order polynomial

$$\xi_j(\mathbf{k}) = \xi_o + \xi_1 k + \xi_2 k^2 \tag{2.3}$$

$A_j(\mathbf{k})$ represents the total jth shell scattering amplitude and can be written (Sandstrom & Lytle, 1979)

$$A_j(\mathbf{k}) = \frac{N_j}{kR_j^2} F_j(\mathbf{k}) \exp - (2\sigma_j^2 k^2) \tag{2.4}$$

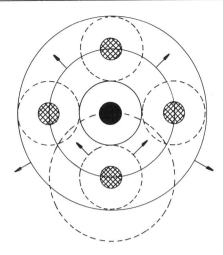

Figure 2.13   A schematic representation of the EXAFS process. An atom (filled circle) absorbs X-rays, emitting a photoelectron wave which is back-scattered by neighbouring atoms (hatched circles). The solid circles denote outgoing electron waves and the broken circles back-scattered electron waves. Constructive or destructive interference can occur when the waves overlap.

N_j and σ_j in equation (2.4) represent the number of atoms in the jth shell and root-mean-square deviation of the interatomic distances over R_j which results both from static and dynamic (thermal) disordering effects respectively. The scattering amplitude, $F_j(\mathbf{k})$ is given by

$$F_j(\mathbf{k}) = f_i(\pi, \mathbf{k})\exp(-2R_j/\lambda) \tag{2.5}$$

where $f_j(\pi,k)$ is the back-scattering amplitude (which is equivalent to an electron back-scattering form factor) and λ is the mean free path of the photoelectron. Equation (2.2) explains why monoatomic gases do not exhibit EXAFS and shows very simply that the absorbance undulations arise from the sine term.

It is at once obvious that Fourier transformation of equation (2.2) should yield information about all the j shells in direct space that contribute to the EXAFS. The R_js so obtained are, however, shortened by ξ_1, the k-dependent part of $\xi_j(\mathbf{k})$. Since the intensity of the outgoing spherical wave decreases very rapidly with increasing R, distant atoms contribute very little to the fine structure. Multiple scattering effects are also relatively unimportant and these have indeed been ignored in the derivation of equation (2.2). EXAFS should contain no information about 'shadowed' or eclipsed atoms, but there are exceptions to this. Other theoretical approaches also use similar effects to explain the EXAFS.

Since the Fourier transformation of equation (2.2) yields only a radial distribution function about the absorber, we note that information obtained from EXAFS is limited to an average, one-dimensional representation of structure. Furthermore, in order that the transform be comparatively free of ripples, the data should extend to at least

~ 500 eV beyond the edge. Other limitations of the Fourier method, such as termination error, are shortcomings that EXAFS shares with other cognate techniques, though use may be made of window functions in order to minimize their effects. The most important parameter involved in the transformation is the total phase-shift function $\xi_j(\mathbf{k})$. Several *ab initio* methods of evaluating $\xi_j(\mathbf{k})$ have been developed and values of $\xi_j(\mathbf{k})$ have been tabulated (Teo & Lee, 1979), but their accuracy is limited by the many approximations involved in the calculations. Often phase-shift corrections are estimated experimentally by making use of structural information available from other techniques on related compounds.

EXAFS data are processed to obtain radial structure functions (RSFs). First, the non-EXAFS components are subtracted from the data. Pre-edge absorption is removed using the Victoreen correction (*International Tables for Crystallography*, 1969) of the form $A\lambda^3 + B\lambda^4$. The monotonic decrease of absorbance beyond the edge, called the photoelectric decay, μ_o, is subtracted out after approximating it either by a second degree polynomial or a spline-function (Eccles, 1978). The normalized $\chi(\mathbf{k})$ is then expressed as

$$\chi(\mathbf{k}) = (\mu - \mu_o)/\mu_o \qquad (2.6)$$

The photoelectron wave-vector k is evaluated using $\hbar^2 k^2 = 2m(E - E_o)$ where E is the energy of the X-ray photon, E_o, a reference energy and m, the mass of the electron. $\chi(\mathbf{k})$ is multiplied by $k^n(n = 2$ or 3 usually) to magnify the faint EXAFS at large k (Lytle *et al.*, 1975); $k^n\chi(\mathbf{k})$ is Fourier transformed to yield the RSF, $\phi(R)$. In the model compound, the first peak at a distance R_1 represents the distance to the nearest-neighbour shell and may be compared to R_1', the known distance. We can then define α^* as $(R_1' - R_1)$, which represents the experimentally determined phase correction. In principle, $2\alpha^*$ should be equal to the theoretically estimated k-dependent part of $\xi_j(\mathbf{k})$, viz. ξ_1, if the identity of the scatterer environment has been correctly assumed. It must be emphasized that wherever scatterer identities are obscure (e.g. in several covalently bonded and disordered systems) use of α^* (and not ξ_1) is advisable. Further, the k-dependence of $\xi_j(\mathbf{k})$ introduces an intrinsic limitation to its quantitative accuracy.

The second stage of improvement in the EXAFS data is achieved through Fourier filtering (Eccles, 1978). Here, a particular peak in $\phi(R)$, corresponding to the ith scattering shell, is transformed back to k-space and the transformed function is fitted to a parameterized expression for EXAFS (Eccles, 1978; Cramer 1978)

$$\chi'(\mathbf{k}) = \frac{k^n C_1 \, \exp \, (-C_2 k^2)}{k C_3} \, \sin \, \{C_4 + (C_5 + 2R_i) \, k + C_6 k^2\} \qquad (2.7)$$

$\chi'(\mathbf{k})$ has six adjustable parameters (the Cs) and consists of both phase and amplitude terms, *cf.* equation (2.3). R_i may be treated as an adjustable parameter. It is now recognized that $\xi_j(\mathbf{k})$ is a property of the absorber–scatterer couple only and is not dependent on details of chemical bonding; $\xi_j(\mathbf{k})$ is therefore transferable from compound to compound. In the Fourier filter method, α^* for the absorber–ith shell couple is given by $\alpha^* = R_i - R_i'$ where R_i is the curve fitted distance to the ith shell and R_i' is the corresponding peak in the RSF. Using this method it is possible to avoid ambiguities due to contributions from different shells to the estimated phase-shifts. The disorder parameter σ_j (represented by C_2) is a measure of the spread of interatomic distances in the jth shell (Lytle $et\ al.$, 1975). In a given shell, j, σ_j is made up of contributions from static disorder, σ_{stat}, and vibrational or dynamic disorder, σ_{vib} (related to the Debye–Waller factor) and can be written $\sigma_j^2 = \sigma_{stat}^2 + \sigma_{vib}^2$. By investigating the temperature dependence of σ_j, estimates of the individual contributions may be obtained. A comparison of the EXAFS expression and the X-ray diffraction (XRD) intensity expression suggests that EXAFS is much more $[\exp(-3\mu_{vib}^2 k^2)$ times] sensitive to disorder than XRD. Distances evaluated from EXAFS can be accurate to ± 0.02 Å but such accuracy is crucially dependent on the quality of the input data. N_j, which represents the coordination number, is estimated by a comparison of C_1 in known and unknown situations. Its accuracy is generally poor and interpretations based solely on such N_j values must be made with caution.

An EXAFS experimental set-up has three primary components: (i) a source of X-rays, (ii) a monochromator (and collimator) and (iii) a detector. Synchrotron radiation is being widely used for EXAFS, but where this facility is not available, a rotating anode source would be suitable. Progress in EXAFS instrumentation has been comprehensively reviewed in the AIP proceedings (1980).

As a powerful technique for the study of short-range order (local structure), EXAFS has been employed effectively to investigate biological substances, amorphous materials, catalysts and various types of materials of interest in solid state chemistry. The subject has been reviewed adequately (Teo, 1980; Parthasarathy $et\ al.$, 1982). We shall cite a few examples related to amorphous materials, catalysts and fast ionic conductors and intercalates. Investigations on superionic (high-temperature α phase) AgI (Fig. 2.14) and β-AgI (the latter is the insulating phase) using Ag_k EXAFS has revealed that Ag ions occupy distorted tetragonal sites in a BCC lattice of iodide ions. Further, Ag ions are displaced in the direction of the most probable path for ion jumps (Boyce $et\ al.$, 1977). Evidence has also been obtained to support the jump–diffusion–conduction model by comparison with theoretically computed EXAFS. A similar investigation of $RbAg_4I_5$ by Stutius $et\ al.$ (1979) has indicated that Ag^+ ions occupy tetrahedral voids in the iodide lattice and that the $Ag^+ - Ag^+$ correlations persist into the superionic conducting (α) phase. In CuI, however, EXAFS shows that the Cu^+ ion is present in both tetrahedral and octahedral locations (Boyce & Hayes, 1979). The favoured path of Cu^+ ion motion appears to be along the [111] direction. The analysis supports an excluded volume model for the Cu—I pair correlation function. Fig. 2.14 gives a schematic diagram of the excluded volume model; also shown are the shapes of the first peak of RSFs at various temperatures for AgI and CuI.

Studies of intercalates have also been reported in recent years. In their EXAFS investigation of Br_2 in graphite, Heald & Stern (1978) have found that while the intercalate retains its molecular structure, the Br–Br distance increases so as to match the periodicity of the graphite lattice. In the series of pseudo-stoichiometric alkali

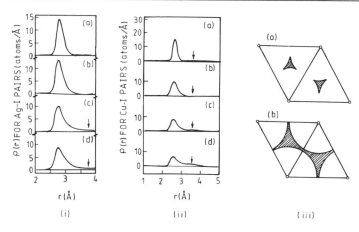

Figure 2.14   (i) Radial distribution function of iodide ions about each Ag in AgI in the β-phase at (a) 293 K (b) 371 K and in the α-phase at (c) 471 and (d) 575 K. Each distribution function is normalized to contain four iodide ions in the first peak. The arrows in (c) and (d) locate the contribution from a silver ion occupying the centre of a face between adjacent tetrahedra. (ii) Radial distribution function of iodide ions about each Cu in CuI in the γ-phase at (a) 295 K; (b) 573 K; (c) 623 K and the α-phase at (d) 743 K. The arrows indicate the location of Cu ions in the centre of a tetrahedral face, the bridging site between the tetrahedra and octahedra. (iii) A schematic two-dimensional representation of the allowed (shaded) and excluded regions for cation centres in the excluded volume model in the (a) insulating and (b) conducting phases. The anion centres are denoted by small circles. Note the connection between allowed regions in the superionic phase. (After Boyce & Hayes, 1979.)

metal–graphite compounds (KC_n, $n = 2, 3, 4$), EXAFS results suggest that the metal is present as a disordered two-dimensional lattice gas (Caswell *et al.*, 1980). Bourdillon *et al.* (1980) have studied alkali metal $NbSe_2$ intercalates and found an increase in the Nb–Nb distance.

A study of the As EXAFS of As_2O_3 glass shows that disorder increases with distance from As, thereby ruling out the existence of As_4O_6 units (Pettifer & McMillan, 1977). In As_2Te_3 glass, EXAFS results show that there is a higher degree of short-range order than in the crystal, as a result of the increased covalency in the glassy state (Pettifer *et al.*, 1977). In As_2Se_3–As_2Te_3 glass system, the non-nearest neighbour distances vary owing to increased ionicity in intermediate compositions (Parthasarathy *et al.*, 1982). Arsenic is found to be present as a substitutional impurity in Si—H films (Knights *et al.*, 1977). Zn EXAFS of $ZnCl_2$ glass through the glass transition has shown that there is no increase in disorder at T_g as in the case of GeO_2 glass (Wong & Lytle, 1980). In amorphous GaAs, Theye *et al.* (1980) find no evidence of chemical disorder; in GaP, however, a fraction of the bonds around Ga appear to be of Ga–Ga type. A number of metallic glasses have been fruitfully investigated by EXAFS (Wong, 1980). The technique has also been employed for the study of ionic glasses. An EXAFS study of Nd^{3+} in BeF_2 glass, for example, has revealed that not only is the coordination number

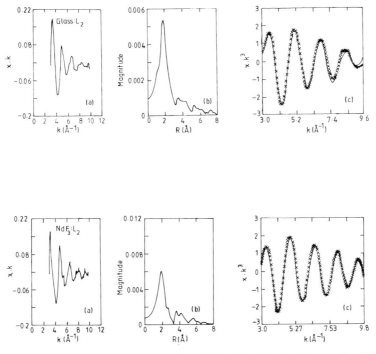

Figure 2.15   Nd L_{II} EXAFS of NdF_3–BeF_2 glass (top) and crystalline NdF_3 (bottom). (a) Normalized EXAFS spectra; (b) Fourier transform of (a); and (c) inverse transform (line) and simulated EXAFS (points) in the region 0.2 to 3.2 Å. (After Rao, K.J. *et al.*, 1983.)

of Nd^{3+} different (seven) in the glass but there is also a decrease in the Nd—F distance as compared to crystalline NdF_3 (Rao *et al.*, 1983). EXAFS of NdF_3—BeF_2 glass and NdF_3 crystal is shown in Fig. 2.15.

EXAFS has been very useful in the study of catalysts, especially in investigating the nature of metal clusters on surfaces of the supported metal catalysts (Kulkarni *et al.*, 1989; Sinfelt *et al.*, 1984). A variety of systems has been examined already and there is still considerable scope for investigation in this area. Since EXAFS gives bond distances and coordination numbers and is absorber-selective, it is possible to study one metal at a time (Fig. 2.12). Thus, an EXAFS investigation of sulphided Co—Mo—Al_2O_3 and related catalysts has shown the nature of the reactive surface species (Kulkarni & Rao, 1991). Cu/ZnO catalysts have revealed certain unusual features suggesting the complex nature of the species involved in methanol synthesis (Arunarkavalli *et al.*, 1993). Time-resolved EXAFS is of considerable value for the study of catalysts (Sankar *et al.*, 1992).

X-ray absorption near edge structure (XANES) is useful in determining the coordination number and the oxidation state of metal ions (Sankar *et al.*, 1983). In Figs. 2.16 and 2.17 we show the XANES of Co and Cu in some compounds as well as catalysts. The $1s$–$3d$ transition of cobalt (or the $1s$–$4d$ transition of Mo) is sensitive to

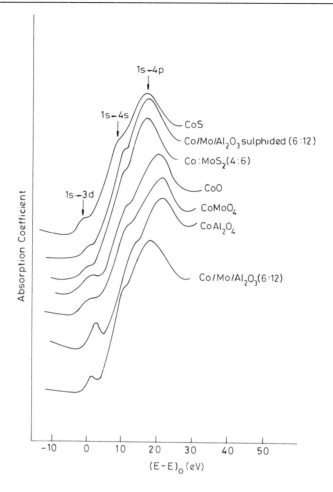

Figure 2.16   XANES of typical cobalt compounds and cobalt catalysts. (After Sankar *et al.*, 1984.)

the coordination, becoming more intense when the coordination is tetrahedral. Accordingly, from Fig. 2.16 it was established that the cobalt in $Co/Mo/Al_2O_3$ catalyst precursor is tetrahedrally coordinated, while in the sulphided form it is octahedrally coordinated (Sankar *et al.*, 1984). The nature of Cu species in cuprate superconductors can be determined by EXAFS (see Fig. 2.17).

Chemical shifts of K or L absorption edges of metals seem to be generally larger than the chemical shifts of binding energies found from X-ray photoelectron spectra. X-ray absorption edge chemical shifts are therefore useful in fixing the oxidation states of metals; the chemical shift increases significantly with the oxidation number (Sarode *et al.*, 1979). The chemical shifts are fruitfully correlated with the effective charge on the metal atoms. Thus, the following relation seems to be valid in many transition-metal compounds

$$\Delta E = aq + bq^2 \tag{2.8}$$

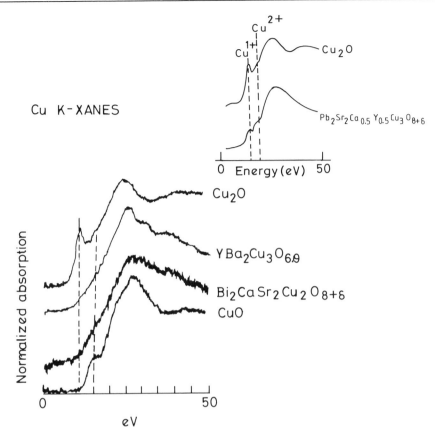

Figure 2.17   XANES of typical cuprate superconductors.
(Courtesy of C.N.R. Rao.)

where q is the effective atomic charge (Sarode *et al.*, 1979). The chemical shifts can be directly used to estimate the oxidation state or the number of d electrons in a system. Absorption edge (and EXAFS) measurements have indeed been used to examine the state of manganese in dispersed MnO_2 catalysts (Brown *et al.*, 1984). The number of unoccupied d electron states in a Pt catalyst has been recently determined by making use of the L_2 and L_3 absorption edge intensity (Mansour *et al.*, 1984); this can be accomplished equally by *electron energy loss spectroscopy* (EELS). EELS carried out in an electron microscope can directly yield information on the L_2 and L_3 intensities which are related to the unoccupied d-electron states of the metal (Rao *et al.*, 1984); besides providing spatial resolution, EELS can yield simultaneous chemical analysis and structure (through electron diffraction) of minute samples ($10^{-12} - 10^{-18}$ g). (Thomas *et al.*, 1985).

2.2.6   *Nuclear magnetic resonance spectroscopy (MASNMR)*
NMR spectroscopy is probably the physical method most widely used by chemists; it provides a powerful means of investigating structure and dynamics. Almost all the

studies hitherto have been on compounds in the liquid or solution state, the spectra in the solid state being generally broad and featureless owing to dipolar interactions. Even the broad bands have been useful to examine structures of solids (e.g. gypsum) by the 'wide-line' NMR method. Measurement of relaxation times has provided valuable information on diffusion, reorientation and similar dynamic aspects. The use of NMR spectroscopy as a routine tool for structural studies on solids was, however, beyond experimental possibility until recently. Advances in techniques and instrumentation have made it possible today to obtain chemical information from NMR spectra of powders and this method will undoubtedly be exploited to the full extent in the near future. Already, there has been significant success in the use of solid state NMR in the investigation of silicates especially zeolites (Fyfe *et al.*, 1983; Klinowski, 1985; Lippmaa *et al.*, 1980; Mägi *et al.*, 1984).

Dipolar interactions, responsible for the broadening of NMR signals, can, in principle, be eliminated in the case of an abundant nucleus such as the proton, but the small range of proton chemical shifts makes the use limited. In the case of dilute nuclei, spectra of fair resolution are readily obtained. The magnetic interaction for such a nucleus is given by

$$H_{tot} = H(\text{Zeeman}) + H_{ii}(\text{dipolar}) + H_{ij}(\text{dipolar}) + H_i(\text{CSA}) \qquad (2.9)$$

where the dipolar interaction H_{ij} is between the nucleus i and the majority nucleus j (e.g. ^{13}C and ^{1}H). The H_{ii} term is negligible because of the low abundance of nucleus i (e.g. ^{13}C). The H_{ij} dipolar interactions can be removed by the application of a strong decoupling field at the resonance frequency. The other broadening contribution is from chemical shift anisotropy $H_i(\text{CSA})$. Random motions of molecules in the solution phase average the chemical shieldings of nuclei resulting from different orientations with respect to the magnetic field. In the solid state, however, a broad envelope characteristic of the chemical environment is obtained. Chemical shift anisotropy can be understood from equation (2.10),

$$H_i(\text{CSA}) = (3\cos^2\theta - 1)\,(\text{other terms}) + (\tfrac{3}{2}\sin^2\theta)\,\sigma I B_o \qquad (2.10)$$

where θ is the angle subtended by the axis of spinning with respect to the magnetic field direction; B_o, the magnetic field; σ, the isotropic chemical shielding and I, the spin angular momentum. When $\theta = 54°\ 44'$, $(3\cos^2\theta - 1)$ becomes zero and we get the isotropic chemical shift. This value of θ is called the 'magic angle' and spinning the sample about an axis inclined at this angle to the magnetic field eliminates CSA, provided the spinning frequency is close to the frequency spread of the signal. If the spinning frequency is smaller, one obtains a central line at the isotropic chemical shift with several sidebands.

We have seen how by *magic-angle spinning* (MAS) and high-power decoupling the resolution is rendered similar to that in the solution phase. The intensity of signals due to less abundant nuclei is enhanced by the use of cross-polarization (CP) pulse sequences. Excellent spectra of a variety of nuclei in the solid state have been obtained

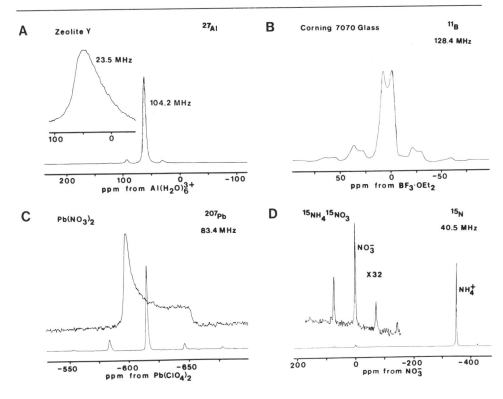

Figure 2.18   MAS NMR spectra of solids: (A) ^{27}Al spectrum of zeolite Y at 23.5 MHz and at 104.2 MHz; note the spectral improvement at higher field; (B) ^{11}B spectrum of Corning 7070 glass at 128.4 MHz; (c) ^{207}Pb spectrum of Pb(NO$_3$)$_2$ at 83.4 MHz (the broader spectrum above is without spinning); (D) ^{15}N spectrum of ^{15}NH$_4$ ^{15}NO$_3$ at 40.5 MHz. The magnified NO$_3$ spectrum is also shown. (After Fyfe et al., 1983).

by employing MAS/CP techniques. It should be noted that it is not essential that a nucleus of interest is always less abundant; the main criterion is that the second term in equation (2.9) be small. Accordingly, spectra of nuclei with 100% abundance (e.g. ^{31}P and ^{27}Al) have been obtained. In solids containing no abundant nucleus such as a proton (e.g. silicates), neither high-power decoupling nor CP is relevant; MAS alone can give the required high-resolution spectra. The subject of high-resolution NMR spectra of solids has been reviewed widely in the literature (Richards & Packer, 1981; Andrew, 1981; Fyfe et al., 1983; Klinowski, 1985).

   In Fig. 2.18 we show MAS NMR spectra of solids with different NMR nuclei; some of the other possible nuclei are ^{17}O, ^{23}Na, ^{31}P, ^{57}Fe, ^{113}Cd and ^{119}Sn. In Fig. 2.19, MAS NMR spectra of aluminium-containing materials are shown to indicate how octahedral and tetrahedral coordination of Al are readily distinguished. In Fig. 2.20, ^{29}Si spectra of several zeolites are shown to point out how silicon atoms coordinated differently (by aluminium atoms) exhibit widely different chemical shifts; the numbers above the peaks represent n in Si(nAl) with five possible values (0 to 4). Si, Al ordering

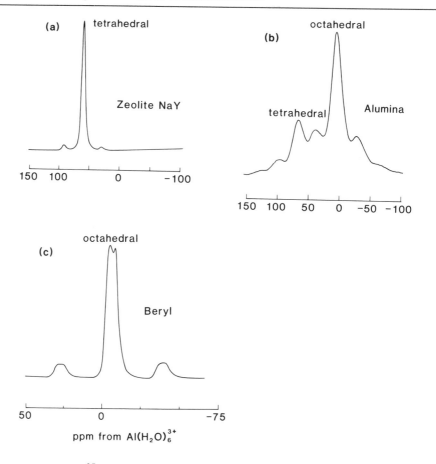

Figure 2.19 ^{27}Al MAS NMR spectra at 104.22 MHz: (a) zeolite NaY; (b) γ-alumina and (c) beryl. (After Fyfe *et al.*, 1983.)

in zeolites cannot be unravelled by X-ray diffraction or other techniques; it is only through the use of MAS NMR that we gain knowledge of this essential aspect of this important class of materials. Silicon and aluminium site distributions in layered clays have also been examined by NMR spectroscopy (Lipsicas *et al.*, 1984). ^{27}Al NMR spectra of aluminosilicate gels have shown how Al atoms are six-coordinated in the alumina-rich gel while they are four-coordinated in the silica-rich gels (Thomas *et al.*, 1983). ^{31}P MAS NMR of solid PCl$_5$ shows two signals due to PCl$_4^+$ and PCl$_6^-$ species. ^{31}P NMR in various phosphates has been examined (Prabakar *et al.*, 1987). ^{19}F NMR of graphite intercalates has revealed the nature of AsF$_x$ species (Moran *et al.*, 1984). NMR studies of silicates have shown that the Si chemical shift is related to the Si—O—Si angle (Smith & Blackwell, 1983). Making use of the observed line shape, distribution functions for the Si—O—Si angle in glassy SiO$_2$, originally estimated from diffuse X-ray scattering (Mozzi & Warren, 1969), have been tested by Dupree & Pettifer (1984). A semiempirical relationship has been derived between the Si chemical shift and the nonbonded Si—T (T = Si or Al) distance in zeolites and other silicates

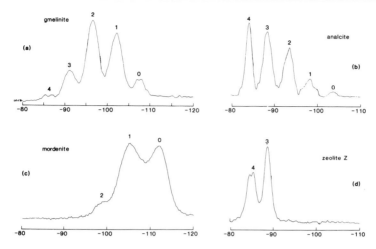

Figure 2.20 ^{29}Si MAS NMR spectra of zeolites at 79.8 MHz: (a) gmelinite; (b) analcite; (c) mordenite and (d) zeolite Z. (After Fyfe *et al.*, 1983.)

(Ramdas & Klinowski, 1984); a relationship between the chemical shift and Si—O distance is also available (Grimmer & Redaglia, 1984). The evolution of crystallinity in a silicate glass has been examined recently by taking advantage of the widely differing relaxation times of the amorphous and crystalline regions (Rao *et al.*, 1985). A distribution of Si—O—Si angles in a glass can also be determined by analysing the NMR profile (Selvaraj *et al.*, 1985). The ^{29}Si chemical shifts in a wide range of silicates (Prabakar *et al.*, 1991a) have been rationalized. The evolution of aluminosilicate and related ceramics from their corresponding gels has been examined by NMR (Prabakar *et al.*, 1991b).

2.2.7   *Electron spectroscopies*

Techniques of electron spectroscopies have emerged to become the principal means for investigating electronic structures of solids and surfaces (Rao, 1985; Mason *et al.*, 1986). Most of these techniques involve the analysis of the kinetic energy of the ejected or scattered electrons. Some of the important techniques of electron spectroscopy used to study solids are photoelectron spectroscopy using X-rays (XPS) or UV radiation (UVPS), Auger electron spectroscopy (AES) and electron energy loss spectroscopy, (EELS). All these techniques are surface-sensitive and probe $\sim 25\,\text{Å}$ or less of solids. Cleanliness of the surfaces and ultra-high vacuum ($\sim 10^{-10} - 10^{-11}$ torr) are therefore prerequisites for studying solids by these techniques. When low-energy electron diffraction (LEED) is used along with these techniques we obtain the two-dimensional structure of the solid surface; LEED patterns also provide information on reconstruction of the surface, if any, when atoms or molecules are adsorbed on surfaces. The various techniques of electron spectroscopy have been extensively reviewed in the literature (Fiermans *et al.*, 1978; Rao & Hegde, 1981; Rao & Sarma, 1982). We shall briefly describe the principles of the techniques and cite a few important applications.

In photoelectron spectroscopy, a monochromatic photon beam impinges on a solid and the resulting photoelectrons are energy-analysed. The binding energy of electrons, E_B, originating from a particular state is obtained by the energy balance equation,

$$hv = E_B + E_k + \phi \qquad (2.11)$$

where E_k is the kinetic energy of the electron and ϕ the work function of the solid. In ultraviolet photoelectron spectroscopy, HeI (21.22 eV) and HeII (40.8 eV) are the most common radiations employed. AlK$_\alpha$ (1486.6 eV) and MgK$_\alpha$ (1253.6 eV) are the common X-radiations used in XPS.

In Auger spectroscopy, an impinging electron beam causes the ionisation of a core electron and this core-level vacancy is filled by a nonradiative electron jump from a higher energy level with the simultaneous ejection of an Auger electron from a higher level. Thus, the energy of a KL_1L_{23} Auger transition in the case of an atom with an effective atomic number Z is given by the equation

$$E_{KL_1L_{23}}(Z) = E_K(Z) - E_{L_1}(Z) - E_{L_{23}}(Z + \Delta) - \phi \qquad (2.12)$$

Here the final state is left with two core holes with reference to the neutral atom and the L_{23} level will have an effective atomic number higher than the singly ionized atom by Δ. The Auger transition energy is independent of the primary electron energy. Auger spectroscopy is particularly useful in the determination of oxidation states of elements and characterization of elements on solid surfaces in the presence or absence of gas adsorption. Scanning Auger microscopy complements scanning electron microscopy for elemental mapping in materials.

In electron energy-loss spectroscopy, one measures the loss in energy of incident electrons due to impact with matter. Depending on the primary beam energy, one can obtain information regarding vibrational and electronic transitions involving adsorbed species; valuable information on surface electronic states of the adsorbent (substrate) can also be obtained. The information obtained from EELS on the vibrational spectra of adsorbed molecules is similar to that obtained from reflection–absorption infrared spectroscopy. For vibrational EELS, the energy of the primary electrons is of the order of 4 eV while for electronic excitation studies the energy used is 100–200 eV.

Photoelectron spectroscopy provides valuable information on both the valence bands and core levels of solids. The advent of synchrotron radiation has made photoelectron spectroscopy a more powerful technique, since the wide range of energies of the radiation provides a range of escape depths. One can distinguish electron states of the surface and the bulk solid using the right radiation. Electron energy-loss spectroscopy (EELS) affords the study of both vibrational and electronic excitations in surface species (both the adsorbate and the adsorbent) depending on the energy of the primary electron beam. Low-energy EELS is very much like the IR or Raman spectrum and corresponds to the vibrational spectrum of the solid and the adsorbate. EELS has been particularly useful in the study of gas–solid interactions of relevance to catalysis. Recent studies of surfaces clearly bring out the need to employ

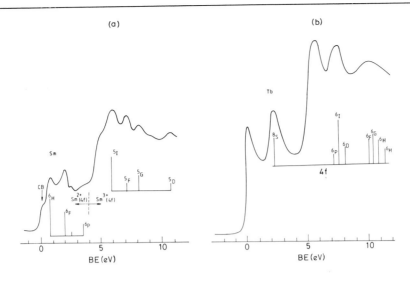

Figure 2.21   UVPS (HeII spectra) in the valence-band region showing final-state effects: (a) Sm; (b) Tb. The valence band of Sm shows the coexistence of both divalent and trivalent species due to valence instability (mixed valence). (After Rao & Sarma, 1982.)

several techniques to obtain useful and worthwhile information. Thus, to study the surface oxidation of metals one can use not only UVPS, XPS, AES and LEED, but also other techniques like electron- or photon-stimulated desorption. Similarly, to examine the interaction of a gas like CO with a metal, one employs EELS and thermal desorption spectra besides XPS, UVPS, AES and LEED.

Photoelectron spectroscopy has been extensively used to examine valence bands of metals, alloys and other inorganic solids. Besides providing the experimental density of states, UVPS and XPS show final state effects in the valence-band regions of transition and rare earth metals and compounds. The structure due to final-state effects is useful in the identification of mixed valence and spin-states (see Fig. 2.21 and Fig. 2.22). Core-level spectra (in the XPS) of metal oxides and other compounds show binding energies characteristic of the oxidation state of the metal. The XPS binding energy increases with the oxidation state of the metal just like the chemical shifts of the X-ray (K or L) absorption edges; the two quantities are proportional to each other. Characteristic spin-orbit splittings of (np) and other levels are found in the XPS of transition metal and rare earth compounds. The ns orbitals show splittings due to exchange interaction (with the d or the f orbitals), the interaction becoming maximum when the d or the f orbital is half full. Other electronic interactions that manifest themselves in core-level spectra are configurational interactions involving the $3s$ state of transition metals, $4d$–$4f$ interactions in rare-earth compounds and charge-transfer (metal–ligand) interactions. Satellites next to metal core levels (and sometimes also next to ligand core levels) are found owing to charge-transfer interactions and the intensity of the satellites crucially depends on the metal–ligand orbital overlap. Both shake-up

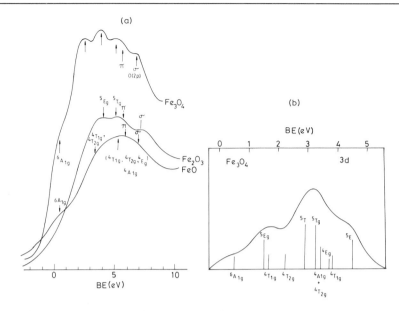

Figure 2.22 (a) UVPS (HeII spectra) of FeO, Fe_3O_4 and Fe_2O_3 showing final-state structures of $3d$ levels. (After Vasudevan *et al.*, 1979); (b) theoretical spectrum of Fe_3O_4 calculated in terms of the final states of Fe^{2+} and Fe^{3+} ions is shown. (After Alvarado *et al.*, 1976.)

and shake-down satellites (appearing at higher or lower energies relative to the main core-level peak) are found. In Fig. 2.23 typical satellites are shown for the purpose of illustration.

Core-level spectra are useful in studying mixed valence (valence instability or interconfigurational fluctuation) in rare-earth systems (e.g. SmS, Ce) which arises when $E_{exc} = E_n - (E_{n-1} + E_e) \approx 0$ where $(E_n - E_{n-1})$ is the energy difference between the $4f^n$ and $4f^{n-1}$ states and E_e is the energy of the promoted electron. The time scale involved in these fluctuations is less than 10^{-11}s, considerably shorter than that in Mössbauer studies, but larger than in XPS ($< 10^{-16}$s). XPS therefore shows the presence of distinct species not found by Mössbauer spectroscopy and other techniques (Rao *et al.*, 1979; Rao & Sarma, 1980). Accordingly, Mössbauer spectroscopy does not show Fe^{3+} and Fe^{2+} states of Fe_3O_4 at room temperature but photoelectron spectroscopy does (see Fig. 2.22 for an example). Metal–insulator transitions in solids can be readily examined by photoelectron spectroscopy since the metallic phase shows a sharp cut-off in intensity at the Fermi level. A change in the spin-state of a metal ion is generally accompanied by changes in the valence band and in the core levels. XPS has been used for compositional analysis (with the knowledge of cross-sections). Angle-resolved XPS (together with X-ray photoelectron diffraction) also yields valuable structural informa-tion and elemental analysis (Evans *et al.*, 1979). Auger electron spectroscopy has been used widely for compositional analysis, especially surface microanalysis. Chemical shifts of Auger signals as well as the ratios of the intensities of the ligand (say oxygen) and the metal Auger lines have been used to examine surface oxidation of metals. It has

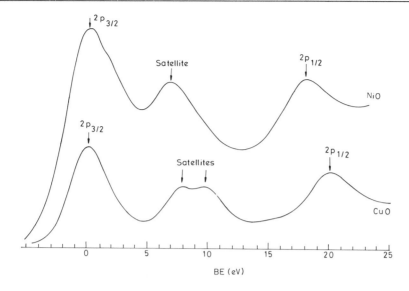

Figure 2.23   XPS of CuO and NiO in the metal $2p$ region showing the presence of satellites. (After Sarma *et al.*, 1983.)

been shown that the ratios of the intensities of the metal Auger lines are related to the number of valence electrons (see Fig. 2.24) and can therefore be used to investigate surface oxidation or to determine the oxidation state of a metal. In Fig. 2.25 we show how the ratios of metal Auger intensities of iron and its oxides are related to the number of valence electrons (Rao *et al.*, 1980; Rao, 1986). Similar relations hold good for compounds of the second and third row transition metals as well (Yashonath *et al.*, 1983). In addition to changes in intra-atomic Auger processes, metal oxides and related systems also show inter-atomic Auger transitions which can be of diagnostic value. We shall discuss the use of photoelectron spectra and related information in understanding the electronic structures of solids later in Chapter 6.

EELS of adsorbed species has thrown considerable light on catalytic reaction mechanisms. For example, CO adsorbed on metals gives characteristic stretching frequencies due to bridged and linear species; in addition, metal–carbon stretching vibrations are seen at lower frequencies. Adsorption of acetylene and ethylene on metals is accompanied by rehybridization as evidenced by C—H and C—C stretching frequencies in regions normally found with saturated hydrocarbons. Studies on adsorbed species by EELS are supplemented by the vibrational spectra of metal cluster compounds containing similar molecular fragments. Other techniques employed in surface characterization are secondary ion mass spectroscopy, ion-scattering spectroscopy and thermal desorption technique, the last giving desorption behaviour of adsorbed species as a function of temperature. Most practitioners use more than one technique to investigate surfaces, and commercial spectrometers offer various combinations of techniques such as UVPS–XPS–AES, AES–SIMS–XPS, XPS–EELS–LEED etc. The efficacy of the different electron spectroscopies and

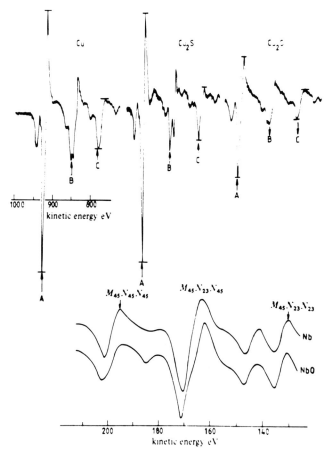

Figure 2.24   Auger spectra of copper and niobium compounds. (A) $Cu(L_3VV)$; (B) $Cu(L_3M_{23}V)$ and (C) $Cu(L_3M_{23}M_{23})$. (After Yashonath *et al.*, 1983.)

cognate techniques has been reviewed by Riviere (1982). Bremsstrahlung isochromat spectroscopy is getting to be used to investigate unoccupied electronic levels (Laubschat *et al.*, 1984).

2.2.8   *Other spectroscopic techniques*

Absorption and fluorescence spectroscopies in the UV and visible regions, infrared and Raman spectroscopy as well as the various resonance spectroscopies are widely used in the characterization of solids. Spectroscopy in the UV and visible regions probes electronic transitions of specific atoms or ions in the solid providing information about the oxidation states of ions, their local symmetry and defect centres. The method is especially suitable when the absorbing ion is present in dilute concentrations in a nonabsorbing host. A typical example is that of Cr^{3+} in oxide lattices; this ion in Al_2O_3 host (ruby) gives a rich red colour, in $Be_3Al_2(SiO_4)_3$ (emerald) a beautiful green colour and in $LiNbO_3$ a dull green colour. The ion occupies octahedral sites in all three cases and the differences in optical absorption arise because of differing crystal fields. Optical

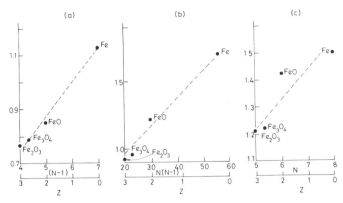

Figure 2.25   Relations between metal Auger line intensity ratios of iron and its oxides and the number of valence electrons N. Z is the oxidation state. (a) $L_3M_{45}M_{45}/L_3M_{23}M_{45}$; (b) $L_3M_{45}M_{45}/L_3M_{23}M_{23}$; (c) $L_3M_{23}M_{45}/L_3M_{23}M_{23}$. (After Rao & Sarma, 1982.)

absorption and luminescence of broad band semiconductors such as GaAs are useful in characterizing the optical band gap, excitons, recombination centres and so on. Infrared and Raman spectroscopies provide essential information on vibrations of atoms and molecules in solids. Complete analysis of the vibrational spectrum of a solid in terms of normal modes and selection rules can complement structure determination by diffraction methods. In addition, infrared and Raman spectroscopy can be used for the study of phase transitions (soft mode behaviour, if any), order–disorder phenomena, adsorbed species (by SERS), IRAS and such special techniques and for the identification of phases in a mixture. *Raman microscopy* may in fact be useful in materials science, especially in the study of glasses.

Electron spin resonance (ESR) detects the presence of unpaired electrons that are localized on paramagnetic ions or a paramagnetic defect in the solid. ESR is mainly applicable to insulators with low dielectric loss. From the observed splitting of the resonance, information about crystal field and spin-orbit coupling of the paramagnetic ion can be obtained. With good resolution, it is possible to see hyperfine and even superhyperfine splittings of the resonance. Hyperfine splitting arises from interaction between the electron spin and the nuclear spin while the superhyperfine splitting is due to interaction of the electron spin with the nuclear spins on neighbouring atoms. ESR studies are generally carried out on single crystals containing a dilute concentration of paramagnetic ions. Ions such as Mn^{2+} and Gd^{3+} having S ground states are useful probes to study various types of structural changes (antiferroelectric, ferroelectric or Jahn–Teller transitions) in solids. Organic spin probes can be used to investigate glasses, plastic crystals and other condensed systems composed of organic molecules.

2.2.9   *Scanning tunnelling microscopy*

Scanning tunnelling microscopy (STM) and related techniques are unique tools available today for a real space view of the atomic structure and electron states of solids.

They have been found most beneficial in studying diverse systems such as fullerenes, cuprate superconductors, adsorbed molecules and layered sulfides. STM, invented in the early 1980s by Binnig and Rohrer et al., 1982 and Binnig & Rohrer, 1984, gives images of the topography and the electronic structure of solid surfaces with atomic resolution. The technique is based on electron tunnelling between a sharp metal tip and the surface of a conducting (or a semiconducting) solid. When the tip is brought close to the sample surface (say 5 to 10 Å), the wave functions of the tip and sample overlap and electrons tunnel between the tip and surface of the solid. If sufficient voltage is applied to the solid sample, there will be a net flow of electrons across the gap which varies exponentially with the separation between the tip and the sample. The flow of electrons gives the tunnelling current. The direction of electron flow depends on the sign of the bias voltage applied. When it is negative, electrons tunnel from the filled electron states (in the sample) to the empty states in the tip. When the bias voltage is positive, electrons tunnel from the filled states (in the tip) to the empty states (in the solid). Depending on the sign of the bias voltage, the filled or the empty electronic states in the solid sample can be examined.

A map of the solid surface is generated by scanning the tip over the surface from side to side in parallel lines, keeping a constant tunnelling current. Features in the map correspond to vertical displacements of the tip as it follows a contour of constant density of states at the surface of the sample. Such a surface may or may not correspond directly to the positions of the surface atoms. A STM image contains information on the topography as well as the electronic structure of the solid sample. The two can be delineated by comparing images obtained at different bias voltages or by combining STM images with tunnelling spectroscopy measurements. The latter directly provide information on the density of states in solids. STM together with the sister technique, AFM or atomic force microscopy (Meyer, 1992), is increasingly being employed to study a variety of solid state chemistry problems.

One of the early applications of STM was in understanding surfaces of silicon (e.g. dangling bonds). Many applications of the technique of direct relevance to solid state chemistry have since been described in the literature (Dai & Lieber, 1993; Lieber, 1994). We shall illustrate these applications by means of a few figures. In Fig. 2.26 we show the image of a C_{70}-C_{60} solid solution which clearly delineates the two molecules and shows the presence of orientational disorder. Fig. 2.27 shows the STM image of a chalcogenide exhibiting incommensurate CDW modulation. In Fig. 2.28 we show the STM and AFM images of carbon nanotubes of different diameters. STM has been employed effectively to investigate metal clusters and their metallicity (Rao et al., 1993).

2.2.10   Concluding remarks

We have briefly covered some of the important developments in structural characterization techniques. There are many other techniques such as Mössbauer spectroscopy, positron annihilation and Rutherford backscattering which have wide applications. Mössbauer spectroscopy is specially useful to investigate different oxidation states, spin-states and coordinations of metal ions, phase transitions, magnetic ordering,

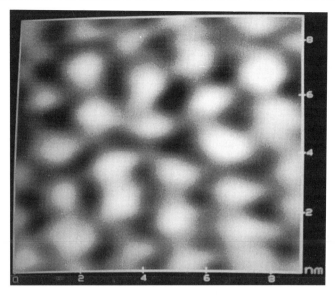

Figure 2.26   STM image of a solid solution of C_{70} and C_{60} (Courtesy H.N. Aiyer and C.N.R. Rao)

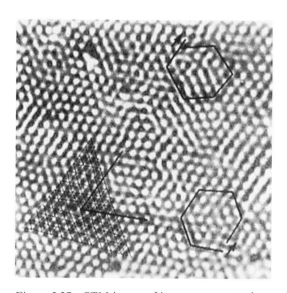

Figure 2.27   STM image of incommensurate charge density wave (CDW) state in $Nb_{0.04}Ta_{0.96}S_2$. Black lines highlight the insertion of extra rows of CDW maxima in the lattice (From Dai & Lieber, 1993).

Debye–Waller factors, etc. Applications of this versatile tool in solid state chemistry have been extensively reviewed in the literature.

It is important to point out that a combination of techniques is often necessary to understand a phenomenon fully or solve a structural problem in solid state chemistry. All the tools of solid state chemistry generally employed to investigate crystalline solids

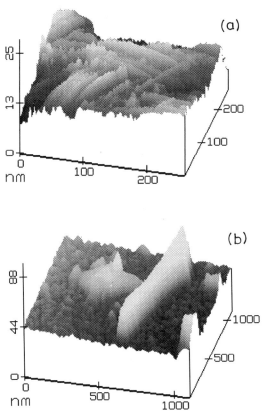

Figure 2.28   Images of carbon nanotubes (Courtesy G. Raina and C.N.R. Rao): (a) STM, (b) AFM

are also useful to investigate amorphous solids, although some of the techniques are specially suited to and more effective for the study of the amorphous state. Thus, X-ray scattering gives radial distribution functions of amorphous materials, which are useful, but EXAFS gives bond distances and coordination numbers; NMR spectroscopy, light scattering and neutron scattering are other important techniques for studying these solids. *Compton scattering* (Williams & Thomas, 1983) can also provide information on the nature of the short-range order; this technique has shown, for example, that amorphous carbon is more nearly graphitic. A variety of techniques is used to study clays, silicates, and catalysts to understand the relation between structural chemistry and reactivity (see, for example, Thomas, 1984). Photoacoustic spectroscopy (Rao *et al.*, 1986) has been useful to study a variety of problems related to solids, surfaces and interfaces.

2.3   Characterization of composition and purity

Chemical characterization of solids generally involves the characterization of the major phase, which includes the determination of stoichiometry, homogeneity and oxidation state if a multivalent element is present and the characterization of minor phases and impurities. The important analytical techniques that are available for this purpose together with their precision, sensitivity and area of application are given in Table 2.2.

One of the main problems in the preparation of solids is to find out whether the product obtained from a solid state reaction is a single phase or a mixture of phases. When the product is single-phase, it is easy to characterize it by chemical and X-ray diffraction methods. Solid state reactions often yield polycrystalline mixtures of phases. Gross chemical analysis and X-ray powder diffraction alone are of little avail in identifying phases. Examination under a polarizing microscope could reveal the presence of a mixture of phases. Growing single crystals is an alternative available for complete characterization purposes but it is not always possible to grow single crystals of all the phases in a mixture. X-ray microanalysis in the electron microscope (analytical electron microscopy), referred to in section 2.2, is most useful in the characterization of polycrystalline mixtures obtained by solid state reaction involving multicomponents (Cheetham & Skarnulis, 1981; Jefferson, 1982). The method provides a means of identifying the various phases present and for determining the composition of individual crystallites in a polycrystalline mixture. The technique is similar to electron microprobe analysis except that thin crystals are used in this method. In the thin crystal limit, the concentration ratio of two elements is directly proportional to the intensity ratio of the characteristic X-ray emissions from the elements:

$$C_x/C_y = k_{xy}\frac{I_x}{I_y} \tag{2.13}$$

where C_x and C_y are the concentrations of elements x and y in the crystal and I_x and I_y are the characteristic X-ray emission intensities of the two elements. This approximation, known as the ratio method, is particularly amenable for characterization using a transmission electron microscope (including STEM). All that is needed is to make use of well-characterized compounds containing elements x and y as standards to determine the constant k_{xy}; using this value, the concentration ratio C_x/C_y in an unknown compound can be determined. The method has been applied to a variety of solid state systems such as oxide solid solutions, mixed metal sulphides, ternary oxides and alloys. A limitation of the method is that it cannot generally be used to determine elements lighter than sodium; use of sealed detectors, however, circumvents this problem and one can estimate, for example, oxygen in oxides. Electron energy loss spectroscopy in the electron microscope can be used for qualitative and quantitative analysis of solids and the method is 'absolute' in compositional analysis, unlike X-ray emission analysis, where a reference material is required (Egerton, 1982). The efficacy of EELS as a quantitative analytical tool (especially when carried out in an electron microscope) is, however, not entirely established. The present-day EDAX attachments to electron microscopes have a minimum scan diameter of 0.5 μm with a sensitivity of much less than 1% (> 200 ppm). Wavelength dispersive X-ray analysis permits higher sensitivity.

Table 2.2. *Compositional characterization of solids*

Method	Application	Sensitivity	Precision
1. (i) Gravimetry	Major- and minor-phase concentration	100 mg–1 mg	0.003–0.01%
		1–10 mg	0.1%
(ii) Titrimetry	Same as above; also impurities	10^{-2} M in solution	0.01%
2. Electrochemical methods			
(i) Coulometry	Major-phase concentration		0.001–0.005%
(ii) Polarography	Major and minor phases;	All in solution	
	impurities	10–100 ppm	0.1–2%
		0.001 ppm	5–10%
3. Thin layer chromatography	Minor constituents and trace impurities	10–1000 μg	5–50%
4. Optical spectroscopy			
(i) Spectrophotometry	Impurities	Down to 0.005 ppm in solution	5–10%
(ii) Emission spectroscopy	Impurities	0.1–100 ppm	5–10%
(iii) Flame emission, atomic absorption and atomic fluorescence	Impurities	0.1–10 ppm in solution	0.5–5%
		0.002–0.1 ppm in solution	5–10%
5. Mass spectrometry	Impurities	0.001–0.1 ppm	5–20%
6. Neutron activation analysis	Impurities	0.001–0.01 ppm	2–10%
7. X-ray fluorescence	Major and minor constituents	20–200 ppm generally;	0.1–0.5%
		0.1 ppm with preconcentration	2–10%
8. Electron probe microanalysis and X-ray emission analysis in electron microscope	Homogeneity of major and minor phases	0.01–0.1% average	0.5%
		1–5 μm scan diameter	
9. Above with electron energy loss spectroscopy		Minimum detectable mass 10^{-18}–10^{-20} g	Minimum detectable fraction, 0.1–3%

Chemical purity is an important criterion in solid state science and the realization of its importance has stimulated great efforts to extend conventional purification methods and to develop new ones. It must, however, be clear that no single set of techniques will be suitable for all purification problems. The exact technique required will depend upon the chemistry of the material and the particular impurity that needs to be removed. Purification techniques can either be physical or chemical in nature. Physical methods include sublimation, evaporation of volatile impurities, recrystallization from melt, liquid extraction and chromatography, while the chemical methods are ion exchange, electrolysis of liquids or the solid itself and purification using a chemical reaction. The most important of the physical methods for purification of solids is zone refining; the method is based on the fact that impurity solubility is different in the solid and the liquid phases. Depending on the purity range, an appropriate method (see Table 2.2) is employed to determine the impurity concentrations quantitatively. For example, electrical resistivity can be a measure of a dopant concentration in an otherwise insulating material.

An elementary technique useful in determining stoichiometry and the oxidation state is that of redox titrations. For example, iodometric titrations can be employed to determine Ni^{3+} in La_2NiO_4 or Mn^{4+} in $LaMnO_3$ and hole concentrations in cuprate superconductors. Characterization of nonstoichiometric compositions often involves determining the stoichiometry by techniques such as thermogravimetry and then determining the structure by electron microscopy, X-ray diffraction and related techniques to find out whether the defects (responsible for anion excess or anion deficiency) are ordered. Measurements of the electromotive force using solid electrolyte cells and of electrical conductivity at different partial pressures of oxygen or any other gaseous species involved in the defect equilibrium (see Chapters 5 and 6 for details) are also employed commonly to characterise nonstoichiometric compounds. Characterization of nonstoichiometric oxides has been reviewed by Sørensen (1981).

References

AIP (1980) Conference Proceedings No. 64, *Laboratory EXAFS Facilities* (ed. Stern, E. A.) AIP, New York.

Allpress, J. G. & Sanders, J. V. (1973) *J. Appl. Crystallogr.* **6**, 165.

Alvarado, F., Erbudak, M. & Munz, P. (1976) *Phys. Rev.* **B14**, 2740.

Anderson, J. S. (1978) *Proc. Indian Acad. Sci.* **87A**, 295.

Andrew, E. R. (1981) *Int. Rev. Phys. Chem.* **1**, 195.

Arunarkavalli, T., Kulkarni, G. U. & Rao, C. N. R. (1993) *Catal. Lett.* **20**, 259.

Beanfils, J. P., Crowley, T., Rayment, T., Thomas, R. K. & White, J. W. (1981) *Molec. Phys.* **44**, 1257.

Beer, M., Carpenter, R. W., Eyring, L., Lyman, C. E. & Thomas, J. M. (1981) *Chem. Engg News* **59**, 40; see also Jefferson, D. A., Thomas, J. M. & Egerton, R. F. (1981) *Chem. Brit.* **17**, 514.

Binnig, G. & Rohrer, H. (1984) *Physica*, **127B**, 37.

Binnig, G., Rohrer, H., Gerber, Ch & Weibel, E. (1982) *Phys. Rev. Lett.* **49**, 57.

Bomchil, G., Hüller, A., Rayment, T., Roser, S. J., Smalley, M. V., Thomas, R. K. & White, J. W. (1980) *Phil. Trans. Roy. Soc. London* **B290**, 537.

Bourdillon, A. J., Pettifer, R. F. & Marseilha, E. A. (1980) *J. Phys.* **C12**, 3889.

Boyce, J. B., Hayes, T. M., Stutius, W. & Mikkelsen Jr, J. C. (1977) *Phys. Rev. Lett.* **38**, 1362.

Boyce, J. B. & Hayes, T. M. (1979) in *Physics of Superionic Conductors* (ed. Salamon, M. B.) Springer-Verlag, Berlin.

Brown, N. M. D., McMonagle, J. B. & Greaves, N. (1984) *J. Chem. Soc. Faraday. I*, **80**, 589.

Caswell, N., Solin, S. A., Hayes, T. M. & Hunter, S. H. (1980) *Physica* B + C, **99**, 463.

Cheetham, A. K. (1986) *Proc. Ind. Nat. Sci. Acad.* **52A**, 25.

Cheetham, A. K. & Taylor, J. C. (1977) *J. Solid State Chem.* **21**, 253.

Cheetham, A. K. & Skarnulis, A. J. (1981) *Anal. Chem.* **53**, 1060.

Cheetham, A. K. & Day, P. (1987) *Solid State Chemistry –Techniques*, Clarendon, Oxford.

Clark, S. M. & Miller, M. C. (1990) *Rev. Sci. Instrum.* **61**, 2253.

Cramer, S. P. (1978) *Stanford Synchrotron Res. Lab. Rep.* 78/07, Stanford, USA.

Dai, H. & Lieber, C. M. (1993) *Ann. Rev. Phys. Chem.* **44**, 237.

Dupree, E. & Pettifer, R. F. (1984) *Nature* **308**, 523.

Eccles, T. K. (1978) Ph.D Thesis, Stanford University, Stanford, USA.

Egerton, R. F. (1982) *Phil. Trans. Roy. Soc. London* **A305**, 521.

Evans, S., Adams, J. M. & Thomas, J. M. (1979) *Phil. Trans. Roy. Soc. London* **A292**, 563.

Fiermans, L., Vennik, J. & Dekeyser, W. (eds) (1978) *Electron and Ion Spectroscopy*, Plenum Press, New York.

Fyfe, C. A., Thomas, J. M., Klinowski, J. & Gobbi, G. C. (1983) *Angew. Chem. Int. Edn (Engl.)* **22**, 259.

Ganguly, P. & Rao, C. N. R. (1984) *J. Solid State Chem.* **53**, 193.

Goodman, P. & Moodie, A. F. (1974) *Acta Crystallogr.* **A33**, 701.

Grimmer, A. R. & Radaglia, R. (1984) *Chem. Phys. Lett.* **106**, 262.

Heald, S. M. & Stern, E. A. (1978) *Phys. Rev.* **B17**, 4069.

Honig, J. M. & Rao, C. N. R. (eds) (1981) *Preparation and Characterization of Materials*, Academic Press, New York.

Hutchison, J. L., Anderson, J. S. & Rao, C. N. R. (1977) *Proc. Roy. Soc. London* **A355**, 301.

International Tables for Crystallography. Vol. III (1969) Kynoch Press, Birmingham, England.

Jefferson, D. A. (1982) *Phil. Trans. Roy. Soc. London* **A305**, 535.

Kihlborg, L. (1979) *Direct Imaging of Atoms in Crystals and Molecules*, Nobel *Symposium* 47, The Royal Swedish Academy of Sciences.

Klinowski, J. (1985) *Solid State Ionics* **16**, 3.

Knights, J. C., Hayes, T. M. & Mikkelsen Jr., J. C. (1977) *Phys. Rev. Lett.* **39**, 712.

Kovatcheva, D., Hewat, A. W., Rangavittal, N., Manivannan, V., Guru Row T. N. & Rao, C. N. R. (1991) *Physica C* **173**, 444.

Kulkarni, G. U., Sankar, G. & Rao, C. N. R. (1989) *Z. Phys. B* **73**, 529.

Kulkarni, G. U. & Rao, C. N. R. (1991) *Catal. Lett.* **9**, 427.

Laubschat, C., Kaindl, G., Sampathkumaran, E. V. & Schneider, W. D. (1984) *Solid State Commun.* **49**, 339.

Lieber, C. M. (1994) *Chem. Eng. News*, April Issue, 28.

Lippmaa, E., Mägi, M., Samoson, A., Engelhardt, G. & Grimmer, A. R. (1980) *J. Amer. Chem. Soc.* **102**, 4889.

Lipsicas, M., Raythatha, R. H., Pinnavaia, T. J., Johnson, I. D., Geise Jr, R. F., Costanzo, P. M. & Robert, J. L. (1984) *Nature* **309**, 604.

Lovesey, S. W. & Springer, T. (eds) (1977) *Dynamics of Solids and Liquids by Neutron Scattering, Topics in Current Physics*, Vol. 3, Springer-Verlag, Berlin.

Lovesey, S. W. & Sterling, G. C. (1984) *Physica* **127B**, 306.

Lytle, F. W., Sayers, D. E. & Stern, E. A. (1975) *Phys. Rev.* **B11**, 4825.

Mägi, M., Lippmaa, E., Samoson, A., Engelhardt, G. & Grimmer, A. R. (1984) *J. Phys. Chem.* **88**, 1518.

Manohar, H. (1983) *Curr. Sci.* (India) **52**, 39.

Mansour, A. N., Cook Jr, J. W. & Sayers, D. E. (1984) *J. Phys. Chem.* **88**, 2330.

Mason, R., Rao, C. N. R., Sheppard, N. & Roberts, M. W. (eds) (1986) *Phil. Trans. Roy. Soc. London*.

Meinke, W. W. (1973) in *Treatise on Solid State Chemistry* (ed. Hannay, N. B.) Vol. 1, Plenum Press, New York.

Meyer, G. (1992) *Prog. Surface Sci.* **41**, 3.

Moran, M. J., Miller, G. R., DeMarco, R. A. & Resing, H. A. (1984) *J. Phys. Chem.* **88**, 1580.

Morris, R. E., Owen, J. J., Stalick, J. K. & Cheetham, A. K. (1994) *J. Solid State Chem.*, **111**, 52.

Mozzi, R. L. & Warren, B. E. (1969) *J. Appl. Crystallogr.* **2**, 164.

Newnham, R. E. & Roy, R. (1973) in *Treatise on Solid State Chemistry* (ed Hannay, N. B.), Vol. 1, Plenum Press, New York.

Olejnik, S., Stirling, G. C. & White, J. W. (1970) *Special Disc. Faraday Soc.* **1**, 1.

Parthasarathy, R., Sarode, P. R., Rao, K. J. & Rao, C. N. R. (1982) *Proc. Ind. Natl. Sci. Acad.* **48A**, 119.

Pettifer, R. F. & McMillan, P. W. (1977) *Phil. Mag.* **35**, 871.

Pettifer, R. F., McMillan, P. W. & Gurman, S. J. (1977) in *The Structure of Non-Crystalline Materials* (eds. Gaskell, P. H.), Taylor & Francis, London.

Prabakar, S., Rao, K. J. & Rao, C. N. R. (1987) *Chem. Phys. Lett.* **139**, 96.

Prabakar, S., Rao, K. J. & Rao, C. N. R. (1991a) *Chem. Phys. Lett.* **183**, 176.

Prabakar, S., Rao, K. J. & Rao, C. N. R. (1991b) *J. Mater. Res.* **6**, 592.

Ramanan, A., Gopalakrishnan, J., Uppal, M. K., Jefferson, D. A. & Rao, C. N. R. (1984) *Proc. Roy. Soc. London* **A395**, 127.

Ramdas, S. & Klinowski, J. (1984) *Nature* **308**, 521.

Rao, C. N. R. (1985) *Proc. Ind. Nat. Sci. Acad.* **51A**, 1.

Rao, C. N. R. (1986) *Phil. Trans. Roy. Soc. London* **317A**, 37.

Rao, C. N. R., Sarma, D. D., Vasudevan, S. & Hegde, M. S. (1979) *Proc. Roy. Soc. London* **A367**, 239.

Rao, C. N. R., Sarma, D. D. & Hegde, M. S. (1980) *Proc. Roy. Soc. London* **A370**, 269.

Rao, C. N. R. & Sarma, D. D. (1980) in *Science and Technology of Rare Earth Materials* (eds Subbarao, E. C. & Wallace, W. E.), Academic Press, New York.

Rao, C. N. R. & Hegde, M. S. (1981) in *Preparation and Characterization of Materials* (eds Honig, J. M. & Rao, C. N. R.) Academic Press, New York.

Rao, C. N. R. & Sarma, D. D. (1982) *J. Solid State Chem.* **45**, 14.

Rao, C. N. R., Sparrow, T. G., Thomas, J. M. & Williams, B. G. (1984) *J. Chem. Soc. Chem. Comm.*, 1238.

Rao, C. N. R., Thomas, J. M., Selvaraj, U., Klinowski, J., Rao, K. J. & Ramdas, S. (1985) *Angew. Chem. Int. Ed. (Engl)* **24**, 61.

Rao, C. N. R., Ganguly, P. & Somasundaram, T. (1986) *J. Indian Chem. Soc.* **63**, 1.

Rao, C. N. R., Vijayakrishnan, V., Aiyer, H. N., Kulkarni, G. U. & Subbanna, G. N. (1993) *J. Phys. Chem.*, **97**, 11157.

Rao, K. J., Wong, J. & Weber, M. J. (1983) *J. Chem. Phys.*, **78**, 6228.

Richards, R. & Packer, K. J. (eds) (1981) *Phil. Trans. Roy. Soc. London* **A229**, 475.

Rietveld, H. M. (1969) *J. Appl. Crystallogr.* **2**, 65.

Riviere, J. C. (1982) *Phil. Trans. Roy. Soc. London* **A305**, 545.

Sandstrom, D. R. & Lytle, F. W. (1979) *Ann. Rev. Phys. Chem.* **30**, 215.

Sankar, G., Sarode, P. R. & Rao, C. N. R. (1983) *Chem. Phys.*, **76**, 435.

Sankar, G., Sarode, P. R., Srinivasan, A., Rao, C. N. R., Vasudevan, S. & Thomas, J. M. (1984) *Proc. Indian Acad. Sci. (Chem. Sci.)* **93**, 321.

Sankar, G., Thomas, J. M., Waller, D., Couves, J. W., Catlow, C. R. A. & Greaves, G. N. (1992) *J. Phys. Chem.*, **96**, 7485.

Sarma, D. D., Kamath, P. V. & Rao, C. N. R. (1983) *Chem. Phys.*, **73**, 71.

Sarode, P. R., Ramasesha, S., Madhusudan, W. H. & Rao, C. N. R. (1979) *J. Phys.*, **C12**, 2439.

Satyamurthy, N. S., Dasannacharya, B. A. & Chakravarty, R. (1981) in *Preparation and Characterization of Materials* (eds Honig, J. M. & Rao, C. N. R.) Academic Press, New York.

Selvaraj, U., Rao, K. J., Rao, C. N. R., Klinowski, J. & Thomas, J. M. (1985) *Chem. Phys. Lett.*, **114**, 24.

Sinfelt, J. H., Via, G. H. & Lytle, F. W. (1984) *Catalysis Reviews* **26**, 81.

Smith, J. V. & Blackwell, C. S. (1983) *Nature* **303**, 223.

Somorjai, G. A. (1981) *Chemistry in Two Dimensions: Surfaces*, Cornell University Press, Ithaca.

Sorensen, O. T. (ed.) (1981) *Nonstoichiometric Oxides*, Academic Press, New York.

Sparrow, T. G., Williams, B. G., Thomas, J. M., Jones, W., Herley, P. J. & Jefferson, D. A. (1983) *J. Chem. Soc. Chem. Comm.* 1432.

Stern, E. A., Rinaldi, S., Callen, E. & Heald, S. M. (1978) *J. Mag., Mat.* **7**, 188.

Stutius, W., Boyce, J. B. & Mikkelsen Jr., J. C. (1979) *Solid State Comm.*, **31**, 539.

Teo, B. K. (1980) *Acc. Chem. Res.*, **13**, 412.

Teo, B. K. & Lee, P. A. (1979) *J. Amer. Chem. Soc.*, **101**, 2815.

Theye, M. L., Launois, H. & Gheorghiu, A. (1980) *J. Phys.*, **C13**, 6139.

Thomas, J. M. (1982) *Ultramicroscopy* **8**, 13.

Thomas, J. M. (1984) *Phil. Trans. Roy. Soc. London* **A311**, 271.

Thomas, J. M., Klinowski, J., Wright, P. A. & Roy, R. (1983) *Angew Chem.*, **22**, 614.

Thomas, J. M., Williams, B. G. & Sparrow, T. G. (1985) *Acc. Chem. Res.*, **18**, 324.

Treacy, M. M. J., Howie, A. & Wilson, C. J. (1978) *Phil. Mag.*, **A38**, 569.

Vasudevan, S., Hegde, M. S. & Rao, C. N. R. (1979) *J. Solid State Chem.*, **29**, 253.

Vasudevan, S., Thomas, J. M., Wright, C. J. & Sampson, C. (1982) *J. Chem. Soc. Chem. Comm.* 418.

West, A. R. (1985) *Solid State Chemistry and its Applications*, John Wiley, Chichester.

Williams, B. G. & Thomas, J. M. (1983) *Int. Rev. Phys. Chem.* **3**, 39.

Wong, J. (1980) in *Metallic Glasses* (eds Güntherodt, H. J. & Beck, H.) *Topics in Applied Physics* **46**, Springer-Verlag, Berlin.

Wong, J. & Lytle, F. W. (1980) *J. Non-Cryst. Solids* **37**, 273.

Yashonath, S., Sen, P., Hegde, M. S. & Rao, C. N. R. (1983) *J. Chem. Soc. Faraday I*, **79**, 1229.

3 Preparative strategies

3.1 Introduction

Availability of pure, well-characterized solid samples is crucial to all solid state studies. A knowledge of the various experimental methods available for the preparation of solids therefore becomes an important and integral part of solid state chemistry (Corbett, 1987; Hagenmuller, 1972; Honig & Rao, 1981; Rao, 1994). A brief reflection on the development of solid state science reveals that, in many cases, it is the synthesis of a novel compound that has triggered off a new line of research. Tables 3.1 and 3.2 provide a few examples to illustrate the point. To many solid state scientists, preparation of solids may mean preparation of single crystals of elements or simple compounds (e.g. Si, Ge, III–V semiconductors, alkali halides, etc.) for a study of a specific property or for technical applications. Preparation of solids is, however, a much more general activity, particularly amenable to chemists. A variety of strategies are adopted to prepare solids and to grow crystals. Technological advances have enabled solid state chemists to employ a broad range of conditions for preparative purposes. Ultra-rapid quenching of materials from very high temperatures, irradiation heating by intense laser beams, melting of solids by electron-beam heating or by the skull method and use of high pressures have become common procedures. Thus, by employing high-power CO_2 lasers ($> 1200\,W$) several entropy-stabilized metastable (e.g. α-$CaCr_2O_4$, $BaNi_2In_8O_{15}$) and mixed-valent oxides (e.g. $Sr_7Nb_2^{IV}Nb_4^{V}O_{21}$ and $Ba_2Ti_{12}^{III}Ti^{IV}O_{22}$) have been synthesized (Möhr & Müller-Buschbaum, 1995). When noncrystalline solids are required, they are prepared by sputtering, quenching and a number of other methods. To meet the challenge of producing materials which have hitherto required high temperatures, advantage is often taken of simple chemical methods.

Five aspects of the preparation of solids can be distinguished: (i) preparation of a series of compounds in order to investigate a specific property, as exemplified by a series of perovskite oxides to examine their electrical properties or by a series of spinel ferrites to screen their magnetic properties; (ii) preparation of unknown members of a structurally related class of solids to extend (or extrapolate) structure–property relations, as exemplified by the synthesis of layered chalcogenides and their intercalates or derivatives of TTF–TCNQ to study their superconductivity; (iii) synthesis of a new class of compounds (e.g. sialons, $(Si, Al)_3(O, N)_4$, or doped polyacetylenes), with novel structural properties (iv) preparation of known solids of prescribed specifications (crystallinity, shape, purity, etc.) as in the case of crystals of Si, III–V compounds and

Table 3.1 *Some early solid state preparations that have led to major developments*[a]

Prototype	First reported by (year)	Subsequent development
InP	Thiel (1910)	III–V semiconductors
$ZrO_2(CaO)$	Ruff (1929)	Solid electrolyte, oxygen sensor
$Na\beta$-alumina	Stillwell (1926)	Solid electrolyte, Na–S battery
V_3Si	Wallbaum (1939)	A15 high temperature superconductors
$BaTiO_3$	Tammann (1925)	Ferroelectrics, ceramic capacitors
$LiNbO_3$	Süe (1937)	Nonlinear optics
$BaFe_{12}O_{19}$	Adelsköld Schrewelius (1938)	Ferrites, memory devices
$LnNi_5$ (Ln = rare earth)	Klemm (1943)	Strong magnets, hydrogen storage materials
Amorphous Si	Konig (1944)	Solar cells
$Ca_5(PO_4)_3X: Sb^{3+}, Mn^{2+}$	McKeag, Ranby Jenkins (1949)	Fluorescent lamp phosphor
ZnS/CdS	Kröger (1940)	Cathode ray tube phosphor

[a] Partly taken from the US Panel Report on New Materials published by Centre for Materials Research, Stanford University (1979).

alkali halides and (v) discovering new strategies for the synthesis of known as well as new solids. Chemists are really at home with the last aspect.

An understanding of the crystal chemistry is of the essence in designing new materials possessing desired properties. It is only when we have the correct prescription of structure (in relation to the properties of interest) that we can start synthetic efforts. As discussed in Chapter 1, even in metal oxides alone, we have several structures such as perovskites, spinels, bronzes and pyrochlores; new families have been identified during characterization and structural elucidation of oxides. Accordingly, several of the homologous series of oxides (see Table 3.3) have been identified through X-ray crystallography or electron microscopy rather than by chemical analysis, the precise compositions often being identified by structural considerations. Preparative solid state chemistry becomes most rewarding when there is close interaction between preparation, characterization (including structure determination) and study of properties. In what follows, we shall discuss the methods of preparation of solids in various states of aggregation ranging from the amorphous and the microcrystalline states to polycrystalline powders and finally to single crystals.

It is helpful to make a distinction between the preparation of *new solids* and the

Table 3.2. *Some recently synthesized materials of technological relevance*

Sialon $(Si, Al)_3 (O, N)_4$	High temperature ceramics
$Sm_{0.4}Y_{2.6}Ga_{1.2}Fe_{3.8}O_{12}$	Bubble memory devices
$M_xMo_6Se_8$ and related Chevrel phases	High field superconductors
$LnRh_4B_4$	Coexistence of superconductivity
(Ln = rare earth)	and magnetism
Aluminosilicates	Catalysis (e.g. methanol–
(including ZSM–5)	gasoline)
$Pb_2Ru_{2-x}Pb_x^{4+}O_{7-y}$	Electrocatalyst (for use in
	oxygen electrodes)
$YBa_2Cu_3O_7$	High T_c superconductivity
Polyacetylene (n–and p–type doping)	Solid state batteries
Diacetylene polymers	Nonlinear optic materials
	(better than $LiIO_3$)
Organic liquid crystals	Display devices

Table 3.3. *Some homologous series of metal oxides*

Ti_nO_{2n-1}	$4 \leqslant n \leqslant 9$ or $16 < n < 36$
Mo_nO_{3n-1}	$8 \leqslant n \leqslant 12$
W_nO_{3n-2}	$n = 20, 24, 25$, etc.
Pr_nO_{2n-2}	$n = 4, 7, 9, 10, 11$ and 12
$Bi(WO_3)_n$	$n = 6, 8, 15$, etc.
$Bi_2W_nO_{3n+3}$	$n = 1, 2$ and 3
$La_nNi_nO_{3n-1}$	$n = 2$ and above
$La_nCo_nO_{3n-1}$	$n = 2$ and above
$(Bi_2O_2)[A_{n-1}B_nO_{3n+1}]$	A = Ba, Bi, etc.
	B = Ti, Nb, W, Fe, Cr
	$n = 1–8$
$A_{n+1}B_nO_{3n+1}$	A = Sr; B = Ti
	A = La; B = Ni
	$n = 1, 2, 3$, etc.
$A_nB_nO_{3n+2}$	A = Na, Ca
	B = Nb
	$n^a = 4, 5, 6$ and 7

[a] Between $n = 4$ and 4.5, a large number of coherent intergrowth phases with long periodicities are known in this system (Portier *et al.*, 1975).

preparation of solids by *new methods*. Preparation of a new solid need not necessarily involve a new method. A number of instances can be cited where preparation of new solids with novel structures and properties is achieved by routine procedures. Typical of them are $Na_3Zr_2PSi_2O_{12}$ (NASICON), $Na_{1+x}Al_{11}O_{17+x/2}$ (β-alumina), $BaFe_{12}O_{19}$ and MMo_6S_8 (M = Cu, Pb etc.) Chevrel phases. All these solids are prepared by the ordinary method of reacting readily available constituents at elevated temperatures (the *ceramic method*). In the last case, the reaction is carried out in closed ampoules for obvious reasons, but in others even this precaution is unnecessary. The importance of such preparations lies not in the method, but in selecting the right constituents in the right proportion, bearing in mind the chemistry as well as the structure and properties desired in the new phase. This aspect is illustrated by the synthesis of $Na_3Zr_2PSi_2O_{12}$, a fast sodium ion conductor, where the composition was chosen keeping in mind the coordination preferences of the atoms involved, the stability of their oxidation states, the nature of the network that would be formed and whether the network would permit three-dimensional mobility of sodium ions (Goodenough *et al.*, 1976). Another example of innovative synthesis by design is the preparation of synthetic bone material for prosthetic application (Roy, 1977). Here the problem is not only the synthesis of the chemical compound constituting the human bone, viz. calcium hydroxyapatite, but also having it with the 100% connectivity and porosity of the natural bone. The synthesis could be accomplished because of two realizations: certain marine corals ($CaCO_3$) possess the same porosity and connectivity as the human bone and the aragonite form of $CaCO_3$ can be topotactically converted to calcium hydroxyapatite in a hydrothermal reaction. Thus, the reaction of marine corals with phosphoric acid under hydrothermal conditions produces synthetic bone material.

Synthesis of new solids is not always achieved by design. Preparative solid state chemistry has its share of serendipitous discovery of new materials. The discovery of $NaMo_4O_6$, the prototype of metal cluster chain compounds, is an example of this category (Torardi & McCarley, 1979). The discovery resulted from a most innocent experiment aimed at preparing $NaZn_2Mo_3O_8$, the sodium-analogue of the lithium-containing cluster compound, $LiZn_2Mo_3O_8$. For this purpose, a mixture of Na_2MoO_4, ZnO, MoO_2 and Mo was sealed in a molybdenum tube and heated at 1370 K. The product turned out to be shiny needles of $NaMo_4O_6$ instead of the expected $NaZn_2Mo_3O_8$. The crystal structure of $NaMo_4O_6$ turned out to be one of the most unique oxide structures, consisting of infinite chains of Mo_6 octahedral clusters sharing opposite edges (Fig. 1.26).

3.2   Preparation of crystalline inorganic materials

3.2.1   *Ceramic methods*

The most common method of preparing inorganic solids is by the reaction of the component materials at elevated temperatures. If all the components are solids, the method is called the *ceramic method*. If one of the constituents is volatile or sensitive to the atmosphere, the reaction is carried out in sealed evacuated capsules. Ternary oxides, sulphides, phosphides, etc. have been prepared by this method. A knowledge of the phase diagram is generally helpful in fixing the desired composition and conditions for synthesis. Some caution is necessary in deciding the choice of the container; platinum, silica and alumina containers are generally used for the synthesis of metal oxides, while graphite containers are suitable for sulphides and other chalcogenides as well as pnictides.

Ceramic preparations often require relatively high temperatures (up to 2300 K) which can generally be attained by resistance heating. Electric arc and skull techniques can give temperatures up to 3300 K while high-power CO_2 lasers can give temperatures up to 4300 K.

Ceramic methods suffer from several disadvantages: (i) When no melt is formed during the reaction, the entire reaction has to occur in the solid state, first by a phase boundary reaction at the points of contact between the components and later by diffusion of the constituents through the product phase. With the progress of the reaction, diffusion paths become longer and longer and the reaction rate slower and slower; the reaction can be speeded up to some extent by intermittent grinding between heating cycles. (ii) There is no way of monitoring the progress of the reaction in a ceramic method. It is only by trial and error that one decides appropriate conditions that lead to the completion of the reaction. Because of this difficulty, one often ends up with mixtures of reactants and products by the ceramic technique. Separation of the desired product from such mixtures is difficult, if not impossible. (iii) Frequently, it becomes difficult to obtain a compositionally homogeneous product by a ceramic technique even where the reaction proceeds almost to completion.

Despite these limitations, ceramic techniques have been widely used for solid state synthesis. Mention must be made, among others, of the successful use of this technique for the synthesis of rare earth monochalcogenides such as SmS and SmSe. The method (Jayaraman et al., 1975) consists of heating the elements, first at lower temperatures (870–1170 K) in evacuated silica tubes; the contents are then homogenised, sealed in tantalum tubes and heated at ~ 2300 K by passing a high current through the tube. Metal-rich halides of lanthanides and other early transition elements have been synthesized using tantalum containers (Corbett, 1980, 1981). Several halides and chalcogenides possessing condensed metal clusters have also been prepared by the ceramic method (Simon, 1981). Metal-rich suboxides of alkali metals have been prepared by reacting the metal with required amounts of oxygen until the gas is entirely absorbed ($p < 10^{-5}$ torr) (Simon, 1975). Examples of such phases are Rb_6O, Rb_9O_2, $Cs_{11}O_3$, Cs_3O, Cs_4O and Cs_7O. Extreme air sensitivity and low temperatures of decomposition/melting render the preparation and investigation of the phases very difficult. These oxides are therefore investigated in situ.

Various modifications of the ceramic technique have been employed to overcome some of the limitations. One of these relates to decreasing the diffusion path lengths. In a polycrystalline mixture of reactants, the individual particles are of approximately 10 μm size, which represents diffusion distances of roughly 10 000 unit cells. By using freeze-drying, spray-drying, coprecipitation and sol–gel and similar techniques, it is possible to bring down the particle size to a few hundred ångstroms (see Section 3.3) and thus effect a more intimate mixing of the reactants. In spray-drying, suitable constituents dissolved in a solvent are sprayed in the form of fine droplets into a hot chamber. The solvent evaporates instantaneously leaving behind an intimate mixture of reactants, which on heating at elevated temperatures gives the product. In freeze-drying, the reactants in a common solvent are frozen by immersing in liquid nitrogen and the solvent removed at low pressures. In coprecipitation, the required metal cations are coprecipitated from a common medium, usually as hydroxides, carbonates, oxalates, formates or citrates, which are subsequently heated at appropriate temperatures to yield the final product. These methods have been used for the preparation of polycrystalline samples of several oxides such as ferrites, perovskites and β-aluminas. The *sol–gel process* (Johnson, 1981) involves forming a concentrated sol of the reactant oxides or hydroxides and converting it to a semirigid gel by removing the solvent. The solvent is removed by passing fine droplets of the sol through a column of an aliphatic alcohol (e.g. 2-ethyl-1-hexanol). The dehydrated gel is heated at an appropriate temperature to obtain the product. Several complex oxide materials such as NASICON, $PbTi_{1-x}Zr_xO_3$, $ThO_2 - UO_2$ solid solutions and $Zr_2(PO_4)_2SO_4$ modifications (Alamo & Roy, 1984) have been prepared by this method. The sol–gel technique has also become important for the preparation of noncrystalline solids (Scholze, 1984).

The sol–gel route allows the fabrication of multi-component ceramics in the form of pellets, fibres, thin films or abrasive coatings (Roy, 1987). Use of heterometallic alkoxides as precursors for the synthesis of multi-component ceramics is a new development. For instance $LiNbO_3$ and $MgAl_2O_4$ have been prepared by the sol–gel route through the hydrolysis of the bimetallic alkoxides, $LiNb(OEt)_6$ and $Mg[Al(OR)_4]_2$ respectively. The sol–gel process is highly amenable for incorporating organic, inorganic and even biological species into insulating gel matrices. These hybrid nanocomposite materials have opened up new avenues for application in non-linear optics, solar energy conversion, biotechnology and so on. For example, a key problem in solar energy conversion and artificial photosynthesis using redox systems is to retard the back electron transfer between the charge separated donor and acceptor. This problem has, to a great extent, been circumvented by immobilizing the donor and the acceptor in a porous silica sol–gel glass. Here pyrene is the photo-sensitized electron donor and methyl viologen, the electron acceptor. The spatial separation between the donor and the acceptor inhibits the back electron transfer to produce long life times (up to a few hours) of the charge separated pair (Slama-Schwok *et al.*, 1992). An inorganic sol–gel matrix which contains both an immobilized donor and a mobile redox relay appears to be the solution for this important problem (Castellano *et al.*, 1994). The sol–gel technique has been used to produce new biomaterials by encapsulation of enzymes and other proteins in optically transparent porous silicate glasses, typical examples being bacteriorhodopsin (Wu *et al.*, 1993) and *Leishmania*, a parasitic protozoa (Livage, 1994). Interestingly, the encapsulated biomolecules retain their characteristic properties and enzymatic activities inside the glass matrix.

Thermodynamic control of solid state synthesis is possible by a judicious choice of reagents and reactions as shown by Wiley & Kaner (1992). A typical strategy is to make use of highly exothermic metathetical reactions between covalent metal halides and alkali metal chalcogenides and pnictides. Synthesis of MoS_2 by the reaction between $MoCl_5$ and Na_2S is one such example (Bonneau *et al.*, 1991).

$$MoCl_5 + 5/2Na_2S \rightarrow MoS_2 + 5\,NaCl + \tfrac{1}{2}S$$

Since the reaction is highly exothermic ($\Delta H = -213\,\mathrm{Kcal\,mol^{-1}}$), it occurs with explosive violence yielding the products within seconds. In this respect, this strategy is similar to the *self-propagating high-temperature synthesis* (SHS) developed by Mer-zhanov and coworkers (1972, 1993). Many oxide materials including superconducting cuprates, have been prepared by SHS, also known as the *combustion method* (Sekar & Patil, 1993; Mahesh *et al.*, 1992). The method employs mixtures of metal nitrates with a fuel (e.g. urea, glycine) which when dry undergo spontaneous combustion to yield oxide products.

3.2.2   *Chemical strategies*

There have been significant attempts to overcome the limitations of the ceramic method, resulting in alternative routes for solid state synthesis. The new methods, which rely on the knowledge of the structural chemistry and the reactivity patterns of solids, have not only enabled the synthesis of known solids in a state of high purity and homogeneity at far lower temperatures than by the conventional methods, but, in addition, have resulted in the synthesis of new phases (Rao, 1994). Among these alternative strategies, the following three methods have proved significant: the solid state precursor method and methods based on topochemical redox reactions and topochemical ion-exchange reactions. The emphasis in all three methods is in achieving the synthesis at low temperatures so that the products obtained are in a finely divided state with large surface areas – a feature essential for catalysis and other applications. More importantly, synthesis at temperatures considerably lower than the sintering temperatures of solids preserves the essential features of the parent structure with mimimal structural reorganization (topochemical methods). Synthesis by topochemi-cal methods often yields metastable phases that cannot be obtained by conventional methods.

Over the years, chemical methods are being increasingly employed for the synthesis of novel solids, especially metastable ones (Rao, 1994; Stein *et al.*, 1993; Go-palakrishnam, 1995). In some instances, a thermodynamically stable phase is first prepared by an ordinary route followed by its transformation to a metastable phase by an appropriate soft-chemical route. Soft-chemistry routes (*chimie douce*) are indeed becoming popular (Rao, 1994; Rouxel *et al.*, 1994). Some of the examples of synthesis of this kind are: ReO_3 -like MoO_3 from its hydrate, $MoO_3 \cdot 1/3 \; H_2O$ (Figlarz, 1989), novel phases of WO_3 and MoO_3 by deintercalation of amine intercalates (Ayyappan *et al.*, 1995), TiO_2 (B) from layered titanes, $A_2Ti_nO_{2n+1}$ (A = Na, K, Cs) (Tournoux *et al.*, 1986; Feist and Davies, 1992), $TiO_2(H)$ from $K_{0.25}TiO_2$ (Latroche *et al.*, 1989), 1T-MoS_2 from $KMoS_2$ (Wypych and Schöllhorn, 1992), NASICON framework $V_2(PO_4)_3$ from $Na_3V_2(PO_4)_3$ (Gopalakrishnan & Kasthuri Rangan, 1992) and superconducting $La_2CuO_{4+\delta}$. In the last example, synthesis is achieved by the chemical (Rudolf & Schöllhorn, 1992) or the electrochemical oxidation of the parent La_2CuO_4 (Grenier *et al.*, 1991). Such synthesis at relatively low temperatures, involves mild solid state reactions such as dehydration, decomposition, ion-exchange, redox insertion/extraction and so on. Accordingly, the synthesis is topochemically-control-led in the sense that the nature of the solid product and the course of the reaction are influenced by the structure of the parent solid. A fine example of the influence of structure is provided by the proton exchange of Li in $LiAlO_2$ (Poeppelmeier & Kipp,

1987). This oxide crystallizes in three different structures, α, β and γ, of which only the α-modification, with the rocksalt superstructure, has both Li and Al in octahedral coordination and undergoes facile exchange of Li^+ with H^+ (in molten benzoic acid) forming spinel-like $HAlO_2$. The β and γ-forms of $LiAlO_2$, where Li occurs in tetrahedral oxygen coordination, do not exhibit the ion-exchange reaction. The reactivity of $LiAlO_2$ as well as of the layered $K_2Ti_4O_9$ towards proton exchange has been examined with the help of semiempirical electronic structure calculations, throwing light on the acidity and basicity of atoms in metal oxides (Dronskowski, 1992, 1993).

Not all soft-chemical synthesis is achieved by topochemical reactions. For example, *acid-leaching*, which is well-known in geology as a means of transforming rocky silicates to clays and soil minerals (Casey *et al.*, 1993), has been exploited for the synthesis of solids. An early example of such a synthesis is that of PrO_2 from Pr_6O_{11} or of TbO_2 from Tb_4O_7, which has attracted renewed attention (Sastry *et al.*, 1966; Kang & Eyring, 1988; Gasgnier *et al.*, 1993). Acid-leading of $AVMoO_6$ (A = Li, Na) and $LiVWO_6$ brannerites leads to new V-substituted molybdenum and tungsten oxides (Feist and Davies, 1991; Gopalakrishnan *et al.*, 1994). Acid-leaching of amine intercalates of WO_3 and MoO_3 gives novel phases of these oxides (Ayyappan *et al.*, 1995). The following general principle seems to govern acid-leaching. A binary or ternary phase containing two or more cations (of the same or different elements) that differ in their basicity leaches out the cation of higher basicity into aqueous acid, leaving behind a new solid phase containing the more acidic cations.

Among the various other syntheses achieved by mild chemical routes, mention must be made of the synthesis of hydrides of metals and alloys using borohydride reduction in aqueous medium (Murphy *et al.*, 1993). The reaction seems to proceed as follows:

$$M + BH_4^- + 3H_2O \rightarrow M - H + H_2BO_3^- + 3.5H_2$$

It is significant that a hydriding power equivalent to 20–30 atm. of H_2 is obtained in a reaction of this kind using aqueous BH_4^- in an open beaker.

Synthesis of inorganic materials by soft-chemical methods that mimic biomineralization is another interesting development (Mann, 1993a&b). Biomineralization essentially involves crystallization at inorganic/organic interfaces via specific molecular recognition (templating) and supermolecular organization. The strategy has been employed in the nanoscale assembly of inorganic materials (e.g. Fe_3O_4, CdS) crystal engineering of inorganic–organic composites and in the fabrication of inorganic materials with controlled microstructure. The last is achieved through the use of functionalized organic/biomolecular matrices as organized frameworks for *in-situ* formation of inorganic materials.

The ideal condition for carrying out a solid state reaction in order to obtain a homogeneous product in the shortest time at the lowest possible temperature is to ensure homogeneous mixing of the reactants on an atomic scale. This, however, cannot be achieved in the ceramic method or its modifications. The only way to achieve this is to prepare a single phase (a chemical compound) in which the reactants are present in the required stoichiometry. Such a solid phase, called a *precursor*, on heating gives the desired product in a stoichiometric and homogeneous state. The *precursor method* has been employed for the synthesis of ternary oxides (Wold, 1980). Spinel-type ferrites, MFe_2O_4 (M = Mg, Mn, Co, Ni), have been prepared by thermal decomposition of acetate precursors of the general formula $M_3Fe_6(CH_3COO)_{17}O_3OH\cdot12C_5H_5N$. The advantage of this method is that the precursors can be prepared as crystalline solids in a state of high purity, possessing the desired M/Fe ratio. Moreover, the products are homogeneous and formed at fairly low temperatures. Other examples of synthesis of metal oxides through the compound precursor route are preparation of chromites, MCr_2O_4, by the decomposition of dichromates, $(NH_4)_2M(CrO_4)_2\cdot6H_2O$, preparations of $BaTiO_3$ and $LiCrO_2$ by the decomposition of $Ba[TiO(C_2O_4)_2]$ and $Li[Cr(C_2O_4)_2(H_2O)_2]$ and of $LaFeO_3$ by the decomposition of $La[Fe(CN)_6]$. Another example is the synthesis of ceramic oxides by the hydrolysis of metal alkoxides.

It is not always possible to find suitable single precursor compounds for the synthesis of all the desired compositions, because the stoichiometry of the precursor may not correspond to that of the product. A solid-solution precursor method which retains all the advantages of single compound precursors has been described (Horowitz & Longo, 1978; Longo & Horowitz, 1981; Vidyasagar *et al.*, 1984, 1985). The strategy is to make use of isostructural compounds containing a common anion so that they form a continuous series of solid solutions. The method has been used for the synthesis of several ternary oxides in the Ca–Mn–O system, such as $Ca_2Mn_3O_8$, $CaMn_3O_6$, $CaMn_4O_8$ and $CaMn_7O_{12}$ by the thermal decomposition of carbonate solid solutions, $Ca_{1-x}Mn_xCO_3$, of calcite structure. $Ca_{1-x}Fe_xCO_3$ has been used to prepare unusual ferrites; $Ca_2Fe_{2-x}Mn_xO_5$ with three types of transition metal–oxygen polyhedra have also been prepared in this manner. The method has been extended to hydroxide, oxalate and nitrate precursor solid solutions. Using hydroxides of the $La(OH)_3$ structure, compounds like $LaNiO_3$ and quaternary oxides have been prepared. Using nitrate precursors, $BaPbO_3$, Ba_2PbO_4 and $BaSrPbO_4$ have been prepared. Solid solution precursors have several advantages: (i) The reacting cations are uniformly blended together, thereby avoiding diffusion problems (the diffusion distance in this method is ~ 10 Å) and compositional inhomogeneities in the final product. (ii) The final product is formed at a much lower temperature than in a conventional ceramic synthesis, thus permitting an examination of subsolidus regions of the phase diagram which would otherwise by inaccessible. The products formed at lower temperatures will have large specific surface areas, which is an important requirement in catalyst preparations. The solid solution precursor method has been used for the low-temperature synthesis of Mo–W alloys (Cheetham, 1980) by hydrogen reduction of $(NH_4)_6[Mo_{7-x}W_xO_{24}]$. $Mo_{1-x}W_xO_3\cdot H_2O$ solid solutions constitute ideal precursors, not only for the synthesis of ReO_3-like $Mo_{1-x}W_xO_3$, but also for $Mo_{1-x}W_xO_2$ solid solutions. Direct reduction of $Mo_{1-x}W_xO_3\cdot H_2O$ at 750°C in hydrogen affords a convenient route for the preparation of monophasic Mo-W alloys (Rao *et al.*, 1986).

Chevrel compounds of the general formula $A_xMo_6S_8$ with A = Cu, Pb, La etc. are difficult to prepare in very pure state by the ceramic procedure. A one-step reduction of a precursor sulphide has been found to yield these phases (Nanjundaswamy *et al.*, 1987):

$$2A_x(NH_4)_yMo_3S_9 + 10\ H_2 \rightarrow A_{2x}Mo_6S_8 + 10\ H_2S + 2y\ NH_3 + y\ H_2$$

Metal thiolates, thiocarbonates and dithiocarbonates are good precursors to prepare sulphides such as CdS and ZnS. A variety of organometallic precursors (usually metal alkyls) have been employed to prepare semiconducting materials (e.g. GaAs, InP).

Vapour phase decomposition of precursors at relatively low temperatures has been employed to produce solids. Thus SiC is formed by the decomposition of CH_3SiH_3 or $(CH_3)_2SiCl_2$, while Si_3N_4 is produced by the reactive decomposition of $SiCl_4$ or SiH_4 and NH_3. Sialon powders are made by the reaction of metakaolin with NH_3 vapour. There has been considerable effort recently to design the right precursors to prepare carbides, nitrides and other ceramics such as SiC, Si_3N_4, TiC, TiN and B_4C (Rao, 1994). Transition-metal sulphides are prepared by the reaction of precursor oxides, sulphates and chlorides with H_2S or CS_2. Pyrochlores of the type $Pb_2[Ru_{2-x}Pb_x^{4+}]O_{7-y}$ and $Bi_2[Ru_{2-x}Bi_x^{5+}]O_{7-y}$ are prepared by a low-temperature method employing strongly alkaline media (Horowitz *et al.*, 1981). The high-temperature phase ($> 1300\,K$) of ZrO_2 is obtained at room temperature by the hydrolysis of zirconium salts. The cubic (δ) form of Bi_2O_3 can be stabilized at room temperature by incorporating additives such as Y_2O_3 and WO_3. γ-Fe_2O_3 is obtained by the dehydration of γ-FeOOH by organic bases (Desiraju & Rao, 1982). *Fused salt electrolysis* (Banks & Wold, 1974) is employed to synthesize oxides such as blue Mo bronzes.

Use of molten salts as fluxes or as high-temperature solvents for the growth of crystals has been known for some time (Elwell & Scheel, 1975). The role of the flux is generally to enhance the rate of diffusion in solid state reactions and promote the crystallization of the final product at lower temperatures. The use of alkali metal polychalcogenides as *reactive fluxes* for the synthesis of novel chalcogenides is a recent development (Keane *et al.*, 1991; Kanatzidis, 1990). The method enables the synthesis of solids at temperatures (~ 470–$770\,K$) intermediate between those accessible by the hydrothermal route and the conventional ceramic route. The method essentially consists of reacting transition metals in low-melting alkali metal polychalcogenide (A_2Q_n; $Q = S$, Se, Te) fluxes. The polychalcogenide not only functions as a classical flux in the sense of promoting the synthesis at lower temperatures but also acts as a source for the alkali metal and chalcogen/polychalcogen in the final product. In a typical synthesis, an intimate mixture of the desired transition metal, alkali metal chalcogenide, A_2Q and elemental Q are sealed in a pyrex or quartz tube under vacuum. The temperature is slowly raised to a level where A_2Q and Q melt to form A_2Q_x where x is between 3 and 6. After the reaction, the product is isolated by washing out the excess A_2Q_x with water or other polar solvents. Sunshine *et al.* (1987) first reported the use of alkali metal polychalcogenides for the synthesis of metal chalcogenides and pointed out the potential of these melts for synthesis of solids. The $K_nCu_{3-n}NbSe_4$ series is an illustrative example of synthesis achieved by this method. In this series, the dimensionality of the structure changes from 0 to 3 as n increases from 0 to 3. Kanatzidis and coworkers have employed polychalcogenide melts to synthesize several new phases including the microporous open frameworks, $(Ph_4P)[M(Se_6)_2]$, where M = Ga, In, Tl (Dhingra & Kanatzidis, 1992).

For the synthesis of oxide materials, alkali metal hydroxides (NaOH and KOH) are used. These are relatively low melting solids forming an eutetic at 448 K. The alkali hydroxide melt contributes to the synthesis in two different ways. The equilibrium that exists in the melt, viz., $2OH^- \rightleftharpoons H_2O + O^{2-}$, enables the dissolution of metal oxides in wet (acidic) fluxes and their recrystallization in dry (basic) fluxes. The highly electropositive environment obtaining in molten hydroxides generally stabilizes high oxidation states of metals. Both these attributes have been exploited in the synthesis and growth of superconducting oxides such as $La_{2-x}Na_xCuO_4$ (M = Na, K) (Ham et $al.$, 1988), $Ba_{1-x}K_xBiO_3$ (Schneemeyer et $al.$, 1988), twin-free $EuBa_2Cu_3O_{7-x}$ (Marquez et $al.$, 1993), and oxide pyrochlores such as $La_2Pb_2O_7$, $La_2Bi_2O_7$ and La_2PbBiO_7 (Uma & Gopalakrishnan, 1993). KOH melt has also been used for the electrochemical synthesis of $KBiO_3$ (Nguyen et $al.$, 1993). Keller et $al.$ (1994) have employed molten NaOH (670 K) to prepare the rare-earth cuprates, $LnCu_2O_4$ (Ln = La, Nd, Sm or Eu).

Electrochemical oxidation yields oxides with high oxidation states of the transition metals (Rao, 1994). For example, La_2CuO_4 is rendered oxygen-excess electrochemically (Grenier et $al.$, 1991). Metallic $NdNiO_3$ and ferromagnetic $LaMnO_3$ (containing $\sim 40\%$ Mn^{4+}) have been prepared electrochemically (Mahesh et $al.$, 1995). Demourgues et $al.$ (1993) have prepared $La_8Ni_4O_{17}$, containing unusual oxidation states of Ni, by electrochemical means. Oxides of the type $SrFeO_3$ and $SrCoO_3$ can also be prepared by electrochemical oxidation.

Transition-metal oxides and chalcogenides, MX_n (M = metal), possessing layered or chain structures can be intercalated at room temperature with lithium and other alkali metals to give the reduced phases, A_xMX_n (A = Li, Na or K). Formation of such phases was first reported by Rüdorff (1959). The process has several features which make it attractive both as a preparative technique and for the application of the phases as battery cathodes (Murphy & Christian, 1979): (i) the reaction is reversible and can be brought about chemically or electrochemically; (ii) the reaction is topochemical in nature, occurring with minimal structural reorganization of the host, MX_n and (iii) A cations as well as electrons transferred to the host possess considerable mobility in A_xMX_n phases, rendering them mixed ionic–electronic conductors. It is this last feature that makes them useful as cathode materials in solid state batteries. The best-known example of alkali metal intercalation is that of lithium into TiS_2 to give Li_xTiS_2 ($0 < x \leqslant 1.0$) (Whittingham, 1978; Whittingham & Chianelli, 1980). The intercalation can be carried out by two methods. In the first, a chemical reagent, n-butyl lithium dissolved in a hydrocarbon solvent such as hexane, is employed as the lithiating agent:

$$xC_4H_9Li + TiS_x \rightarrow Li_xTiS_2 + \frac{x}{2}C_8H_{18}.$$

The second method for synthesizing lithium intercalates involves electrochemical reduction of TiS_2. A polycrystalline sample of TiS_2 bonded into an electrode form is immersed in a polar organic solvent (e.g. dioxolane) in which lithium chlorate (VII) is dissolved. A sheet of lithium metal or LiAl serves as the anode. On shorting the two electrodes, lithium ions intercalate TiS_2, the charge-compensating electrons passing through the external circuit. Electrochemical intercalation is particularly advantageous because the rate of the reaction can be controlled by imposing an external voltage across the cell; when the voltage exceeds the value corresponding to the free energy change ΔG of the intercalation reaction, the reverse reaction – namely deintercalation – occurs.

Intercalation of lithium has been achieved in a variety of MX_n hosts. *Deintercalation* of lithium has also been done both electrochemically and chemically. Deintercalation of lithium from $LiMX_n$ using mild oxidizing agents such as I_2/CH_3CN and $Br_2/CHCl_3$ offers a low-temperature route for the synthesis of Li_xMX_n and MX_n phases that are otherwise impossible to prepare, such as Li_xVS_2 and Li_xVO_2. Similarly Mo_6S_8 can be prepared by acid leaching of $Cu_xMo_6S_8$ (Chevrel *et al.*, 1974). In Table 3.4 we list representative examples of lithium intercalation and deintercalation. From the table, it is seen that lithium intercalation occurs in a wide variety of oxide and sulphide hosts possessing one-, two-, and three-dimensional structures. In most cases, the gross structural features of the host are retained. In some cases, there are specific structural changes which can be accounted for within the framework of a topochemical mechanism, e.g. $LiReO_3$ and Li_2ReO_3 (Cava *et al.*, 1982). In both $LiReO_3$ and Li_2ReO_3, the ReO_3 host undergoes significant change on lithium insertion without breaking bonds. The 12-coordinated cavities in the ReO_3 framework each become two octahedral cavities which are occupied by lithium. Lithium insertion in close-packed oxides such as Fe_2O_3, Fe_3O_4 and TiO_2 results in interesting structural changes. Lithium insertion changes the anion array of Fe_2O_3 from hexagonal to cubic close-packing. Lithium insertion suppresses the cooperative Jahn–Teller distortion of Mn_3O_4. Lithium-inserted anatase, $Li_{0.5}TiO_2$, transforms to the superconducting spinel $LiTi_2O_4$ around 770 K. Deintercalation of lithium by electrochemical methods or with iodine from Li_2FeS_2 leads reversibly to $LiFeS_2$, which has the $LiTiS_2$ structure.

Intercalation of sodium and potassium differs from that of lithium. In layered A_xMX_n, lithium is always octahedrally coordinated, while sodium and potassium occupy octahedral or trigonal prismatic sites; octahedral coordination is favoured by large values of x and low formal oxidation states of M. For smaller x and larger oxidation states of M, the coordination of sodium or potassium is trigonal prismatic. Intercalated caesium in MX_n is always trigonal prismatic. Intercalation of sodium and potassium in layered MX_2 oxides and sulphides results in structural transformations involving a change in the sequence of anion layer stacking.

Superconducting Li_xNbO_2 and Na_xNbO_2 have been prepared by the deintercalation of alkali metals from $LiNbO_2$ and $NaNbO_2$ using I_2 or Br_2/CH_3CN. (Geselbracht *et al.*, 1990; Rzeznik *et al.*, 1993). Besides the conventional redox reagents, Cl_2, NO_2^+ and MoF_6 as well as alkali metal iodides and borohydrides have been found useful in such deintercalation/intercalation reactions (Wizansky *et al.*, 1989; Kanatzidis & Marks, 1987). Deintercalation of amines from the intercalates of WO_3 and MoO_3 yield unusual metastable structures (Ayyappan *et al.*, 1995). The recently developed lithium-ion 'rocking chair' batteries make use of lithium-insertion oxides, $LiMn_2O_4$ or $LiCoO_2$, as cathode and a graphitic carbon as anode (Guyomard & Tarascon, 1994). Use of $VO_2(B)$ in place of graphite has enabled the use of aqueous $LiNO_3$ as electrolyte

Table 3.4. *Intercalation and deintercalation of lithium into MX_n hosts*

MX_n host	Remarks	References
TiS_2	Li_xTiS_2: $0 < x \leqslant 1$. Phase is homogeneous over the composition	Whittingham (1978)
VS_2	Li_xVS_2: $0 < x \leqslant 1$. Phases obtained by deintercalation of lithium from $LiVS_2$ using I_2/CH_3CN. Three different phase regions: $0.25 \leqslant x \leqslant 0.33$; $0.48 \leqslant x \leqslant 0.62$ and $0.85 \leqslant x \leqslant 1$ apart from VS_2	Murphy *et al.* (1977)
$NbS_2(3R)$	$Li_{0.5}NbS_2$ and $Li_{0.70}NbS_2$	Whittingham (1978)
TiS_3	Li_2TiS_3 and $Li_{2+x}TiS_3(0 < x \leqslant 1)$	Whittingham (1978)
MoS_3	$Li_xMoS_3(0 < x \leqslant 4)$	Jacobson *et al.* (1979)
MO_2 (rutile)	$Li_xMO_2(x \geqslant 1)(M = Mo, Ru, Os$ or $Ir)$	Murphy *et al.* (1978)
TiO_2 (anatase)	$Li_xTiO_2(0 < x \leqslant 0.7)$. $Li_{0.5}TiO_2$ transforms irreversibly to $LiTi_2O_4$ spinel at $500°C$	Murphy *et al.* (1982)
CoO_2	$Li_xCoO_2(0 < x < 1)$: Phases obtained by electrochemical delithiation of $LiCoO_2$	Mizushima *et al.* (1980)
VO_2	$Li_xVO_2(0 < x > 1)$: Phases obtained by chemical delithiation of $LiVO_2$ using $Br_2/CHCl_3$	Vidyasagar & Gopalakrishnan (1982)
$VO_2(B)$	$Li_xVO_2(0 < x < 2/3)$: Chemical lithiation using *n*-butyl lithium	Murphy & Christian (1979)
Fe_2O_3	$Li_xFe_2O_3(0 < x < 2)$: Anion array transforms from hcp to ccp on lithiation	Thackeray *et al.* (1982)
Fe_3O_4	$Li_xFe_3O_4(0 < x < 2)$: Fe_2O_4 subarray of the spinel structure remains intact	Thackeray *et al.* (1982)
Mn_3O_4	$Li_xMn_3O_4(0 < x < 1.2)$: Lithium insertion suppresses tetragonal distortion of Mn_3O_4	Thackeray *et al.* (1983)
MoO_3	$Li_xMoO_3(0 < x < 1.55)$	Dickens & Pye (1982)
V_2O_5	$Li_xV_2O_5(0 < x < 1.1)$: Intercalation of lithium by using LiI	Dickens *et al.* (1979)
ReO_3	$Li_xReO_3(0 < x \leqslant 2)$: Three phases $0 < x \leqslant 0.35$; $x = 1$ and $1.8 \leqslant x \leqslant 2$	Murphy *et al.* (1981) Cava *et al.* (1982)

for the first time, making the batteries safe and cost-effective (Li *et al.*, 1994).

Organic molecules such as aniline, pyrrole and 2,2′-bithiophene have been interca-lated and polymerized within the galleries of clay minerals, FeOCl, V_2O_5 gel and other layered hosts to yield multilayered inorganic/organic polymer nanocomposites

(Kanatzidis *et al.*, 1989a, 1989b; Mehrotra & Giannelis, 1992; Liu *et al.*, 1993). A distinction has to be made between two kinds of intercalation polymerization. When the host material contains a reducible transition-metal cation as in FeOCl and V_2O_5 gel, intercalation and oxidative polymerization occurs simultaneously, resulting in synergistic modification of host–guest properties. A typical example is the polymerization of aniline in V_2O_5 gel (Kanatzidis, 1989a).

$$C_6H_5NH_2 \rightarrow 1/n\,(-C_6H_4NH-)_n + 2H^+ + e$$

In 2:1 clays and α-Zr(HPO$_4$)$_2$ and such hosts materials, monomer intercalation occurs first; subsequent polymerization is achieved through an external oxidizing agent (Cu^{2+}, $(NH_4)_2S_2O_8$ or oxygen). Intercalation–polymerization of aniline in layered MoO_3 has been carried out using $(NH_4)_2S_2O_8$ (Bissessur *et al.*, 1993a). Such polymerization has also been achieved in zeolites and mesoporous MCM-41, yielding conducting filaments of polyaniline (molecular wires) encapsulated in the channels of the insulating host (Enzel & Bein, 1989; Wu & Bein, 1994).

Exfoliation of layered materials to yield single-layer colloidal dispersions and their *restacking* in the presence of guest molecules afford a novel strategy to synthesize new materials (Divigalpitiya *et al.*, 1989; Jacobson, 1994). For example, MoS_2 disperses into single layers on reaction of Li$_x$MoS$_2$ with water. Restacking of these single-layer MoS_2 in the presence of organic molecules such as styrene, polystyrene and organometallics such as ferrocene yields highly oriented films of the composite material. A similar restacking of MoS_2 in presence of polyaniline, polyvinylpyrrolidone, polyethylene oxide, etc., has yielded inorganic/organic polymer composites exhibiting novel electronic properties. For instance, (PPG)$_x$MoS$_2$ (PPG = polypropylene glycol) and (PEO)$_x$MoS$_2$ (PEO = polyethylene oxide), exhibit metallic conductivity above 50 K and a metal–insulator transition at low temperatures (Bissessur *et al.*, 1993b). Intercalation of organic molecules, having a large second-order hyperpolarizability, such as donor–acceptor substituted stilbenes into layered MPS$_3$ (M = Cd or Mn) has resulted in new non-linear optical materials possessing large SHG efficiency (Lacroix *et al.*, 1994). The high efficiency has been attributed to a spontaneous polar alignment of guest molecules in the interlayer region of the host. The manganese derivative shows spontaneous magnetization below 40 K.

Pillaring is an intercalation reaction that has been successfully used for the synthesis of novel materials. Pillaring refers to the intercalation of robust thermally stable molecular species into layered hosts to prop the layers apart so as to convert the two-dimensional interlayer space into micropores of molecular dimension, somewhat similar to the pores present in zeolites. Smectite clays, layered zirconium (IV) phosphates, MoO_3, TaS_2 and layered double hydroxides (LDHs) are among the hosts that have been pillared by cationic/anionic species such as alkylammonium ions, polyoxocations (e.g., $Al_{13}O_4(OH)_{24}(H_2O)_{12}^{7+}$) and isopoly- and heteropoly-ions (e.g., $Mo_7O_{24}^{6-}$, $V_{10}O_{28}^{6-}$ and $PV_3W_9O_{40}^{6-}$) (Pinnavaia, 1983; Narita *et al.*, 1993; Wang *et al.*, 1992; Alberti *et al.*, 1993; Nazar *et al.*, 1991; Lerf *et al.*, 1992).

A number of inorganic solids having layered or a three-dimensional network structure exhibit fast cation transport. Sodium β- and β''-aluminas are typical examples. Sodium ions in these solids move rapidly in layers which provide a number of empty sites and easy pathways for ionic motion. Diffusion coefficients are typically of the order of 10^{-7} cm^{2}s^{-1}. Fast cation conductors such as β-aluminas are good ion-exchangers. The exchange can be carried out easily at room temperature both in aqueous and molten salt conditions. Thus, sodium β-alumina has been exchanged with H_3O^+, NH_4^+ and other monovalent and divalent cations, giving rise to different β-aluminas (Farrington & Briant, 1978; Tofield, 1982). Ion-exchange in inorganic solids is a general phenomenon, not restricted to fast ion conductors alone. The kinetic and thermodynamic aspects of ion-exchange in inorganic solids have been examined by England et al. (1983). Their results reveal that ion-exchange is quite a widespread phenomenon, occurring even when the diffusion coefficients are as small as $\sim 10^{-12}$ cm^{2}s^{-1}, at temperatures far below the sintering temperatures of solids. Ion-exchange can occur at a considerable rate even in stoichiometric solids; mobile ion vacancies (introduced by nonstoichiometry or doping) are not required. Since the exchange occurs topochemically, it enables preparation of metastable phases that are inaccessible by high-temperature reactions.

England et al. (1983) have shown that a variety of metal oxides having layered, tunnel or close-packed structures can be ion-exchanged in aqueous solutions or molten salt media to produce new phases. Typical examples are:

$$\alpha\text{-NaCrO}_2 \xrightarrow[300\,^\circ\text{C; 24 h}]{\text{LiNO}_3} \alpha\text{-LiCrO}_2$$

$$\text{KAlO}_2 \xrightarrow{\text{AgNO}_3(\text{l})} \beta\text{-AgAlO}_2$$

$$\alpha\text{-LiFeO}_2 \xrightarrow{\text{CuCl}(\text{l})} \text{CuFeO}_2$$

The structure of the framework is largely retained during the exchange except for minor changes to accommodate the structural preference of the incoming ion. Thus, when α-LiFeO$_2$ is converted to CuFeO$_2$ by exchange with molten CuCl, the structure changes from that of α-NaCrO$_2$ to that of delafossite to provide a linear anion coordination for Cu$^+$. CuCoO$_2$ in the delafossite structure is similarly obtained by the reaction of CuCl with LiCoO$_2$. Similarly when KAlO$_2$ is converted to β-AgAlO$_2$ by ion-exchange, there is a structure change from cristobalite-type to ordered wurtzite-type. The change probably occurs to provide a tetrahedral coordination for Ag$^+$. AgNiO$_2$ has been prepared by the interaction of AgNO$_3$ with LiNiO$_2$ or NaNiO$_2$ (Shin et al., 1993); the material is semimetallic. An interesting ion-exchange reaction reported by Rice & Jackel (1982) is the conversion of LiNbO$_3$ and LiTaO$_3$ to HNbO$_3$ and HTaO$_3$, respectively, by treatment with hot aqueous acid. The exchange of Li$^+$ by protons is accompanied by a topotactic transformation of the rhombohedral LiNbO$_3$ structure to the cubic perovskite structure of HNbO$_3$. The mechanism suggested for the transformation is the reverse of the transformation of cubic ReO$_3$ to rhombohedral LiReO$_3$ and Li$_2$ReO$_3$ (Cava et al., 1982), involving a twisting of the octahedra along the [111] cubic direction so as to convert the 12-coordinated perovskite tunnel sites to two

6-coordinated sites in the rhombohedral structure. Another interesting structural change accompanied by ion-exchange is reported in $Na_{0.7}CoO_2$ by Delmas *et al.* (1982). In $Na_{0.7}CoO_2$, the anion layer sequence is ABBAA; cobalt ions occur in alternate interlayer octahedral sites and sodium ions in trigonal prismatic coordination in between the CoO_2 units. When this material is ion-exchanged with LiCl, a metastable form of $LiCoO_2$ with the layer sequence ABCBA is obtained. The phase transforms irreversibly to the stable $LiCoO_2$ (ABCABC) around 520 K.

A variety of inorganic solids has been exchanged with protons to give new phases, some of which exhibit high protonic conduction; typical of them are $HTaWO_6 \cdot H_2O$ (Groult *et al.*, 1982), $HMO_3 \cdot xH_2O$ (M = Sb, Nb, Ta) pyrochlores (Chowdhry *et al.*, 1982) and $HTiNbO_5$ layer phase (Rebbah *et al.*, 1982). Ion-exchange has also been reported in metal sulphides. For example, $KFeS_2$ undergoes topochemical exchange of potassium in aqueous solutions with alkaline-earth metal cations to give new phases in which the $[FeS_{4/2}]_\infty$ tetrahedral chain is preserved (Boller, 1978).

3.2.3   *High-pressure methods*

The use of high pressure for solid state synthesis has acquired special significance after the synthesis of diamond by the General Electric Co. in 1955. Early investigations in this area were concerned with pressure–temperature phase diagrams of elements like Si, Ge and Bi. With the development of high-pressure technology, commercial equipment permitting simultaneous use of both high-pressure and high-temperature conditions has become available since the 1960s. Reviews presenting the experimental aspects and research progress in this area are available (Goodenough *et al.*, 1972; Joubert & Chenavas, 1975; Pistorius, 1976).

Experimental facilities required for high-pressure synthesis may be divided into two major categories depending on the range of pressures involved. For the 1–10 kbar pressure range, the *hydrothermal method* is often employed. In this method, the reaction is carried out either in an open or a closed system. In the open system, the solid is in direct contact with the reacting gases (F_2, O_2 or N_2) which also serve as pressure-intensifiers. Normally, a gold container is used in this type of synthesis. The method has been used for the synthesis of transition-metal compounds such as RhO_2, PtO_2 and Na_2NiF_6 where the transition metal is in a high oxidation state. Hydrothermal high-pressure synthesis under closed system conditions has also been used for the preparation of higher-valence metal oxides; here, an internal oxidant like $KClO_3$ is added to the reactants, which on decomposition under reaction conditions provides the necessary oxygen pressure. For example, oxide pyrochlores of palladium(IV) and platinum(IV), $Ln_2M_2O_7$, have been prepared by this method (970 K, 3 kbar). $(H_3O)Zr_2(PO_4)_3$ (Subramanian *et al.*, 1984) and a family of zero thermal expansion ceramics (e.g. $Ca_{0.5}Ti_2P_3O_{12}$) (Roy *et al.*, 1984) have also been prepared hydrothermally.

Zeolites are generally prepared under hydrothermal conditions in the presence of alkali (Barrer, 1982). The alkali, the source of silicon and the source of aluminium are mixed in appropriate proportions and heated (often below 370 K). A common reactant mixture is a hydrous gel composed of an alkali (alkali or alkaline-earth metal

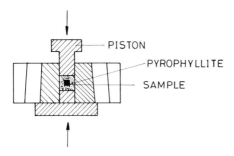

Figure 3.1   Piston-cylinder apparatus.

hydroxide, organic quaternary bases, etc.), fresh $Al(OH)_3$ and silica sol or alkali, soluble aluminate, and silica sol. Under alkaline conditions, Al is present as $Al(OH)_4^-$ anions. The OH^- ions act as mineralizing catalyst while the cations present in the reactant mixture determine the kinds of zeolite formed. Besides water, some inorganic salts are also encapsulated in some zeolites. Zeolitization in the presence of organic bases has been developed and this route is useful for synthesizing silica-rich zeolites. Silicate with a tetrahedral framework enclosing a three-dimensional system of channels (defined by 10 rings wide enough to absorb molecules up to 0.6 nm in diameter) has been synthesized by the reaction of tetrapropylammonium (TPA) hydroxide and a reactive form of silica between 370 and 470 K. The precursor crystals have the composition $(TPA)_2O.48SiO_2 \cdot H_2O$ and the organic cation is removed by chemical reaction or thermal decomposition to yield microporous silicalite which may be considered to be a new polymorph of SiO_2 (Flanigen et al., 1978). The clathrasil (silica analogue of a gas hydrate), dodecasil-1H, is prepared from an aqueous solution of tetramethoxysilane and $N(CH_3)_4OH$; after the addition of aminoadamantane, the solution is treated hydrothermally under nitrogen for four days at 470 K (Groenen et al., 1983).

The hydrothermal method has been employed in recent years to synthesize a variety of solids that include aluminium phosphates (ALPOs) and other microporous transition-metal phosphates and transition-metal polychalcogenides (Davis & Lobo, 1992; Haushalter & Mundi, 1992; Liao & Kanatzidis, 1990, 1992). Unlike zeolites, synthesis of ALPOs requires acidic or mildly basic conditions and no alkali metal cations. A typical synthetic mixture for making ALPO consists of alumina, H_3PO_4, water and an organic material such as a quaternary ammonium salt or an amine. The hydrothermal reaction occurs around 373–573 K. The use of fluoride ions, instead of hydroxide ions as mineralizer, allows synthesis of novel microporous materials under acidic conditions (Estermann et al., 1991; Ferey et al., 1994).

The role of organic additives in hydrothermal synthesis of zeolitic solids has been discussed adequately in the literature (Davis & Lobo, 1992). Besides templating, the organic species can act as space fillers and structure-directing agents. Structure direction implies that a specific structure is synthesized using a particular organic additive. An example of such structure direction is provided by the synthesis of hexagonal faujasite using 1,4,7,10,13,16-hexaoxacyclooctane (18-crown-6) (Arhancet & Davis, 1991). A good example of templating is provided by $C_{18}H_{36}N^+$-triquaternary amine which is specific for the synthesis of ZSM-18 (Lawton & Rohrbaugh, 1990). Tetraalkylammonium salts are common structure-directing agents used in the synthesis of high-silica zeolites. On the basis of the accumulated knowledge on various aspects of hydrothermal synthesis, a rational design of microporous solids is becoming possible. The recent synthesis of mesoporous silicates with adjustable pore sizes of 15 to 100 Å using quaternary ammonium ions, $C_nH_{2n+1}(CH_3)_3N^+$, which form micelles in aqueous solution, is a step in that direction (Kresge et al., 1992). By recognizing a correlation between the minimum framework density and the smallest ring size, Annen et al. (1991) have prepared a zincosilicate (VPI.7) that is topologically related to lovdarite, containing three-membered rings and a low framework density. Chen et al. (1994) have recently prepared an open framework cobalt (II) phosphate, hydrothermally. (See 1.5.1 for synthesis of porous solids).

Hydrothermal methods have enabled the synthesis of solid materials that contain tetrahedra and octahedra or other polyhedra, the variety of phosphates of molybdenum and vanadium synthesized by Haushalter and coworkers being typical examples. An interesting instance is the hydrothermal synthesis of an inorganic double helix, $[(CH_3)_2NH_2]K_4[V_{10}O_{10}(H_2O)_2(OH)_4(PO_4)_7]\cdot 4H_2O$ (Soghomonian et al., 1993). This solid, containing double helices formed by VO_6 octahedra, VO_5 square pyramids and PO_4 tetrahedra, crystallizes in the space group $P4_3$ (or its enantiomorph $P4_1$). The crystals are therefore enantiomorphic and the unit cell contents are chiral.

Hydrothermal and solvothermal (using organic polar solvents) conditions have enabled the synthesis of novel chalcogenides, examples being $K_{12}Mo_{12}Se_{56}$, $K_2Mo_3Se_{18}$ and $K_8Mo_9Se_{40}\cdot 4H_2O$ (Liao & Kanatzidis, 1990, 1992). Besides the synthesis of new solids, hydrothermal methods are employed for processing ceramics and other materials, which involves hydrothermal sintering, oxidation, crystallization, hot pressing and so on (Sōmiya, 1989).

Pressures in the range 10–150 kbar, commonly used for solid state synthesis, are attained with three different kinds of apparatus. (i) In the *piston–cylinder apparatus* (Fig. 3.1), consisting of a tungsten carbide chamber and a piston assembly, the sample is contained in a suitable metal capsule surrounded by a pressure-transducer (pyrophyllite). Pressure is generated by moving the piston through the blind hole in the cylinder. An internal microfurnace made of graphite or molybdenum can be incorporated in the design. Pressures up to 50 kbar and temperatures up to 1800 K are readily reached in a volume of 0.1 cm^3 using this design. (ii) In the *anvil apparatus* (Fig. 3.2), first designed by Bridgman, the sample is subjected to pressure by simply squeezing it between two opposed anvils. Massive support to the anvils is provided by the surrounding rings. Although pressures of ~ 200 kbar and temperatures up to 1300 K can be reached in this technique, it is not popular for solid state synthesis since only milligram quantities can be handled. An extension of the opposed anvil principle is the tetrahedral anvil design, where four massively supported anvils disposed tetrahedrally ram towards the centre where the sample is located in a pyrophyllite medium together with a heating arrangement. The multi-anvil design has been extended to cubic geometry, where six anvils act on the faces of a pyrophyllite cube located at the centre. (iii) The '*belt*' *apparatus* (Fig. 3.3) provides probably the most

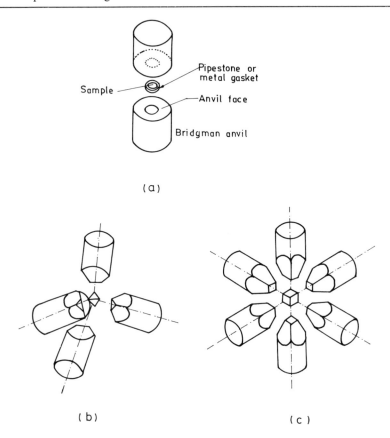

Figure 3.2   Anvil designs: (a) opposed anvil; (b) tetrahedral anvil; and (c) cubic anvil.

successful high pressure–high temperature combination for solid state synthesis. This apparatus, which has been used for the synthesis of diamonds in the US is in a way a combination of the piston–cylinder and the opposed anvil designs; massive support is provided to the anvil as well as the cylinder. The apparatus consists of two conical pistons made of tungsten carbide, which ram through a specially shaped chamber from opposite directions. The chamber and pistons are laterally supported by several steel rings making it possible routinely to reach fairly high pressures ($\sim$ 150 kbar) and high temperatures ($\sim$ 2300 K).

In the 'belt' apparatus, the sample is contained in a noble metal capsule (a BN or MgO container is used for chalcogenides) and surrounded by pyrophyllite and a graphite sleeve, the latter serving as an internal heater. In a typical high-pressure run, the sample is loaded, the pressure raised to the desired value and then the temperature increased. After holding the pressure for about 30 minutes, the sample is quenched (400 K s^{-1}) while still under pressure. The pressure is then released after the sample has cooled to room temperature.

High-pressure research requires suitable calibration and measurement of pressures.

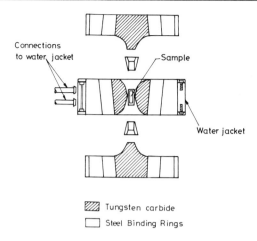

Connections
to water jacket

Sample

Water jacket

Tungsten carbide
Steel Binding Rings

Figure 3.3   The 'belt' apparatus.

Calibration is carried out by making use of standard substances which are known to undergo structure transitions at definite pressures.

Solids generally undergo transformations at high pressures to more close-packed structures, with an increase in coordination number. For example, on application of pressure, Si and Ge transform from a four to a six-coordinated (white tin) structure, iron transforms from an eight- to a twelve-coordinated structure, while NaCl goes to the CsCl structure. The simple rule that high pressures result in lower volume and higher coordination number is based on the hard sphere model, which assumes constant radius and compressibility for atoms in different structures. The only generalization we can make is that pressure-induced transitions proceed in the direction of greater packing efficiency; less efficiently packed structures have a greater probability of undergoing pressure-induced structural changes. Pressure-induced phase transitions of solids may be quenchable or nonquenchable. Many high-pressure forms of solids can be retained under ambient conditions as metastable phases, a typical case being that of diamond (quenchable). In general, transformations belonging to the reconstructive type, involving breaking and making of bonds, can be quenched; transformations involving only a slight movement of atoms are difficult to quench. This is the case with the monoclinic–tetragonal transition of ZrO_2, as also the tetragonal (red PbO type)–wurtzite transition of SnO. A large class of materials having quenchable high-pressure phases is the class of perovskite-related polytypes of ABX_3 (X = F, O) compositions (Goodenough et $al.$, 1972). In these cases, pressure has the effect of decreasing the tolerance factor, t, and hence stabilizing the cubic close-packing of the AX_3 layers (B ion octahedra share corners) as against hexagonal close-packing (B ion octahedra share faces). In transition-metal hydroxides, MOOH, the normal-pressure boehmite structure gives way to diaspore and InOOH-type structures at high pressures.

High-pressure methods have been used for the synthesis of new solids that cannot possibly be made otherwise. In general, the formation of a new compound from its components requires that the new composition has a lower free energy than the sum of the free energies of the components. Pressure can aid in lowering of free energy in different ways (Goodenough et al., 1972): (i) Pressure can delocalize outer d electrons in transition-metal compounds by increasing the magnitude of coupling between the d electrons on neighbouring cations, thereby lowering the free energy. Typical examples are provided by the synthesis of $ACrO_3$ (A = Ca, Sr, Pb) perovskites and CrO_2. (ii) Pressure can stabilize higher-valence states of transition metals, thus promoting the formation of a new phase. For instance, in the Ca–Fe–O system only $CaFeO_{2.5}$ (brownmillerite) is stable under ambient pressures. Under high oxygen pressures, iron is oxidized to the 4 + state and hence $CaFeO_3$ with the perovskite structure is formed. (iii) Pressure can suppress the ferroelectric displacement of cations and this aids the synthesis of new phases. The synthesis of A_xMoO_3 bronzes, for example, requires populating the empty d orbitals centred on molybdenum; at ambient pressures, MoO_3 is stabilized by a ferroelectric distortion of MoO_6 octahedra up to the melting point. (iv) Pressure can alter site-preference energies of cations, thus facilitating the formation of new phases. For example, it is not possible to synthesize $A^{2+}Mn^{4+}O_3$ (A = Mg, Co, Zn) ilmenites because of the strong tetrahedral site preference of the divalent cations and one obtains a mixture of $A[AMn]O_4$(spinel) + MnO_2(rutile) under atmosphere pressure instead of a single phase $AMnO_3$. However, the latter is formed at high pressures with a corundum-type structure in which both A and Mn ions are in octahedral coordination. (v) Pressure can suppress the $6s^2$ core polarization in oxides containing Tl^+, Pb^{2+}, Bi^{3+} isoelectronic cations. For example, a perovskite-type $PbSnO_3$ cannot be made at atmospheric pressure because a mixture of $PbO + SnO_2$ is more stable than a perovskite.

Stabilization of unusual oxidation states and spin-states of transition metals is of considerable interest (e.g. $La_2Pd_2O_7$). Such stabilization can be rationalized by making use of correlations of structural factors with the electronic configuration. Six-coordinated high-spin iron(IV) has been stabilized in $La_{1.5}Sr_{0.5}Li_{0.5}Fe_{0.5}O_4$, which has the K_2NiF_4 structure (Demazeau et al., 1982a). The elongated FeO_6 octahedra and the presence of ionic Li—O bonds resulting from the K_2NiF_4 structure favour the high-spin Fe(IV) state. The Li and Fe ions in this oxide are ordered in the ab-plane as evidenced by supercell spots in the electron diffraction pattern. Such an oxide is prepared under oxidizing conditions. $CaFeO_3$ and $SrFeO_3$ prepared under oxygen pressure also contain octahedral Fe(IV); while Fe(IV) in $SrFeO_3$ is in the high-spin state with the e_g electron in the narrow σ^* band down to 4K, Fe(IV) in $CaFeO_3$ disproportionates to Fe(III) and Fe(V) below 290 K (Takano & Takeda, 1983). La_2LiFeO_6 prepared under high oxygen pressure has the perovskite structure with the iron in the pentavalent state (Demazeau et al., 1982a).

Nickel in the 3 + state is present in the perovskite $LaNiO_3$, which can be prepared at atmospheric pressure; other rare-earth nickelates can, however, be prepared at high oxygen pressures. $La_2Li_{0.5}Ni_{0.5}O_4$ and $MLaNiO_4$(M = Sr or Ba) with the K_2NiF_4 structure are other examples of oxides with Ni(III) (Demazeau et al., 1982b). $MNiO_{3-x}$ (M = Ba or Sr) prepared under high pressure contain Ni(IV) as well (Takeda et al., 1976). In $La_2Li_{0.5}Co_{0.5}O_4$, there is evidence for the transformation of the low-spin

Co(III) to the intermediate- as well as high-spin states. The Li and Co ions are ordered in the ab-plane of this oxide of K_2NiF_4 structure; the highly elongated CoO_6 octahedra seem to stabilize the intermediate-spin state (Mohanram $et\ al.$, 1983). Intermediate-spin Co ions are reported to be formed in Sr_4TaCoO_8 and Sr_4NbCoO_8 as well. Oxides in perovskite and K_2NiF_4 structures with trivalent Cu have been prepared under high oxygen pressure (Demazeau $et\ al.$, 1982b). High F_2 pressure has been employed to prepare Cs_2NiF_6 and other fluorides (Hagenmuller, 1985).

As noted earlier, solid state reactions are generally slow under ordinary pressures even when the product is thermodynamically stable. Pressure has a marked effect on the kinetics of the reaction, reducing the reaction times considerably and at the same time resulting in more homogeneous and crystalline products. For instance, $LnFeO_3$, $LnRhO_3$ and $LnNiO_3$ (Ln = rare earth) are prepared in a matter of hours at high pressure – high temperature conditions, whereas under ambient pressures the reactions require several days ($LnFeO_3$ and $LnRhO_3$) or they do not occur at all ($LnNiO_3$). In several $(AX)(ABX_3)_n$ series of compounds, the end members ABX_3 and A_2BX_4, having the perovskite and K_2NiF_4 structures respectively, are formed at atmospheric pressures but not the intermediate phases such as $A_3B_2X_7$ and $A_4B_3X_{10}$. Pressure facilitates synthesis; for instance, $Sr_3Ru_2O_7$ and $Sr_4Ru_3O_{10}$ are formed in 15 min at 20 kbar and 1300 K. TaS_3, $NbSe_3$ and such solids are prepared in 30 min at 2 GPa and 970 K.

3.2.4   Arc techniques

Use of the electric arc for the preparation of materials as well as growth of crystals of refractory solids has been reviewed by Loehman $et\ al.$ (1969) and Müller-Buschbaum (1981). Basically, an arc for synthetic purposes is produced by passing a high current from a tungsten cathode to a crucible anode which acts as the container for the material to be synthesized (Fig. 3.4). The cathode tip is ground to a point in order to sustain a high current density. Typical operating conditions involve currents of the order of 70 amp at 15 volts. The arc is maintained in inert (He, Ar, N_2) or reducing (H_2)

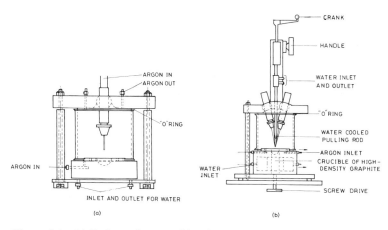

Figure 3.4   (a) D.C. arc furnace; (b) Triarc.

atmospheres. Even traces of oxygen attack the tungsten electrode and therefore the gases are freed from oxygen (by gettering with heated titanium sponge) before passing into the arc chamber. The arc can be maintained in an oxygen atmosphere using graphite electrodes instead of tungsten. The crucible (anode) is made from a cylindrical copper block and is water-cooled during operation.

For the synthesis of materials, the reactants are placed in the copper crucible. An arc is struck by allowing the cathode to touch the anode. The current is raised slowly while the cathode is simultaneously withdrawn so as to maintain the arc. The arc is then positioned so that it bathes the sample in the crucible. The current is increased until the reactants melt. When the arc is turned off, the product solidifies in the form of a button. Because of the enormous temperature gradient between the melt and the water-cooled crucible, a thin solid layer of the sample usually separates the melt from the copper hearth; in this sense, the sample forms its own crucible and hence contamination with copper does not take place. Contamination of the sample by tungsten vaporizing from the cathode can be avoided by using water-cooled cathodes. The arc method has been successfully used for the synthesis of various oxides of Ti, V and Nb. A number of lower-valence rare-earth oxides, $LnO_{1.5-x}$, have been prepared by arc fusion of Ln_2O_3 with Ln metal.

A sophisticated version of the arc method is the *triarc technique* (Fig. 3.4) developed by Fan & Reed (1972) for growing single crystals of high-melting solids. The main differences between the single and triarc furnace lie in the use of three symmetrically arranged thorated tungsten cathodes and a graphite crucible as anode, together with a water-cooled pulling rod. For crystal growth using the triarc furnace, the polycrystalline sample is loaded in the crucible, all the three arcs are struck simultaneously and the sample is brought to just-melting condition. The pulling rod is then lowered down until the tip just touches the melt. By slowly pulling the rod up, a 'neck' is formed between the frozen solid and the melt in the crucible. This is equivalent to using a single seed crystal in the conventional pulling technique for crystal growth. Careful driving of the pulling rod while maintaining the triarc produces a single crystal boule. The method has been used to grow single crystals of Ti_3O_5, Ti_2O_3 and NbO.

A further modification of the arc method is the *arc transport (transfer) technique* in which the material to be grown as a single crystal is transported by an electric arc. A vertical arc between the cathode (filled with the melt) and the anode carrying the seed crystal is kept at constant length by progressively raising the anode. The fused material is transported across the arc from cathode to anode where it grows on the seed crystal. NiO crystals have been prepared by this technique by using Ni electrodes (Honig & Rao, 1981).

3.2.5   *Skull melting*

This technique is useful not only for preparing metal oxides, but also for growing crystals of these oxides. The technique involves coupling of the material to a radio frequency electromagnetic field (200 kHz–4 MHz, 20–50 kW) (Harrison *et al.*, 1980). The material is placed in a container consisting of a set of water-cooled cold fingers set in a water-cooled base (all made of copper), the space between the fingers being large enough to permit penetration of the electromagnetic field into the interior, but small enough to avoid leakage of the melt. The process is crucibleless and a thin solid skull separates the melt from the water-cooled container. Single crystals of very large sizes can be grown by this method and the mass of the starting materials can be up to 1 kg. Temperatures up to 3600 K are reached in this technique, permitting growth of crystals of materials like ThO_2 and stabilized ZrO_2. The stoichiometry of the oxide is readily controlled by the use of an appropriate ambient gas (CO/CO_2 mixtures, air or oxygen). Large crystals of CoO, MnO and Fe_3O_4 have been grown by the skull method and the technique will undoubtedly find many applications in the future. In CoO and MnO, trivalent metal ions were eliminated by heating the crystals in an appropriate CO/CO_2 mixture. Recently, crystals of La_2NiO_4 and Nd_2NiO_4 have been grown by the skull method.

3.2.6   Chemical vapour deposition (CVD)

Deposition of a solid on a substrate by means of a chemical reaction between vapours is called chemical vapour deposition (CVD). The process involves (i) formation of reactants in the vapour state, (ii) transport of the vapour to the deposition region and (iii) deposition of the solid from the vapour. CVD has become an important technique in solid state synthesis and finds application in the synthesis of a broad spectrum of solids ranging from amorphous and polycrystalline deposits to epitaxial growth of single-crystal films used in devices (Tietjen, 1973). As a synthetic technique, CVD has several advantages. The method permits preparation of virtually any material in almost any geometry. Since product formation is not limited by kinetic factors and diffusion, it is possible to synthesize solids at relatively low temperatures; homogeneity and stoichiometry are also easily controlled. An additional advantage is that incorporation of selected dopants in controlled concentration during the synthesis is achieved.

CVD is widely used for the preparation of silicon and III–V semiconductors for device purposes. There are two different CVD processes for the preparation of silicon: (i) hydrogen reduction of $SiCl_4$ and (ii) pyrolysis of SiH_4. The deposition is carried out either homoepitaxially on silicon or on an insulating substrate such as sapphire. Several methods have been developed for the synthesis of III–V compounds. For example, GaAs is made by CVD using $AsCl_3$ and Ga vapour in a hydrogen atmosphere; here, $AsCl_3$ acts not only as the source of As but as the transporting agent for Ga as well. Another method for the synthesis of this class of compounds makes use of the reaction between a volatile hydride of a group V element with a group III element in the presence of HCl gas:

$$AsH_3 + Ga + HCl \rightarrow GaAs + \tfrac{3}{2}H_2 + HCl$$

A third method makes use of hydrides of group V elements together with volatile organometallic compounds of group III elements such as trimethylgallium and trimethylaluminium.

Insulating materials such as SiO_2, Si_3N_4 and Al_2O_3 have also been made by CVD methods. Si_3N_4, a high-temperature ceramic, is formed by the reaction of NH_3 with

SiH_4 or $SiCl_4$. The reaction using SiH_4 has the advantage that it occurs at lower temperatures. CVD has been particularly useful in the synthesis of superconducting and magnetic materials. One such material Nb_3Si is an important A15 superconductor. Preparation of this phase by the reaction of the elements at elevated temperatures yields only a nonsuperconducting phase which has the Ti_3P structure. It is only by deposition from vapours below 1120 K that the desired phase with the A15 structure is obtained. Rare-earth garnets, useful as bubble-domain magnetic materials, have been prepared as single crystals by deposition from volatile metal halides.

3.3   Organic solids

Synthesis is the main preoccupation of organic chemists and one cannot possibly deal with the varied aspects of organic synthesis here. Synthesis of organic solids exhibiting properties of interest to the solid state scientists however involves certain procedures which are noteworthy and we shall discuss a few typical systems of current interest. TTF–TCNQ and such donor–acceptor complexes and polyacetylene and other conducting organic polymers are areas of active study in solid state science today (Hatfield, 1979; Grossel & Weston, 1994). Synthetic methods for the preparation of TTF, TCNQ and related donor and acceptor molecules are described in the literature (Narita & Pittman, 1976; Wheland & Martin, 1975; Grossel & Weston, 1994). The charge-transfer salts are usually prepared by mixing hot acetonitrile solutions of the donor and the acceptor. What is important for solid state chemistry is not the synthesis *per se* of these organic compounds, but the design of donors and acceptors so that the charge-transfer solids possess the desired structure and properties. Certain guidelines based on molecular stacking, stoichiometry and degree of charge-transfer have been evolved for the design of donor–acceptor complexes exhibiting metallic properties (Torrance, 1979). Thus, 1:1 charge-transfer salts possessing uniform segregated stacking and incomplete charge-transfer show metallic properties. There have been attempts to link the three factors to the molecular properties of donors and acceptors. Segregated stacking seems to be favoured when the donor–acceptor overlap is particularly weak. The degree of charge-transfer is chiefly determined by the ionization energy of the donor (reduction potential of the cation). A plot of the room-temperature conductivity against the reduction potential of cation for the TCNQ family of donor–acceptor solids reveals that only in the narrow region of the reduction potential close to zero are the solids metallic. When the potential is too negative, complete charge transfer occurs giving ionic salts and when the potential is positive, neutral complexes are obtained (Torrance, 1979). Factors that determine the stoichiometry of donor– acceptor complexes are not definitely known. In the family of TTF-halide salts, the observed nonstoichiometries have been accounted for in terms of Madelung energy, showing that ionization potential of the donor and electron affinity of the acceptor are important factors determining stoichiometry. Synthesis of organic materials for molecular electronic applications has been recently reviewed by Grossel & Weston (1994) who discuss TTF and related compounds, TCNQ and related compounds as well as other aspects in detail.

Attempts have also been made to design one-dimensional organic superconductors based on donor–acceptor interaction (Bechgaard & Jerome, 1982). For this purpose it is necessary to avoid Peierls distortion, which results in the loss of metallic property (see Section 6.8). This is accomplished by preparing charge-transfer complexes of selenium analogues of TTF such as TMTSF(Tetramethyltetraselenafulvalene). Charge-transfer solids of the type $TMTSF_2X(X = ClO_4, ReO_4, PF_6$ and AsF_6) indeed show superconductivity, the perchlorate salt showing the superconducting transition at 1.2 K and atmospheric pressure and the others at high pressures.

Polyacetylene,$(CH)_x$, is the prototype of conducting organic polymers (Etemad et al., 1982) whose electrical conductivity can be varied by doping over twelve orders of magnitude from that of an insulator ($< 10^{-9}$ ohm^{-1} cm^{-1}) through semiconductor to a metal ($> 10^3$ ohm^{-1} cm^{-1}). Polyacetylene can be prepared by admitting acetylene gas into a glass reactor vessel whose inside wall has been wetted with a methylbenzene solution of $(C_2H_5)_3Al$ and $(n-C_4H_9O)_4Ti$ Ziegler catalyst (Shirakawa & Ikeda, 1971; Shirakawa et al., 1973). A cohesive film of polyacetylene grows on all surfaces which have been wetted by the catalyst solution. Polymerization at room temperature produces a mixture of 80% cis and 20% trans isomers. When the temperature is around 350 K, nearly 100% cis-$(CH)_x$ is produced; at 420 K (decane solvent) the product is 100% trans. The cis–trans isomerization occurs at $\sim$ 470 K. As formed $(CH)_x$ films consist of randomly oriented fibrils, each fibril being a few hundred angstroms in diameter. The room-temperature conductivity of $(CH)_x$ depends upon the cis and trans content, the value varying between 10^{-5} ohm^{-1} cm^{-1} for the trans polymer ($E_g \sim 0.8$ eV) and 10^{-9} ohm^{-1} cm^{-1} for the cis polymer ($E_g \sim 1.0$ eV). As a result of the large overall band-width arising from the conjugated π-electrons, $(CH)_x$ is different from the common organic semiconductors made up of weakly interacting molecules (e.g. anthracene) or from saturated polymers with no π-electrons (e.g. polyethylene).

Like inorganic semiconductors, polyacetylene can be doped with a variety of donors and acceptors to give n-type or p-type semiconductors. Doping to high levels (above $\sim 1\%$) results in a semiconductor–metal transition giving a new class of metals with a wide range of electronegativity (MacDiarmid & Heeger, 1979; Mammone & MacDiarmid, 1984). Doping can be done chemically or electrochemically. Chemical doping can be done by exposure of the $(CH)_x$ film to a known vapour pressure of a volatile dopant, e.g. I_2, AsF_5 etc., until the desired conductivity is obtained. Removal of the dopant vapour essentially 'freezes' the conductivity at that stage. Doping can also be done by treatment of the polymer film with a solution of the dopant in an appropriate solvent (e.g. I_2 in pentane, sodium napthalide in THF) or by treatment with molten Na or K. Polyacetylene can act as an electron-source or as an electron-sink according to whether it is oxidized (doped p-type) or reduced (doped n-type). $(CH)_x$ begins to lose electrons at an applied potential of + 3.10 volts and gains electrons at an applied potential of + 1.75 volts versus a Li/Li^+ electrode. Electrochemical doping can be carried out by using strips of $(CH)_x$ as anode and cathode. For example, when the electrodes, immersed in a solution of $LiClO_4$ in propylene carbonate, are connected to a d.c. source (4V), oxidation of $(CH)_x$ occurs at the anode giving $[(CH)^{y+}(ClO_4^-)_y]_x$ and reduction occurs at the cathode giving $[Li_y^+(CH)^{y-}]_x$. The possibility of reversible redox reactions of polyacetylene in an electrochemical cell forms the basis of light weight, high power-density, rechargeable storage batteries (Nigrey et al., 1981).

A number of polyheterocycles have been prepared. For instance, oxidative electrochemical polymerization of pyrrole produces polypyrrole, invariably containing dopant anions from the electrolyte. A polypyrrole film containing AsF_6, of about $10\,\mu m$ thickness, is produced when a mixture of 0.2 m pyrrole and 0.004 M tetra-*n*-butylammonium hexafluoroarsenate in dimethyl sulphate is electrolysed using indium–tin oxide anode and platinum cathode at a current density of $2\,mA\,cm^{-2}$ for 20 minutes. The polymer grows on the indium–tin oxide surface (Hotta *et al.*, 1983). The doped polymers so obtained can be converted into neutral polymers by immediate reversal of the polarity after electrochemical polymerization. Polythienylene and poly(3-methylthienylene) are prepared by similar methods. Organometallic polymers possessing high electrical conductivity have also been prepared (Sheats *et al.*, 1984).

3.4   Small particles and clusters (nanomaterials)

There is a distinct region of small aggregates or clusters which falls between the atomic (or molecular) domain and that of condensed matter. These small particles and clusters possess unique properties and have several technological applications. The formation of these particles involves a vapour–solid, a liquid–solid, a solid–solid or a vapour–liquid–solid type of phase change governed by nucleation and it is important that the size of the growing nucleus is controlled (Multani, 1981; Hadjipanyas & Siegel, 1994).

Microclusters of metals up to 500 atomic mass units (AMU) have been obtained by inert-gas vaporization–condensation; the metal vapour effuses from a hot cell through a narrow orifice. The mean particle size by this technique varies in the range 5–200 Å, the microclusters containing anywhere between 1 and 500 atoms (e.g. Sb_1–Sb_{500}; Bi_1–Bi_{280}). Mass spectroscopy has been used to determine sizes of such clusters. Granqvist and Buhrman (1976) have described an apparatus to synthesize particles in the range 30–60 Å diameter. The arc-plasma method has been employed to prepare microclusters of ceramic materials such as tungsten carbide (50–80 Å) and silicon carbide (100–200 Å). Biphasic ceramic–metal composites of Al_2O_3, SiO_2 or ZrO_2 with Cu, Pt, Sn or Ni have been prepared by a sol–gel process (Roy & Roy, 1984). The xerogel consists of a microcrystalline or a noncrystalline ceramic matrix in which the metallic component is dispersed as small islands.

Microcrystals made up of relatively larger particles (still in the submicron range) are useful in applications involving solid state diffusion (or reaction) and sintering. These are synthesized by any of the following techniques: spray–dry, freeze–dry, sol–gel, pyrogel (high-temperature spray) or liquid drying. Thus, the freeze–dry technique has been used to prepare α- and γ-Fe_2O_3 (150–200 Å) while the sol–gel technique has been used to prepare $Y_3Fe_5O_{12}$ and $PbZr_{0.52}Ti_{0.48}O_3$ (PZT). In all these techniques, one starts with aqueous solutions or mixed sols of the appropriate stoichiometry; the solutions are atomized and heat or mass transfer is achieved in milliseconds. The temperature of the sink (source) is around 300 K for sol-gelling (in a bath of 2-ethylhexanol and liquid drying in a bath of acetone), at 78 K for freeze-drying, around 600 K for spray-drying and around 1200 K for the pyrogel technique. The preparations enable homogeneity at molecular level, small grain size, low porosity, large surface area and theoretical density of sintered compacts.

Transmission and scanning electron microscopy are employed for a direct study of microclusters while the distribution of sizes (or average diameter) is provided by sedimentation and other techniques. The average particle diameter is obtained by the Brunauer–Emmett–Teller (BET) surface-area method and by X-ray line broadening.

Microcrystals exhibit properties distinctly different from those of bulk solids. The fractional change in lattice spacing has been found to increase with decreasing particle size in Fe_2O_3. Magnetic hyperfine fields in α-Fe_2O_3 and Fe_3O_4 are lower in the microcrystalline phase compared to those of the bulk crystalline phases. The tetragonality (i.e. the departure of the axial ratio from unity) of ferroelectric $BaTiO_3$ decreases with decrease in particle size; in PZT, the low-frequency dielectric constant decreases and the Curie temperature increases with decreasing particle size. The small particle size in microcrystals cannot apparently sustain low-frequency lattice vibrations.

Particles of metals, semiconductors or ceramics having diameters in the range 1–50 nm constitute nanoclusters. Physical properties of such clusters correspond neither to those of the free atoms or molecules (making up the particle) nor to those of bulk solids of the same composition. The clusters are characterized by a large surface-to-volume ratio which implies that a large fraction of the atoms reside at the grain boundary. Metastable structures can therefore be generated in nanoclusters and these are different from those of the bulk. The science of nanoscale clusters has attracted considerable attention in recent years (Andres et al., 1989; Jena et al., 1986, 1992; Hadjipanyas & Siegel, 1994). Several techniques have been employed to prepare various types of clusters and to investigate their properties. Most of the investigations have been carried out on clusters prepared by gas-phase condensation which provides a substrate-free configuration for the nanoscale particles. Clusters grown in liquid and solid media have also been investigated.

The various methods of preparation employed to prepare nanoscale clusters include evaporation in inert-gas atmosphere, laser pyrolysis, sputtering techniques, mechanical grinding, plasma techniques and chemical methods (Hadjipanyas & Siegel, 1994). In Table 3.5, we list typical materials prepared by inert-gas evaporation, sputtering and chemical methods. Nanoparticles of oxide materials can be prepared by the oxidation of fine metal particles, by spray techniques, by precipitation methods (involving the adjustment of reaction conditions, pH etc) or by the sol–gel method. Nanomaterials based on carbon nanotubes (see Chapter 1) have been prepared. For example, nanorods of metal carbides can be made by the reaction of volatile oxides or halides with the nanotubes (Dai et al., 1995).

Nanophase materials are prepared by compacting the nanosized clusters generally under high vacuum. Synthesis of such nanomaterials has been reported in a few systems. The average grain sizes in these materials range from 5 to 25 nm.

Chemical methods of preparing nanoparticles offer many possibilities – the case of metal particles being illustrative. Microemulsions and micelles can be employed as the media to produce small particles of sulfides and oxides of 1.5–10 nm diameter with narrow size distribution (e.g. CdS). The sol–gel technique also gives small particulates of many oxidic materials. Recently, homogeneous nanoparticles of ZnO and of the

Table 3.5. *Typical nanoclusters prepared by physical and chemical methods*[a]

Material	Cluster size (nm)	Method of characterization
(i) *By inert-gas evaporation*		
CdS	20–250	Optical
CuCl	17–60	Optical
Fe	5–40	Mössbauer
Er	10–70	ESR
Au, Pd	1–10	Microscopy (HREM, STM) Tunnelling conductance
(ii) *By sputtering*		
Al–Cu	10–100	Microstructure
CdS	1.5–2.6	Optical
GaAs	1.6–3.2	Optical
$Co_{0.4}Fe_{0.4}B_{0.2}$	5	Electrical
Ni	1–10	Magnetic
Au, Pd	1–10	Microscopy (HREM, STM)
(iii) *By chemical methods*		
$MnFe_2O_4$ (by coprecipitation in alkali medium)	5–35	Magnetic, Mössbauer, EXAFS
HgSe (Colloidal preparation)	2–10	Optical
PbI_2 (Colloidal preparation)	1.2–3.6	Optical
Ag (Colloidal preparation)	4	Microstructure
ZrO_2, TiO_2 (Gel precipitation, sol-gel)	5–20	Structure
Al_2O_3–ZrO_2 (Chemical polymerization and precipitation)	20–80	Structure
$PbTiO_3$ (Sol-gel)	12	Ferroelectricity
Nd-Fe-Zr-B (Glass ceramics)	20–40	Microstructure
Fe-Co (Pyrolysis of organometallics	10–50	Magnetic

[a]Clusters of carbon, metals and metal carbides (e.g. M_8C_{12}, M = V, Zr, Hf or Ti) etc. are prepared by laser ablation.

oxalate precursor of $YBa_2Cu_3O_7$ have been prepared by coprecipitation in aqueous cores of water-in-oil microemulsions (Hingorani *et al.*, 1993; Kumar *et al.*, 1993). Nanoparticles of oxide materials are also obtained by nebulized spray pyrolysis (Vallet-Regi *et al.*, 1993).

Metal particles. Chemical preparation of metal colloids was initiated by Michael Faraday long ago. A number of procedures have been employed for the preparation of metal sols. In general, the preparation involves the treatment of a metal salt solution with a suitable reducing agent (e.g., $NaBH_4$, hydroxylamine) in the presence of a protective agent; the latter is to prevent the coagulation of the colloidal particles. The undesired end-products are sometimes removed by dialysis. The size distribution of the particles depends on the method employed for the preparation. Large metal particles (micrometer dimensions) have also been prepared by chemical methods involving the reduction of metal salts (Fievet et al., 1989; Matijevic, 1991). In the polyol process, the metal salts are reduced by glycerol, ethylene glycol etc. Monodispersed metal particles of different shapes have been obtained in this manner. Sols containing small gold particles (~ 6 nm) have been synthesized employing phosphorus in ether as a reducing agent. Duff et al. (1992) have recently suggested a much safer method involving the reduction of $HAuCl_4$ by using tetrakis-dihydroxymethyl-phosphonium chloride in an alkaline medium. This procedure yields particles in the range of ~ 1 nm.

A variety of reducing agents have been used in the preparation of platinum sols. The citrate reduction of chloroplatinic acid gives Pt particles as small as 2 nm while dimethylamine borane produces particles with a wide size distribution (5–20 nm). Small particles (< 3 nm) can also be obtained using sodium borohydride as the reducing agent in the presence of polyvinyl pyrrolidone (Van Rheenen et al., 1987). In the absence of such a protective agent, the reaction mixture yields strand-like aggregates composed of ~ 6 nm diameter links. Synthesis using hydroxylamine hydrochloride in the presence of PVP at 277K gives unsintered Pt particles of a mean diameter of ~ 2 nm. The reducing power of the solution has been increased by adding glucose.

A more general procedure has been suggested by Tsai & Dye (1993) for the preparation of sols of a number of metals. Alkali metals solubilized in aprotic solvents such as dimethylether or tetrahydrofuran along with a suitable cation complexant (a crown ether or a cryptand) has been employed by these workers as an attractive reducing medium. Homogeneous reduction of various metal salts with alkalides or electrides in dimethyl ether or THF at 223K or below produces 2–15 nm diameter particles of metals and alloys. The reactions are rapid, complete and applicable to a wide range of elements from Ti to Te. Synthesis involving highly reactive metals (3d metals) requires an inert atmosphere. Organic as well as inorganic precursors can be effectively used to synthesize mono- and bimetallic particles. Nanoscale Ag–Pd and Cu–Pd alloys have been prepared recently (Vasan & Rao, 1995).

3.5   Amorphous materials

It is only in recent years that there has been the realization that most materials can be rendered amorphous. Referring specifically to glasses, it may be said that most materials, if cooled fast enough (and far enough) from the liquid state, can be made in the form of glasses. The various techniques available for the preparation of glasses (Angell, 1981, Elliott et al., 1985) are cooling of supercooled liquid phases, vapour deposition, shock disordering, radiation disordering, desolvation and gelation (Fig. 3.5). All these techniques with the exception of the first produce materials which cannot easily be characterized, especially with regard to their entropies relative to those of the corresponding crystals.

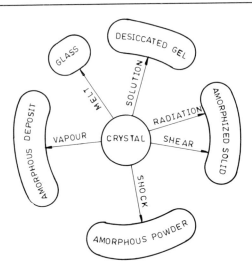

Figure 3.5   Various routes for preparing amorphous solids.

A variety of glasses, possessing different types of bonding, have been prepared. In Table 3.6, we list a few of them along with their glass transition temperatures to illustrate the universality of glass formation. The rate of cooling has to be very fast to prepare glasses from metallic materials; the temperature drop required is around 1000 degrees in a millisecond ($\sim 10^6 \, K \, s^{-1}$). Melt-spinning, where the metallic glass is spun off a copper rotor as a ribbon at a rate exceeding a kilometre per minute is employed for the purpose. Most other glasses prepared by quenching melts require slower rates; SiO_2 requires a cooling rate of hardly $0.1 \, K \, s^{-1}$ and this is attained by allowing the melt to cool freely. *Splat-quenching* techniques with a range of cooling rates (10^5–$10^8 \, K \, s^{-1}$) have also been employed to prepare metallic glasses. Of special interest is the technique of *laser-glazing* where the surface of a crystalline material is made glassy by exposing it to a moving focussed laser beam so that the subjacent solid acts as heat sink; the quenching rate in this technique has been estimated at $10^{11} \, K \, s^{-1}$.

The vapour deposition method is widely used to obtain amorphous solids. In this technique, atoms, molecules or ions of the substance (in dilute vapour phase) are deposited on to a substrate maintained at a low temperature. Most vapour-deposited amorphous materials crystallize on heating, but some of them exhibit an intervening second-order transition (akin to the glass transition). Amorphous solid water and methanol show such transitions. The structural features of vapour-deposited amorphous solids are comparable to those of glasses of the same materials prepared by melt-quenching.

Very low density amorphous SiO_2 prepared by the sol–gel technique (Vacher *et al.*, 1988) exhibits low thermal conductivity and fracton–phonon crossover in the vibrational density of states. Novel glasses with low T_g values have been prepared by making use of organic hosts with other organic and inorganic materials (Blair *et al.*, 1992). Mixed metal carboxylates appear to be good hosts for making such glasses.

Table 3.6. *Typical glassy amorphous solids*

Substance	Nature of bonding	Glass transition, K
SiO_2	Covalent	1450
GeO_2	Covalent	820
As_2S_3	Covalent	470
Se	Covalent (Polymeric)	320
Polyvinylacetylene	Covalent (Polymer)	310
BeF_2	Ionic	570
$75\%AgI-25\%Ag_2SeO_4$	Ionic	75
$Au_{0.8}Si_{0.1}Ge_{0.1}$	Metallic	295
Propylene carbonate	van der Waals	160
Isopentane	van der Waals	65
H_2O	Hydrogen bonded	139(?)
C_2H_5OH	Hydrogen bonded	90
$Fe_{80}B_{20}$	Metallic	crystallizes

Disordering of crystalline materials by subjecting them to shock (e.g. as in the case of quartz and other minerals) or by large dosages of radiation (neutrons or alpha particles) also produces the amorphous state. The *metamict* form (Pabst, 1952) produced by irradiation is unstable with respect to the vitreous form. Substances such as magnesium acetate are rendered glassy by a desolvation process. In the gelation method, the material is first produced in gel form by hydrolysis of an organic derivative (e.g. ethyl silicate); the gel is then dried and collapsed into the glassy state.

Solid state reactions may also produce amorphous solids starting from crystalline solids. For instance, Yeh *et al.* (1983) have found that absorption of hydrogen by crystalline Zr_3Rh transforms it to a hydrided amorphous material.

3.6   Crystal growth

Crystals are essential both for fundamental studies of solids and for fabrication of devices. The ideal requirements are large size, high purity and maximum perfection (freedom from defects). It may also be necessary to incorporate selective impurities (dopants) during growth in order to achieve required electronic properties. A number of methods are available for growing crystals (Table 3.7) and the subject has been reviewed extensively in the literature (Laudise, 1970; Banks & Wold, 1974; Mroczkowski, 1980; Honig & Rao, 1981).

The most common methods of growing crystals involve solidification from the melt (in the case of one-component systems) or crystallization from solution. Some of the methods for growing crystals from melt are described schematically in Fig. 3.6. In the *Czochralski method*, commonly known as the pulling technique, the material is melted by induction or resistance heating in a suitable nonreactive crucible. The melt temperature is adjusted to slightly above the melting point and a seed crystal is dipped into the melt. After thermal equilibration is attained, the seed is slowly lifted from the

Table 3.7. *Methods of crystal growth*

I. *Growth in one-component systems*
    (a) *Solid–solid*
        Strain-annealing, devitrification or polymorphic phase change
    (b) *Liquid–solid*
        1. Directional solidification (Bridgman–Stockbarger)
        2. Cooled seed (Kyropoulos)
        3. Pulling (Czochralski and also triarc)
        4. Zoning (Horizontal zone, vertical zone)
        5. Flame fusion (Verneuil)
        6. Slow cooling in skull melter
    (c) *Gas–solid*
        Sublimation–condensation or sputtering
II. *Growth involving more than one component*
    (a) *Solid–solid*
        Precipitation from solid solution
    (b) *Liquid–solid*
        1. Growth from solution (evaporation, slow cooling and temperature
           differential)
           (i) Aqueous solvent; (ii) Organic solvents; (iii) Molten solvents; (iv)
           Hydrothermal
        2. Growth by reaction
           (i) Chemical reaction; (ii) Electrochemical reaction
        3. Growth from melt; e.g. congruently melting intermetallic compounds
    (c) *Gas–solid*
        1. Growth by reversible reaction (chemical vapour transport)
        2. Growth by irreversible reaction (epitaxial process).

melt. As the seed is pulled, continuous growth occurs at the interface. The diameter of the growing crystal can be controlled by adjusting the rate of pulling, the rate of melt-level drop and the heat fluxes into and out of the system. The technique has the advantage that the growth interface does not come into contact with the walls of the crucible and hence formation of unwanted nuclei is avoided. The method has been used for the growth of silicon, germanium, III–V semiconductors, ceramic oxides like Al_2O_3, rare-earth perovskites such as $LnAlO_3$, $LnFeO_3$, garnets, scheelites, etc. Oxide materials can be grown in air while others require closed systems and atmosphere control. In the triarc method (described earlier in Section 3.2.4) the crystal is pulled from the melt. The *Kyropoulos method* is similar to the Czochralski method, but instead of pulling the seed crystal, the crystal–liquid interface moves into the melt as crystallization proceeds. In the *Bridgman–Stockbarger method*, a sharp temperature gradient is provided across the melt which results in nucleation in the colder region. The conical geometry of the crucible at the bottom limits the number of nuclei formed.

Another method for growing crystals from melt is the *floating zone technique* in which a section of the starting material, held vertically in the form of a rod, is melted by

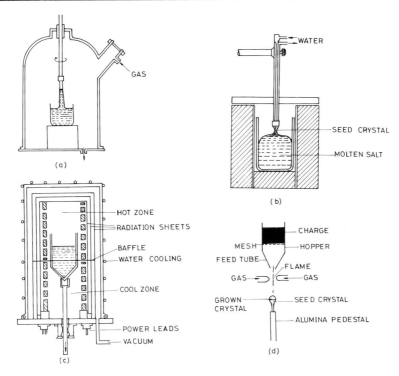

Figure 3.6   Methods for crystal growth from melt; (a) Czochralski method; (b) Kyropoulos method; (c) Bridgman–Stockbarger method and (d) Verneuil method.

suitable heating. As the molten zone is moved along the rod, progressive melting of the sample at one end of the zone and crystal growth at the other end occur. If a seed crystal is provided at one end, the whole rod can be converted into a single crystal. The method has the advantage that there is no contamination from the crucible. The method is similar to purification by *zone-refining* and has been used routinely to grow single crystals of silicon. In the *flame fusion (Verneuil) method*, the powder sample is directly fed into an oxy-hydrogen flame and the melt allowed to drip on a seed crystal. As crystallization occurs on the top, the growing seed is lowered slowly, facilitating the growth of large crystals. The method has been used for the growth of high-melting oxides such as ruby and sapphire. A variant of the flame fusion technique is the *plasma torch method* in which the powder is dropped through a hot plasma generated and maintained by high-frequency current. The skull technique (see Section 3.2.5) has emerged to become a useful method for growing large crystals of several oxides.

Frequently, growth of crystals from melt involves more than one component, such as impurities, intentionally added dopants, etc., in addition to the major component. In these cases, it is essential to know the distribution of the second component between the growing crystal and the melt. This distribution occurs according to the phase diagram relating the equilibrium solubilities of the second component (impurity) in the liquid and the solid phases.

A number of crystal-growth methods making use of the solubility of a solute in a suitable solvent are known. Crystallization requires supersaturation, which can be provided by a temperature difference between the dissolution and growth zones, by solvent-evaporation or by a chemical reaction. In the solution methods, it is difficult to avoid contamination of the product by other components in the solution or flux. The flux technique has nevertheless been used to grow crystals of GeO_2, SiO_2, $BaTiO_3$, $KTaO_3$, α-Al_2O_3, $GdAlO_3$ and so on. The role of the solvent is to depress the melting point of the solute to be grown as a crystal. Although supersaturation is required for crystal growth, experimentally it is found that growth rate from solution is enormously faster than expected, considerable growth occurring even at low degrees of supersaturation (1%). This discrepancy was explained by Frank and Cabrera in terms of the effect of dislocations on crystal growth. Whereas the growth of a perfect crystal requires nucleation of a new layer on a perfect surface after completion of the previous layer, in the presence of a screw dislocation, growth does not require nucleation of a new layer. The dislocation provides a stepped surface where growth can occur even at low degrees of supersaturation. Spiral growth patterns arising from screw dislocations have been observed on a number of crystals grown from solution and vapour supporting the Frank model.

Among the solution-growth methods, crystallization from aqeous solution is well known. Materials with low water solubility may be brought into solution by the use of complexing agents (mineralizers). Selenium, for instance, can be grown from aqueous sulphide solutions, by making use of the reaction

$$2Se + S^{2-} \rightleftharpoons Se_2S^{2-}$$

An important method for growth of crystals with low water solubility is the hydrothermal method discussed earlier in Section 3.2.3. Hydrothermal synthesis is generally carried out with an autoclave of the type shown in Fig. 3.7. The autoclave consists of a lower nutrient region and an upper growth region separated by a baffle. The solute is placed in the nutrient region and a few seed crystals are suspended in the growth region. The vessel is filled with water to a predetermined volume, mineralizers are added if necessary, then it is closed and the temperature raised while providing a gradient. Since the solubility of most substances increases with increasing temperature, the nutrient region is maintained at a higher temperature than the growth region. The solvent saturates at the dissolving zone, moves by convection to the cooler growth zone where it is supersaturated and deposits the solute on the seed crystal. The solution is undersaturated when it reaches the nutrient region, thus dissolving more solute and the cycle goes on. Large quartz crystals are grown by this method. If the solute happens to have a retrograde solubility, as in the case of $AlPO_4$, the temperature gradient has to be reversed between the growth and the nutrient regions. In hydrothermal synthesis, the degree of filling the autoclave with solvent is an important factor; it has to be more than 32% in order to provide a high density of the solvent. Conventionally, hydrothermal experiments are carried out in alkaline medium, the OH^- ion acting as complexing agent. Hydrothermal synthesis has been used to stabilize unusual oxidation states in

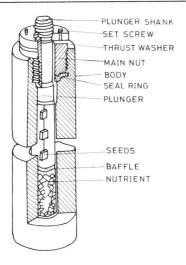

Figure 3.7   Apparatus for hydrothermal synthesis.

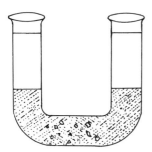

Figure 3.8   Gel method for growth of water-insoluble ionic solids.

transition-metal compounds and to synthesize low-temperature and metastable phases (Rabenau, 1985).

Insoluble ionic solids such as $CaCO_3$ and $BaSO_4$ cannot be grown by conventional solution methods or even by controlled chemical reaction because of instantaneous precipitation of the product when the reactants are mixed. The *gel method* is useful for the growth of such solids; this simple method depends on controlled diffusion of the reagents through the gel. A U-tube is filled with silica gel produced by acidifying a solution of sodium metasilicate (Fig. 3.8). The reagents are added to the two arms of the U tube; they diffuse slowly towards one another. In local regions where the concentration exceeds the solubility product, nuclei are formed and grow further into large crystals. Henisch and coworkers (1965) have made use of this method to grow crystals of calcium tartrate, calcium tungstate and lead iodide and have even incorporated selective impurity ions (Mn^{2+}, Cu^{2+}, Cr^{3+} etc.) during growth.

Solids such as KF, PbO, PbF_2 and B_2O_3 are powerful solvents (flux) in the molten state for many inorganic substances and hence can be used as media for the growth of crystals. The usual technique is to dissolve the solute in a suitable combination of flux contained in a platinum crucible while maintaining the temperature slightly above the saturation point. After the crucible has been cooled at a programmed rate, the flux is poured off or leached away by mineral acid, leaving behind the crystals. Yttrium iron garnet and its isomorphs are typical examples of solids grown by the *flux method* using PbO–PbF_2 fluxes. A recent development relating to growth of crystals from molten media is the use of molten metals as solvents from which crystals of intermetallic compounds are grown. Crystals of rare-earth borides, LnB_4 and LnB_6, have been grown from liquid aluminium and rare-earth rhodium stannides from molten tin (Fisk *et al.*, 1972; Espinosa, 1980).

Electrolysis of molten salt solutions has been applied for preparation and crystal growth (Banks & Wold, 1974; Wold & Bellavance, 1972; Feigelson, 1980). The *electrolytic method* involves reduction, usually, of a cation and deposition of a product containing the reduced cation at the cathode. The technique was used by Andrieux and coworkers for the synthesis of a variety of transition-metal compounds. Typical examples are: (i) growth of vanadium spinels, $MV_2O_4(M = Mn, Fe, Co, Zn$ and $Mg)$, from $Na_2B_4O_7$ and NaF melts using carbon crucibles and a carbon cathode (iron crucible and iron cathode for the iron compound), (ii) synthesis of $NaMO_2$ ($M = Fe$, Co, Ni) by the electrolysis of $NaOH$ melts contained in alumina crucibles and M electrodes, (iii) synthesis of alkali metal tungsten bronzes, A_xWO_3, by the electrolysis of alkali metal tungstates using platinum electrodes. Although most of the early studies of fused-salt electrolysis were empirical, aimed at obtaining large single crystals by manipulating melt composition and current density, later studies as exemplified by the work of Whittingham & Huggins (1972) on the preparation of alkali metal tungsten bronze, have shown that if the potential difference across the cell rather than the current density is kept constant during electrolysis, crytals of fixed composition can be grown. The electrolysis method has been employed for the preparation and growth of crystals of a wide variety of solids such as borides, carbides, silicides, phosphides, arsenides, and sulphides.

The experimental cell assembly for electrolytic growth can be exceedingly simple or complex, depending on the system studied. For example, electrolytic growth of alkali metal tungsten bronzes requires a Gooch crucible set inside a bigger Gooch crucible, the inner crucible serving as the anode chamber and the outer one as the cathode chamber; electrodes of platinum or gold are used. No inert atmosphere is necessary since atmospheric oxygen has no influence upon the current–potential relationship. In contrast to this, the electrolytic cell used by Didchenko & Litz (1962) for the synthesis of CeS and ThS is quite elaborate, consisting of graphite and alundum crucibles (serving as electrode compartments) together with molybdenum electrodes and involving a provision for inert atmosphere.

Growth of crystals from vapour may be divided into two categories depending on whether the change, vapour → crystal, is physical or chemical. When the composition of the vapour and the crystal is the same, the process is physical; examples are sublimation–condensation and sputtering. The process is termed chemical when a chemical reaction occurs during the growth; in such a case, the composition of the solid is different from the vapour. The use of chemical vapour deposition (CVD) as a preparative technique was discussed earlier in this chapter (see Section 3.2.6). We shall now briefly discuss the related *chemical vapour transport* (CVT) which is useful both for preparation of new solids as well as growing them into crystals. The method has been treated extensively by Schäfer (1972) and Rosenberger (1981). In CVT, a condensed phase reacts with a gas to form volatile products. An equilibrium exists between the reactants and products:

$$iA(s) + kB(g) + \cdots \rightleftharpoons jC(g) + \cdots$$

CVT makes use of the temperature dependence of the above heterogeneous equilibrium to transport solid A through the vapour phase by means of gaseous intermediate(s) C. That the process involves true transport and not just evaporation and condensation is evident from the fact that solid A does not possess an appreciable vapour pressure at the experimental temperature; moreover, transport of A is not observed without the transporting agent B.

Chemical transport is normally carried out by maintaining a temperature difference between the charge-end and growth-end in a closed system (Fig. 3.9). The forward reaction occurs at the charge-end forming gaseous products and the reverse reaction occurs at the growth-end depositing crystals. The temperatures T_1 and T_2 chosen at the growth-end and charge-end respectively depend on whether the reaction is endothermic or exothermic. For endothermic reactions, transport requires that $T_2 > T_1$; the reverse is true for exothermic reactions. The factors that determine growth of crystals by this method are, the choice of the chemical reaction chosen for transport (the CVT equilibrium should not lie at extremes lest the reversal becomes difficult), and the magnitude of T_1 and T_2 chosen and the concentration of transport agent used. The examples listed in Table 3.8 clearly suggest that the method can be used to prepare or grow crystals of almost any type of solid material provided a suitable transporting agent that can give volatile products can be found. Metals have been transported using the volatile halides.

CVT is indeed a popular technique and has been employed to grow crystals of oxide phases such as V_8O_{15} free from contamination by the neighbouring Magnéli phases, V_7O_{13} and V_9O_{17}. The transporting agents commonly used are I_2, $TeCl_4$ and Cl_2. When a metal oxide can be volatilized readily (e.g. $SnO_2 \xrightarrow{\text{heat}} SnO + \frac{1}{2}O_2$), the vapour species formed can be recombined to form single crystals at the cooler end.

Molecular beam epitaxy is an important technique for the preparation of semiconductors (III–V compounds). The finesse and sophistication of modern preparative solid

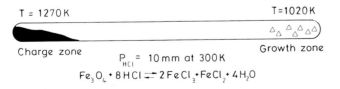

T = 1270 K   T = 1020 K

Charge zone   Growth zone

P_{HCl} = 10 mm at 300 K

$Fe_3O_4 + 8HCl \rightleftharpoons 2FeCl_3 + FeCl_2 + 4H_2O$

Figure 3.9   Chemical vapour transport in a closed system.

Table 3.8. *Some crystals grown by chemical vapour transport method*

Starting material	Final product (crystals)	Transport agent	Temperature, K
SiO_2	SiO_2	HF	$470 \rightarrow 770$
Fe_3O_4	Fe_3O_4	HCl	$1270 \rightarrow 1070$
Cr_2O_3	Cr_2O_3	$Cl_2 + O_2$	$1070 \rightarrow 870$
$MO + Fe_2O_3$ (M = Mg, Co, Ni)	MFe_2O_4	HCl	—
$Nb + Nb_2O_5$	NbO	Cl_2	—
$NbSe_2$	$NbSe_2$	I_2	$1100 \rightarrow 1050$

state science is exemplified by this method which employs reactions of multiple molecular beams with a single-crystal substrate.

When a polycrystalline solid sample is heated for a sufficient length of time below its melting point, an increase in the average size of crystals is observed. The driving force for this change is the lowering of outer crystalline area. The number of crystals obtained depends on the number of active nuclei capable of growing at the expense of the polycrystalline matrix. Metals are normally grown as crystals in the solid state by the so-called 'strain-anneal' procedure involving recrystallization by annealing out strain. Growth in the solid state is particularly useful for solids exhibiting incongruent melting or showing a polymorphic transition below the melting point.

Thin films of complex metal oxides have been deposited on solid substrates by employing a variety of techniques such as metal–organic chemical vapour deposition (MOCVD), sol–gel method, pulsed laser ablation and nebulized spray pyrolysis. The last technique provides a truly simple inexpensive chemical means of obtaining films starting with organic precursors (Raju *et al.*, 1995). It is noteworthy that films obtained by some of these methods are epitaxial and have single-crystal character. For example, epitaxial films of $YBa_2Cu_3O_7$, infinite-layer $Ca_{1-x}Lu_xCuO_2$ and other cuprate superconductors have been deposited on $SrTiO_3$ and other oxide substrates. What is still more interesting is the deposition of atomic scale oxide superlattices (thin film heterostructures) by pulsed laser deposition, typical examples being $SrTiO_3/BaTiO_3$, $SrCuO_2/CaCuO_2$ and $BaCuO_2/SrTiO_3$ (Kawai *et al.*, 1993, Shaw *et al.*, 1994, Gupta *et al.*, 1995). This is comparable to superlattices of semiconductor materials prepared by molecular beam epitaxy.

References
Alamo, J. & Roy, R. (1984) *J. Solid State Chem.* **51**, 270.

Alberti, G., Constantino, U., Marmottini, F., Vivani, R. & Zappelli, P. (1993) *Angew. Chem. Int. Ed. (Engl.)* **32**, 1357.

Andres, R. P., Averback, R. S. & Brown, W. L. (1989) *J. Mater. Res.* **4**, 704.

Angell, C. A. (1981) in *Preparation and Characterization of Materials* (eds Honig, J. M. & Rao, C. N. R.) Academic Press, New York.

Annen, M. J., Davis, M. E., Higgins, J. B. & Schlenker, J. L. (1991) *J. Chem. Soc. Chem. Commun.* 1175.

Arhancet, J. P. & Davis, M. E. (1991) *Chem. Mater.* **3**, 567.

Ayyappan, S., Subbanna, G. N. & Rao, C. N. R. (1995) *Chem. Euro J.* **1**, 165.

Banks, E. & Wold, A. (1974) in *Solid State Chemistry* (ed. Rao, C. N. R.) Marcel Dekker, New York.

Barrer, R. M. (1982) *Hydrothermal Chemistry of Zeolites*, Academic Press, London.

Bechgaard, K. & Jerome, D. (1982) *Scientific American* **247**, 50.

Bissessur, R., DeGroot, D. C., Schindler, J. L., Kannewurf, C. R. & Kanatzidis, M. G. (1993a) *J. Chem. Soc. Chem. Commun.* 687.

Bissessur, R., Kanatzidis, M. G., Schindler, J. L. & Kannewurf, C. R. (1993b) *J. Chem. Soc. Chem. Commun.* 1582.

Blair, J. A., Duffy, J. A. & Wardell, J. L. (1992) *Phys. Chem. Glasses*, **93**, 191.

Boller, H. (1978) *Monatsh. Chem.* **109**, 975.

Bonneau, P. R., Jarvis, Jr., R. F. & Kaner, R. B. (1991) *Nature* **349**, 510.

Casey, W. H., Westrich, H. R., Banfield, J. F., Ferruzzi, G. & Arnold, G. W. (1993) *Nature* **366**, 253.

Castellano, F. N., Heimer, T. A., Tandhasetti, M. T. & Meyer, G. J. (1994) *Chem. Mater.* **6**, 1041.

Cava, R. J., Santoro, A., Murphy, D. W., Zahurak, S. & Roth, R. S. (1982) *J. Solid State Chem.* **42**, 251.

Cheetham, A. K. (1980) *Nature* **288**, 469.

Chen, J., Jones, R. H., Natarajan, S., Hursthouse, B. M. & Thomas, J. M. (1994) *Angew. Chem. Int. Ed. (Engl.)* **33**, 639.

Chevrel, R., Sergent, M. & Prigent, J. (1974) *Mater. Res. Bull.* **9**, 1487.

Chowdhry, U., Barkley, J. R., English, A. D. & Sleight, A. W. (1982) *Mater. Res. Bull.* **17**, 917.

Corbett, J. D. (1980) in *Solid State Chemistry-Techniques: A Contemporary Overview* (eds. Holt, S. L., Milstein, J. B. & Robbins, M.) American Chemical Society, Washington, D.C.

Corbett, J. D. (1981) *Acc. Chem. Res.* **14**, 239.

Corbett, J. D. (1987) in *Solid State Chemistry* (eds. Cheetham, A. K. & Day, P.), Clarendon Press, Oxford.

Dai, H., Wong, E. W., Lu, Y. Z., Fan, S. & Lieber, C. M. (1995) *Nature* **375**, 769.

Davis, M. E. & Lobo, R. F. (1992) *Chem. Mater.* **4**, 756.

Delmas, C., Braconnier, J. J. & Hagenmuller, P. (1982) *Mater. Res. Bull.* **17**, 117.

Demazeau, G., Buffat, B., Pouchard, M. & Hagenmuller, P. (1982a) *J. Solid State Chem.* **45**, 881.

Demazeau, G., Buffat, B., Pouchard, M. & Hagenmuller, P. (1982b) *Z. Anorg. Allgem. Chem.* **491**, 60.

Demourgues, A., Weill, F., Darriett, B., Wattiaux, A., Grenier, J. C., Gravereau, P. & Pouchard, M. (1993) *J. Solid State Chem.* **106**, 317, 330.

Desiraju, G. R. & Rao, M. (1982) *Mater. Res. Bull.* **17**, 443.

Dhingra, M. & Kanatzidis, M. G. (1992) *Science* **258**, 1769.

Dickens, P. G., French, S. J., Hight, A. T. & Pye, M. F. (1979) *Mater. Res. Bull.* **14**, 1295.

Dickens, P. G. & Pye, M. F. (1982) in *Intercalation Chemistry* (eds Whittingham, M. S. & Jacobson, A. J.) Academic Press, New York.

Didchenko, R. & Litz, L. M. (1962) *J. Electrochem. Soc.* **109**, 247.

Divigalpitiya, W. M. R., Frindt, R. F. & Morrison, S. R. (1989) *Science* **246**, 369.

Dronskowski, R. (1992) *J. Amer. Chem. Soc.* **114**, 7230.

Dronskowski, R. (1993) *Inorg. Chem.* **32**, 1.

Duff, D. G., Baiker, A. & Edwards, P. P. (1992) *J. Chem. Soc. Chem. Commun.* 96.

Elliott, S. R., Rao, C. N. R. & Thomas, J. M. (1985) *Angew. Chem. Int. Ed. (Engl.)* **25**, 31.

Elwell, D. & Scheel, H. J. (1975) *Crystal Growth from High Temperature Solutions*, Academic Press, London.

England, W. A., Goodenough, J. B. & Wiseman, P. J. (1983) *J. Solid State Chem.* **49**, 289.

Enzel, P. & Bein, T. (1989) *J. Phys. Chem.* **93**, 6270.

Espinosa, G. P. (1980) *Mater. Res. Bull.* **15**, 791.

Estermann, M., McCusker, L. B., Baerlocher, C., Merrouche, A. & Kessler, H. (1991) *Nature* **352**, 320.

Etemad, S., Heeger, A. J. & MacDiarmid, A. G. (1982) *Ann. Rev. Phys. Chem.* **33**, 443.

Fan, J. C. C. & Reed, T. B. (1972) *Mater. Res. Bull.* **7**, 1403.

Farrington, G. C. & Briant, J. L. (1978) *Mater. Res. Bull.* **13**, 763.

Feigelson, R. S. (1980) in *Solid State Chemistry: A Contemporary Overview* (eds Holt, S. L., Milstein, J. B. & Robbins, M.) American Chemical Society, Washington, D.C.

Feist, T. P. & Davies, P. K. (1991) *Chem. Mater.* **3**, 1011.

Feist, T. P. & Davies, P. K. (1992) *J. Solid State Chem.* **101**, 275.

Ferey, G., Loiseau, T. & Riou, D. (1994) *Materials Science Forum* **152–153**, 125.

Fievet, F., Lagier, J. P. & Figlarz, M. (1989) *MRS Bulletin* **14**, 29.

Figlarz, M. (1989) *Prog. Solid State Chem.* **19**, 1.

Fisk, Z., Cooper, A. S., Schmidt, P. H. & Castellano, R. N. (1972) *Mater. Res. Bull.* **7**, 285.

Flanigen, E. M., Bennett, J. M., Grose, R. W., Cohen, J. P., Patton, R. L., Kirchner, R. M. & Smith, J. V. (1978) *Nature* **271**, 512.

Gasgnier, M., Derouet, J., Albert, L., Beaury, L., Caro, P. & Deschamps, M. (1993) *J. Solid State Chem.* **107**, 179.

Geselbracht, M. J., Richardson, T. J. & Stacy, A. M. (1990) *Nature* **345**, 324.

Goodenough, J. B., Kafalas, J. A. & Longo, J. M. (1972) in *Preparative Methods in Solid State Chemistry* (ed. Hagenmuller, P.), Academic Press, New York.

Goodenough, J. B., Hong, H. Y-P. & Kafalas, J. A. (1976) *Mater. Res. Bull.* **11**, 203.

Gopalakrishnan, J. (1995) *Chem. Mater.* **7**, 1265.

Gopalakrishnan, J. & Kasthuri Rangan, K. (1992) *Chem. Mater.* **4**, 745.

Gopalakrishnan, J., Bhuvanesh, N. S. P. & Raju, A. R. (1994) *Chem. Mater.* **6**, 373.

Granqvist, C. G. & Buhrman, R. A. (1976) *J. Appl. Phys.* **47**, 2200.

Grenier, J.-C., Wattiaux, A., Lagueyte, N., Park, J. C., Marquestaut, E., Etourneau, J. & Pouchard, M. (1991) *Physica C* **173**, 139.

Groenen, E. J. J., Alma, N. C. M., Bastein, A. G. T. M., Hays, G. R., Huis, R. & Kortbeek, A. G. T. G. (1983) *J. Chem. Soc. Chem. Comm.* 1360.

Grossel, M. C. & Weston, S. C. (1994) *Contemp. Org. Synthesis*, **1**, 367.

Groult, D., Pannetier, J. & Raveau, B. (1982) *J. Solid State Chem.* **41**, 277.

Gupta, A., Shaw, T., Chern, M. Y., Hussey, B. W., Guloy, A. M. & Scott, B. A. (1995) *J. Solid State Chem.* **114**, 190.

Guyomard, D. & Tarascon, J-M. (1994) *Adv. Mater.* **6**, 408.

Hadjipanyas, G. & Siegel, R. W. (1994) *Nanophase Materials: Synthesis, Properties and Applications*, Kluwer, Holland.

Hagenmuller, P. (ed.) (1972) *Preparative Methods in Solid State Chemistry*, Academic Press, New York.

Hagenmuller, P. (ed.) (1985) *Solid Inorganic Fluorides*, Academic Press, New York.

Ham, W. K., Holland, G. F. & Stacy, A. M. (1988) *J. Amer. Chem. Soc.* **110**, 5214.

Harrison, H. R., Aragon, R. & Sandberg, C. J. (1980) *Mater. Res. Bull.* **15**, 571.

Hatfield, W. E. (ed.) (1979) *Molecular Metals*, Plenum Press, New York.

Haushalter, R. C. & Mundi, L. A. (1992) *Chem. Mater.* **4**, 31.

Henisch, H. K., Dennis, J. & Hanoka, J. I. (1965) *J. Phys. Chem. Solids* **26**, 493.

Hingorani, S., Pillai, V., Kumar, P., Multani, M. S. & Shah, D. O. (1993) *Mater. Res. Bull.* **28**, 1303.

Honig, J. M. & Rao, C. N. R. (eds) (1981) *Preparation and Characterization of Materials*, Academic Press, New York.

Horowitz, H. S. & Longo, J. M. (1978) *Mater. Res. Bull.* **13**, 1359.

Horowitz, H. S., Longo, J. M. & Lewandowski, J. T. (1981) *Mater. Res. Bull.* **16**, 489.

Hotta, S., Hosaka, T. & Shimotsuma, W. (1983) *Syn. Met.* **6**, 319.

Jacobson, A. J., Chianelli, R. R., Rich, S. M. & Whittingham, M. S. (1979) *Mater. Res. Bull.* **14**, 1437.

Jacobson, A. J. (1994) *Materials Science Forum* **152–153**, 1.

Jayaraman, A., Dernier, P. D. & Longinotti, L. D. (1975) *High Temp. High Pr.* **7**, 1.

Jena, P., Rao, B. K. & Khanna, S. N. (eds.) (1986) *Physics and Chemistry of Small Clusters*, Plenum Press, New York.

Jena, P., Rao, B. K. & Khanna, S. N. (eds) (1992) *Physics and Chemistry of Finite Systems: From Clusters to Crystals*, Kluwer, Dordrecht.

Johnson, Jr, D. W. (1981) *Amer. Ceram. Soc. Bull.* **60**, 221.

Joubert, J. C. & Chenavas, J. (1975) in *Treatise on Solid State Chemistry*, Vol. 5 (ed. Hannay, N. B.) Plenum Press, New York.

Kanatzidis, M. G. (1990) *Chem. Mater.* **2**, 253.

Kanatzidis, M. G. & Marks, T. J. (1987) *Inorg. Chem.* **26**, 783.

Kanatzidis, M. G., Wu, C. G., Marcy, H. O. & Kannewurf, C. R. (1989a) *J. Amer. Chem. Soc.* **111**, 4139.

Kanatzidis, M. G., Marcy, H. O., McCarthy, W. J., Kannewurf, C. R. & Marks, T. J. (1989b) *Solid State Ionics* **32/33**, 594.

Kang, Z. C. & Eyring, L. (1988) *J. Solid State Chem.* **75**, 60.

Kawai, M., Liu, Z., Sekine, R. & Koinuma, H. (1993) *Jpn. J. Appl. Phys. C* **32**, 1208.

Keane, P. M., Lu, Y. J. & Ibers, J. A. (1991) *Acc. Chem. Res.* **24**, 223.

Keller, S. W., Carlson, V. A., Sandford, D., Stenzel, F., Stacy, A. M., Kwei, G. H. & Alario-Franco, M. (1994) *J. Amer. Chem. Soc.* **116**, 8070.

Kresge, C. T., Leonowicz, M. E., Roth, W. J., Vartuli, J. C. & Beck, J. S. (1992) *Nature* **359**, 710.

Kumar, P., Pillai, P., Bates, S. R. & Shah, D. O. (1993) *Mater. Lett.* **16**, 68.

Lacroix, P. G., Clement, R., Nakatani, K., Zyss, J. & Ledoux, I. (1994) *Science* **263**, 658.

Latroche, M., Brohan, L., Marchand, R. & Tournoux, M. (1989) *J. Solid State Chem.* **81**, 78.

Laudise, R. A. (1970) *The Growth of Single Crystals*, Prentice-Hall, New Jersey.

Lawton, S. L. & Rohrbaugh, W. J. (1990) *Science* **247**, 1319.

Lerf, A., Lalik, E. Kolodziejski, W. & Klinowski, J. (1992) *J. Phys. Chem.* **96**, 7389.

Li, W. U., Dahn, J. R. & Wainwright, D. S. (1994) *Science* **264**, 1115.

Liao, J-H. & Kanatzidis, M. G. (1990) *J. Amer. Chem. Soc.* **112**, 7400.

Liao, J-H. & Kanatzidis, M. G. (1992) *Inorg. Chem.* **31**, 431.

Liu, Y. J., DeGroot, D. C., Schnidler, J. L., Kannewurf, C. R. & Kanatzidis, M. G. (1993) *J. Chem. Soc. Chem. Commun.*, 593.

Livage, J. (1994) *Materials Science Forum* **152–153**, 43.

Loehman, R. E., Rao, C. N. R., Honig, J. M. & Smith, C. E. (1969) *J. Sci. Ind. Res.* (India) **28**, 13.

Longo, J. M. & Horowitz, H. S. (1981) in *Preparation and Characterization of Materials* (eds. Honig, J. M. & Rao, C. N. R.) Academic Press, New York.

MacDiarmid, A. G. & Heeger, A. J. (1979) in *Molecular Metals* (ed. Hatfield, W. E.) Plenum Press, New York.

Mahesh, R., Pavate, V. A., Parkash, O. & Rao, C. N. R. (1992) *Supercond. Sci. Tech.* **5**, 174.

Mahesh, R., Kannan, K. R. & Rao, C. N. R. (1995) *J. Solid State Chem.* **114**, 294.

Mammone, R. J. & MacDiarmid, A. G. (1984) *Syn. Met.* **9**, 143.

Mann, S. (1993a) *Nature* **365**, 499.

Mann, S. (1993b) *Science* **261**, 1286.

Marquez, L. N., Keller, S. W., Stacy, A. M., Fendorf, M. & Gronsky, R. (1993) *Chem. Mater.* **5**, 761.

Matijevic, E. (1991) *Faraday Discussion* **92**, 229.

Mehrotra, V. & Giannelis, E. P. (1992) *Solid State Ionics* **51**, 115.

Merzhanov, A. G. & Borovinskaya, I. P. (1972) *Dokl. Akad. Nauk SSSR* (Engl. Transl.) **204**, 429.

Merzhanov, A. G. (1993) in *Chemistry of Advanced Materials* (ed. C. N. R. Rao), Blackwell, Oxford.

Mizushima, K., Jones, P. C., Wisemann, P. J. & Goodenough, J. B. (1980) *Mater. Res. Bull.* **15**, 783.

Mohanram, R. A., Singh, K. K., Madhusudan, W. H., Ganguly, P. & Rao, C. N. R. (1983) *Mater. Res. Bull.* **18**, 703.

Möhr, S. & Müller-Buschbaum, H. (1995) *Angew. Chem. Int. Ed.* (*Engl.*) **34**, 365.

Mroczkowski, S. (1980) *J. Chem. Ed.* **57**, 537.

Müller-Buschbaum, H. K. (1981) *Angew. Chem. Int. Edn.* (*Engl.*) **20**, 22.

Multani, M. (1981) in *Preparation and Characterization of Materials* (eds Honig, J. M. & Rao, C. N. R.) Academic Press, New York.

Murphy, D. W., Cros, C., DiSalvo, F. J. & Waszczak, J. V. (1977) *Inorg. Chem.* **16**, 3027.

Murphy, D. W., DiSalvo, F. J., Carides, J. N. & Waszczak, J. V. (1978) *Mater. Res. Bull.* **13**, 1395.

Murphy, D. W. & Christian, P. A. (1979) *Science* **205**, 651.

Murphy, D. W., Greenblatt, M., Cava, R. J. & Zahurak, S. M. (1981) *Solid State Ionics* **5**, 327.

Murphy, D. W., Greeblatt, M., Zahurak, S. M., Cava, R. J., Waszczak, J. V. & Hutton, R. S. (1982) *Rev. Chim. Miner.* **19**, 441.

Murphy, D. W., Zahurak, S. M., Vyas, B., Thomas, M., Badding, M. E. & Fang, W. C. (1993) *Chem. Mater.* **5**, 767.

Nanjundaswamy, K. S., Vasanthacharya, N. Y., Gopalakrishnan, J. & Rao, C. N. R. (1987) *Inorg. Chem.* **26**, 4286.

Narita, M. & Pittman, C. U. (1976) *Synthesis* 489.

Narita, E., Kaviratna, P. D. & Pinnavaia, T. J. (1993) *J. Chem. Soc. Chem. Commun.* 60.

Nazar, L. F., Liblong, S. W. & Yin, X. T. (1991) *J. Amer. Chem. Soc.* **113**, 5889.

Nguyen, Tu, N., Giaquinta, D. M., Davis, W. M. & zur Loye, H. C. (1993) *Chem. Mater.* **5**, 1273.

Nigrey, P. J., McInnes, D., Nairns, D. P., MacDiarmid, A. G. & Heeger, A. J. (1981) *J. Electrochem. Soc.* **128**, 1651.

Pabst, A. (1952) *Amer. Mineral.* **37**, 137.

Pinnavaia, T. J. (1983) *Science* **220**, 365.

Pistorius, C. W. F. T. (1976) *Prog. Solid State Chem.* **11**, 1.

Poeppelmeier, K. R. & Kipp, D. O. (1987) *Inorg. Chem.* **27**, 766.

Portier, R., Carpy, A., Fayard, M. & Galy, J. (1975) *Phys. Stat. Solidi.* **A30**, 683.

Rabenau, A. (1985) *Angew, Chem. Int. Ed. (Engl.)* **24**, 1026.

Raju, A. R., Aiyer, H. N. & Rao, C. N. R. (1995) *Chem. Mater.* **7**, 225.

Rao, C. N. R. (1994) *Chemical Approaches to the Synthesis of Inorganic Materials*, Wiley, New York.

Rao, C. N. R., Gopalakrishnan, J., Vidyasagar, K., Ganguli, A. K., Ramanan, A. & Ganapathi, L. (1986) *J. Mater. Res.* **1**, 280.

Rebbah, H., Pannetier, J. & Raveau, B. (1982) *J. Solid State Chem.* **41**, 57.

Rice, C. E. & Jackel, J. L. (1982) *J. Solid State Chem.* **41**, 308.

Rosenberger, F. (1981) in *Preparation and Characterization of Materials* (eds Honig, J. M. & Rao, C. N. R.) Academic Press, New York.

Rouxel, J., Tournoux, M. & Brec, R. (1994) *Soft Chemistry Routes to New Materials*, Trans Tech Publications, Switzerland.

Roy, R. (1977) *J. Amer. Ceram. Soc.* **60**, 350.

Roy, R. (1987) *Science* **238**, 1664.

Roy, R., Agarwal, D. K., Alamo, J. & Roy, R. A. (1984) *Mater. Res. Bull.* **19**, 471.

Roy, R. A. & Roy, R. (1984) *Mater. Res. Bull.* **19**, 169.

Rudolf, P. & Schöllhorn, R. (1992) *J. Chem. Soc. Chem. Commun.* 1158.

Rüdorff, W. (1959) *Angew. Chem.* **71**, 127, 487.

Rzeznik, M. A., Geselbracht, M. J., Thompson, M. S. & Stacy, A. M. (1993) *Angew. Chem. Int. Ed. (Engl.)* **32**, 254.

Sastry, R. L. N., Mehrotra, P. N. & Rao, C. N. R. (1966) *J. Inorg. Nucl. Chem.* **28**, 2167.

Schäfer, H. (1972) in *Solid State Chemistry*, NBS Special Publication 364 (eds Roth, R. S. & Schneider Jr., S. J.) National Bureau of Standards, USA.

Schneemeyer, L. F., Thomas, J. K., Siegrist, T., Batlogg, B., Rupp, L. W., Opila, R. L., Cava, R. J. & Murphy, D. W. (1988) *Nature* **335**, 421.

Scholze, H. (ed.) (1984) Proc. Second Int. Workshop on Glasses and Glass Ceramics from Gels, *J. Non-Cryst. Solids* **63**, 1–300.

Sekar, M. M. & Patil, K. C. (1993) *Mater. Res. Bull.* **28**, 485.

Sheats, J. E., Pittman Jr, C. U. & Carraher Jr, C. E. (1984) *Chem. Brit.* **20**, 709.

Shaw, T. M., Gupta, A., Chern, M. Y., Batson, P. E., Laibowitz, R. B. & Scott, B. A. (1994) *J. Mater. Res* **9**, 2566.

Shin, Y. J., Doumerc, J. P., Dordor, P., Delmas, C., Pouchard, M. & Hagenmuller, P. (1993) *J. Solid State Chem.* **107**, 303.

Shirakawa, H. & Ikeda, S. (1971) *Polym. J.* **2**, 231.

Shirakawa, H., Ito, T. & Ikeda, S. (1973) *Polym. J.* **4**, 460.

Simon, A. (1975) in *Crystal Structure and Chemical Bonding in Inorganic Chemistry* (eds. Rooymans, C. J. M. & Rabenau, A.) North-Holland, Amsterdam.

Simon, A. (1981) *Angew. Chem.* **20**, 1.

Slama-Schwok, A., Ottolenghi, M. & Avnir, D. (1992) *Nature* **355**, 240.

Soghomonian, V., Chen, Q., Haushalter, R. C., Zubieta, J. & O'Connor, C. J. (1993) *Science* **259**, 1596.

Sōmiya (ed.) (1989) *Hydrothermal Reactions for Materials Science and Engineering*, Elsevier Applied Science, London.

Stein, A., Keller, S. W. & Mallouk, T. E. (1993) *Science* **259**, 1558.

Subramanian, M. A., Roberts, B. D. & Clearfield, A. (1984) *Mater. Res. Bull.* **19**, 1471.

Sunshine, S. A., Kang, D. & Ibers, J. A. (1987) *J. Amer. Chem. Soc.* **109**, 6202.

Takano, M. & Takeda, Y. (1983) *Bull. Inst. Chem. Res.* (Kyoto University, Japan) **61**, 406.

Takeda, Y., Kanamaru, F., Shimada, M. & Koizumi, M. (1976) *Acta Crystallogr.* **B32**, 2464.

Thackeray, M. M., David, W. I. F. & Goodenough, J. B. (1982) *Mater. Res. Bull.* **17**, 785.

Thackeray, M. M., David, W. I. F., Parks, P. G. & Goodenough, J. B. (1983) *Mater. Res. Bull.* **18**, 461.

Tietjen, J. J. (1973) *Ann. Rev. Mater. Sci.* **3**, 317.

Tofield, B. C. (1982) in *Intercalation Chemistry* (eds Whittingham, M. S. & Jacobson, A. J.) Academic Press, New York.

Torardi, C. C. & McCarley, R. E. (1979) *J. Amer. Chem. Soc.* **101**, 3963.

Torrance, J. B. (1979) in *Molecular Methods* (ed. Hatfield, W. E.) Plenum Press, New York.

Tournoux, M., Marchand, R. & Brohan, L. (1986) *Prog. Solid State Chem.* **17**, 33.

Tsai, K. L. & Dye, J. L. (1993) *Chem. Mater.* **5**, 540.

Uma, S. & Gopalakrishnan, J. (1993) *J. Solid State Chem.* **105**, 545.

Vacher, R., Woignier, T., Pelous, J. & Courtens, E. (1988) *Phys. Rev.* **B37**, 6500.

Vallet-Regi, M., Ragel, V., Roman, J., Martinez, J., Labeau, M., Varela, A. & Gonzalez-Calbet, J. M. (1993) *Solid State Ionics*, **63–65**, 164.

Van Rheenen, P. R., McKelvy, M. J. & Glaunsinger, W. S. (1987) *J. Solid State Chem.* **67**, 151.

Vasan, H. N. & Rao, C. N. R. (1995) *J. Mater. Chem.* **5**, 1755.

Vidyasagar, K. & Gopalakrishnan, J. (1982) *J. Solid State Chem.* **42**, 217.

Vidyasagar, K., Gopalakrishnan, J. & Rao, C. N. R. (1984) *Inorg. Chem.* **23**, 1206. Also see *JCS Chem. Commun.* (1986) 449.

Vidyasagar, K., Gopalakrishnan, J. & Rao, C. N. R. (1985) *J. Solid State Chem.* **58**, 29.

Wang, J., Tian, Y., Wang, R-C. & Clearfield, A. (1992) *Chem. Mater.* **4**, 1276.

Wheland, R. C. & Martin, E. L. (1975) *J. Org. Chem.* **28**, 3101.

Whittingham, M. S. (1978) *Prog. Solid State Chem.* **12**, 41.

Whittingham, M. S. & Huggins, R. A. (1972) in *Solid State Chemistry*, NBS Special Publication 364 (eds Roth, R. S. & Schneider Jr, S. J.) National Bureau of Standards, USA.

Whittingham, M. S. & Chianelli, R. R. (1980) *J. Chem. Ed.* **57**, 569.

Wiley, J. B. & Kaner, R. B. (1992) *Science* **255**, 1093.

Wizansky, A. R., Rauch, P. E. & DiSalvo, F. J. (1989) *J. Solid State Chem.* **81**, 203.

Wold, A. (1980) *J. Chem. Ed.* **57**, 531.

Wold, A. & Bellavance, D. (1972) in *Preparative Methods in Solid State Chemistry* (ed. Hagenmuller, P.) Academic Press, New York.

Wu, C-G. & Bein, T. (1994) *Science* **264**, 1757.

Wu, S., Ellerby, L. M., Cohan, J. S., Dunn, B., El-Sayed, M. A., Valentine, J. S. & Zink, J. I. (1993) *Chem. Mater.* **5**, 115.

Wypych, F. & Schöllhorn, R. (1992) *J. Chem. Soc. Chem. Commun.* 1386.

Yeh, X. L., Samwer, K. & Johnson, W. L. (1983) *Appl. Phys. Lett.* **42**, 242.

4   Phase transitions

4.1   Introduction

Many solids undergo transformations from one crystal structure to another as the temperature or pressure is varied and this phenomenon is popularly referred to as polymorphism. Whereas polymorphism normally refers to phase transitions involving changes in the atomic configurations in crystals, there are also transitions where the electronic or spin configuration undergoes changes. The subject of phase transitions is not only of great academic interest but also of technological importance. Phase transitions are exhibited by a wide variety of systems (Table 4.1) from simple metals and alloys to complex inorganic and organic materials. The subject has grown enormously in the last two decades with new types of transitions as well as new approaches to explain the phenomena. Traditionally, the subject has been of vital concern to metallurgists (Porter & Easterling, 1981) but there are many aspects of great importance to solid state chemistry (Rao, 1984). Varied aspects of phase transitions such as critical phenomena, soft modes, mechanisms and changes in properties at phase transitions have been treated in a unified manner by Rao & Rao (1978). We shall deal with some highlights of the subject and examine some classes of transitions in this chapter.

4.2   Thermodynamics

During a phase transition at the equilibrium temperature, the free energy of the solid remains continuous, but thermodynamic quantities like entropy, volume and heat capacity undergo discontinuous changes. Depending on which derivative of the Gibbs free energy shows a discontinuous change at the transition, phase transitions have been classified as first-order, second-order and so on. In a *first-order phase transition*, the first derivatives of the Gibbs free energy (S and V) exhibit changes:

$$\left(\frac{\partial G}{\partial T}\right)_P = -S: \quad \left(\frac{\partial G}{\partial P}\right)_T = V \tag{4.1}$$

In a *second-order transition*, the second derivatives of the free energy show changes:

$$\left(\frac{\partial^2 G}{\partial T^2}\right)_P = -\frac{C_p}{T}; \quad \left(\frac{\partial^2 G}{\partial P^2}\right)_T = -V\beta$$

$$\left(\frac{\partial^2 G}{\partial P \partial T}\right) = V\alpha \tag{4.2}$$

168

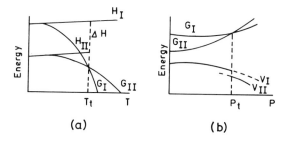

Figure 4.1   Variation of (a) the free energy and enthalpy with temperature and (b) the free energy and volume with pressure in a first-order transition.

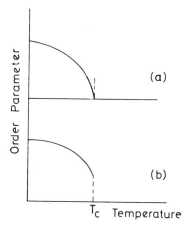

Figure 4.2   Variation of order parameter (a) in a second-order transition and (b) in a first-order transition.

Here C_p, α and β are the heat capacity, volume thermal expansivity and compressibility respectively. First-order transitions involving discontinuous changes in entropy and volume are depicted in Fig. 4.1. In this figure curves G_I, G_{II} represent variations in free energies of phases I and II respectively, while H_I, H_{II} and V_I, V_{II} represent variations in enthalpy and volume. We see that the free energy curves intersect ($\Delta G = 0$) sharply at the transition temperature, T_t, or the transition pressure, P_t.

In a second-order transition, the $G(P,T)$ curves do not intersect sharply and we see a continuous variation in enthalpy or volume through the transition. The heat capacity generally shows changes in such transitions. Second-order transitions are almost always associated with some sort of disordering process (Fig. 4.2). The long-range order parameter varies with temperature in a second-order transition, having a maximum value of unity when there is perfect order and a value of zero when there is perfect disorder. There are several transitions where the heat capacity tends towards infinity at the transition temperature (giving rise to a lambda-shaped C_p–temperature curve) and such transitions are referred to as λ-transitions. First-order transitions are associated with considerable hysteresis, and hysteresis effects are often used to characterize the 'order' of a transition. The magnitude of thermal hysteresis depends primarily on the

Table 4.1. *Typical reversible phase transitions*

(i) Thermal transitions[a]

		$T_t(\text{K})$	$\Delta V_t(\text{cm}^3\,\text{mol}^{-1})$	$\Delta H_t(\text{kJ}\,\text{mol}^{-1})$
Quartz	α–β	848	1.33	0.36
CsCl	CsCl(8:8)–NaCl(6:6)	752	10.3	2.42
AgI	hexagonal–cubic	427	– 2.2	6.15
NH_4Cl	CsCl(8:8)–NaCl(6:6)	469	7.1	4.47
$NaNO_2$	orthorhombic–orthorhombic	439	1.66	1.05
$RbNO_3$	trigonal–CsCl	439	6.0	3.97
	CsCl–hexagonal	461	3.12	2.72
	hexagonal–NaCl	551	3.13	2.72

(ii) Pressure transitions

		$P_t(\text{k bar})$	$\Delta V_t(\text{cm}^3\,\text{mol}^{-1})$	$\Delta H_t(\text{kJ}\,\text{mol}^{-1})$
KCl	NaCl(6:6)–CsCl(8:8)	19.6	– 4.11	8.03
RbCl	NaCl(6:6)–CsCl(8:8)	5.7	– 6.95	3.39
SiO_2	quartz–coesite	18.8	– 2.0	2.93
	coesite–stishovite	93.1	– 6.6	57.27
ZnO	wurtzite–NaCl	88.6	– 2.55	19.23
$CdTiO_3$	ilmenite–perovskite	40.4	– 2.90	15.88
CdS	wurtzite–NaCl	17.4	– 7.2	– 10.45
$FeCr_2O_4$	spinel–Cr_3S_4 type	36.0	– 6.5	– 30.51

[a]All except the transitions of quartz and $NaNO_2$ show appreciable thermal hysteresis. ΔV_t and ΔH_t are respectively the volume change and enthalpy change associated with the transition at temperature T_r.

volume change accompanying the transition, but is also affected by other factors (Rao & Rao, 1966; Natarajan *et al.*, 1969).

The Clausius–Clapeyron equation describes the thermodynamics at a first-order transition:

$$\frac{dP}{dT} = \frac{\Delta S}{\Delta V} = \frac{\Delta H}{T\Delta V} \tag{4.3}$$

In the case of second-order transitions, ΔV and ΔS have zero values, and equation (4.3) will have a 0/0 indeterminacy. We can, however, derive analogues of the Clausius–Clapeyron equation for these transitions, such as equation (4.4):

$$\frac{dP}{dT} = \frac{\Delta C_p}{VT\Delta\alpha} \tag{4.4}$$

In Table 4.1, thermodynamic data on the phase transitions of a few typical substances are listed.

Every phase transition is associated with a change in symmetry as well as in order. To facilitate quantitative analysis, Landau introduced the concept of an *order parameter*, ξ, and expressed the free energy, ϕ, as follows:

$$\phi(P, T, \xi) = \phi_o(P, T) + a\xi + b\xi^2 + c\xi^3 + d\xi^4 + \cdots \tag{4.5}$$

where $\phi_o(P,T)$, a, b, c, and d are constants. Here, $\xi = 1$ in the completely ordered phase (at low temperatures) and $\xi = 0$ in the completely disordered phase after the transition. Since changing the sign of ξ should not alter the value of ϕ, the coefficients of odd powers of ξ should be equal to zero. Equation (4.5) then becomes

$$\phi(P, T, \xi) = \phi_o(P, T) + b\xi^2 + d\xi^4 + \cdots \tag{4.6}$$

The equilibrium value of the long-range order parameter, ξ, is obtained by the conditions

$$\left(\frac{\partial\phi}{\partial\xi}\right)_{P,T} = \xi(b + 2d\xi^2) = 0$$

$$\left(\frac{\partial^2\phi}{\partial\xi^2}\right)_{P,T} = (b + 6d\xi^2) > 0$$

From these relations we obtain solutions $\xi = 0$ and $\xi^2 = -b/2d$. Since $\xi = 0$ in the disordered state, $b > 0$ at temperatures above the transition temperature; we then find that $b < 0$ for the ordered phase. That is, b changes sign through a second-order transition. Assuming b to vary linearly with temperature, $b(P,T) = B(T - T_c)$ where T_c is the transition temperature (or critical temperature), we obtain

$$\xi^2 = -b/2d = -B(T-T_c)/2d \tag{4.7}$$

It can be readily shown that the change in heat capacity at the transition is given by $B^2 T_c/2d$. We thus see that Landau's theory provides the basis for second-order transitions. The most important point to note is that the average value of the order parameter, $\langle \xi \rangle$, vanishes above T_c in a second-order transition and is non-vanishing below T_c; in a first-order transition, the change in $\langle \xi \rangle$ is discontinuous (Fig. 4.2). Second- and higher-order transitions are often referred to as continuous transitions. The identification of the order parameter is easy for some transitions and not as straightforward in others. For example, in iron, the order parameter associated with the ferromagnetic–paramagnetic transition is magnetization; in ferroelectrics such as $BaTiO_3$, polarization is the order parameter. In the phase transition of $SrTiO_3$ at 108 K, the angle of rotation of the oxygen octahedra (Fig. 4.3) is the order parameter.

Another important contribution by Landau is related to symmetry changes accompanying phase transitions. In second-order or structural transitions, the symmetry of the crystal changes discontinuously, causing the appearance (or disappearance) of certain symmetry elements, unlike first-order transitions, where there is no relation between the symmetries of the high- and low-temperature phases. If $\rho(x, y, z)$ describes the probability distribution of atom positions in a crystal, then ρ would reflect the symmetry group of the crystal. This means that for $T > T_c$, ρ must be consistent with

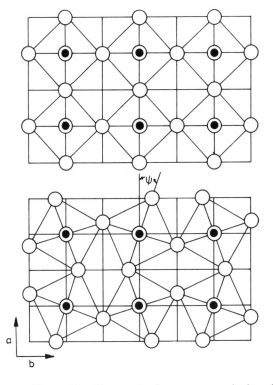

Figure 4.3   Changes in the oxygen octahedra of $SrTiO_3$ at the 108 K transition.

the symmetry group G of the high-temperature (high-symmetry) phase. When $T < T_c$, ρ must be consistent with the symmetry groups G_o of the low-temperature (low-symmetry) phase. We can write

$$\sigma = \rho_o + \delta\rho = G_o + \Sigma C_i \psi_i^\lambda(\Gamma_\lambda) \tag{4.8}$$

where ψ_i^λ are basic functions transforming according to the irreducible representation Γ_λ, and G_o, ρ_o corresponds to the symmetry of G_o and ρ to the symmetry of G; $\delta\rho$ has the same symmetry as ρ. Equation (4.6) is actually written as

$$\phi = \phi_o(P,T) + b\xi^2 + \xi^4 \sum_\alpha V_\alpha f_\alpha^4(\gamma_i) \tag{4.9}$$

where $\xi^2 = \sum C_i^2$, $C_i = \gamma_i \xi$ and $\sum \gamma_i^2 = 1$. The quantity f_α^4 is an invariant of the 4th order constructed from γ_i. A transition from high-symmetry G to low-symmetry G_o is accordingly accompanied by the appearance of an order parameter $\langle \xi \rangle$, when b changes sign.

4.3   Soft modes

The fact that the order parameter vanishes above T_c does not mean that Nature does not have an inkling of things to come well below (or above) T_c. Such indicators are indeed found in many instances in terms of the behaviour of certain vibrational modes. As early as 1940, Raman and Nedungadi discovered that the α–β transition of quartz was accompanied by a decrease in the frequency of a totally symmetric optic mode as the temperature approached the phase transition temperature from below. Historically, this is the first observation of a soft mode. Operationally, a soft mode is a collective excitation whose frequency decreases anomalously as the transition point is reached. In Fig. 4.4, we show the temperature dependence of the soft-mode frequency. While in a second-order transition the soft-mode frequency goes to zero at T_c, in a first-order transition the change of phase occurs before the mode frequency is able to go to zero.

Prediction of soft modes was made independently in 1959 by Cochran and Anderson, who pointed out that phase transitions in some ferroelectrics may result from lattice dynamical instability. Indeed, Born and Huang had earlier shown that a crystal lattice becomes unstable if one of its normal mode frequencies becomes purely imaginary. Cowley in 1962 discovered from neutron-scattering studies that one of the optic modes of $SrTiO_3$ exhibited softening behaviour. Fig. 4.5 shows the temperature dependence of one of the vibrational modes of $SrTiO_3$ associated with the angular oscillations of the oxygen octahedra. In recent years, soft modes have been found in a variety of systems including superconductors and organic solids (Blinc & Zeks, 1974; Rao & Rao, 1978; Scott, 1974). Zone-centre and zone-boundary phonons as well as acoustic phonons can show soft-mode behaviour (Fig. 4.6).

We can understand the relation between soft modes and order parameters in phase

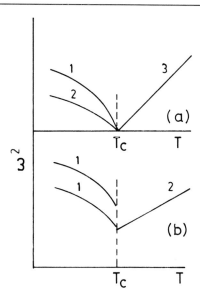

Figure 4.4   Variation of soft-mode frequency, ω^2, with temperature in (a) second-order transitions and (b) first-order transitions. The numbers next to the curves represent degeneracies.

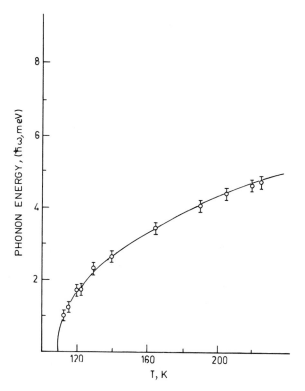

Figure 4.5   Variation with temperature of the soft-mode frequency in $SrTiO_3$. (After Shirane & Yamada, 1969.)

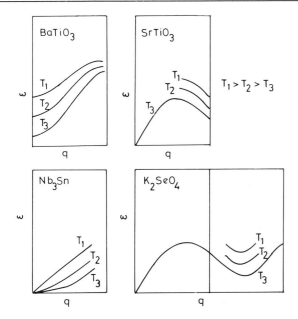

Figure 4.6 Various kinds of soft modes observed in phase transitions. Softening can occur anywhere in the Brillouin zone. (After Venkataraman, 1979.)

transitions as follows. The order parameter, ξ, is not a static quantity although $\langle \xi \rangle$ is; ξ can fluctuate locally and these fluctuations can be derived by a suitable superposition of the atomic displacements associated with soft-mode vibrations. We accordingly find that the square of the vibration frequency, $\omega^2 = B(T - T_c)$ just as in eqn.(4.7). Not every phase transition is associated with a soft mode. Irrespective of whether there is a soft mode or not, the fluctuation spectrum near T_c shows a characteristic behaviour. Fluctuations are thus a key feature associated with a phase transition. Soft-mode behaviour under pressure has been examined by Samara (1984). If there is considerable anharmonicity, the soft-mode behaviour is given by, $\omega^2 = AT + A' (T - T_c)$.

4.4   Central peaks

One of the most striking features of the scattered spectrum for either neutrons or light in the vicinity of a phase transition is the appearance of a divergent elastic or quasielastic peak centred near zero frequency shift that lies entirely outside the quasiharmonic soft-mode description of the dynamics (Fleury & Lyons, 1983). The first observation of a divergence in scattered intensity is due to Yakovlev et $al.$, (1956), who observed the phenomenon in the α–β transition of quartz. The scattered intensity increases dramatically, sometimes by a factor of 10000 near T_c, and the maximum value of line width of the diverging feature is itself rather small (~ 1 cm^{-1}). In Fig. 4.7, typical central peaks are shown for the purpose of illustration.

It was thought for some time that central peaks were due to impurities, defects and other such extrinsic or intrinsic factors. A number of models and mechanisms based on entropy fluctuations, phonon density fluctuations, dielectric relaxation, molecular

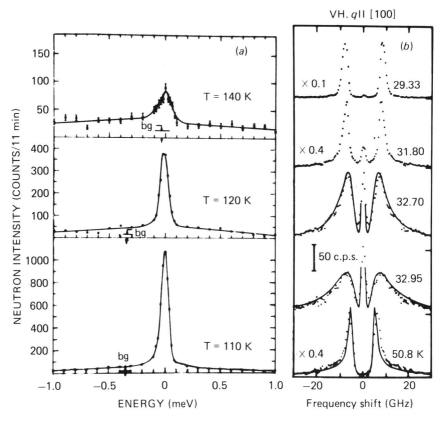

Figure 4.7   Central peaks: (a) scattered neutron specta of $KMnF_3(T_c = 90 \, K)$. (After Shapiro *et al.*, 1974.); (b) quasielastic light scattering of $TbVO_4(T_c \sim 32 \, K)$. (After Harley *et al.*, 1980.)

reorientation, solitons, phasons (fluctuations in the phase of the wave along the **k** vector), dynamic and static clusters etc. have been proposed. It is now recognized that singular central peaks at temperatures near T_c can arise from either dynamic or static phenomena. Measurements of line shapes and line widths have permitted definitive mechanisms, but no single phenomenon appears to be applicable for the central peaks in all the systems.

4.5   Critical phenomena

Many physical properties diverge near T_c (i.e. show large values as T_c is approached from either side). Interestingly, divergences of similar quantities in different phase transitions are strikingly similar as shown by the typical C_p–temperature curves in Fig. 4.8. These divergences can be quantified in terms of the so-called critical exponents. A critical point exponent is given by

$$\lambda = \lim_{\varepsilon \to 0} \left| \frac{\ln \, f(\varepsilon)}{\ln \, \varepsilon} \right| \tag{4.10}$$

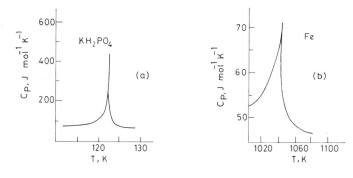

Figure 4.8   Comparison of specific heats of two second-order transitions: (a) KH_2PO_4; (b) Fe.

where $\varepsilon = (T - T_c)/T_c$ and $f(\varepsilon)$ is the function whose exponent is λ. The most important exponents are those associated with the specific heat (α), the order parameter (β), the magnetic susceptibility (γ) and the correlation length (v). The correlation length refers to the range over which individual constituents (like atoms, atomic moments, etc.) are correlated. It so happens that the individual exponents for different transitions are roughly similar (e.g. $\beta \approx 0.33$). More interesting is the fact that relations such as $\alpha + 2\beta + \gamma = 2$ hold for transitions independent of the detailed nature of the system.

Such a universality in critical point exponents had intrigued scientists for a long time, but is now understood in the light of Kadanoff's concept of *scale invariance* (Kadanoff, 1966) associated with fluctuations near T_c. Kadanoff made use of the 'self-similarity' amongst the fluctuations near T_c and established that the Gibbs potential must be a homogeneous function close to T_c. Wilson & Fisher (1972) employed the *renormaliz-ation group* method to calculate the exponents precisely. All phase transitions can be characterized in terms of the dimensionality of the system, d, and the dimensionality of the order parameter, n (see Fig. 4.9). Wilson and Fisher treated the dimensionality of the system as a continuously varying parameter, $\varepsilon = (4 - d)$, where ε varies continuously. For developments in critical phenomena see Domb, (1985). Renormalization group has been applied to many situations: tricritical points, transitions in 2D systems, generaliz-ation of the Ising model to more than two components and the concept of fractals (Mandelbrot, 1983).

4.6   Structural changes in phase transitions

On the basis of our knowledge of crystal chemistry, we can predict the nature of structural changes in the phase transitions of simple ionic solids. In making such predictions, we take advantage of the fact that in thermal transitions the high-temperature form has higher symmetry than the low-temperature form. In pressure transitions, the high-pressure form will be more closely packed (higher coordination number) and will have a smaller volume. Thus, on application of pressure, NaCl-type solids would give rise to CsCl-type solids. Heating CsCl-type solids, on the other hand, would give rise to NaCl-type solids. In AB_2-type solids, we could expect a distorted rutile or some other low-symmetry structure to transform to the rutile or the fluorite

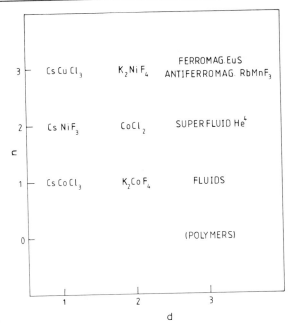

Figure 4.9   Some observed transitions in the *n–d* plane. The three Cs salts are structurally one-dimensional while the polymers are structurally three-dimensional but there is no magnetic ordering. (From Rao, 1984.)

structure at high temperatures. A distorted perovskite would be expected to transform to the cubic structure at high temperatures. The Born model of ionic solids with the appropriate repulsive and van der Waals parameters can explain the relative stabilities of crystal structures; in partly covalent solids, an ionicity parameter would have to be used to predict the preferred crystal structure (see Chapter 1, Section 1.3).

Buerger (1951) classified phase transitions on the basis of structural changes involving primary or higher coordination as follows:

(i)   Transformations involving first coordination (e.g. CsCl–NaCl, aragonite–calcite)
  (a) reconstructive (sluggish)
  (b) dilatational (rapid).
(ii) Transformations involving second or higher coordination ($BaTiO_3$; α–β quartz)
  (a) reconstructive (sluggish)
  (b) displacive (rapid)

In displacive transitions only small changes in the arrangement of coordination polyhedra occur. *Reconstructive transitions* would require the breaking and making of bonds, but the same can be accomplished by a simple dilatational mechanism. Buerger proposed such a mechanism for the transformation from the CsCl structure to the NaCl structure (Fig. 4.10). Such deformational relations are known to exist between

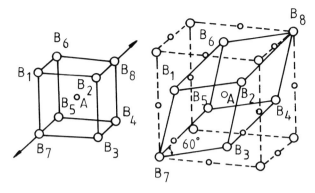

Figure 4.10   Dilatational mechanism for the transformation from the CsCl structure to the NaCl structure. (After Buerger 1951.)

different structures (Rao & Rao, 1978) and they are important in understanding mechanisms of phase transitions. In displacive transitions, only small changes in the arrangement of coordination polyhedra occur. In order–disorder transitions, large atomic displacements occur.

Structural transitions are also classified into ferrodistortive (no change in the number of formula units in the unit cell) or antiferrodistortive (number of formula units in the unit cell changes) transitions. *Ferrodistortive transitions* can be displacive (e.g. $BaTiO_3$) or of order–disorder type (e.g. NH_4Cl). *Antiferrodistortive transitions* can also be displacive (e.g. $SrTiO_3$) or of order–disorder type (e.g. NH_4Br). Ferrodistortive displacive transitions are exhibited only by ferroelectric materials; antiferrodistortive displacive transitions are shown by both ferroelectrics and antiferroelectrics. In materials such as KH_2PO_4, there is a strong coupling between order–disorder and displacive types of variables. The pressure transition of SnI_4 is unique in that the high pressure phase is amorphous (Sugai, 1985). Does the amorphous state arise because of topological frustration in packing tetrahedra under pressure?

An interesting aspect of many structural phase transitions is the coupling of the primary order parameter to a secondary order parameter. In transitions of molecular crystals, the order parameter is coupled with reorientational or libration modes. In Jahn–Teller as well as ferroelastic transitions, an optical phonon or an electronic excitation is coupled with strain (acoustic phonon). In antiferrodistortive transitions, a zone-boundary phonon (primary order parameter) can induce spontaneous polarization (secondary order parameter). Magnetic resonance and vibrational spectroscopic methods provide valuable information on static as well as dynamic processes occurring during a transition (Owens *et al.*, 1979; Iqbal & Owens, 1984; Rao, 1993). Complementary information is provided by diffraction methods.

4.7   Mechanisms of phase transitions

Phase transitions in solids are also fruitfully classified on the basis of the mechanism. The important kinds of transitions normally encountered are: (i) nucleation-and-growth transitions; (ii) order–disorder transitions and (iii) martensitic transitions.

In the *nucleation-and-growth transitions*, nuclei of the new phase possessing a critical size have to be first formed in the parent phase. The change in free energy, ΔG, due to the formation of spherical nuclei is given by

$$\Delta G = -\tfrac{4}{3}\pi r^3 \Delta G_v + 4\pi r^2 \gamma \tag{4.11}$$

where the first term on the right-hand side represents the decrease in bulk free energy and the second term represents the increase in surface free energy. When the radius of the spherical nucleus, r, is small, the second term will dominate and the nucleus will therefore be thermodynamically unstable. However, if the nucleus attains a critical radius, r_c, ΔG will begin to decrease and further growth will become thermodynamically favourable. In Fig. 4.11, the variation of ΔG with r is shown at different temperatures. We see that at the thermodynamic transition temperature, T_o, ΔG is always positive and no nucleation can occur. A little supercooling or a little superheating is always necessary for the transformation to occur. By setting $(\partial \Delta G / \partial r)$ equal to zero, we can obtain r_c and ΔG_c from equation (4.11)

$$r_c = 2\gamma / \Delta G_v$$

$$G_c = \frac{16\pi\gamma^3}{3\Delta G_v^2}$$

The rate of nucleation can be shown to be given by

$$R = A \exp[-(\Delta G_c + E_a)/RT] \tag{4.12}$$

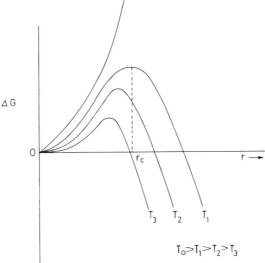

$$T_o > T_1 > T_2 > T_3$$

Figure 4.11   Variation of nucleation barrier, ΔG, with the radius, r, of the nucleus at various temperatures.

where A is the pre-exponential factor and E_a is the activation energy for the jump of atoms across the nucleus–bulk interface. Once nucleation has occurred, phase transitions proceed towards completion by the growth (or the propagation) of the critical-sized nuclei. The growth step will also be associated with a free energy of activation.

Empirical relations are often used to analyse kinetic data on nucleation-growth transitions. One of the common relations is that established by Avrami

$$x = 1 - \exp(-kt^n) \tag{4.13}$$

where x is the volume fraction of the transformed phase and t is the time. Equation (4.13) adequately describes kinetics of transitions, the value of n varying (anywhere between 1 and 5) depending on the nature of the transition. In Fig. 4.12, Avrami plots for the transformation of grey tin are shown (Burgers & Green, 1957). The value of n for the white $\rightarrow$ grey tin transition is 3, while for the grey $\rightarrow$ white transition it is between 1.5 and 2.0. The values of k and n in eqn.(4.13) are obtained by plotting log log $[1/(1 - x)]$ against log t to obtain straight lines. The transformation of the anatase form of TiO_2 to the rutile form is another example of a nucleation-growth transition.

Order–disorder transitions are generally associated with (i) positional disordering, (ii) orientational disordering or (iii) disordering of electronic (or nuclear) spin states. The configurational entropy due to disordering is given by

$$\Delta S = R \ln(\Omega_{II}/\Omega_I) \tag{4.14}$$

where Ω_{II} and Ω_I are the total number of configurations (or orientations) in the product and parent phases respectively. Two familiar examples of positional disordering are the transitions of β-brass and AgI. AgI undergoes a transition from a hexagonal to a cubic

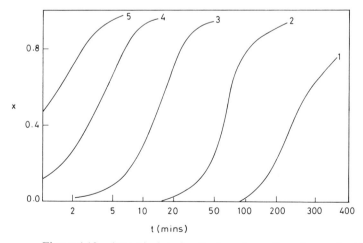

Figure 4.12   Avrami plots for the transformation of grey tin: 1298 K; 2300.5 K; 3303 K; 4305.5 K; 5308 K.

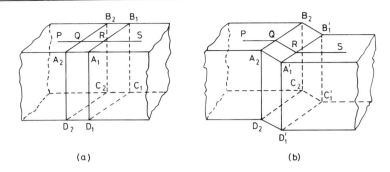

Figure 4.13   Representation of a plate of martensite (b) formed from a rectangular block of the parent crystal (a). Notice that the plane surfaces $A_2B_2C_2D_2$ and $A_1'B_1'C_1'D_1'$ remain undistorted and unrotated during the transition.

phase, the latter exhibiting high ionic conductivity owing to the randomization of Ag^+ ions. The increase in entropy at this transition ($16.74\,J\,deg^{-1}\,mol^{-1}$) is even greater than the expected increase in configurational entropy ($R\,\ln 12/2 = 14.86\,J\,deg^{-1}$ mol^{-1}), suggesting that the Ag^+ ions in the high-temperature cubic phase explore many more sites than the available 12 tetrahedral sites.

There are many examples of orientational disorder in inorganic solids. Most solids containing oxyanions, or any di- or polyatomic ions, exhibit orientational order–disorder transitions (Parsonage & Staveley, 1978). Thus in the III–II transition of KCN at 83 K, ΔS is around $R\ln 2$, indicating that CN^- ions gain two possible orientations. In the II–I transition of KCN at 168 K, the ΔS is $R\ln 4$, indicating that in the cubic phase I, the CN^- ions orient in any of the eight possible directions. Similarly, in the III–II transition of NH_4Cl at 243 K, ΔS is $0.7\,R$, showing that in the disordered phase II, NH_4^+ ions utilize two possible orientations (and that there is no free rotation).

Although *martensitic transitions* were originally discovered in steel, they are now considered to provide the mechanism for transitions in a variety of inorganic (and possibly even organic) solids. A martensitic transition is a structural change caused by atomic displacements (and not by diffusion) corresponding to a homogeneous deformation which gives rise to an invariant strain plane through which the parent and product phases are related by a substitutional lattice correspondence, an irrational habit plane and a precise orientational relationship (see Fig. 4.13). The systematic change in atomic positions over macroscopic distances has led to the metaphoric description of these transitions as 'military' transformations (Christian, 1965). Martensitic transitions often occur with high velocities (of the order of sound velocity). It is interesting that the CsCl–NaCl transition for which Buerger ingeniously proposed a dilatational mechanism (Fig. 4.10) is a martensitic transition with orientational relations between the two phases. The actual deformation mechanism differs only slightly from that in Fig. 4.10. Other examples of solids undergoing martensitic transitions are ZrO_2, $KTa_{0.65}Nb_{0.35}O_3$ and superconducting A-15 compounds such as V_3Si.

The phase transitions discussed hitherto occur without any change in composition.

Spinodal decompositions, often observed in binary solid solutions of metals and in glasses, on the other hand, arise from thermodynamic instabilities caused by composition (Cahn, 1968). A special feature of this type of solid state transformation is the absence of any nucleation barrier. There is a class of transformation called *eutectoid decomposition* in which a single phase decomposes into two coupled phases of different compositions, the morphology generally consisting of parallel lamellae or of rods of one phase in the matrix of the other.

4.8   Organic solids

It was believed until a few years ago by many workers that there are no structural relations between phases in organic crystal transformations, and the transformations were considered to be somewhat similar to melting. Many interesting transitions of organic solids, specially of the displacive type, have been investigated (e.g. triazine, benzil, chloranil) by employing spectroscopic methods (Rao, 1993). Jones *et al.* (1975) have examined the stress-induced phase transition of 1,8-dichloro-10-methylanthracene which proceeds by a diffusionless, displacive mechanism (somewhat similar to a martensitic transition) with definite orientational relationship. The irrational habit plane seems to be composed of close-packed planes, and the properties of the interface could be formulated in terms of slip dislocations. The reversible topotactic phase transition of 5-methyl-1-thio-5-azoniacyclooctane-1-oxide perchlorate has been explained by Parkinson *et al.* (1976) in terms of recurrent glissile partial dislocations; it was earlier thought that the transition involved a cooperative inversion and rotation of half of the molecular cations (Paul & Go, 1969). The γ-phase of *p*-dichlorobenzene is characterized by unusually high intramolecular vibration frequencies and the α–γ transition shows athermal nucleation behaviour (Ganguly *et al.*, 1979); the α–β transition is associated with disordering. Benzothiophene exhibits an order–disorder transition (Rao *et al.*, 1982). The transformation of the yellow form of 2-(4'-methoxyphenyl)-1,4-benzoquinone to the red form involves nucleation and migration of well-defined fronts; the molecules in the two phases are diastereoisomers (Desiraju *et al.*, 1977). The phase transition of *p*-terphenyl involving rotational disorder has been elucidated by evaluating pairwise interaction between nonbonded atoms (Ramdas & Thomas, 1976).

Investigations of the phase transitions of several organic solids by employing magnetic resonance and vibrational spectroscopic techniques have yielded valuable information on the mechanisms of the transitions. (Owens *et al.*, 1979; Iqbal & Owens, 1984; Rao, 1993). While the magnetic resonance methods are generally more appropriate to examine local dynamics, inelastic scattering and infrared absorption spectroscopy are useful to study collective dynamics. The vibrational spectroscopic methods generally employed are inelastic neutron scattering, IR absorption and light scattering, the last including Brillouin scattering and Raman scattering. Typical displacive transitions studied are those of sym-triazine, chloranil and benzil. Dunitz (1991) has reviewed some interesting features of the phase transitions in organic solids. A puzzling aspect of the crystallization of many organic solids is the disappearance of elusive polymorphs due to loss of control over the process. Molecules that can adopt different shapes particularly give rise to such disappearing polymorphs. The nature of this phenomenon has been discussed by Dunitz & Bernstein (1995).

Phase transitions of hydrogen-bonded solids such as ferroelectric hydrogen phosphates and Rochelle salt have been investigated widely in the literature. The phase transitions of alkanedioic acids have been investigated by employing vibrational spectroscopy (Rao *et al.*, 1982). The transition of malonic acid is specially interesting. At ordinary temperatures, the unit cell of malonic acid contains two cyclic dimeric rings orthogonal to each other; above its phase transition at 360 K, the two hydrogen-bonded rings become similar, as evidenced by IR and Raman spectra (de Villepin *et al.*, 1984). Hydrogen bonds in the high-temperature phase are, on average, weaker than those in the low-temperature phase. The phase transition occurs at a higher temperature (366 K) in the fully deuterated acid and the vibrational bands show a positive deuterium isotope effect. It appears that the transition is governed by vibrational and torsional modes of the hydrogen-bonded rings (around 90 and 50 cm^{-1} respectively below the transition temperature), which show a tendency to soften. Layered compounds of the type $(RNH_3)_2MCl_4(M = Mn, Fe, Cd$ or $Cu)$ exhibit several phase transitions related to the hydrogen bonding of the RNH_3 groups, which gives rise to ordered low-symmetry structures at low temperatures. In the methylammonium compounds, for example, the $(CH_3NH_3)^+$ ion attains C_{3v} symmetry only in the high-temperature phase; infrared spectra clearly reveal the nature of the transitions through changes in the vibrational modes of the alkylammonium groups (Rao *et al.*, 1981).

4.9   Incommensurate phases

Solids generally crystallize in regular structures with a periodic arrangement of atoms. In certain solids, interaction between electrons and ion cores or the presence of two modes of transformation of different symmetry renders the regular array of atoms unstable with respect to small distortions. The stable state would then be one where the charge density or the atomic positions exhibit long-period modulations (Riste, 1977; Bak, 1982). The period of the modulations may be commensurate or incommensurate with the underlying lattice, the incommensurate phase showing no true periodicity but having two unrelated periods. Generally speaking, incommensurate phases occur in systems with competing periodicities; while one of the periods is that of the basic lattice, the other period can be that of another atomic lattice with its own periodicity, giving rise to a misfit when the two sublattices are packed together. Such a misfit is responsible for the incommensurate two-dimensional structures found in many of the layered solids such as the graphite intercalation compounds (Makovicky & Hyde, 1981). Besides such compositional incommensurability, phases where the incommensurate quantity is a periodic distortion of an otherwise regular lattice (displacive incommensurability) is commonly encountered both in conducting (e.g. TaS_2) and insulating (e.g. $NaNO_2$, K_2SeO_4) materials. Incommensurate phases are encountered commonly in metal oxides, sulphides and other materials where point defects (vacancies) order themselves, giving rise to superstructures (see Chapter 5, Section 5.5). Incommensurate magnetic structures have been known for some time; the spiral magnets analysed by

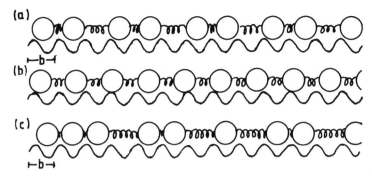

Figure 4.14   One-dimensional model showing interaction between a row of atoms (circles) and a periodic potential (wavy lines) giving rise to (a) commensurate structure, (b) incommensurate structure and (c) chaotic structure. (Following Bak, 1982.)

Dzyaloshinskii (1964) belong to this category. The term *modulated structures* has been used to describe incommensurate perturbation (i.e. one in which the ratio of the imposed periodicity to that of the unit cell is irrational). A broader definition of modulated structure can be used to describe any periodic or partly periodic perturbation of a cystal structure with a repetition distance appreciably greater than the basic unit cell dimensions; such a definition would include a variety of superstructures as well (Cowley *et al.*, 1979).

The formation of incommensurate phases can be visualized by considering a one-dimensional array of atoms (Fig. 4.14). The array of atoms possessing period *a* interacts with a potential of period *b*. We can distinguish three possible cases: (a) When the interaction is strong enough, the lattice is commensurate with the underlying potential and the lattice period *a* is a simple rational fraction of the period *b* of the potential. (b) When *a* and *b* are not related by a simple rational ratio, the structure becomes incommensurate. (c) The third possibility is a *chaotic structure* where the atoms are distributed in a random way among the potential minima. The incommensurate wave is made up of an amplitude wave (amplitudon) and a phase wave (phason). The presence of higher harmonics of the incommensurate modulation leads to the so-called 'devil's staircase'.

In conducting solids, the conduction electron density is spatially modulated, forming *charge density waves* (CDW); the periodic distortion accompanying the CDW (due to interaction between the conduction electron and the lattice) is responsible for the incommensurate phase (Overhauser, 1962; Di Salvo & Rice, 1979; Riste, 1977). The occurrence of CDW and the periodic distortion can be understood in terms of the model proposed by Peierls and Fröhlich for one-dimensional metals. Let us consider a row of uniformly spaced chain of ions (spacing $= a$) associated with conduction electrons of energy $E(k)$ and a wave vector k. At 0 K, all the states are filled up to the Fermi energy, $E_F = E(k_F)$. If the electron density is sinusoidally modulated as in Fig. 4.15 such that

$$\rho(x) = \rho_o(x)[1 + \rho_1 \cos(\vec{Q}\cdot\vec{x} + \phi)] \qquad (4.15)$$

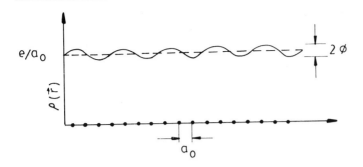

Figure 4.15   CDW formation in a one-dimensional metal. Sinusoidally modulated electron density is shown above. The dots along the x-axis represents the ions. (Following Di Salvo, 1977.)

where ρ_1 is the amplitude of charge modulation and $\bar{Q}$ its wave vector, the phase factor ϕ describes the position of the CDW relative to the ions of the underlying lattice. Displacing the wave by a distance $\bar{u}$ would then be equivalent to a change in phase from ϕ to $\phi + \bar{Q}\cdot\bar{u}$ and the charge modulation causes a modulation of ionic positions as given by

$$U_n = U_o \sin(n\,\bar{Q}\,a + \phi) \tag{4.16}$$

where n is an integer defining the position of the ion. The periodic distortion of the lattice with wave vector $\bar{Q} = 2k_F$ opens up a gap in the energy spectrum at E_F. The states below E_F which are pushed down in energy are occupied while those that are raised are empty. The total energy is therefore lowered. If the gain in energy is larger than the potential energy cost of the distortion, a CDW will spontaneously appear. At 0 K, such a distortion will always occur in a one-dimensional system no matter how weak the electron–lattice coupling is. As the temperature is raised, electrons will be thermally excited across the gap, thus reducing the gain in energy. Finally, at the transition temperature, T_t, the gap vanishes. Above the transition temperature, the processes that give rise to crystal and charge distortion are still active. Although they can no longer give rise to a stable distortion of the lattice, they do produce an anomaly in the behaviour of longitudinal acoustic phonons, called a *giant Kohn anomaly*; the anomaly is in the energy of the phonons, which shows a decrease as the wave number is increased through $2k_F$. According to the simple model, the phonon energy at $2k_F$ should decrease to zero as the temperature approaches T_t. Below T_t, the phonon is frozen into a static distortion.

Besides electron–phonon coupling, the shape of the Fermi surface is important in deciding CDW formation. Fermi surfaces of systems containing linear or planar arrays of atoms reflect their one- or two-dimensional character. Since the Fermi surfaces connect many states with the same wave vector, $\bar{Q}$, a periodic distortion having the wave vector $\bar{Q}$ will produce gaps at those portions connected by $\bar{Q}$. The energy gained by creating such gaps around the Fermi surface may overcome the cost of potential energy due to the lattice distortion, thereby rendering the distortion stable.

Dichalcogenides of group Va transition metals provide classic examples of metallic conductors exhibiting CDW-induced periodic lattice distortions (Wilson *et al.*, 1975; Williams, 1976). These solids crystallize in several polytypic structures, 1T (CdI$_2$ type), 2H (MoS$_2$) and 4H, where the transition metal has octahedral, trigonal prismatic and both types of coordination respectively. The distorted low-temperature phases show satellite (superstructure) spots in the electron-diffraction patterns (Fig. 4.16), which are separated from the Bragg spots by $m \vec{Q}$, where m is an integer. The location of the satellite spots helps to determine the period of CDW. Many of these dichalcogenides exhibit two transitions, one at T_{IC} and the other at T_C. At T_{IC}, the superlattice spots are incommensurate with the basic hexagonal lattice, the wave vector $\vec{Q}$ of the distortion being equal to $(a^*/n) + \delta$. The fraction δ becomes zero at T_C, with Q exactly equal to $a^*/3$. Below T_C, the phase is commensurate possessing a unit cell which is n times as big as the basic cell. In the temperature interval $T_C < T < T_{IC}$, the incommensurate structure exists with a $\delta (\neq 0)$ that decreases with temperature and drops discontinuously to zero at T_C. For the incommensurate phase, a unit cell cannot be defined; no unit cell can contain an exact period of both the wave and the underlying crystal structure. Accompanying the phase transitions, anomalies in electrical resistivities of the chalcogenides are observed (Fig. 4.17).

Rouxel (1992) has nicely reviewed the interesting properties of niobium and tantalum chalcogenides possessing chain structures. NbSe$_3$ contains irregular MSe$_3$ chains with a Se–Se bond running parallel to the b-axis of the unit cell. Three types of chains with Se–Se distances 2.37, 2.49, and 2.91 Å have been found, and the nearest-neighbour NbSe$_3$ chains are shifted by $b/2$. NbSe$_3$ exhibits two charge density waves (CDW) at 145 and 59 K, and is an archetypal solid for the study of CDW. The two CDWs are localized on the two chains with the shortest Se–Se bond distances as confirmed by the observation of superstructures below the CDW transitions. NbSe$_3$ is not a perfect one-dimensional conductor because it is metallic below the CDW transitions, suggesting that only a portion of the Fermi surface is nested. The analogous TaS$_3$ however exhibits increased one-dimensional character including a metal–insulator transition at 210 K. The chains in this compound behave more like independent units.

Simple chains occur in tri- and tetrachalcogenides of niobium and tantalum. In the tetrachalcogenides, only telluride chains exist independently. The selenide chains need to be stabilized by counter ions or chains. More complex low-dimensional structures are formed by the condensation of different chains, as in FeNb$_3$Se$_{10}$, by the one-dimensional intergrowth of different units, as in Nb$_6$Se$_6$Br or by multicomponent fibres, as in Ta$_4$SiTe$_4$. Mixed polyhedra with a sort of supermolecular architecture are found in P-Nb(Ta)-S phases; Ta$_4$P$_4$S$_{29}$ has double intertwined helices. The study of low-dimensional solids is especially interesting because it is at the interface between inorganic coordination chemistry and solid state chemistry.

K$_2$[Pt(CN)$_4$]Br$_{0.3}$·xH$_2$O (KCP) is an example of a one-dimensional compound (see Chapter 6 for a discussion of low-dimensional solids) consisting of parallel stacks of

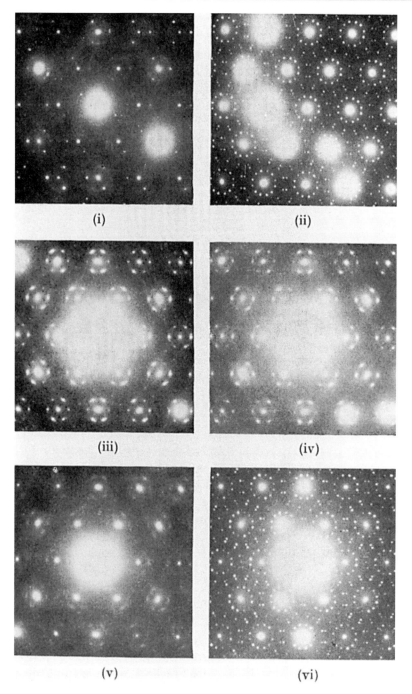

Figure 4.16   Electron diffraction patterns of 1T TaSe$_2$ and 4H$_b$ TaS$_2$. (i) 1T TaSe$_2$ at 500 K; (ii) 4H$_b$ TaS$_2$ at 300 K showing 13a$_0$ Superlattice; (iii) 4H$_b$ TaS$_2$ at 320 K; (iv) 4H$_b$ TaS$_2$ at 350 K; (v) 4H$_b$ TaS$_2$ at 460 K (notice that the pattern is similar to (i); (vi) 4H$_b$ TaS$_2$ cooled back to 300 K, 13 a$_0$ Superlattice is reestablished. (After Wilson *et al.*, 1975.)

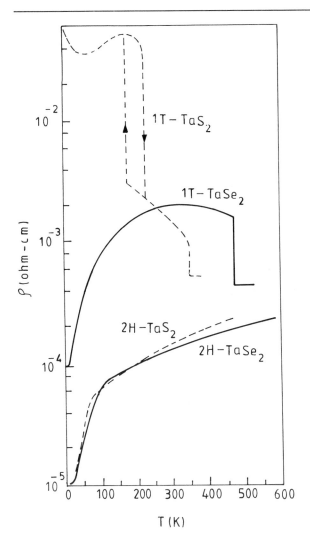

Figure 4.17   Electrical resistivity parallel to the layers of transition-metal dichalcogenides. (After Wilson *et al.*, 1975.)

[Pt(CN)$_4$] forming linear chains of platinum atoms. Within each chain, the metal atoms are equally spaced with a Pt–Pt distance of 2.88 Å (slightly higher than the value in platinum metal, 2.77 Å) as shown in Fig. 4.18. The stacks are separated from one another by 9.87 Å. Whereas in the insulating K$_2$[Pt(CN)$_4$]·xH$_2$O the highest occupied band (comprising Pt (d_z^2) orbitals) is completely filled, in partly oxidized KCP, 0.3 Br removes an equivalent number of electrons from the platinum, leaving the d_z^2 band 85% filled. KCP undergoes a broad metal–insulator transition near 200 K and is a good insulator at low temperatures. X-ray diffraction patterns of KCP at room temperature show diffuse satellite lines along with the Bragg spots (Comes *et al.*, 1973). The diffuse lines reveal a displacement of atoms parallel to the chain with wave vector $\vec{Q} = 0.3C^*$

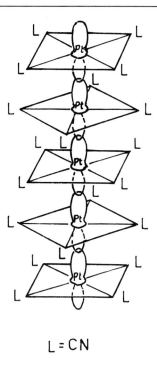

L = CN

Figure 4.18   Schematic structure of $K_2[Pt(CN)_4]Br_{0.3} \cdot 3H_2O$ (KCP). A schematic arrangement of the $[Pt(CN)_4]$ groups in the chain is shown.

where C^* is the reciprocal lattice constant along the metal chain direction. On lowering the temperature to 77 K, the diffuse satellites become sharper. Neutron scattering studies (Carneiro et al., 1976) have established the existence of a $2k_F$ Kohn anomaly in KCP. Low-temperature results reveal three-dimensional ordering of the lattice distortion on neighbouring chains, but the transverse correlation length is not very large. Salts of tetrathiafulvalene (TTF) have crystal structures composed of two incommensurate sublattices (Johnson & Watson, 1976).

An exotic, anisotropically conducting solid showing incommensurate phases is $Hg_{3-x}AsF_6$, consisting of chains of Hg atoms and a host lattice of AsF_6 (Pouget et al., 1978; Emery & Axe, 1978). The Hg atoms are well ordered along the chain direction and the intrachain Hg distance is incommensurate with the host lattice parameter along the chain direction. There is no phase order amongst the neighbouring Hg chains at room temperature, but progressive phase ordering occurs on cooling. There is a periodicity in the position of the Hg chains imposed by the regular network of open-channels of the host AsF_6 sublattice. At 120 K, the competing orthogonal chain–chain interaction leads to three-dimensional order and phase ordering phase transition. The three-dimensional low-temperature phase is accompanied by an incommensurate modulation of the Hg chains, indicating an interaction with the host lattice. $Hg_{3-x}AsF_6$ is indeed a curious solid where, roughly speaking, the one-dimensional columns of Hg atoms do not fit into an integral number of AsF_6 cells.

Besides its anisotropic electrical and optical properties, this solid exhibits superconductivity at low temperatures.

As mentioned earlier, many insulating solids exhibit incommensurate phases, the phase transitions generally occurring in the order normal → incommensurate → commensurate as the temperature is lowered. The normal–incommensurate phase transition is referred to as the incommensurate transition. For example, $K_2Pb[Cu(NO_2)_6]$ shows the incommensurate transition at 281 K (T_{IC}) and the transition to the commensurate phase at 273 K (T_C). The corresponding transitions in K_2SeO_4 are at 129.5 K and 93 K, the latter being associated with ferroelectricity. Several tetraalkylammonium halide derivatives of the type $[N(CH_3)_4]_2MCl_4$, where M = Cd, Zn, etc., show such incommensurate transitions as well. Heine & McConnell (1981, 1984) have proposed a model to account for incommensurate transitions in insulators. According to this model, if there are two different modes of transformation ψ and ϕ (with different symmetries) from the high-temperature undistorted (disordered) phase, they cannot be present in a uniform, commensurate phase. They can, however, interact in a modulated (incommensurate) phase to lower the free energy if their symmetries satisfy a certain connecting relationship. One of the modes dominates at low temperatures, giving rise to the commensurate phase. These authors use Landau's theory of soft-mode phase transitions, expanding the free energy on the high-temperature side and looking for an incommensurate instability at T_{IC}. In $NaNO_2$, the NO_2 groups order with $q = 0$ below T_c (all pointing in the same direction along the b axis) with an accompanying displacement of Na^+ ions. In the incommensurate phase, the NO_2 groups point positively along the b axis with a sinusoidal probability P, and along $-b$ with a probability $1 - P$, where $P = \frac{1}{2} + \psi \cos qx$ and q varies from 0.1 to 0.12 times a^*. The other transformation is a local shear of the unit cell as described by $\varepsilon = \phi \sin qx$. In the Jahn–Teller system, $K_2Pb[Cu(NO_2)_6]$, the operative pair of modes are the equivalent $x^2 - y^2$ and $3z^2 - r^2$ types of distortion of the octahedral complex (McConnell & Heine, 1982).

Let us briefly look at the phase transitions of $K_2Pb[Cu(NO_2)_6]$ which has a cubic (F3m) structure at room temperature (Joesten $et\ al.$, 1977). This solid undergoes phase transitions at 281 K and 273 K. The intermediate incommensurate phase is tetragonal while the lowest temperature-commensurate phase has a symmetry lower than orthorhombic, with superlattice reflections at $(h + \frac{1}{2}, k + \frac{1}{2}, l + \frac{1}{2})$ in cubic indices. The diffraction data are compatible with an antiferrodistortive structure consisting of two different sublattices; in one sublattice the a-axis is in the direction of tetragonal elongation of the $Cu(NO_2)_6$ octahedron, while in the other the b-axis becomes the direction of elongation. The incommensurate phase is more interesting. The remarkable feature is that it shows satellite reflections at $(h \pm 0.425, k \pm 0.425, 0)$ in cubic indices, indicating that it possesses an incommensurate structure. The model proposed for the static structure of this phase involves local Jahn–Teller distortion which is modulated along [110] with $|k_o| = 0.425$. The structure based on this model is shown in Fig. 4.19. In terms of the pseudospin model, the commensurate phase is characterized as a canted spin structure, while the incommensurate phase is described as a 'fan' spin structure, wherein the orientation of the pseudospin changes from site to site in a plane

Z = 0 layer

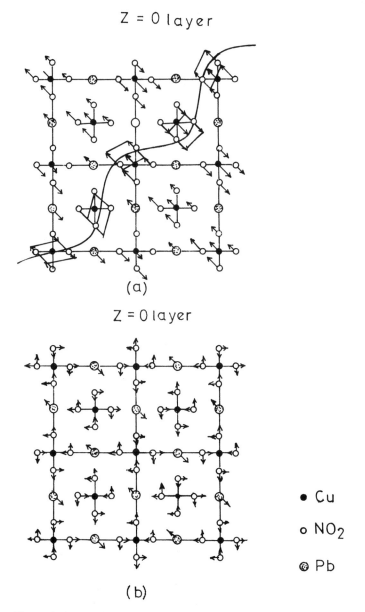

(a)

Z = 0 layer

(b)

● Cu

○ NO$_2$

◉ Pb

Figure 4.19   Model structures of (a) incommensurate and (b) commensurate phases of K$_2$Pb[Cu(NO$_2$)$_6$]. The displacement pattern of Jahn–Teller active phonons is shown by arrows. In (a) the phonon mode has wave-vector $k = (0.425, 0.425, 0)$ and in (b), wave-vector of the phonon mode is $k = (\frac{1}{2}, \frac{1}{2}, \frac{1}{2})$. (After Yamada, 1977.)

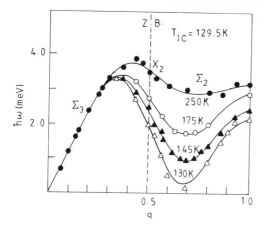

Figure 4.20   Dispersion relation of the soft mode in K_2SeO_4. (After Iizumi et al., 1977.)

describing a 'fan'-type arrangement. In summary, the phase transitions in $K_2Pb[Cu(NO_2)_6]$ are characterized as normal → incommensurate → commensurate structures which are accompanied by cubic → pseudotetragonal → pseudomonoclinic structural changes. The change in the electronic state of the Jahn–Teller active Cu^{2+} ion is characterized as para → fan → canted spin ordering. Appearance of the incommensurate phase is due to propagation of the local Jahn–Teller mode with an incommensurate wave vector k.

Light scattering and neutron scattering have been employed effectively for investigating incommensurate phases (Klein, 1983; Axe, 1980). Soft modes associated with incommensurate transitions show interesting features. For example, in K_2SeO_4 softening occurs at a somewhat odd value of q, as shown in Fig. 4.20. Fig. 4.21 shows a portion of the reciprocal lattice of the high-temperature (normal) orthorhombic phase; the corresponding lattice of the ferroelectric (commensurate) phase is also shown in the figure. We see from Fig. 4.20 that the minimum of the soft branch changes with temperature though it is close to $q = (\frac{1}{3}, 0, 0)$. The displacement δ from this point (Fig. 4.21) decreases with temperature. In the incommensurate phase, a satellite Bragg reflection develops in the region of reciprocal space and gives a measure of δ down to T_C (93 K); at T_C, δ becomes zero discontinuously, leading to an incommensurate-commensurate phase transition. Iizumi et al. (1977) propose an interaction term of the type $C[\psi^3(q) + \psi^3(-q)]P$ to account for the phase transitions. SeO_4^- produced by γ-irradiation has also been used to study the K_2SeO_4 transition by ESR spectroscopy. NMR spectroscopy has been used to investigate the transitions in such solids as Rb_2ZnCl_4.

Makovicky & Hyde (1981) have reviewed incommensurate misfit structures in graphite intercalation compounds, brucite-type compounds, sulphides and related layered systems. A simple two-dimensional incommensurate system is provided by graphite with adsorbed rare gas monolayers. At low densities and high temperatures,

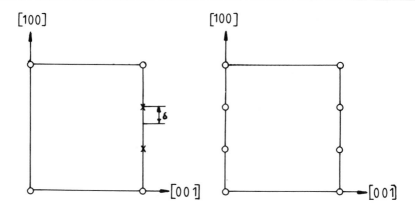

Figure 4.21   Portion of the (010) plane in the reciprocal lattice of K_2SeO_4. On the left is the lattice for the high-temperature phase and on the right that for the ferroelastic phase. In the high-temperature phase, a softening occurs at a point displaced by δ from $q = (\frac{1}{3}, 0, 0)$, shown by a cross. In the incommensurate phase a satellite reflection develops at X.

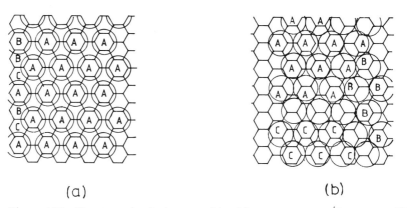

(a)   (b)

Figure 4.22   Krypton adsorbed on graphite. (a) commensurate $\sqrt{3}$ structure. The krypton atoms occupy $\frac{1}{3}$ of the honeycomb cells; (b) incommensurate structure. (Following Bak, 1982).

the monolayers form a two-dimensional fluid phase. At low densities and low temperatures, a commensurate phase occurs (Fig. 4.22). Certain aspects of incommensurate structures are yet to be explored (e.g. pressure effects of incommensurate transitions). It is not clear why these incommensurate transitions generally occur around laboratory temperatures.

Misfit layered compounds of the type, $(RX)_n(TX_2)_m$, where R is a rare earth, Pb or Bi, T = Ti, V, Nb or Ta and X = S or Se, are typical incommensurate phases that have been investigated in detail (Rouxel & Meerschaut, 1994). The incommensuration here arises from a structural misfit between the rocksalt-like RX and layered TX_2 dichalcogenide units which are stacked alternately. The interplay of structure and electron-transfer from the RX to TX_2 units gives rise to novel electronic properties. For example, $LnSe(NbSe_2)_2$ is a superconductor ($T_C = 5.3$ K) but does not exhibit a CDW instability.

4.10   Cooperative Jahn–Teller effect

According to the Jahn–Teller theorem, if the ground state of an ion in a crystal is orbitally degenerate, with no other perturbations present, the crystal will distort to one of lower symmetry in order to remove the degeneracy. There are any number of solids exhibiting tetragonal distortion due to the presence of Jahn–Teller ions such as Mn^{3+} and Cu^{2+}. The distortion preserves the centre of gravity of the e_g level of the cation and the ion can therefore attain equal stabilization through the distortion to tetragonal symmetry with $c/a > 1$ or < 1. If there is coupling between the electronic states and vibrational modes, we have a situation referred to as the *dynamic Jahn–Teller effect*. Let us briefly examine the Hamiltonian of a JT system. The JT contribution to the Hamiltonian is written as AQS^2, where Q is the nuclear coordinate, S an electronic operator and A a measure of the strength of coupling. If we include the elastic energy, $\frac{1}{2}m\omega^2Q^2$, the potential energy exhibits two minima where $E_{JT} = A^2/2m\omega^2$ at $Q = \pm Q_o$, where $Q_o^2 = A/m\omega^2$. There are four possible values of Q for a given value of energy, and the values of Q as well as the range of energies change if the value of A is changed. If $\hbar\omega \ll E_{JT}$, then at low temperatures $kT \ll E_{JT}$ and the system will be situated at the bottom of the potential well and the presence of a distortion should be noticeable (as in static JT effect). At higher T when $kT \gtrsim E_{JT}$, thermally induced fluctuations from one potential well to another can occur (dynamic case). If $\hbar\omega \gtrsim E_{JT}$, even at the lowest temperature, there will be no stabilization of one distortion. When the electronic states are coupled to doubly degenerate lattice modes, the energy would depend on two linearly independent normal modes in the form $(Q_1^2 + Q_2^2)^{\frac{1}{2}}$ and there would be no minima in the potential energy surface. The Hamiltonian for the interaction of a JT ion with a local distortion is written in the form $H = \sum VQ_\Gamma^m O_\Gamma^m$ where the local distortion Q_Γ^m transforms like the mth component of the irreducible representation Γ of the site symmetry group. The strength of coupling is given by V, and O_Γ^m is the electronic operator; Q_Γ^m is expressed in terms of the normal modes of the crystal.

Since the mechanism of the JT effect involves interaction between ligand displacements (and therefore the JT ions) due to the elastic properties of the lattice, the entire crystal can become unstable with respect to distortion if the concentration of JT ions is large. Phase transitions arising from such interactions can result in a parallel or some other alignment of all distortions. The cooperative Jahn–Teller effect (CJTE) is such a phase transition driven by the interaction between the electronic states of one of the ions in the crystal and the phonons. The transition can be first- or second-order and in either case there will be symmetry-lowering distortion and splitting of electronic levels. CJTE in many spinel compounds was discussed some years ago by Englman (1972). Our understanding of CJTE today, however, owes much to the investigations of CJTE in rare-earth arsenates and vanadates of zircon structure. These solids, being optically transparent, have enabled spectroscopic studies of the energy levels involved in the phase transition. Measurements of heat capacities, elastic constants and other properties have thrown considerable light on the phenomenon (Gehring & Gehring,

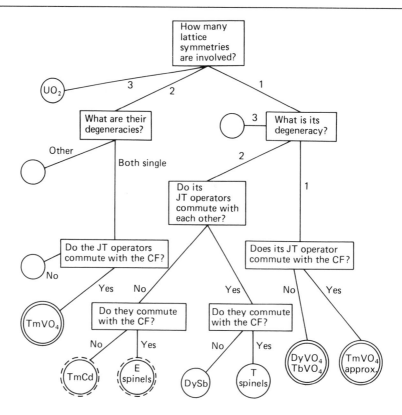

Figure 4.23   Various degrees of complexity that can arise in JT systems. (After Gehring & Gehring, 1975.) CF stands for crystal field. Single circle indicates first-order transition and a double circle indicates a second-order transition. A single circle with a broken circle indicates a transition which is first-order because of the existence of anharmonic forces.

1975). The elastic excitation mode (strain mode) is the soft mode in many of the second-order CJTE transitions.

Most of the experimental results on CJTE can be explained on the basis of molecular field theory. This is because the interaction between the electron strain and elastic strain is fairly long-range. Employing simple molecular field theory, expressions have been derived for the order parameter, transverse susceptibility, vibronic states, specific heat, and elastic constants. A detailed discussion of the theory and its applications may be found in the excellent review by Gehring & Gehring (1975). In Fig. 4.23 various possible situations of different degrees of complexity that can arise in JT systems are presented.

As mentioned earlier, rare-earth zircons are ideal cases for the study of CJTE. In $TmVO_4$, at high temperatures, the lowest electronic state of Tm^{3+} ion is an E doublet. A pseudo-JT effect occurs if there is accidental or near degeneracy. $DyVO_4$ is such an example where two Kramers doublets are separated by $9\,cm^{-1}$; another example is $TbVO_4$. The rather unusual features of the low-lying electronic states of rare-earth

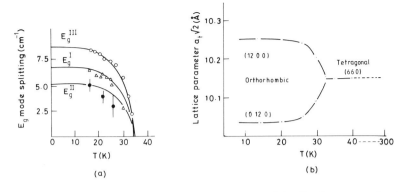

Figure 4.24   (a) Temperature dependence of the E_g phonon mode of $TbVO_4$ in Raman scattering. (After Elliott *et al.*, 1972.); (b) temperature variation of lattice parameters of $TbVO_4$. (After Will *et al.*, 1972.)

zircons have been discussed by Pytte & Stevens (1971) and Elliott *et al.* (1972). Phonon modes and elastic constants of these materials are discussed in the review by Gehring & Gehring (1975). In Fig. 4.24 we show the temperature dependence of the splitting of the E_g phonon mode of $TbVO_4$ as well as the temperature variation of the lattice parameters for purposes of illustration. In cubic spinels, AB_2O_4, where A and B are both transition-metal ions, the JT effect can arise from either A or B site ions or both simultaneously. The lowest electronic state can be a doublet (E spinel) or a triplet (T spinel). Since the electrons responsible for the JT effect are on the outside of the ions, they interact strongly with the lattice, giving rise to structural phase transitions at fairly high temperatures. This should be contrasted with the behaviour of rare-earth zircons, which show transitions at very low temperatures. Rare-earth pnictides (of NaCl structure) show first-order phase transitions involving magnetic ordering and structural distortions; in these systems there is an interplay of magnetic and JT interactions.

Gehring & Gehring (1975) have listed a number of CJTE materials along with literature references. We shall briefly examine the transitions of $PrAlO_3$, which are quite fascinating. $PrAlO_3$ undergoes three phase transitions, starting from the high-temperature cubic perovskite structure through trigonal and orthorhombic phases to a tetragonal structure at $T \sim 0\,K$. These transitions can be understood in terms of rotations of the AlO_6 octahedra coupled to the electronic levels of the Pr^{3+} ion through electron–phonon interaction (Harley *et al.*, 1973). The orthorhombic–monoclinic transition (second-order) at 151 K is driven by the coupling of the lowest-lying exciton to phonons, the first-order trigonal–orthorhombic transition at 205 K is also similarly due to electron–phonon interaction (Birgeneau *et al.*, 1974). For such systems, order parameters can be derived from the splitting of the electronic levels, the macroscopic strain, and the appropriate internal displacements, and all these order parameters should in principle show the same temperature dependence. All the above three order parameters have been derived for the second-order transition of $PrAlO_3$ at 151 K; splitting of electronic states from optical absorption, fluorescence, and electronic Raman spectroscopy, internal displacements by ESR spectroscopy, and macroscopic

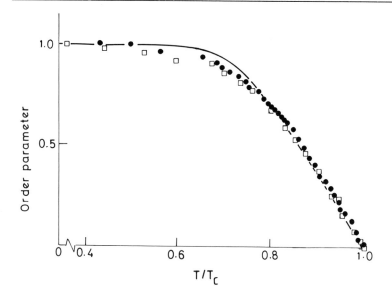

Figure 4.25   Comparison of the different order parameters for the 151 K second-order transition of $PrAlO_3$. (After Sturge et al., 1975.) Unbroken line is the smooth curve through the internal displacement order parameter, $\cos^2 2\phi$, from ESR measurements. Black circles represent the electronic order parameter from optical absorption studies. Squares represent the reduced macroscopic strain from elastic neutron scattering.

strain by neutron scattering. The three order parameters have been compared and found to show fairly good agreement (Fig. 4.25). The order parameters also follow the $(T_c - T)^{1/2}$ behaviour within 0.2° of T_c (Sturge et al., 1975).

4.11   Spin-state transitions

Transition-metal ions having electronic configurations d^4, d^5, d^6 or d^7 can exist in the low-spin or the high-spin ground state depending on the strength of the octahedral ligand field. When the field strength is close to the exchange energy corresponding to the cross-over of the ground state terms (Tanabe & Sugano, 1954), a transition can occur between the two spin states (rate constant $\sim 10^6 \text{s}^{-1}$ or more.) Spin-state transitions are accompanied by changes in the magnetic susceptibility and other properties. Low-spin to high-spin transitions in complexes of iron and cobalt have been investigated extensively in the last few years by employing magnetic susceptibility, heat capacity and spectroscopic studies. Mössbauer spectroscopy is specially effective for examining the transitions in Fe^{2+} complexes, the low- and the high-spin states being associated with distinctly different isomer shifts. Spin-state transitions in complexes have been reviewed (Gütlich, 1981a; König et al., 1985), and the topic is of considerable relevance to the properties of cytochromes. In some complexes, the transition appears to be continuous while in others it is abrupt. Structural changes and enthalpy changes accompany abrupt spin-state transitions in many of the complexes (Fig. 4.26) and

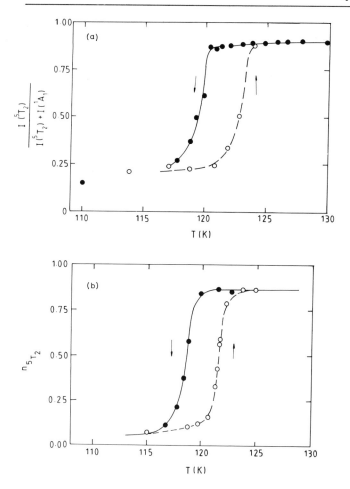

Figure 4.26   Temperature dependence of (a) the relative X-ray diffraction intensity of high-spin ($^5T_2$) and low-spin ($^1A_1$) phases of [Fe(4, 7-$(CH_3)_2$ phen)$_2$ (NCS)$_2$] and (b) the high-spin fraction (n^5T_2) estimated from Mössbauer spectra. In (a) the ordinate represents relative diffraction intensity of lines at θ ($^5T_2$) = 4.92° and θ ($^1A_1$) = 4.95° (CuK α radiation). (After König et al., 1979.)

several explanations and models have been proposed to account for the varied features of these transitions. Spin-state transitions are by no means unique to transition-metal complexes. They have been found in extended solids such as metal oxides, sulphides and other systems as well. Those who work with transition-metal complexes, however, seem to be unaware of the observations reported on metal oxide systems by solid state chemists (see, for example, Ramasesha et al., 1979), and yet the transitions in the two types of systems bear many similarities. Rao (1985a) has discussed the subject in a unified manner and has reviewed the various models for spin-state transitions. We shall briefly present some of the important features of these transitions to illustrate how such phase transitions are investigated experimentally and theoretically.

The discontinuous spin-state transitions in complexes generally seem to be first-order phase transitions with appreciable thermal hysteresis (Fig. 4.26). Discontinuities in X-ray Debye–Waller factors also accompany the structural changes in these transitions. The exact nature of the structural change (symmetry, unit cell dimensions, etc.) has not been investigated in most of the complexes exhibiting spin-state transitions. A change in unit cell parameters would be expected because of the difference in size of the high- and the low-spin ions, but a change in symmetry need not necessarily occur. Anomalous changes in heat capacity are found in complexes such as $[Fe(Phen)_2(NCS)_2]$ undergoing abrupt changes in spin-state population (Sorai & Seki, 1974). Some of the gradual transitions (see, for example, König et al., 1983) could be thermodynamically second-order. Some complexes show a plateau in the plots of the magnetic moment of the susceptibility against temperature. For example, in Fe(III) complexes, instead of the μ_{eff} value varying between $5.9\,\mu_B(^6A_1)$ and $2.0\,\mu_B(^2T_2)$, a plateau is seen over a wide temperature range at an intermediate value of μ_{eff}. Grinding makes a transition gradual, leaving more low-spin molecules at high temperatures; this may be because grinding could produce the same effects as the application of high pressure.

Spin-state transitions are sensitive to doping by other metal ions. In complexes such as $[Fe_xZn_{1-x}(2\text{-pic})_3]\,Cl_2 \cdot C_2H_5OH$, the nature of the spin transition depends on x, indicating the importance of the cooperative interaction between the electronic state of the metal ion and the lattice (Gütlich, 1981b). What is more interesting is the observation that doping affects the transition temperature markedly in some cases. Here again, pressure effects of dopant metal ions have to be borne in mind in understanding the observations. In the case of $[Fe(3\text{-}OCH_3\text{-Sal Een})_2]\,PF_6$, doping with Cr^{3+} has only a small effect on the transition temperature, but substitution with Co^{3+} drastically increases the transition temperature (Haddad et al., 1981). This is probably because the radius of Co^{3+} (0.53 Å) is considerably smaller than that of Fe^{3+} and would be expected to have a positive pressure effect. Such pressure effects of dopant metal ions on phase transitions of solids are indeed well-known (Rao & Rao, 1978). For example, substitution of Gd^{3+} in place of Sm^{2+} in SmS makes it metallic; the solid solution on cooling explosively transforms to a low-density structure. Substitution of Cr^{3+} in place of V^{3+} in V_2O_3 has a negative pressure effect on the metal–insulator transition of V_2O_3. Partial substitution of Fe^{2+} by Mn^{2+} in $[Fe\,(Phen)_2\,(NCS)_2]$ drastically lowers the transition temperature. Mn^{2+} has a much larger radius (0.82 Å) than Fe^{2+} and would have a negative pressure effect; large Mn^{2+} concentration does indeed make the transition gradual. Co^{2+} with a radius of 0.72 Å, on the other hand, has less marked effect on the transition; Zn^{2+} which has the same radius as Fe^{2+} has no effect. Residual paramagnetism at low temperatures in $[Fe_{1-x}M_x(Phen)_2\,(NCS)_2]$ can also be interpreted in terms of the relative size of the M^{2+} ion, the larger ionic size increasing the paramagnetism (Gütlich, 1981b). Such pressure effects of dopant metal ions are also found in Mn^{2+} substituted FeS_2. Substitution of Fe in FeS_2 by Mn stabilizes the high-spin state of Fe^{2+}; ordinarily, Fe in this sulphide is in the low-spin state. Application of pressure on $Fe_{1-x}Mn_xS_2$ results in a high-spin to low-spin transition.

Spin-state transitions in metal oxides are by and large limited to those containing

Co^{3+} ions. The rare-earth cobaltates of the general formula LnCoO$_3$ (Ln = La, Y or any other lanthanide element) have been extensively studied oxides with respect to the spin-state equilibria (Raccah & Goodenough, 1967; Bhide *et al.*, 1972; Madhusudan *et al.*, 1980). In these oxides, Co^{3+} is in the low-spin t_{2g}^6 configuration at low temperatures and transforms to the high-spin state ($t_{2g}^4 e_g^2$) with increase in temperature (say up to 300–400 K). At higher temperatures, the behaviour depends on the rare-earth ion; the lighter rare-earth (La, Nd, etc.) cobaltates show evidence for charge-transfer between the low- and the high-spin states while the heavier rare-earth (Ho, Er) cobaltates do not. We shall be mainly concerned here with the low- to high-spin transition rather than the charge-transfer effects. The spin-state transition in LaCoO$_3$ is quite unique in that the two spin states order themselves, causing a change in symmetry from R$\bar{3}$c to R$\bar{3}$. The ordering of the spin states is evidenced by a plateau in the inverse magnetic susceptibility–temperature curve (Fig. 4.27), such a behaviour having been noticed earlier in MnAs$_{1-x}$P$_x$. The lighter rare-earth cobaltates show thermal anomalies due to the spin-state transitions, with ΔH values of around 400 J mol^{-1} or more; at the phase transition temperatures, changes in the Debye–Waller factor, the Lamb–Mössbauer factor and structural features are noticed. It is interesting that at the transition temperatures, the low- to high-spin population ratio is generally unity. The inverse susceptibility–temperature plots of NdCoO$_3$, PrCoO$_3$ and other cobaltates show a maximum (Fig. 4.27) and no plateau, as in the case of LaCoO$_3$; the gradual variation of the magnetic moments in these cobaltates is not unlike that in the complexes discussed

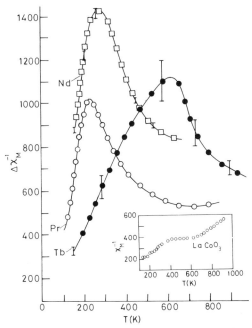

Figure 4.27   Plots of the cobalt contribution to the inverse susceptibility of rare-earth cobaltates. The curve for LaCoO$_3$ is shown in the inset. (After Madhusudan *et al.*, 1980.)

earlier. A study of the solid solutions of the type $La_{1-x}Nd_xCoO_3$ has shown a transition from the $NdCoO_3$-like behaviour (maximum in the $\chi^{-1}-T$ plot) to the $LaCoO_3$-like behaviour (plateau in the $\chi^{-1}-T$ plot). This observation seems to suggest that we may be dealing with two extremes in behaviour of these spin cross-over systems. Furthermore, the spin-state transition temperature increases as we go across the rare-earth series from La onwards. This may be because of the polarizing effect of the rare-earth ion on the crystal field splitting, the latter increasing across the rare-earth series accompanied by a decrease in the unit cell volume. The π^* orbital also would get significantly stabilized as we go across the series, contributing to the increase in crystal field splitting. It must be noted, however, that the energy separation between the low- and the high-spin states in the cobaltates is in the range $100-500\,cm^{-1}$.

Chenavas & Joubert (1971) have reported a synthesis of Co_2O_3 under high pressure. The high-pressure phase seemed to consist of only low-spin Co^{3+} ions which on heating transformed to the high-spin state with a positive volume change of 6.7%. The results are suggestive of a first-order transition, but it is not entirely certain that a pure phase of Co_2O_3 was indeed prepared.

Spin-state transitions have been noticed in quasi-two-dimensional oxides of the K_2NiF_4 structure. La_4LiCoO_8 is one such oxide (Mohanram et al., 1983). It was first considered to show a maximum in the inverse susceptibility–temperature plot because of a low- to high-spin state transition of the Co^{3+} ions. Demazeau et al. (1980) have interpreted the susceptibility data of La_4LiCoO_8 in terms of the occurrence of low- to intermediate-spin and low- to high-spin transitions. In this oxide, the Li^+ and Co^{3+} ions are ordered in two dimensions (along the ab planes) as revealed by electron diffraction (Mohanram et al., 1983 Fig. 2.1). Sr_4TaCoO_8 and Sr_4NbCoO_8 which possess the K_2NiF_4 structures show a susceptibility behaviour reminiscent of the plateau in the $\chi^{-1}-T$ plot of $LaCoO_3$. The Co^{3+} ions seem to transform from the low- to the intermediate-spin ($t_{2g}^5e_g^1$) state in these oxides.

Demazeau et al. (1982) reported that Ni^{3+} is in the low-spin $t_{2g}^6e_g^1$ state in $LaSrNiO_4$, but in the high-spin $t_{2g}^5e_g^2$ state in $LaBaNiO_4$. A magnetic susceptibility study of $LaSr_{1-x}Ba_xNiO_4$ has shown the e_g electrons in this system to be always in extended states forming a $\sigma^*_{x^2-y^2}$ band; with increase in x, the band-width decreases, accompanying an increase in the unit cell volume. While a small proportion of high-spin Ni^{3+} ions may be formed with increasing x, there appears to be no transition (Mohanram et al., 1983).

The first model for spin-state transitions was proposed by Chestnut (1964), who gave a phenomenological description of systems containing triplet excitons such as TCNQ radical ion salts and Würster's blue perchlorate, in terms of the relatively high exciton concentrations. By treating the free energy as quadratically dependent on exciton concentration, Chestnut showed the occurrence of phase transitions in such systems. König & Kremer (1971) analysed changes in the spin-state population semiempirically by assuming a temperature-dependent energy separation between the high- and low-spin states and a tetragonal distortion. Slichter & Drickamer (1972) provided an explanation for pressure-induced changes in the high-spin fraction of Fe(II) complexes by a treatment similar to that of Guggenheim's regular solution theory. Based on similar considerations, Sorai & Seki (1974) attempted to explain first-order spin-state transitions. The Gibbs free energy at a temperature was written by them as

$$G = xG_{HS} + (1 - x)G_{LS} + NkT\{x\ln x + (1 - x)\ln(1 - x)\} \tag{4.17}$$

where G_{HS} and G_{LS} are the Gibbs free energies of the high-temperature (high-spin) and low-temperature (low-spin) phases. The equilibrium mole fraction of the high-spin state, x, is given by

$$x = 1/\{1 + \exp(\Delta G/NkT)\} \tag{4.18}$$

where $\Delta G(= G_{HS} - G_{LS})$ includes the vibrational term. Sorai and Seki showed the spin-state transition to be cooperative, occurring through coupling between the electronic state of the metal ion and the phonon system; the strength of the coupling would determine the shape of the spin-state conversion curve, a weak coupling giving rise to a gradual conversion.

Spiering *et al.* (1982) have developed a model where the high-spin and low-spin states of the complex are treated as hard spheres of volume V_{HS} and V_{LS} respectively and the crystal is taken as an isotropic elastic medium characterized by bulk modulus and Poisson constant. The complex is regarded as an inelastic inclusion embedded in spherical volume V_c. The decrease in the elastic self-energy of the incompressible sphere in an expanding crystal leads to a deviation of the high-spin fraction from the Boltzmann population. Pressure and temperature effects on spin-state transitions in Fe(II) complexes have been explained based on such models (Usha *et al.*, 1985).

Kambara (1979) has proposed a microscopic model in which the coupling between the *d*-electrons and the lattice has been given a definitive meaning. Assuming that there is Jahn–Teller coupling between the *d*-electrons and the local intramolecular distortion, the Hamiltonian of the system is written as

$$H = \sum_{i=1}^{N} h_i + \tfrac{1}{2}NM\omega^2 Q^2 \tag{4.19}$$

where h_i is the Hamiltonian for a Fe^{2+} ion in the ith molecule and Q the displacement of the six ligands in the normal coordinate with the E_g symmetry of the O_h point group, M the effective mass and ω the frequency of the mode. The Hamiltonian h_i is given by

$$h_i = h_o + V_o + V_{so} + (\partial V/\partial Q)_{oQ} + \tfrac{1}{2}(\partial^2 V/\partial Q^2)_{oQ_2} \tag{4.20}$$

where h_o is the Hamiltonian of the free ion, V_o the octahedral ligand field, V the ligand field due to the distorted molecule and V_{so} the spin-orbit interaction.

The free energy per iron ion is represented as

$$F = -kT\sum_{i=1}^{3} g_i\exp(-E_i/kT) \tag{4.21}$$

and equilibrium Q values are found by minimizing F with respect to Q. Terms containing Q or Q^2 appear in the eigenvalues. The energy separation between the $^5T_{2g}$ and the A_{1g} states (500–$1000\,\mathrm{cm}^{-1}$) in the undistorted phase is given by $E_o = 20Dq$–$5B$–$8C$, where B and C are the Racah parameters. The Jahn–Teller coupling parameter, a, appearing in the expression for Q, is found by the relation $E_{JT} = \frac{1}{2}a^2$, where $E_{JT} = 100$–$1000\,\mathrm{cm}^{-1}$ and a is therefore between 0 and $50\,\mathrm{cm}^{-1/2}$. In the linear coupling case (last term in equation 4.20 is zero), a first-order transition, with Q as the order parameter, occurs over a wide range of the coupling strength, a. If the second-order coupling is introduced, three possible kinds of the transition, with $Q > 0 \rightleftharpoons Q = 0$, $Q > 0 \rightleftharpoons Q < 0$ and $Q < 0 \rightleftharpoons Q = 0$, can occur. The temperature variation of μ_{eff} calculated from equation (4.22) is in fair agreement for discontinuous spin-state transitions, when reasonable values of E_o and a are employed.

$$\mu_{eff}^2 = \frac{kT}{N}(\chi_{11} + 2\chi_1) \tag{4.22}$$

where χ is the magnetic susceptibility; the expression for χ includes E_i values, which in turn are determined by E_o, a, Q and constants that relate the force constants and vibrational frequencies.

Kambara (1981) has also treated intramolecular distortions as dynamical variables and taken intermolecular coupling into account. The effect of coupling between transition-metal ions and a lattice strain has also been studied; the lattice would then drive the spin-state transition accompanying the cooperative intramolecular distortion even in the absence of strong coupling. When there is no distortion, the lattice strain varies continuously with temperature along with the spin-state. Discontinuous changes in the lattice strain occur with the cooperative intramolecular distortion over a narrow range of coupling strength. The results of the treatment with intermolecular coupling are similar to those from the static treatment (Kambara, 1979), the only difference being in the meanings of the different parameters. Sasaki & Kambara (1982) have shown that a cooperative molecular distortion associated with a spin-state transition can be induced by a magnetic field. A small shift in the transition temperature of $[\mathrm{Fe(Phen)_2(NCS)_2}]$ seems to occur in the presence of a field of $5.5\,\mathrm{T}$(Qi et al., 1983). This aspect is worthy of further investigation.

Bari & Sivardiere (1972) had proposed a model for spin-state transitions based on Chestnut's model for the singlet–triplet transition in radical ion salts (Chestnut, 1964), which included the displacement of the breathing mode, Q, and the elastic constant in the Hamiltonian:

$$H = N\xi Q^2 + \sum_i (\Delta - VQ)n_i \tag{4.23}$$

Here, $(\Delta - VQ)$ represents the energy difference between the low- and the high-spin states and $n_i = 0$ and 1 for the low-spin and the high-spin states respectively. This model showed a first-order phase transition from the low- to the high-spin state with increasing temperature for $0.43 < (V^2/4\xi\Delta) < 1$. Bari & Sivardiere (1972) extended the model to include exchange interactions between the high-spin states leading to the prediction of heat of magnetization; magnetic ordering has, however, not been observed in any of the systems showing spin-state transitions. Bari (1972) proposed a two-sublattice model to explain the structural transition accompanying the spin-state transition in $LaCoO_3$. The Hamiltonian of Bari is given by

$$H = N\xi Q^2 + \sum_{iA}(\Delta - VQ)n_i + \sum_{iB}(\Delta + VQ)n_i \tag{4.24}$$

where the displacement on sublattice A is opposite to the displacement on sublattice B. This model shows a first- or second-order transition due to ordering followed by a spin-state transition of first- or second-order. Experimentally, the spin-state transition occurs before the ordering transition as one would expect. This study, however, points to the importance of elastic properties of the solid and of the coupling to a lattice distortion.

The models of Bari and Sivardiere are static and can be solved when the dynamics of the lattice are included. Ramasesha et al. (1979) have examined this problem and find that the dynamic model does not show a phase transition; this is true of the two-sublattice model as well. The dynamics of the lattice seem to wash away any phase transition arising from a static and linear coupling of the displacement to the spin-states. A dynamic two-sublattice model where the high-spin and the low-spin states are mixed by an ion-cage mode (with the transition-metal ion moving off-centre with respect to the octahedral cage), however, shows a non-zero population of the high-spin state at low temperatures and a plateau in the susceptibility curve. The coupling in this treatment is taken to be quadratic in displacement and the Hamiltonian is written as

$$H = [(P_Q^2/2M) + \tfrac{1}{2}M\omega^2Q^2]I + \Delta\sigma_z + \tfrac{1}{2}aQ^2M\omega^2\sigma_x \tag{4.25}$$

where ω is the vibrational frequency and σ represents the spin matrix. A coupling of the spin-states to the cube of the sublattice displacement gives a first- or a second-order transition depending on the interaction parameter, but the high-spin state population is always zero at low temperatures (Ramasesha et al., 1979). It is possible to explain the different features of spin-state transition systems by including couplings of the spin-states to the displacements of a breathing mode and an ion-cage mode, the former being linear and the latter quadratic. The ion-cage mode would mix the spin-states.

The preceding discussion should indicate how the chemically interesting phenomenon of spin cross-over becomes quite a complex one to understand in the solid state. It must be pointed out that in spite of the extensive studies, there are many aspects of spin-state transitions (e.g. critical behaviour, soft modes) that are yet to be explored.

Table 4.2. *Structure and thermodynamic properties of some plastic crystals*[a]

Substance	$a(\text{Å})$	$T_t(\text{K})$	$T_m(\text{K})$	$\Delta S_t(\text{R})$	$\Delta S_m(\text{R})$	$E_a(\text{kJ mol}^{-1})$
Neopentane	8.78	140	257	2.22	1.46	4.2
Cyclohexane	8.76	186	280	4.34	1.10	6.4
Hexamethyl disilane	8.47	222	288	5.28	1.25	6.3
t-Butyl chloride	8.62	183	245	4.38	0.95	3.0
Camphor	10.10	238	449	3.84	1.40	11.9

[a]All the plastic phases listed in the table possess FCC structure: T_t, crystal–plastic crystal transition temperature; ΔS_t, entropy change at T_t; ΔS_m, entropy change at T_m; E_a, activation energy for molecular reorientation obtained from NMR spectroscopy.

4.12   The plastic crystalline state

Ordinarily, crystals possess both translational and orientational order. Crystals of certain substances (especially those formed by spherical or globular molecules) exhibit orientational disorder while retaining the long-range translational order. Such an orientationally disordered crystalline state is referred to as the *plastic crystalline state*. The transition from the crystalline to the plastic crystalline state is exhibited by molecular crystals as the molecules acquire orientational degrees of freedom with higher temperature. The characteristic thermodynamic feature of plastic crystals, which led to the recognition of their existence by Timmermanns, is that the entropy change in the crystal–plastic crystal transition is far greater than the entropy change accompanying the melting of the plastic crystal (Table 4.2). In addition to the low entropy of melting, plastic crystals usually melt at a relatively high temperature; for example, plastic crystalline neopentane melts at 257 K while *n*-pentane, which does not form the plastic state, melts at 132 K. In some cases, plastic crystals directly sublime without melting, camphor, sulphur hexafluoride and hexachlorethane being typical examples.

Because of the orientational freedom, plastic crystals usually crystallize in cubic structures (Table 4.2). It is significant that cubic structures are adopted even when the molecular symmetry is incompatible with the cubic crystal symmetry. For example, *t*-butyl chloride in the plastic crystalline state has a fcc structure even though the isolated molecule has a three-fold rotation axis which is incompatible with the cubic structure. Such apparent discrepancies between the lattice symmetry and molecular symmetry provide clear indications of the rotational disorder in the plastic crystalline state. It should, however, be remarked that molecular rotation in plastic crystals is rarely free; rather it appears that there is more than one minimum potential energy configuration which allows the molecules to tumble rapidly from one orientation to another, the different orientations being random in the plastic crystal.

As the name suggests, plastic crystals are generally soft, frequently flowing under their own weight. The pressure required to produce flow of a plastic crystal, as for instance to extrude through a small hole, is considerably less (2–14 times) than that required to extrude a regular crystal of the same substance. *t*-Butyl alcohol, pivalic acid and *d*-camphor provide common laboratory examples of plastic crystals. The subject of plastic crystals has been reviewed fairly extensively (Aston, 1963; Sherwood, 1979) and we shall restrict our discussion to the nature of the orientational motion (Rao, 1985*b*).

Existence of a high degree of orientational freedom is the most characteristic feature of the plastic crystalline state. We can visualize three types of rotational motions in crystals: free rotation, rotational diffusion and jump reorientation. Free rotation is possible when interactions are weak, and this situation would not be applicable to plastic crystals. In classical rotational diffusion (proposed by Debye to explain dielectric relaxation in liquids), orientational motion of molecules is expected to follow a diffusion equation described by an Einstein-type relation. This type of diffusion is not known to be applicable to plastic crystals. What would be more appropriate to consider in the case of plastic crystals is collision-interrupted molecular rotation.

The rotational diffusion model was generalized by Gordon (1966) to include molecular reorientation in angular steps of fairly large size. According to this model, the molecules are supposed to rotate freely during the intervals between collisions. Two limiting cases have been discussed by Gordon, both of which involve angular diffusion steps (of arbitrary size). These are commonly known as the J-diffusion and the M-diffusion models. In the former, the direction as well as the magnitude of the angular momentum vector of the molecule follow Boltzmann distribution due to collisions. In the M-diffusion model, the orientation of the angular momentum vector is randomized, but the magnitude is unchanged. In Gordon's model applied to linear molecules, the general diffusion equation follows classical mechanics for short times. For longer times, the rotational frequency is unchanged in successive steps in the M-diffusion model, whereas it follows a Boltzmann distribution in the J-diffusion model.

In the model proposed by Hill (1963) and Wyllie (1971), molecules exhibit librations (perturbed angular oscillations), having been trapped in potential wells. Fluctuations in orientation of neighbouring molecules result in a change in the shape of the potential well and consequently give rise to angular motion of molecules with large amplitudes. It is possible that diffusion of the potential well on the surface of a sphere gives rise to reorientational motion of molecules. If the potential surface is related to the crystal space group, molecular reorientation will be associated with damped librations (Guthrie & McCullough, 1961). Rotations of the crystal space group will give rise to allowed potential wells.

No single model can exactly describe molecular reorientation in plastic crystals. Models which include features of the different models described above have been considered. For example, diffusion motion interrupted by orientation jumps has been considered to be responsible for molecular reorientation. This model has been somewhat successful in the case of cyclohexane and neopentane (Lechner, 1972; De Graaf & Sciesinski, 1970). What is not completely clear is whether the reorientational motion is cooperative. There appears to be some evidence for coupling between the reorientational motion and the motions of neighbouring molecules. Comparative experimental studies employing complementary techniques which are sensitive to autocorrelation and monomolecular correlation would be of interest.

The nature of rotational motion responsible for orientational disorder in plastic crystals is not completely understood and a variety of experimental techniques have been employed to investigate this interesting problem. There can be coupling between rotation and translation motion, the simplest form of the latter being self-diffusion. The diffusion constant D is given by the Einstein relation

$$D = \frac{\langle r^2 \rangle}{6\tau} \tag{4.26}$$

where $\langle r^2 \rangle$ is the mean-square-jump distance and τ is the mean time between successive jumps of a molecule at a lattice site. The diffusion constant depends on the magnitude of molecular velocity as well as the persistence velocity. Both radioactive tracer and NMR methods have been employed to measure the coefficients of self-diffusion in plastic crystals. In many cases, the two methods yield similar results since they refer to a similar process on the time scales of the velocity correlation function. The diffusion coefficients are anomalously high (compared with nonplastic organic crystals), the softest crystals having the highest diffusion coefficients.

NMR spectroscopy provides spin-lattice (T_1) and spin-spin (T_2) relaxation times. Making appropriate assumptions with regard to the magnetic interactions responsible for the relaxation process, these relaxation times can be related to molecular motions. Since nuclear spin relaxation results from all processes which cause a fluctuation in the magnetic field at the nucleus, the correlation function will generally correspond to more than one kind of motion involving all possible interactions. The equations for the relaxation times are generally of the form

$$\frac{1}{T} = \frac{C\tau}{(1 + \omega^2\tau^2)} \tag{4.27}$$

where the constant of proportionality, C, as well as τ are determined by the nature of the nucleus and the different interactions. If the relaxation process is essentially due to dipolar interactions, then NMR spectroscopy gives information on rotational diffusion. If nuclear quadrupole relaxation is dominant, then we obtain information on molecular reorientation. In Table 4.2, activation energies for molecular reorientation obtained from NMR spectroscopy are listed.

Thermal or low-energy neutron scattering experiments have been most valuable in throwing light on molecular motion in plastic crystals. These experiments measure changes in the centre of mass of a molecule. Diffusion constants obtained from neutron experiments differ widely from those obtained from tracer experiments since neutron scattering is mainly determined by rotational diffusion. The scattering function has the form

$$S(\mathbf{Q}, \omega) = \iint G(\mathbf{R}, t) \exp (i\mathbf{Q}^r - i\omega t) \, d\mathbf{Q}.d\omega \tag{4.28}$$

where $G(\mathbf{R}, t)$ is the rotational correlation function for a scattering centre, $\mathbf{Q}$ the scattering vector, ω the angular frequency, and $\mathbf{R}$ the equilibrium position vector. The inelastic part of the scattering provides information on the rotational motion while the elastic part is related to the translational motion. If the reorientational motion is diffusive, we can employ a diffusion equation similar to that mentioned earlier, the width of the quasielastic peak being related to rotational diffusion. One often employs different functions to see if rotational motion occurs through jumps. A good example of a neutron scattering study of plastic crystals is that on t-butyl cyanide by Frost et al. (1980).

One of the most direct methods of examining reorientational motion of molecules is by far infrared absorption spectroscopy or dielectric absorption. In the absence of vibrational relaxation, the relaxation times obtained by IR and dielectric methods are equivalent. In both these techniques we obtain the correlation function, $\langle \mu(t_o) . \mu(t_o + t) \rangle$, for the motion about a specific molecular axis with respect to the principal rotation axes and the direction of vibration. Rayleigh scattering as well as Raman scattering are most useful in the study of molecular orientational processes. With the aid of isotropic and anisotropic Raman scattering measurements, orientational processes can be delineated from other mechanisms which cause line broadening. Analysis of infrared and Raman band shapes provides information on the short-term and long-term behaviour of the correlation function.

Correlation times and activation energy parameters obtained from different techniques may or may not agree with one another. Comparison of these data enables one to check the applicability of the model employed and examine whether any particular basic molecular process is reflected by the measurement or whether the method of analysis employed is correct. In order to characterize rotational motion in plastic crystals properly it may indeed be necessary to compare correlation times obtained by several methods. Thus, values from NMR spectroscopy and Rayleigh scattering enable us to distinguish uncorrelated and correlated rotations. Molecular disorder is not reflected in NMR measurements; to this end, diffraction studies would be essential.

It is important to understand whether molecular motion in most plastic crystals is correlated or uncorrelated. Since the stability of plastic crystals depends crucially on the strength and shape of the intermolecular potential, it is of interest to examine whether phenomenological atom–atom potentials can be employed to describe the plastic crystalline state. Bounds et al. (1980) have carried out a molecular dynamics study of the plastic state of methane while Yashonath & Rao (1983) have carried out a Monte Carlo study of the plastic state as well as the glassy state obtained by quenching the plastic state. Nose & Klein (1983a) have studied CF_4 by molecular dynamics (also see Rao, 1985b). Both Monte Carlo and molecular dynamics simulations have been employed to investigate the plastic states as well as the glassy crystalline states of C_{60} and C_{70} (Chakrabarti et al., 1993; Rao & Seshadri, 1994).

4.13   The liquid crystalline state

The liquid crystalline state is another intermediate state of matter which several molecular crystals pass through before they become isotropic liquids on melting. The term *mesophase* is often used to describe such a state of matter. Mesophases share the characteristics of both liquids and crystals. In liquid crystals, there is no translational order but there is orientational order. The nature of the mesophase depends on the molecular shape; spherical or globular molecules gain rotational freedom easily and form plastic crystals. Rod-shaped molecules, on the other hand, gain translational freedom more easily than rotational freedom and give rise to liquid crystals. Liquid crystals have become a vast subject of study, and we shall only introduce the various types of transitions exhibited by liquid crystals in this section. Newer varieties of liquid crystals and transitions are being reported at a rapid pace and there are many good reference works on the subject (Gray & Winsor, 1974; Chandrasekhar, 1994; Hilsum & Raynes, 1983; Frank, 1983).

Liquid crystals can broadly be classified into lyotropic and thermotropic. *Lyotropic* liquid crystals are formed by amphiphilic substances in the presence of a solvent. The

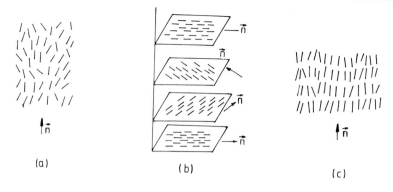

Figure 4.28   Schematic representation of (a) nematic; (b) chiral nematic and (c) smectic A liquid crystals.

majority of liquid crystal systems of current interest are *thermotropic* in origin, wherein the liquid crystal state arises as a result of a temperature change. Both finite molecules as well as polymers form thermotropic liquid crystals. *Nematic, cholesteric* and *smectic* are the three Friedelian classes of liquid crystals formed by rod-like elongated molecules. Since the cholesterics are nothing but nematics formed by chiral molecules, in the present-day classification only nematic and smectic are recognized as distinct classes of liquid crystals. In the nematic phase, the molecules are arranged almost parallel to one another in the direction of the long axis (Fig. 4.28(a)), but there is no long-range positional correlation in the direction of the long axis. Cholesteric phases are spontaneously twisted nematics, the twist arising from the optical activity of the molecules (Fig. 4.28(b)). A number of cholesterol derivatives exhibit this phase and hence the name. Smectic phases are, in the Friedelian sense, lamellar phases with no long-range positional correlation in the individual lamellae (Fig. 4.28(c)). Smectics possess, in addition to the orientational order as in nematics, some positional order as well; the centres of the molecules are on the average in equidistant planes.

Direct evidence for the layered structure of smectics is seen in X-ray diffraction patterns, where a sharp reflection is observed at small Bragg angles, which correspond to layer thickness. Many structural variants (polymorphs) of smectics are known, depending upon how the molecules are arranged in the smectic lamellae, the tilt angle of the molecules with respect to the layer plane, and the degree of correlation of structure from layer to layer. Two major smectic phases are recognized: those without ordered layers and those with ordered layers. In the first group, the distribution of molecular centres within the layers corresponds to that of a two-dimensional liquid. In the second group, the layers are built up regularly so that the positions of the molecular centres lie on a two-dimensional lattice. Further subdivisions to S_A, S_B, S_C, S_F and S_I of the first category and to S_B, S_E, S_G and S_H of the second category are made to specify molecular tilt and packing within layers. The smectic shown in Fig. 4.28(c) corresponds to S_A.

The local direction of alignment of molecules in liquid crystals is described by a unit vector $\vec{n}$, called the director, which gives the direction of the preferred axis at each point in the sample. The concept of director is very valuable in nematics. The degree of order in a nematic phase is described by an order parameter ξ defined as $\xi = \frac{1}{2}\langle 3 \cos^2 \theta - 1 \rangle$ where θ is the angle between the long molecular axis and $\bar{n}$, and the angular brackets denote a statistical average. $\xi = 1$ corresponds to complete orientational ordering and $\xi = 0$ obtains for complete randomness. In nematics ξ has an intermediate value that depends on temperature. In chiral nematics (cholesterics), in addition to the long-range orientational order, there exists a spatial variation (twist) of the director leading to a helical structure. The twist is characterized by a pitch whose value can be in the range of a few hundred nanometres. Comparing the nematic and chiral nematic phases, the following points can be made: (i) a phase transition from nematic to chiral nematic or vice versa is never seen; (ii) both nematic and chiral nematic are completely miscible; addition of a small amount of an optically active compound (whether itself mesogenic or not) changes a nematic to a long-pitch chiral nematic; (iii) a mixture of two compounds of opposite chirality can give a nematic phase; at a certain composition the pitch becomes infinite; (iv) X-ray diffraction patterns of chiral nematic and nematic phases are similar, showing lack of long-range positional order.

The thermodynamics of phase transitions of liquid crystalline systems have been widely investigated. The enthalpy change associated with the crystal-to-liquid crystal (nematic) transition, ΔH_{CN}, is much larger than the heat change associated with the liquid crystal to isotropic liquid transition, ΔH_{NI}. ΔH_{NI} is typically around 2% of ΔH_{CN}. Most of the phase transitions involving liquid crystals seem to be thermodynamically of the first order, except possibly intersmectic transitions. That the nematic to isotropic liquid transition has to be first-order was shown by de Gennes (1971) in terms of a symmetry argument. In the spirit of the Landau theory of phase transitions, the free energy in the vicinity of phase transition can be expanded in a power series in temperature and order parameter

$$G = G_o(\rho, T) + C_2 \xi^2 + C_3 \xi^3 + C_4 \xi^4 + \cdots \tag{4.29}$$

Equation (4.29) is similar to the expression used for second-order phase transitions except for the presence of the cubic term. A second-order transition occurs when C_2 goes through zero at $T = T^* < T_{NI}$, where T_{NI} is the nematic–isotropic liquid transition temperature. In order for a second-order transition to be pre-empted by a first-order transition at a temperature $T_{NI} > T^*$, the conditions are that the free energy show a minimum at a nonzero value of the order parameter ξ_{NI} and that the free energy at that value of ξ_{NI} be the same as in the isotropic phase. The first condition is expressed by $\partial G/\partial \xi$ and the second condition by equating $G(\xi_{NI})$ with $G(\xi = 0)$ in the following equations, disregarding higher-order terms

$$2C_2 + 3C_3 \xi_{NI} + 4C_4 \xi_{NI}^2 = 0 \tag{4.30}$$

$$C_2 + C_3 \xi_{NI} + C_4 \xi_{NI}^2 = 0 \tag{4.31}$$

The two conditions are simultaneously fulfilled when $\xi_{NI} = -C_3/2C_4$. But, in nematic

Table 4.3. *Some examples of organic molecules showing liquid crystalline states* [a]

p-azoxyanisole   C 391 N 408 I

CH₃O— ⟨ring⟩ —N=N(→O)— ⟨ring⟩ —OCH₃

p-heptyl-p'-cyanobiphenyl   C 301.5 N 315 I

C_7H_{15}— ⟨ring⟩—⟨ring⟩ —CN

N-p-ethoxybenzylidene)-p'-   C 288 N* 633 I
(β-methylbutyl) aniline

C_2H_5O— ⟨ring⟩ —CH=N— ⟨ring⟩ —CH₂CH(CH₃)CH₂CH₃

p, p'-dinonylazobenzene   C 311 S_B 314 S_A 327 I

C_9H_{19}— ⟨ring⟩ —N=N— ⟨ring⟩ —C_9H_{19}

p, p'-diheptyloxyazoxybenzene   C 347.5 S_C368.5 N 397 I

$C_7H_{15}O$— ⟨ring⟩ —N=N(→O)— ⟨ring⟩ —OC_7H_{15}

[a]C stands for crystalline phase, N for nematic phase, N* for chiral nematic, S_A,S_B etc. for smectic phases and I for isotropic liquid. The numbers in between are the transition temperatures in Kelvin.

ordering, unlike in other order–disorder transitions, positive and negative values of the order parameter correspond to different physical ordering of the molecules. Hence the coefficients of the cubic and higher-order terms in the free-energy expansion cannot be identically zero and the transition must be first-order.

Representative examples of organic molecules that exhibit liquid crystalline states are listed in Table 4.3. It is seen that they are all elongated, rod-like molecules containing rings and double bonds, the latter preventing the molecules from adopting nonlinear conformations. There has been considerable work during the last few years to elucidate the relationship between molecular structure and the type of liquid crystalline state exhibited, especially to understand the relative smectic and nematic tendencies (Gray, 1983). We shall deal with some of these aspects in Section 6.10.

In the last few years disc-like molecules have been shown to form liquid crystals (Chandrasekhar, 1994). Typical of them are hexasubstituted esters of benzene (I) and certain porphyrin esters (II) (see below). In the liquid crystalline state, the disc-like molecules are stacked aperiodically in columns (liquid-like), the different columns packing in a two-dimensional array (crystal-like). The phases have translational periodicity in two dimensions but liquid-like disorder in the third. In addition to the columnar phase(D), the disc-like molecules also exhibit a nematic phase (N_D). A transition between D and N_D phases has been reported.

Recently, it has been shown that certain spiropyrans exhibit some of the features of liquid crystals, but have structures that are different from conventional liquid crystals; these solids have been called *quasiliquid crystals* (Shvartsman & Krongauz, 1984).

$X = RCOOCH_2CH_2$

(I)   (II)

4.14   Noncrystalline state and the glass transition

It was mentioned earlier (Section 3.4) that all materials can be rendered amorphous. What is often not recognized by chemists is that a large variety of materials – catalysts, catalyst supports, xerographic photoreceptors, optical fibres, large-area solar cells and many biominerals – are noncrystalline (Elliott *et al.*, 1986). Noncrystalline solids possess no long-range order and their structure is akin to that of a frozen liquid. Most

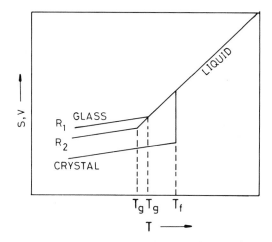

Figure 4.29 Changes in volume or entropy which can occur on cooling a liquid. Crystallization may occur at T_f or, if the liquid is supercooled below T_f, a glass is formed. The temperature corresponding to the break in slopes of V(or S) versus T is termed the glass transition temperature, T_g. The value of T_g varies with the cooling rate, $R(R_1 > R_2)$.

noncrystalline (or amorphous) materials become crystalline on appropriate heat treatment. For example, ordinary glasses devitrify at some temperature; metallic glasses become crystalline on heating ($\sim$ 700 K or so). Crystallization is by no means a universal feature as exemplified by fused quartz. Vitreous carbon is a material employed in high-temperature applications. Crystallization of vitreous B_2O_3 is possible only with severely contrived crystallization procedures.

While crystallization is an important phase transition of noncrystalline materials, the phenomenon of glass transition is characteristic of glasses, but not of all amorphous substances. At the glass transition temperature, T_g, the supercooled liquid departs from the equilibrium behaviour (Fig. 4.29) and the actual value of T_g itself depends on the cooling rate. T_g is generally around $0.67\,T_f$, where T_f is the fusion temperature. The glass transition has some features apparently similar to those of second-order phase transitions. For example, C_p and other second derivatives of Gibbs free energy increase sharply at T_g. According to Fig. 4.29, the entropy of a liquid would become equal to that of a crystal well above 0 K, provided that there is no intervening transition. This problem is circumvented by postulating the existence of a transition at a temperature T_o (thermodynamic transition temperature) that limits the decrease in entropy. In order to avoid the paradoxical situation as to what happens when $T < T_o$ in the absence of a transition, one assumes that the glass is forced to crystallize at $T > T_o$. It is interesting that T_o values estimated from calorimetry (C_p vs. log T plots) are close to those obtained from equation (4.32), which describes the temperature variation of viscosity and other dynamic properties, ψ

$$\psi = \psi_o \exp [A/T - T_o] \qquad (4.32)$$

An ideal glass has been defined as one in which T_g approaches T_o. Nonideality of a glass would then be a measure of the entropy frozen in at T_g (this becoming zero when $T_g = T_o$). The existence of a thermodynamic transition at T_o has not been observed so far, since the relaxation times are too long below T_g. The experimental T_g itself occurs when the relaxation time attains a constant value on the time scale of the experiment. The situation at $T < T_o$ in the absence of a transition at T_o is referred to as the *Kauzmann paradox*; Kauzmann (1948) suggested that glasses would crystallize at a temperature greater than T_o.

A variety of relaxation, spectroscopic and other techniques have been employed to investigate the behaviour of various properties of glasses around T_g, but the nature of the glass transition is not entirely understood (Wong & Angell, 1976; Parthasarathy *et al.*, 1983). Among the various models proposed for the transition, the free volume model (Cohen & Grest, 1981) and the configurational entropy model (Gibbs, 1963) are noteworthy. In the free volume model, dynamic properties are related to the free volume whereas in the configurational entropy model they are related to the configurational entropy as $\psi(T) = A \exp [B/TS_c]$ where A and B are constants. According to the latter model, a solution of the Kauzmann paradox requires a transition at a temperature T_o where the configurational entropy vanishes. The various models for the glass transitions have been reviewed (Parthasarathy *et al.*, 1983).

The entropy frozen below the glass transition has a fairly large nonconfigurational component. The role of communal entropy has been considered and it is suggested that the communal entropy vanishes at T_g, when the liquid-like properties cease to exist (Kanno, 1983). This would require the glass transition to be insensitive to cooling history, but this is contrary to experience. Computer experiments (Hoare & Barker, 1977) as well as electron microscopic studies (Gaskell *et al.*, 1979; Bursill *et al.*, 1981) have indicated the existence of ordered regions or clusters (with noncrystalline motifs) distributed in a tissue material of lower density. When clusters grow in number and size, a congelation to glass occurs, resulting in the elimination of configurational entropy; the tissue material in the intercluster regions could continue to undergo configurational changes, thereby accounting for significant values of configurational entropy. A cluster model of the glass transition has been developed, with the relative size of the cluster as the order parameter (Rao & Rao, 1982). The cluster model predicts the glass transition to be essentially second-order unlike the free volume model which predicts it to be first-order.

The term 'glass transition' is generally used in connection with positionally disordered materials, but the transition is also found in solids that are characterized by other degrees of freedom. For example, orientational disorder in plastic crystals may be quenched to yield *'glassy crystals'*, which exhibit glass-like transitions (Parthasarathy *et al.*, 1984; Suga & Seki, 1980). Disorder in dipole interactions may be frozen-in to yield *dipole glasses* (similar to spin glasses), typical examples being KBr doped with CN^- ions or $KTaO_3$ doped with Li^+ (Hochli *et al.*, 1979; Bhattacharya *et al.*, 1982). Frozen liquid crystals are also found to exhibit glass-like transitions (Johari & Goodby, 1982; Johari, 1982). Thus, the glassy state includes long-range disorder of many types, while the glass transition manifests itself when relaxational and experimental time scales intersect.

Simulation of local structures of liquids has its origins in the classic work of Bernal, who used mechanical assemblies for this purpose. The importance of nonspace-filling symmetries (such as the icosahedral and the pentagonal) to ordered aggregate in the amorphous state has been indicated by some of the studies. The role of such aggregates or clusters in understanding the glass transition was pointed out earlier. Computer simulation studies (Angell et al., 1981) have yielded useful information on the prototype glass transition in simple liquids, although the use of high quenching rates limits the applicability of the results. The high fictive temperatures (the temperature at which the extrapolated liquid and glass curves intersect in a V–T plot) of the simulated glasses suggest high diffusion rates. Both hard and soft spheres can be compacted into amorphous assemblies, but the characteristic of C_p discontinuity is absent in these cases. The hard sphere ensemble shows that the diffusivity does not vary linearly with free volume at low temperatures and V_o^D in equation (4.33)

$$D = A \exp \left[- B/(V - V_o^D) \right] \qquad (4.33)$$

seems to represent Bernal's dense random packing of hard spheres limit. Computer simulation studies are yet to demonstrate clearly the existence of the C_p overshoot, which is the hallmark of the experimental glass transition. Molecular dynamics simulation studies, however, provide valuable insight on ion movements in glasses (Angell, 1981).

Water, the most common substance in nature, can be rendered amorphous in the solid state. Amorphous solid water can be made by vapour deposition of dilute gas molecules on a cold substrate (Sceats & Rice, 1982). The resulting material looks like glass and is amorphous to X-rays. A glass transition has been noticed in such deposits (McMillan & Los, 1965) and the T_g predicted by the extrapolation of data on binary salt solutions (139 K) has been considered to be consistent with the T_g obtained by thermal measurements (Angell & Sare, 1970). The heat capacity of amorphous solid water measured by adiabatic calorimetry shows an increase in the heat capacity, as expected in a glass transition (Sugisaki et al., 1968). The extrapolated T_g was, however, not considered to be compatible with the heat capacity of liquid water. The structure of amorphous solid water prepared by vapour deposition is not different from what is expected for the low-temperature ground state for liquid water. Ultrafast quenching of liquid water produces amorphous samples similar to those prepared by vapour deposition; furthermore, crystallization does not occur during such cooling (Brugeller & Mayer, 1980; Dubocket et al., 1982). Crystallization of amorphous solid water occurs around 162 K. What is rather puzzling about amorphous solid water is the report (MacFarlane & Angell, 1984) that no thermal manifestation of the glass transition occurs in DSC curves anywhere in the temperature region predicted by the extrapolation of the data on binary salt solutions. It appears that water in the amorphous solid state is as difficult to understand as in the liquid state.

4.15   Monte Carlo and molecular dynamics methods

Monte Carlo and molecular dynamics methods are two important techniques employed in the study of fluids and solids (Klein, 1985; Klein & Lewis, 1990). In these techniques, properties of finite system of particles interacting via a known interparticle potential, are evaluated. The principle of the Monte Carlo method is to perform a stochastic averaging of the properties by means of the Metropolis importance sampling technique (Metropolis *et al.*, 1953). Most of the Monte Carlo calculations have been carried out in a cell of constant volume, Gibbs petit-canonical or NVT ensemble. However, since the work of Wood (1968) and McDonald (1972), calculations in the isothermal–isobaric ensemble in which the cell varies in volume during the simulation but is of fixed shape have also appeared frequently in the literature. In molecular dynamic calculations, Newton's equations of motion are solved numerically for a system of particles (Alder & Wainwright, 1957). As the total energy E of the system is conserved, the ensemble generated in the simulation corresponds to the microcanonical or NVE ensemble. The main advantage of the Monte Carlo method is that the isothermal–isobaric calculations are closer to the experimental conditions of fixed temperature and pressure. Dynamical properties, on the other hand, are readily obtained from molecular dynamics (MD).

The constant energy molecular dynamics and the canonical ensemble Monte Carlo techniques cannot be used to study phase transitions resulting from change in temperature or pressure, since the shape of the simulation cell is not permitted to change during the course of the simulation. Andersen (1980) extended the molecular dynamics method to allow for the variation in the size of the cell. Parrinello & Rahman (1980) extended it further to allow for the variation in shape and size of the cell. This has enabled studies of phase transitions by the molecular dynamics method (Nose & Klein, 1983*b*). The Monte Carlo method in the canonical or in the isothermal–isobaric ensemble calculation does not permit variation of the shape of the simulation cell. The Monte Carlo method has been generalized to enable the study of phase transitions and related phenomena (Yashonath & Rao, 1985*a*, *b*) by allowing for the variation in shape of the simulation cell. In molecular systems this has been done by including the orientational coordinates in terms of the Euler angles, making it possible to investigate polymorphic or structural phase transitions in solids. During the simulation of solid tetrachloromethane, it has been found that the cell rotates in space. The rotation of the simulation cell is possible in the variable shape Monte Carlo (as well as in the variable shape MD) studies. MC and MD simulations of phase transitions have been reviewed adequately (Rao & Yashonath, 1987; Klein & Lewis, 1990; Rao *et al.*, 1992).

It is difficult to investigate the relationship between the structure of a solid and the interparticle potential through which the particles interact by the existing theoretical and experimental methods. Yashonath & Rao (1985*b*) have investigated changes in the crystal structure of a monoatomic solid resulting from changes in the interaction potential. The solid interacting via a Lennard–Jones potential has a face-centred cubic structure and when the interaction potential is changed to caesium potential the solid transforms to body-centred cubic structure. The transformation was reversed when the interaction potential was changed to the Lennard–Jones form, just as in the case of a similar molecular dynamics study (Parrinello & Rahman, 1980).

The transition from the crystalline to the plastic crystalline phase has been studied by employing the molecular dynamics technique, in which the shape of the cell could continuously vary during the simulation. A study of the phase transition in tetrachloromethane has been carried out by employing both MC and MD methods (McDonald *et al.*, 1982; Yashonath & Rao, 1985a). The high pressure phase III of tetrachloromethane transforms to a plastic crystal phase at about the same temperature as that found in experiments. Transformation of the ordered monoclinic phase to the plastic crystal phase at atmospheric pressure in carbon tetrafluoride has also been studied by the MC method and the results are in general agreement with those from experiment and from a molecular dynamics study (Nose & Klein, 1983a).

The plastic crystalline states of several other compounds including C_{60}, C_{70}, adamantane and neopentane have been studied by these methods (Rao & Yashonath, 1987; Rao & Seshadri, 1994). In Fig. 4.30, we show the ordered and disordered phases of adamantane. Orientational glasses obtained by quenching plastic crystals have been examined by the Monte Carlo method (Chakrabarti *et al.*, 1993; Yashonath & Rao, 1983). Simulation studies on the structural changes accompanying the vitrification of isopentane have yielded interesting results (Yashonath & Rao, 1986b). It has been found that cooling the liquid results in translational as well as orientational relaxation. The changes in the radial distribution functions associated with the peripheral groups such as CH_3 and the inaccessible groups, such as CH in isopentane (viz., the absence of shift of the peaks towards lower distance of the intermolecular CH_3—CH_3 groups), has enabled one to obtain information regarding the orientational contribution to the structural rearrangement. A similar study has been carried out on *n*-pentane, neopentane and neohexane to understand the relationship between molecular shape and the orientational contribution (Chakrabarti *et al.*, 1992). The results suggest that the orientational contribution is highest for globular geometry. About 24% of the total contribution to the increase in intermolecular interactions comes from the reorientation of molecules during cooling. For molecules such as isopentane, which are not exactly globular, the contribution is large (about 22%), but smaller than for globular geometry. Linear molecules such as *n*-pentane have a small orientational contribution.

Normal alkanes are known to exhibit several phases below their melting point with a significant degree of orientational and conformational disorder. Ryckaert *et al.* (1989) have carried out molecular dynamics simulations on sixty $C_{23}H_{48}$ chain molecules (tricosane) at 311 and 315 K. In spite of the small difference in temperature, they find the 311 K phase to be crystalline wherein the molecules are in the all-*trans* elongated conformation. The crystalline phase transforms to a rotator phase when heated to 315 K (Figure 4.31). Diffraction data predict the transformation to be at 313 K. The simulation suggests that the longitudinal diffusivity increases significantly in the rotator phase, in agreement with NMR studies. Increase in diffusivity is accompanied by an increase in the orientational disorder. Simultaneously, the number of *gauche* defects increases, mostly towards the chain ends. The excellent agreement of the simulation with experiment in this case is due to the good intermolecular potentials available for *n*-alkanes.

Orientational disorder in ionic solids has been investigated by Klein and coworkers

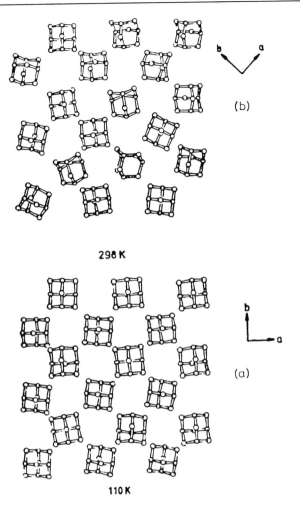

298 K

110 K

Figure 4.30   A view of instantaneous arrangement in adamantane looking down the c axis: (a) tetragonal ordered phase at 110K; (b) orientationally disordered cubic phase at 298K (after Yashonath and Rao, 1986a).

(Sprik *et al.*, 1993; Signorini *et al.*, 1990), a typical example being the orientational disorder associated with NH_4^+ in NH_4Br. Detailed simulations have been reported on $(NaCN)_{1-x}(KCN)_x$ and other mixed alkali halides and alkali cyanides. Other systems studied include potassium and calcium nitrate crystals and their mixtures. The transition from the crystalline to the superionic conductor phase in solid electrolytes has also been successfully investigated. Molecular dynamics studies of AgI were carried out by Parrinello, Rahman & Vashishta (1983). Li_2SO_4 has been investigated by molecular dynamics by Impey *et al.* (1985). Here, the Li^+ ions become mobile at high temperatures. The SO_4^{2-} ions exhibit orientational disorder and the orientational dynamics of the SO_4^{2-} group can be described in terms of time autocorrelation functions appropriate for tetrahedral symmetry. These are reported for two models,

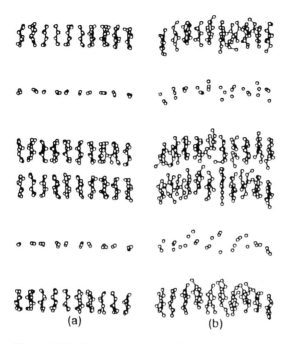

Figure 4.31   Snapshot picture of molecular dynamics of $C_{23}H_{48}$, tricosane, viewed along the a axis. A bilayer is shown for (a) the crystalline phase at 311K and (b) the rotator phase at 315K. For clarity no H atoms, only the central C atoms along with a few terminal C atoms are shown. Note the disorder in the central C atoms in the rotator phase which is predominantly along the longitudinal axis (after Ryckaert *et al.*, 1989).

one with a charge of $-0.5e$ and another $-0.8e$ on the oxygen of SO_4^{2-}. The calculated results of the model with a charge of $-0.8e$ for the oxygen atom are in good agreement with the experimental results.

4.16   Applications of phase transitions

Changes in properties at the phase transition are often of technological interest, and several materials applications of phase transitions have been found. Following Goodenough (1973), applications of phase transitions may be classified into those utilizing (i) the formation and/or motion of mobile boundaries between two or more thermodynamic states coexisting below a critical temperature T_c, (ii) changes in physical properties as the temperature approaches T_c, (iii) changes in properties at T_c and (iv) metastable phases obtained by control of the kinetics of nucleation or diffusion required for transformation to stable phases.

Under the first category (phases at $T < T_c$), where one makes use of changes in the net material properties by control of nucleation or movement of domain boundaries, we have large classes of materials like ferroelectrics, ferroelastics, ferromagnets, superconductors and liquid crystals. In ferromagnets, long-range magnetic order below the Curie temperature induces a spontaneous magnetization, M_s, resulting in many magnetic domains, each with a M_s vector oriented in a direction different from that in adjacent domains (Section 6.3.1). The adjacent domains are separated by domain walls within which the M_s vector rotates from the orientation in one domain to that in the other. An external magnetic field changes the size of the domains, enlarging those of favourable orientation at the expense of others. Ferroelectrics are characterized by different orientational states below T_c, but the spontaneous polarization is induced by a cooperative crystallographic distortion. The domain boundaries in ferroelectrics can be controlled by an external field. In ferroelastics, a cooperative crystal distortion induces a spontaneous strain below T_c. An externally applied stress controls the domain boundaries of a ferroelastic. Magnetic (or electric) and elastic long-range order coexist in many materials, so that spontaneous magnetization (or polarization) is controlled by an applied stress or a spontaneous strain by an applied magnetic or electric field. Coexistence of magnetic and electric long-range order is known in substances like $BiFeO_3$, which is an antiferromagnetic–ferroelectric. Magnetoelectric phenomena have been investigated and exploited (Freeman & Schmid, 1975). Optical properties of ferroelectrics, ferroelastics and ferromagnets have found many applications, a typical example being the ferroelectric–ferroelastic $Gd_2(MoO_4)_3$ with a T_c of 433 K, which is transparent in the visible region.

Liquid crystals have found widespread application in optical display devices as well as in detection of temperature uniformity and impurities. These properties are related to the orientational order of molecules in the temperature region between T_c and the melting point. The possible applications of ferroelectric liquid crystals are promising. Superconductors (type II) can be used to create high magnetic fields at low power; the ability of type I superconductors to trap magnetic flux within the domains of the normal material may also have applications.

Two of the important properties which change near T_c are softening of an optical vibration mode before a displacive transition, and temperature dependence of spontaneous magnetization in ferromagnets below T_c. These properties are used in dielectric and pyromagnetic detectors respectively. Invar alloys have lattice parameters which are essentially temperature-independent below the ferromagnetic Curie temperature, since magnetically induced lattice change is superimposed on the usual thermal expansion in this system. In metallic $Bi_2Ru_2O_{7-x}$ (defect pyrochlore structure), resistivity is nearly temperature-independent in the range $150 < T < 500$ K. This is apparently due to electron coupling to soft vibrational modes above T_c (150 K), where a phase change occurs. Electron–phonon coupling to soft modes in materials will probably afford some significant applications. Device applications of soft-mode anomalies have been discussed by Fleury (1973).

Coming to properties at T_c, we can conceive of uses being made of the latent heat of a first-order transition for storing energy and regulating temperature. First-order magnetic transitions could be used for switching. Semiconductor–metal transitions can be employed in circuit breakers, voltage dividers or optical switches.

The importance of metastable phases which persist at ambient pressure and temperature need not be emphasized. Control of the eutectoidal decomposition of the metastable BCC iron–carbon phase, austenite, below 996 K is essential to the steel industry. Austenite is made by quenching a FCC phase. When the carbon content is greater than $0.2\,wt\%$, cubic austenite transforms to tetragonal martensite by a diffusionless mechanism. Many high pressure phases can be retained at atmospheric pressure. Diamond and boron nitride are examples of such metastable phases. Several metastable phases can be prepared by low-temperature chemical methods (Chapter 3). Cubic $NaSbO_3$ exhibiting fast Na^+ conduction is a metastable phase prepared from $KSbO_3$ by ion exchange. The best-known examples of metastable materials are glasses and other amorphous materials which have wide applications.

References
Alder, B. J. & Wainright, T. W. (1957) *J. Chem. Phys.* **27**, 1208.

Andersen, H. C. (1980) *J. Chem. Phys.* **72**, 2384.

Anderson, P. W. (1959) in *Fizika Dielectrikov* (ed. Shanavi, G. I.) Acad. Nauk SSSR, Moscow.

Angell, C. A. (1981) in *Preparation and Characterization of Materials* (eds Honig, J. M. & Rao, C. N. R.) Academic Press, New York.

Angell, C. A. & Sare, E. J. (1970) *J. Chem. Phys.* **52**, 1058.

Angell, C. A., Clarke, J. H. R. & Woodcock, L. V. (1981) *Adv. Chem. Phys.* **48**, 397.

Aston, J. G. (1963) in *Physics and Chemistry of the Organic Solid State* (eds Fox, F., Labes, M. & Weissberger, A.) Interscience, New York.

Axe, J. D. (1980) *Phil. Trans. Roy. Soc. London* **B290**, 593.

Bak, P. (1982) *Repts Prog. Phys.* **45**, 587.

Bari, R. A. (1972) *Proc. 7th Annual Conf. Magnetism and Magnetic Materials*, Chicago, American Institutes of Physics.

Bari, R. A. & Sivardiere, J. (1972) *Phys. Rev.* **B5**, 4466.

Bhattacharya, S., Nagel, S. R., Fleishman, L. & Susman, S. (1982) *Phys. Rev. Lett.* **46**, 1267.

Bhide, V. G., Rajoria, D. S., Rao, G. R. & Rao, C. N. R. (1972) *Phys. Rev.* **B6**, 1021.

Birgeneau, R. J., Kjems, J. K., Shirane, G. & Van Uitert, L. G. (1974) *Phys. Rev.* **B10**, 2512.

Blinc, R. & Zeks, B. (1974) *Soft-Modes in Ferroelectrics and Antiferroelectrics*, North-Holland, Amsterdam.

Bounds, D. G., Klein, M. L. & Patey, G. N. (1980) *J. Chem. Phys.* **72**, 5348.

Brugeller, P. & Mayer, E. (1980) *Nature* **288**, 569, 579.

Buerger, M. J. (1951) in *Phase Transformations in Solids* (eds Smoluchowski, R., Mayer, J. E. & Weyl, W. A.) John Wiley, New York.

Burgers, W. G. & Groen, L. J. (1957) *Disc. Faraday Soc.* **23**, 183.

Bursill, L. A., Thomas, J. M. & Rao, K. J. (1981) *Nature* **289**, 157.

Cahn, J. W. (1968) *Trans AIME* **242**, 168.

Carneiro, K., Shirane, G., Werner, S. A. & Kaiser, S. (1976) *Phys. Rev.* **B13**, 4258.

Chakrabarti, A., Yashonath, S. & Rao, C. N. R. (1992) *J. Phys. Chem.* **96**, 6762.

Chakrabarti, A., Yashonath, S. & Rao, C. N. R. (1993) *Chem. Phys. Lett.* **215**, 591.

Chandrasekhar, S. (1994) *Liquid Crystals*, Cambridge University Press, Cambridge, 2nd edn.

Chenavas, J. & Joubert, J. C. (1971) *Solid State Comm.* **9**, 1057.

Chestnut, D. B. (1964) *J. Chem. Phys.* **40**, 405.

Christian, J. W. (1965) *Iron and Steel Institute Special Report No. 93.*

Cochran, W. (1959) *Phys. Rev. Lett.* **3**, 412.

Cohen, M. H. & Grest, G. S. (1981) *Adv. Chem. Phys.* **48**, 455.

Comes, R., Lambert, M., Launois, H. & Zeller, H. R. (1973) *Phys. Rev.* **B8**, 571.

Cowley, J. M., Cohen, J. B., Salamon, M. B. & Wuensch, B. J. (eds) (1979) *Modulated Structures.* American Institute of Physics.

de Gennes, P. G. (1971) *Mol. Cryst. Liq. Cryst.* **12**, 193.

De Graaf, L. A. & Sciesinski, J. (1970) *Physica* **48**, 79.

Demazeau, G., Pouchard, M., Thomas, M., Columbet, J. F., Grenier, J. C., Fournes, L., Soubeyroux, J. L. & Hagenmuller, P. (1980) *Mater. Res. Bull.* **15**, 451.

Demazeau, G., Marty, J. L., Buffat, B., Dance, J. M., Pouchard, M., Dordor, P. & Chevalier, B. (1982) *Mater. Res. Bull.* **17**, 37.

Desiraju, G. R., Paul, I. C. & Curtin, D. Y. (1977) *J. Amer. Chem. Soc.* **99**, 1594.

de Villepin, J., Limage, M. H., Novak, A., Toupry, N., Le Postollec, M., Poulet, H., Ganguly, S. & Rao, C. N. R. (1984) *J. Raman Spec.* **15**, 41.

Di Salvo, F. J. (1977) in *Electron-Phonon Interactions and Phase Transitions* (ed. Riste, T.) Plenum Press, New York, p. 107.

Di Salvo, F. J. & Rice, T. M. (1979) *Physics Today* (April issue).

Domb, C. (1985) *Contemp. Phys.* **26**, 49.

Dubocket, J., Lepault, J. & Freeman, R. (1982) *J. Microse,* **128**, 219.

Dunitz, J. (1991) *Pure Appl. Chem.* **63**, 177.

Dunitz, J. D. & Bernstein, J. (1995) *Acc. Chem. Res.* **28**, 193.

Dzyaloshinskii, J. E. (1964) *Soviet Phys. JETP* **19**, 960.

Elliott, R. J., Harley, R. T., Hayes, W. & Smith, S. R. P. (1972) *Proc. Roy. Soc. London* **A328**, 217.

Elliott, S. R., Rao, C. N. R. & Thomas, J. M. (1986) *Angew. Chem. Int. Ed. (Engl.)* **25**, 31.

Emery, V. J. & Axe, J. D. (1978) *Phys. Rev. Lett.* **40**, 1507.

Englman, R. (1972) *The Jahn-Teller Effect in Molecules and Crystals,* John Wiley, London.

Fluery, P. A. (1973) in *Phase Transitions* (ed. Henisch, H., Roy, R. & Cross, L. E.) Pergamon Press, New York.

Fleury, P. A. & Lyons, K. B. (1983) in *Light Scattering near Phase Transitions* (eds Cummins, H. Z. & Levanyuk, A. P.) North-Holland, Amsterdam.

Frank, C. (1983) *Phil, Trans. Roy. Soc. London* **A309**, 71.

Freeman, A. J. & Schmid, H. (eds) (1975) *Magnetoelectric Interaction Phenomena in Crystals,* Gordon & Breach, New York.

Frost, J. C., Leadbetter, A. J. & Richardson, R. M. (1980) *Disc. Faraday Soc.* **69**, 32.

Ganguly, S., Fernandes, J. R., Bahadur, D. & Rao, C. N. R. (1979) *J. Chem. Soc. Faraday II,* **75**, 923.

Gaskell, P. H., Smith, D. J., Catto, C. J. D. & Cleaver, J. R. A. (1979) *Nature* **281**, 465.

Gehring, G. A. & Gehring, K. A. (1975) *Repts. Prog. Phys.* **38**, 1.

Gibbs, J. H. (1963) in *Modern Aspects of the Vitreous State* (ed. Mackenzie, J. D.) Butterworths, London.

Goodenough, J. B. (1973) in *Phase Transitions* (eds Henisch, H., Roy, R. & Cross, L. E.) Pergamon Press, New York.

Gordon, R. G. (1966) *J. Chem. Phys.* **44**, 1830.

Gray, G. W. (1983) *Phil. Trans. Roy. Soc.* London **A309**, 77.

Gray, G. W. & Winsor, P. A. (eds) (1974) *Liquid Crystals and Plastic Crystals*, Ellis Horwood, Chichester.

Guthrie, G. B. & McGullough, J. (1961) *J. Phys. Chem. Solids* **18**, 53.

Gütlich, P. (1981a) *Structure and Bonding* **44**, 83.

Gütlich, P. (1981b) in *Mössbauer Spectroscopy and its Chemical Applications*, Advances in Chemistry Series **194**, American Chemical Society.

Haddad, M. S., Lynch, M. W., Federer, W. D. & Hendrickson, D. N. (1981) *Inorg. Chem.* **20**, 131.

Harley, R. T., Hayes, W., Perry, A. M. & Smith, S. R. P. (1973) *J. Phys.* **C6**, 2382.

Harley, R. T., Lyons, K. B. & Fleury, P. A. (1980) *J. Phys.* **C13**, L447.

Heine, V. & McConnell, J. D. C. (1981) *Phys. Rev. Lett.* **46**, 1092.

Heine, V. & McConnell, J. D. C. (1984) *J. Phys.* **C17**, 1199.

Hill, N. E. (1963) *Proc. Phys. Soc.* London **82**, 723.

Hilsum, C. & Raynes, E. P. (eds) (1983) *Phil. Trans. Roy. Soc.* London **A309**, 69.

Hoare, M. R. & Barker, J. A. (1977) in *The Structure of Non-Crystalline Materials* (ed. Gaskell, P. H.) Taylor & Francis, London.

Höchli, U. T., Weibel, H. E. & Boatner, L. A. (1979) *J. Phys.* **C12**, L563.

Iizumi, M., Axe, J. D., Shirane, G. & Simaoka, K. (1977) *Phys. Rev.* **B15**, 4392.

Impey, R. W., Klein, M. L. & McDonald, I. R. (1985) *J. Chem. Phys.* **82**, 4690.

Iqbal, Z. & Owens, F. J. (1984) *Vibrational Spectroscopy of Phase Transitions*, Academic Press, New York.

Joesten, M. D., Takagi, S. & Lenhert, P. G. (1977) *Inorg. Chem.* **16**, 2680.

Johari, G. P. (1982) *J. Chem. Phys.* **77**, 4619.

Johari, G. P. & Goodby, J. W. (1982) *J. Chem. Phys.* **77**, 565.

Johnson, C. K. & Watson, C. R. (1976) *J. Chem. Phys.* **64**, 2271.

Jones, W., Thomas, J. M. & Williams, J. O. (1975) *Phil. Mag.* **32**, 1.

Kadanoff, L. P. (1966) *Physics* **2**, 263.

Kambara, T. (1979) *J. Chem. Phys.* **70**, 4199.

Kambara, T. (1981) *J. Chem. Phys.* **74**, 4557.

Kanno, H. (1983) *J. Non-Cryst. Solids* **37**, 203.

Kauzmann, W. (1948) *Chem. Rev.* **3**, 219.

Klein, M. L. (1985) *Ann. Rev. Phys. Chem.* **36**, 525.

Klein, M. L. & Lewis, L. J. (1990) *Chem. Rev.* **90**, 459.

Klein, M. V. (1983) in *Light Scattering near Phase Transitions* (eds Cummins, H. Z. & Levanyuk, A. P.) North-Holland, Amsterdam.

König, E. & Kremer, S. (1971) *Theor. Chem. Acta* **20**, 143.

König, E., Ritter, G. & Irler, W. (1979) *Chem. Phys. Lett.* **66**, 336.

König, E., Ritter, G. & Kulshreshta, S. K. & Nelson, S. M. (1983) *J. Amer. Chem. Soc.* **105**, 1924.

König, E., Ritter, G. & Kulshreshta, S. K. (1985) *Chem. Revs.* **85**, 219.

Lechner, R. E. (1972) *Solid State Comm.* **10**, 1247.

Madhusudan, W. H., Jagannathan, K., Ganguly, P. & Rao, C. N. R. (1980) *J. Chem. Soc.* Dalton, 1397.

Makovicky, E. & Hyde, B. G. (1981) *Structure and Bonding* **46**, 101.

Mandelbrot, B. B. (1983) *The Fractal Geometry of Nature*, Freeman & Co. New York.

McConnell, J. D. C. & Heine, V. (1982) *J. Phys.* **C15**, 2387.

McDonald, I. R. (1972) *Mol. Phys.* **23**, 41.

McDonald, I. R., Bounds, D. G. & Klein, M. L. (1982) *Mol. Phys.* **45**, 521.

McFarlane, D. R. & Angell, C. A. (1984) *J. Phys. Chem.* **88**, 759.

McMillan, J. A. & Los, S. C. (1965) *J. Chem. Phys.* **42**, 829.

Metropolis, N. A., Rosenbluth, A. W., Rosenbluth, M. N., Teller, A. H. & Teller, E. (1953) *J. Chem. Phys.* **21**, 1087.

Mohanram, R. A., Singh, K. K., Madhusudan, W. H., Ganguly, P. & Rao, C. N. R. (1983) *Mater. Res. Bull.* **18**, 703.

Natarajan, M., Das, A. R. & Rao, C. N. R. (1969) *Trans. Faraday Soc.* **65**, 3081.

Nose, S. & Klein, M. L. (1983a) *J. Chem. Phys.* **78**, 6928.

Nose, S. & Klein, M. L. (1983b) *Mol. Phys.* **50**, 1055.

Overhauser, A. W. (1962) *Phys. Rev.* **128**, 1437.

Owens, F. J., Poole, C. P. & Farach, H. A. (1979) *Magnetic Resonance of Phase Transition*, Academic Press, New York.

Parkinson, G. M., Thomas, J. M., Williams, J. O., Goringe, M. J. & Hobbs, L. W. (1976) *J. Chem. Soc. Perkin* **II**, 836.

Parrinello, M. & Rahman, A. (1980) *Phys. Rev. Lett.* **45**, 1196.

Parrinello, M., Rahman, A. & Vashista, P. (1983) *Phys. Rev. Lett.* **50**, 1073.

Parsonage, N. G. & Staveley, L. A. K. (1978) *Disorder in Crystals*, Clarendon Press, Oxford.

Parthasarathy, R., Rao, K. J. & Rao, C. N. R. (1983) *Chem. Soc. Revs.* **12**, 361.

Parthasarathy, R., Rao, K. J. & Rao, C. N. R. (1984) *J. Phys. Chem.* **88**, 49.

Paul, I. C. & Go, K. T. (1969) *J. Chem. Soc.* **B**, 33.

Porter, D. A. & Easterling, K. E. (1981) *Phase Transitions in Metals and Alloys*, Van Nostrand-Reinhold, New York.

Pouget, J. P., Shirane, G., Hastings, J. M., Heeger, A. J., Miro, N. D. & MacDiarmid, A. G. (1978) *Phys. Rev.* **B18**, 3645.

Pytte, E. & Stevens, K. W. H. (1971) *Phys. Rev. Lett.* **27**, 862.

Qi, Y., Müller, E. W., Spiering, H. & Gütlich, P. (1983) *Chem. Phys. Lett.* **101**, 503.

Raccah, P. M. & Goodenough, J. B. (1967) *Phys. Rev.* **155**, 932.

Ramasesha, S., Ramakrishnan, T. V. & Rao, C. N. R. (1979) *J. Phys.* **C12**, 1307.

Ramdas, S. & Thomas, J. M. (1976) *J. Chem. Soc. Faraday II*, **72**, 1251.

Rao, C. N. R. (1984) *Acc. Chem. Res.* **17**, 83.

Rao, C. N. R. (1985a) *Int. Rev. Phys. Chem.* **4**, 19.

Rao, C. N. R. (1985b) *Proc. Ind. Acad. Sci.* (*Chem. Sci.*) **94**, 181.

Rao, C. N. R. (1993) *J. Mol. Struc.* **292**, 229.

Rao, C. N. R. & Rao, K. J. (1978) *Phase Transitions in Solids*, McGraw-Hill, New York.

Rao, C. N. R. & Yashonath, S. (1987) *J. Solid State Chem.* **68**, 193.

Rao, C. N. R. & Seshadri, R. (1994) *MRS Bulletin.*

Rao, C. N. R., Ganguly, S., Swamy, H. R. & Oxton, I. A. (1981) *J. Chem. Soc. Faraday II*, **77**, 1825.

Rao, C. N. R., Ganguly, S. & Swamy, H. R. (1982) *Croat. Chem. Acta* **55**, 207.

Rao, C. N. R., Rao, K. J., Ramasesha, S., Sarma, D. D. & Yashonath, S. (1992) *Annual Reports, Section C, The Royal Society of Chemistry*, Cambridge.

Rao, K. J. & Rao, C. N. R. (1966) *J. Mat. Sci.* **1**, 238.

Rao, K. J. & Rao, C. N. R. (1982) *Mater. Res. Bull.* **13**, 1337.

Riste, T. (ed.) (1977) *Electron-Phonon Interactions and Phase Transitions*, Plenum Press, New York.

Rouxel, J. (1992) *Acc. Chem. Res.* **25**, 328.

Rouxel, J. & Meerschaut, A. (1994) *Mol. Cryst. Liq. Cryst.* **244**, 343.

Ryckaert, J. P., McDonald, I. R. & Klein, M. L. (1989) *Mol. Phys.* **67**, 957.

Samara, G. A. (1984) *Solid State Phys.* **38**, 1.

Sasaki, N. & Kambara, T. (1982) *J. Phys.* **C15**, 1035.

Sceats, M. S. & Rice, S. A. (1982) in *Water: A Comprehensive Treatise* (ed. Franks, F.), Vol. 7, Plenum Press, New York.

Scott, J. F. (1974) *Rev. Mod. Phys.* **46**, 83.

Shapiro, S. M., Axe, J. D., Shirane, G. & Riste, T. (1974) in *Anharmonic Lattices, Structural Transitions and Melting* (ed. Riste, T.) Noordhoff, Leiden.

Sherwood, J. N. (ed.) (1979) *The Plastically Crystalline State.* John Wiley, New York.

Shirane, G. & Yamada, Y. (1969) *Phys. Rev.* **177**, 858.

Shvartsman, F. & Krongauz, V. (1984) *Nature* **309**, 608.

Signorini, G. F., Barrat, J. L. & Klein, M. L. (1990) *J. Chem. Phys.* **92**, 1294.

Slichter, C. P. & Drickamer, H. G. (1972) *J. Chem. Phys.* **56**, 2142.

Sorai, M. & Seki, S. (1974) *J. Phys. Chem. Solids* **35**, 555.

Spiering, H., Meissner, E., Köppen, H., Müller, E. W. & Gütlich, P. (1982) *Chem. Phys.* **68**, 65.

Sprik, M., Cheng, A. & Klein, M. L. (1993) *Phys. Rev. Lett.* **69**, 1660.

Sturge, M. D., Cohen, E., Van Uitert, L. G. & Van Stapele, R. P. (1975) *Phys. Rev.* **B11**, 4768.

Suga, H. & Seki, S. (1980) *Faraday Disc. Chem. Soc.* **69**.

Sugai, S. (1985) *J. Phys.* **C18**, 799.

Sugisaki, M., Suga, H. & Seki, S. (1968) *Bull. Chem. Soc. Japan* **41**, 2591.

Tanabe, Y. & Sugano, S. (1954) *J. Phys. Soc. Japan* **9**, 753, 766.

Usha, S., Srinivasan, R. & Rao, C. N. R. (1985) *Chem. Phys.* **100**, 447.

Venkataraman, G. (1979) *Bull. Mater. Sci.* **1**, 129.

Will, G., Gobel, H., Sampson, C. F. & Forsyth, J. B. (1972) *Phys. Lett.* **38A**, 207.

Williams, P. M. (1976) in *Crystallography and Crystal Chemistry of Materials with Layered Structures* (ed. Levy, F.) Reidel, Dordrecht-Holland.

Wilson, J. A., Di Salvo, F. J. & Mahajan, S. (1975) *Adv. Phys.* **24**, 117.

Wilson, K. G. & Fisher, M. E. (1972) *Phys. Rev. Lett.* **28**, 248.

Wong, J. & Angell, C. A. (1976) *Glass Structure by Spectroscopy*, Marcel Dekker, New York.

Wood, W. W. (1968) in *Physics of Simple Liquids* (eds Temperley, H. N. V., Rowlinson, J. N. & Rushbrooke, G. S.), North-Holland, Amsterdam.

Wyllie, G. (1971) *J. Phys.* **C4**, 564.

Yakovlev, I. A., Velichkina, T. S. & Mikheeva, L. F. (1956) *Sov. Phys. Doklady* **1**, 215.

Yamada, Y. (1977) in *Electron-Phonon Interactions and Phase Transitions* (ed. Riste, T.), Plenum Press, New York.

Yashonath, S. & Rao, C. N. R. (1983) *Chem. Phys. Lett.* **101**, 524.

Yashonath, S. & Rao, C. N. R. (1985a) *Chem. Phys. Lett.* **119**, 22.

Yashonath, S. & Rao, C. N. R. (1985b) *Mol. Phys.* **54**, 245.

Yashonath, S. & Rao, C. N. R. (1986a) *J. Phys. Chem.* **90**, 2552.

Yashonath, S. & Rao, C. N. R. (1986b) *J. Phys. Chem.* **90**, 2581.

5 New light on an old problem: defects and nonstoichiometry

5.1 Introduction

That compounds have definite compositions is taken as a matter of faith and yet there are several inorganic solids which exhibit a wide range of compositions or show no simple correspondence between the composition and the detailed structure (or chemical identity). It has been known since the 1920s that stoichiometric $FeO_{1.00}$ does not fall in the stability range of iron(II) oxide ($FeO_{1.05}$–$FeO_{1.15}$). Point defects in crystals such as vacancies and interstitials first described by Schottky, Frenkel and Wagner account for the transport properties of ionic solids, but there are serious difficulties in applying the point-defect formalism to solids possessing a wide stoichiometric range or to those solids exhibiting ordering of defects or extended defects (such as crystallographic shear planes). Although there is no clear-cut transition between the point-defect regime and the regime of highly ordered structural imperfections in nonstoichiometric solids, we can certainly say that the point-defect model is really valid only when the defect concentration (or the deviation from stoichiometry) is extremely small. Only in such dilute point-defect systems can one satisfactorily relate the electronic properties and the nonstoichiometry to the concentration of point defects.

A few general comments related to disorder and nonstoichiometry in crystals would be in order. Solids rarely attain a state of perfect order even when they are cooled close to absolute zero of temperature. At ordinary temperatures, crystalline solids generally depart from perfect order and contain several types of imperfections, which are, indeed, responsible for many important solid state phenomena such as diffusion, electrical conduction, plasticity and so on. The types of disorder that occur in solids fall into one of the following categories: point defects, linear defects, planar defects and volumetric defects. Point defects arise from the absence of atoms (or ions) on lattice sites, the presence of atoms in interstitial positions, the presence of atoms in the wrong positions (not possible in ionic solids) and the presence of alien atoms (Fig. 5.1). Presence of point defects results in the polarization of the surrounding crystal structure, giving rise to small displacements of neighbouring atoms or ions. For example, a cation vacancy in an ionic solid will have effective negative charge and causes displacements of neighbouring anions. The energy of formation of a point defect depends mainly on the atomic arrangement in its immediate neighbourhood since the relaxation effects

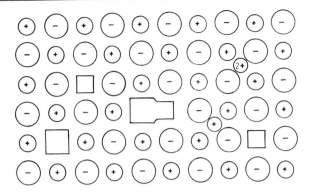

Figure 5.1   Point defects in ionic solids: Schottky defect, vacancy pair, Frenkel defect and aliovalent impurity (for definitions see Section 5.2).

decrease rapidly with distance. Linear defects in crystals are dislocations, corresponding to rows of atoms which do not possess the right coordination. Planar defects are grain boundaries (boundaries between small crystallites), stacking faults, crystallographic shear planes, twin boundaries and antiphase boundaries. Segregation of point defects can give rise to volumetric (three-dimensional) defects.

Compounds exhibiting variable stoichiometry have been referred to as *Berthollides*, whereas those with a nearly fixed composition have been called *Daltonides*, reflecting a controversy over a century ago between Dalton and Berthollet. The term 'nonstoichiometric compound' has also been used, though not always with the same meaning. A *nonstoichiometric compound* can be defined (in terms of operational criteria) as a crystalline compound in equilibrium with its environment, behaving as a thermodynamically bivariant system; although properties of the crystal (e.g. cell dimensions) may change in magnitude with composition, the symmetry remains the same throughout the composition range of stability (Anderson, 1984). It is well to remember that the stoichiometry of a crystal is uniquely defined by the composition of the repeating unit cell. The formula of the unit cell may be complex, with seemingly irrational ratios of the constituent atoms as in $NbO_{2.4906}$, $NbO_{2.4681}$, $NbO_{2.4167}$, $PrO_{1.833}$ and $PrO_{1.714}$ (actually these oxides are the crystallographically well-defined compounds $Nb_{53}O_{132}$, $Nb_{47}O_{116}$, $Nb_{12}O_{29}$, $Pr_{12}O_{22}$ and Pr_7O_{12} respectively); such closely spaced phases are structurally related and throw light on the nature of nonstoichiometric compounds. In nonstoichiometric compounds, the average number of atoms per unit cell is not equivalent to the number of sites and in one of the sublattices (anion or cation), there is a deficiency or an excess of the species.

Nonstoichiometric compounds are mixed-valence compounds with nonintegral electron/atom ratios. Electronic properties of these compounds depend crucially on the nature and magnitude of nonstoichiometry. Electronic conduction in many such compounds occurs by hopping between the cations of different valencies (e.g. Pr^{3+} and Pr^{4+} in $Pr_{12}O_{22}$). Nonstoichiometry with a wide range of compositions is more common in oxides, sulphides, and related materials where the bonding is not completely ionic. In ionic nonstoichiometric compounds, structural rearrangements can occur owing to coulombic interactions between the defects and the altervalent cations present in them. In metallic systems, however, the delocalized electrons can screen centres of local electron deficiency or excess from long-range interactions (Anderson, 1984); such systems may form superstructures, although the probability of structural rearrangement is reduced considerably. Heteroionic solid solutions provide model systems for comparison with bona fide nonstoichiometric compounds. Such solid solutions are formed by a fixed-valency guest ion replacing another fixed-valency ion in a host solid (e.g. $CaF_2–YF_3$, $CdCl_2–NaCl$), by a fixed-valency ion replacing a variable-valency host ion (e.g. $Ni_{1-x}Li_xO$) or by a variable-valency guest ion replacing a fixed-valency host ion (e.g. the valency of Mn ion in oxide matrices can be 2^+, 3^+ or 4^+). Unlike in many nonstoichiometric compounds where the structural defects and the compensating electronic defects (charges) are both involved in ordering processes, the structural defects (and the electronic defects) get localised in heteroionic solid solutions, giving rise to a wider range of nonstoichiometric compositions in the latter.

It was mentioned earlier that defects order themselves, giving superstructures; point defects such as vacancies are eliminated by the formation of planar defects (shear planes) or structural singularities. There is now a wealth of information in the literature on the real structure of defect solids, especially those with irrational cation:anion ratios (Anderson & Tilley, 1974; Tilley, 1980), but rarely has it been necessary to invoke point defects. In fact, much of the recent literature on the chemistry of defect solids does not pertain to the regime of point defects; instead, chemists are mainly concerned with superstructures, shear and block structures, and intergrowths. Defect behaviour in solids actually varies from the entropy-controlled point-defect regime at one end to the enthalpy-controlled systems such as crystallographic shear (cs) planes at the other, superlattice ordering and defect complexes coming in between. These aspects of solid state chemistry truly belong to mainstream inorganic and structural chemistry.

5.2   Point defects

In ionic solids, the common point defects are *Schottky pairs* (pairs of cation and anion vacancies) and *Frenkel defects* (cation or anion interstitial plus a vacancy). When there is a large concentration of Schottky pairs, the measured pyknometric density of the solid is considerably lower than the density calculated from the X-ray unit cell dimensions (e.g. VO). Alkali halides are good examples of solids with Schottky defects; in AgCl and AgBr, we have cation Frenkel defects while in CaF_2 and SrF_2 we have anion Frenkel defects. Replacement of a host cation by an ion with a different valency gives rise to a variety of defect situations. If a small concentration of Sr^{2+} or Cd^{2+} is introduced in NaCl, M^{2+}-cation vacancy pairs are formed. The number of Schottky pairs or Frenkel defects, n, in an ionic solid such as NaCl is estimated by means of a statistical thermodynamic model. In the case of Schottky pairs, n is given by

$$n = N \exp(S_t/2k) \exp(-E_f/2kT)$$

$$= N\left(\frac{v}{v'}\right)^{3z} \exp(-E_f/2kT) \tag{5.1}$$

where N is the total number of lattice sites, v and v' are the vibrational frequencies of the perfect and defect crystals respectively, S_t is the thermal entropy and E_f is the energy of formation of a Schottky pair.

There are certain unusual types of defects in metal systems that are noteworthy. It has been found (Taylor & Doyle, 1972) that in NiAl alloys Al atoms on the Al-rich side do not substitute on the Ni sublattice; instead there are vacancies in the Ni sites. For example, at 55 at.% Al, 18% of Ni sites are vacant while the Al sites are filled. Such vacancies determined by composition are referred to as *constitutional vacancies*. Other alloys have since been found to exhibit such vacancies, typical of these being NiGa and CoGA. Another rather curious aspect of defects is the formation of *void lattices* when metals such as Mo are irradiated with neutrons or more massive projectiles (Gleiter, 1983). Void lattices arise from agglomeration of vacancies and are akin to superlattices. Typically, neighbouring voids in Mo are separated by 200 Å. An explanation for the stability of void lattices on the basis of the continuum theory of elasticity has been proposed (Stoneham, 1971; Tewary & Bullough, 1972).

When divalent cation impurities (e.g. Cd^{2+}, Sr^{2+}) are present in an ionic solid of the type MX consisting of monovalent ions, the negatively charged cation vacancies (created by the divalent ions) are bound to the impurity ions at low temperatures. Similarly, the oppositely charged cation and anion vacancies tend to form neutral pairs. Such neutral *vacancy pairs* are of importance in diffusion, but do not participate in electrical conduction. The interaction energy of vacancy pairs or *impurity–vacancy pairs* decreases with the increase in distance between the two oppositely charged units.

A variety of techniques has been employed to investigate aliovalent impurity–cation vacancy pairs and other point defects in ionic solids. Dielectric relaxation, optical absorption and emission spectroscopy, and ionic thermocurrent measurements have been most valuable; ESR studies of Mn^{2+} in NaCl have shown the presence of impurity-vacancy pairs of at least five different symmetries. The techniques that have provided a wealth of information on the energies of migration, formation and other defect energies in ionic solids are diffusion and electrical conductivity measurements. Electrical conductivity in ionic solids occurs by the motion of ions through vacancies or of interstitial ions. In the case of motion through vacancies, the conductivity, σ, is given by

$$\sigma = \sigma_0 \exp -\left[\left(\frac{E^f}{2} + E^m\right)\bigg/kT\right] \tag{5.2}$$

where E^f and E^m are the energies of formation of vacancies (Schottky pairs) and ion migration (generally cation migration) respectively. If the crystal is doped with aliovalent impurities, the concentration of vacancies is much greater at low temperatures than the intrinsic equilibrium value and the slope of $\ln \sigma$ versus T^{-1} plots directly gives E^m/k; the slope at higher temperatures will be $(E^f/2 + E^m)$. In Fig. 5.2 we show the different regions in the conductivity plot of a typical ionic solid. Diffusion measurements using labelled isotopes give complementary information on defect energies.

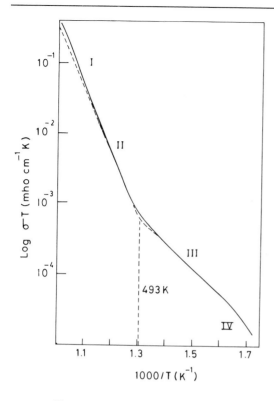

Figure 5.2   The ionic conductivity of 'pure' NaCl as a function of temperature. Intrinsic conduction occurs in stages I and II; stage III corresponds to conduction by cation vacancies present as a result of impurities. Vacancies become associated to form neutral pairs in stage IV.

Direct observation of point defects in metals has been possible by field ion microscopy. Impurity point defects may be usefully investigated by electron microscopy in combination with electron diffraction and electron spectroscopy. Direct observation of the dopant environment in fluorites has been attempted by Catlow *et al.* (1984) by employing EXAFS in conjunction with computer simulation.

Energies associated with the formation, interaction and migration of defects have been determined experimentally in a variety of solids (Corish *et al.*, 1977; Wollenberger, 1983). In noble-gas solids, the vacancy formation energies are of the order of 0.1 eV; in metals, it is around 1 eV (e.g. 0.73, 1.03 and 1.49 eV in Al, Cu and Pt respectively). The vacancy *formation energy* in these solids increases with the melting temperature T_m. In metals, the *migration energy*, E^m, is generally around $7 \times 10^{-4} T_m$, sodium being an exception. Furthermore, the E^m values in metals are close to the E^f values. For example, E^m is 0.65, 0.94 and 1.42 eV respectively in Al, Cu and Pt. Binding energies of divacancies in Au and Ag are found to be around 0.2 eV. The formation energies of Schottky pairs in Na and K halides are in the range 2.0–2.5 eV, while the cation migration energy by the vacancy mechanism is in the 0.6–0.9 eV range. In AgCl and AgBr, the Frenkel defect formation energy is around 1.3 eV and the cation Frenkel

(Ag^+) migration energy is 0.1 eV or less; the Ag^+ ions move without any barrier, probably by an interstitialcy mechanism involving the displacement of an Ag^+ ion in the (000) position (by an interstitial (111) ion) to the (111) position. In the anion Frenkel solids CaF_2 and SrF_2, the formation energy is around 2.5 eV and the anion migration energy is 1 eV. Experimental binding energies of divalent cations and cation vacancies in NaCl and KCl are ~ 0.5 eV. Interaction energies of oppositely charged vacancies to form pairs are also of this order. It should be noted that creation of defects is always an endothermic process. The formation energies of vacancies in ionic solids, for example, are generally 2 eV or more. The intrinsic defect concentration in these solids is therefore extremely small even at high temperatures. The intrinsic (equilibrium) defect concentration in binary solids around $0.8\,T_m$ is $\sim 10^{-5}$.

Formation energies of defects and other defect energy parameters have been estimated theoretically. In the case of metals, the formation energy of a vacancy would be equal to $(E_s - E_r)$, where E_s is the energy required to remove an atom from the surface and E_r is the relaxation energy. A vacancy in an ionic solid carries a virtual charge of opposite sign and the polarization energy therefore becomes significant. The first successful evaluation of vacancy formation energies in ionic solids was made by Mott & Littleton (1938), who evaluated the polarization energy after allowing the lattice to relax. In such defect energy calculations, the displacements and polarization are evaluated using specific ionic interactions in the neighbourhood of the defect while treating the rest of the lattice as a dielectric continuum. The well-known Born potentials of ionic solids are generally satisfactory for such calculations. Theoretical studies of point defects have been reviewed adequately in the literature (Corish *et al.*, 1977; Stoneham, 1975). With the availability of fast computers, several improvements have been incorporated in defect energy calculations. For example, a large number of ions have been included in the defect region and potentials chosen to match dielectric and elastic properties (see, for example, Uppal *et al.*, 1978). Adequate corrections have also been incorporated to account for the alteration of short-range interactions due to polarization by employing the shell model. These new computational techniques have been most useful in understanding defect ordering, extended defects and non-stoichiometry (Catlow & Mackrodt, 1982; Catlow & James, 1980) and we shall examine some of the results later in this chapter. Besides defect energies, the entropy of formation of defects has also been evaluated theoretically (Gillan & Jacobs, 1983).

5.2.1 *Point defect equilibria*

The importance of interactions amongst point defects, at even fairly low defect concentrations, was recognized several years ago. Although one has to take into account the actual defect structure and modifications of short-range order to be able to describe the properties of solids fully, it has been found useful to represent all the processes involved in the intrinsic defect equilibria in a crystal (with a low concentration of defects), as well as its equilibrium with its external environment, by a set of coupled quasichemical reactions. These equilibrium reactions are then handled by the law of mass action. The free energy and equilibrium constants for each process can be obtained if we know the enthalpies and entropies of the reactions from theory or experiment. Such an approach has been used to investigate a variety of defect equilibria (Kröger, 1974). In order to write equilibrium reactions conveniently, we shall use Kröger's notation as adopted by Libowitz (1973). In this scheme A_A means atom A on site A; M_A means atom M on site A. In a solid MX, V_M and V_X stand for a vacancy on metal atom site M and a vacancy on anion site X respectively. The symbol I or i is used to indicate an interstitial. Since defect concentrations are generally low, we shall use concentrations instead of activities. The use of activities and the applicability of Debye–Hückel treatment in defect ionic solids have been examined.

$$K_s = [V_M][V_X] \tag{5.3}$$

where K_s stands for the equilibrium constant. Since MX is in equilibrium with the external phase (X_2 gas), the equilibrium of defects is governed by equations (5.4) and (5.5):

$$X_X \rightleftharpoons \tfrac{1}{2}X_2(g) + V_X$$
$$K_{VX} = p_{X_2}^{1/2}[V_X]/[X_X]$$
$$= p_{X_2}^{1/2}[V_X]/(1 - [V_X]) \tag{5.4}$$

where p_{X_2} is the partial pressure of X_2. If X_2 affects the M vacancy concentration

$$\tfrac{1}{2}X_2(g) \rightleftharpoons X_X + V_M + \alpha V_I$$
$$K_{VM} = [V_M][V_I]^\alpha/p_{X_2}^{1/2} \tag{5.5}$$

$[X_X]$ being unity. Furthermore, neutral defects can undergo ionization

$$V_X \rightleftharpoons V_X^+ + e^-$$

and

$$V_M \rightleftharpoons V_M^- + h^+$$

where e^- and h^+ stand for electrons in the conduction band and holes in the valence band respectively. The equilibrium constants for the above reactions may be written as

$$K_{VX}^i = [V_X^+][e^-]/[V_X] \tag{5.6}$$

and

$$K_{VM}^i = [V_M^-][h^+]/[V_M] \tag{5.7}$$

Whenever electrons and holes are generated in a solid through ionization, another equilibrium is set up: $0 \rightleftharpoons e^- + h^+$. The corresponding intrinsic equilibrium constant is

$$K_i = [e^-][h^+] \equiv [n][p] \tag{5.8}$$

where $[n]$ and $[p]$ represent concentrations of electrons and holes respectively. Electronic defects such as electrons and holes are created in thermal equilibrium in solids and their equilibrium concentrations are also governed by the law of mass action. The magnitude of K_i is determined by the band-gap energy. Electrical neutrality requires that

$$[n] + [V_M^-] = [p] + [V_X^+] \qquad (5.9)$$

Thus we have seven unknowns, $[V_M]$, $[V_X]$, $[V_M^-]$, $[V_X^+]$, $[n]$, $[p]$ and p_{X_2}, and the seven equations (5.3)–(5.9). All these equations, except (5.9), can be expressed in the form of linear equations in ln (concentration) terms as follows:

$$\ln K_s = \ln[V_M] + \ln[V_X] \qquad (5.3)'$$
$$\ln K_{VX} = \tfrac{1}{2}\ln p_{X_2} + \ln[V_X] \qquad (5.4)'$$
$$\ln K_{VM} = \ln[V_M] - \tfrac{1}{2}\ln p_{X_2} \qquad (5.5)'$$
$$\ln K_{VX}^i = \ln[n] + \ln[V_X^+] - \ln[V_X] \qquad (5.6)'$$
$$\ln K_{VM}^i = \ln[p] + \ln[V_M^-] - \ln[V_M] \qquad (5.7)'$$
$$\ln K_i = \ln[n] + \ln[p] \qquad (5.8)'$$

In the above equations necessary simplifications such as $[V_X] \ll 1$, $[V_I] \approx [X_X] = 1$ etc. have been introduced. Although it is possible in principle to obtain solutions for these equations once the various K values are known, the form of equation (5.9) makes it very tedious. Brouwer made the approximation that for large negative deviations from stoichiometry, $[V_M^-]$ and $[p]$ become negligible:

$$[n] = [V_X^+] \qquad (5.10)$$

or

$$\ln[n] = \ln[V_X^+]$$

Using relation (5.10) it is possible to express all the concentrations in terms of $\ln p_{X_2}$. We can now introduce a quantity R, by making use of equation (5.5)':

$$R = K_{VM}p_{X_2}^{1/2} = [V_M] \qquad (5.11)$$

We thus have $[V_X] = K_s/R$ from equation (5.3)' and $[n] = [V_X^+] = K_{VX}^i[V_X]^{1/2} = K_{VX}^i(K_s/R)^{1/2}$ from equation (5.6)'. Since the Schottky equilibrium constants are known for ionized defects, it is profitable to express K_s in terms of K_s^i using the relation

$$K_s = K_s^i K_i / K_{VM}^i K_{VX}^i \qquad (5.12)$$

Now we see readily that when ln (concentration) of any of the defects is plotted as a function of $\ln R$, the slope is $\pm \tfrac{1}{2}$ or ± 1.

At large positive deviations from stoichiometry, we can use the approximation

$$\ln[V_M^-] = \ln[p] \tag{5.13}$$

and relations between R and concentrations of various defects are derived in the same manner as above. In the vicinity of the stoichiometric composition, defect concentrations depend upon the intrinsic defect equilibrium constants K_s and K_i. For materials with large band gaps such as alkali halides, $K_s \gg K_i$ so that equation (5.9) leads to the relation

$$\ln[V_M^-] = \ln[V_X^+] \tag{5.14}$$

With this relation we can set the boundaries for the regions corresponding to large negative deviation and large positive deviation from stoichiometry. We designate the regions as Region I (large negative deviation), Region II (close to stoichiometric composition) and Region III (large positive deviation). From the relation (5.14), the R value corresponding to the boundary $R_{I,II}$ is governed by the relation

$$(RK_s^i K_{VM}^i/K_i)^{1/2} = (K_s^i K_i/K_{VM}^i R)^{1/2}$$

or

$$R_{I,II} = K_i/K_{VM}^i \tag{5.15}$$

From similar considerations

$$(K_{VM}^i R)^{1/2} = K_s^i/(K_{VM}^i R)^{1/2}$$

or

$$R_{II,III} = K_s^i/K_{VM}^i \tag{5.16}$$

The width of Region II may be found as follows:

$$\ln R_{II,III} - \ln R_{I,II} = \ln(K_s^i/K_i) \tag{5.17}$$

Using relations (5.14)–(5.16), along with equations (5.3)′–(5.8)′, we can determine the concentrations of various defects in Region II on both sides of the stoichiometric composition. The stoichiometry is rigorously defined by the condition

$$[V_M] + [V_M^-] = [V_X] + [V_X^+].$$

Since $K_s^i \gg K_i$ in Region II, $[V_M^-] \approx [V_X^+]$ and $[V_M]$ and $[V_X]$ are negligible. Therefore, $[n] \approx [p]$ and R_o, corresponding to the stoichiometric composition, is equal to $(K_i K_s^i)^{1/2}/K_{VM}^i$. Furthermore, both $[V_M^-]$ and $[V_X^+]$ are quite insensitive to R or to large changes in p_{X_2}. The concentrations of various defects in the three different regions are

Table 5.1. *Defect equilibria in an MX crystal exhibiting Schottky defects*[a]

Defect	Region I	Region II		Region III
		$K_s^i \gg K_i$	$K_i \gg K_s^i$	
n	$\left(\dfrac{K_s^i K_i}{K_{VM}^i}\right)^{1/2} R^{-1/2}$	$\dfrac{(K_s^i)^{1/2} K_i}{K_{VM}^i R}$	$(K_i)^{1/2}$	$\dfrac{K_i}{(K_{VM}^i R)^{1/2}}$
p	$\left(\dfrac{K_i K_{VM}^i}{K_s^i}\right)^{1/2} R^{1/2}$	$\dfrac{K_{VM}^i R}{(K_s^i)^{1/2}}$	$(K_i)^{1/2}$	$(K_{VM}^i R)^{1/2}$
$[V_M]$	R	R	R	R
$[V_M^-]$	$\left(\dfrac{K_s^i K_{VM}^i}{K_i}\right)^{1/2} R$	$(K_s^i)^{1/2}$	$\left(\dfrac{K_{VM}^i}{K_i^{1/2}}\right) R$	$(K_{VM}^i R)^{1/2}$
$[V_X]$	$\dfrac{K_s^i K_i}{K_{VX}^i K_{VM}^i R}$	$\dfrac{K_s^i K_i}{K_{VX}^i K_{VM}^i R}$	$\dfrac{K_s^i K_i}{K_{VX}^i K_{VM}^i R}$	$\dfrac{K_s^i K_i}{K_{VX}^i K_{VM}^i R}$
$[V_X^+]$	$\left(\dfrac{K_s^i K_i}{K_{VM}^i R}\right)^{1/2}$	$(K_s^i)^{1/2}$	$\dfrac{K_i^{1/2} K_s^i}{K_{VM}^i R}$	$\dfrac{K_s^i}{(K_{VM}^i R)^{1/2}}$

[a] $R = K_{VM} p_{X_2}^{1/2}$

given in Table 5.1. A schematic plot of $\ln[D]$ vs. $\ln R$ is shown in Fig. 5.3, where $[D]$ represents the concentration of defects.

The situation in semiconductors is just the opposite to that of the wide band-gap insulators discussed hitherto. The condition that holds good for semiconductors is $K_i \gg K_s^i$ and one can develop expressions for the concentrations of various defects following the methodology outlined above.

Analysis of defect equilibria in KBr is a typical example (Kröger, 1974). Since K_i and K_s^i are 3×10^{-35} and 8×10^{-14} respectively, Region II in Fig. 5.3 extends over ~ 43 decades (equation 5.7) and hence the partial pressure of Br_2, which is proportional to R^2 (see equation 5.5'), does not affect the defect concentrations. The deviation from stoichiometry, δ, given by $\Sigma[V_M] - \Sigma[V_X]$ is also rather small ($\sim 10^{-3}$) over the same region.

Partial pressure of oxygen controls the nature of defects and nonstoichiometry in metal oxides. The defects responsible for nonstoichiometry and the corresponding oxidation or reduction of cations can be described in terms of quasichemical defect reactions. Let us consider the example of transition metal monoxides, $M_{1-x}O$ (M = Mn, Fe, Co, Ni), which exhibit metal-deficient nonstoichiometry. For the formation of metal vacancies in $M_{1-x}O$, the following equations can be written:

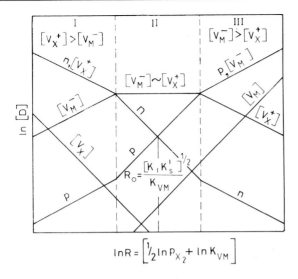

$$\ln R = \left[\tfrac{1}{2}\ln P_{X_2} + \ln K_{VM}\right]$$

Figure 5.3   Plots of defect concentrations in a solid MX against the partial pressure of $X_2 (K_s^l > K_l)$.

$$\tfrac{1}{2}O_2 \rightleftharpoons V_M + O_o$$
$$V_M + e^- \rightleftharpoons V_M^- + h^+$$
$$V_M^- + e^- \rightleftharpoons V_M^{2-} + h^+$$
$$2M_M + 2h^+ \rightleftharpoons 2M_M^+$$

The total reaction is,

$$\tfrac{1}{2}O_2 \rightleftharpoons V_M^{2-} + 2M_M^+ + O_o \tag{5.18}$$

Here, V_M^{2-} denotes a doubly ionized metal vacancy, M_M^+ the association of M^{2+} and h^+ (which may be regarded as a M^{3+} ion) and h^+ a hole in the valence band. Using the law of mass action, introducing the neutrality condition, $[h^+] = 2[V_M^{2-}]$, and expressing the cation vacancies as x in $M_{1-x}O$, the equilibrium constant for (5.18) can be written as

$$K_{V_M^{2-}} = 4x^3 p_{O_2}^{-1/2} \tag{5.19}$$

The composition of the oxide therefore depends on the oxygen partial pressure as given by

$$x \propto p_{O_2}^{1/6} \tag{5.20}$$

If the defects are singly ionized vacancies, V_M^-, then

$$x \propto p_{O_2}^{1/4} \tag{5.21}$$

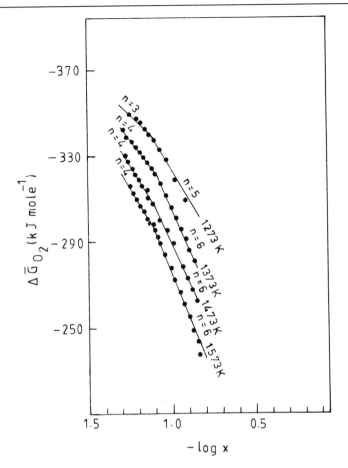

Figure 5.4   Relative partial molar free energy of oxygen. $\Delta \bar{G}_{O_2}$, of $Fe_{1-x}O$ as a function of composition (log x). (Taken from the data given in Sørensen, 1981.)

It is convenient to discuss the thermodynamics of nonstoichiometry of oxides in terms of the relative partial molar-free energy of oxygen, $\Delta \bar{G}_{O_2}$, which is defined (Sørensen, 1981) as

$$\Delta \bar{G}_{O_2} = \mu_{O_2} - \mu_{O_2}^0 = RT \ln p_{O_2} \qquad (5.22)$$

where μ_{O_2} is the chemical potential of oxygen in the solid and $\mu_{O_2}^0$ is the chemical potential of oxygen in its standard state; p_{O_2} is the equilibrium partial pressure of oxygen in the atmosphere around the nonstoichiometric oxide.

For $M_{1-x}O$, the relation between $\Delta \bar{G}_{O_2}$ and composition, x, can be generally expressed as

$$\Delta \bar{G}_{O_2} = nRT \ln x \qquad (5.23)$$

A plot of $\Delta \bar{G}_{O_2}$ versus $\ln x$ would be linear, the slope giving n; the value of n gives information about the types of defects involved. For doubly ionized metal vacancies, V_M^{2-}, $n = 6$; for V_M^-, $n = 4$ and so on. Isothermal $\Delta \bar{G}_{O_2} - \ln x$ plots can be constructed from equilibrium thermogravimetric data or electrochemical measurements. A typical $\Delta \bar{G}_{O_2} - \log x$ plot for the $Fe_{1-x}O$ system is shown in Fig. 5.4, where n values corresponding to various regions are indicated. We see that $n = 6$ applies when $x \geqslant 0.09$, while $n = 5$, corresponding to $(V_M V_M)^{4-}$ pairs, also applies close to this region; values of $n = 3$ and 4 observed for $x < 0.09$ can be explained in terms of $[4V_M^{2-} - Fe_i^{3+}]^{5-}$ and $[16V_M^{2-} - 5Fe_i^{3+}]^{17-}$ defect complexes, respectively. There is structural evidence for the presence of both of these defect complexes in $Fe_{1-x}O$. The 16:5 complex corresponds to an element of Fe_3O_4 structure. It is, however, significant that thermodynamic data do not support the existence of the Koch–Cohen complex, $[9V_M^-, 4V_M^{2-} - 4Fe_i^{3+}]^{5-}$, in $Fe_{1-x}O$ (Section 5.5.3). On the other hand, thermodynamic data show clear evidence for the presence of the Koch–Cohen complex in the $Mn_{1-x}O$ system (Sørensen, 1981). The results underline the importance of simultaneous thermodynamic and structural investigations for the characterization of defect structures of oxides.

5.2.2   Paraelectric and molecular impurities in ionic solids

Dipolar ions like CN^- and OH^- can be incorporated into solids like NaCl and KCl. Several small dopant ions like Cu^+ and Li^+ ions get stabilized in off-centre positions (slightly away from the lattice positions) in host lattices like KCl, giving rise to dipoles. These dipoles, which are present in the field of the crystal potential, are both polarizable and orientable in an external field, hence the name *paraelectric impurities*. Molecular ions like S_2^-, Se_2^-, N_2^- and O_2^- can also be incorporated into alkali halides. Their optical spectra and relaxation behaviour are of diagnostic value in studying the host lattices. These impurities are characterized by an electric dipole vector and an elastic dipole tensor. The dipole moments and the orientation direction of a variety of paraelectric impurities have been studied in recent years. The reorientation movements may be classical or involve quantum–mechanical tunnelling.

Ions like S_2^-, Se_2^- and SSe^- are found to align along the $\langle 110 \rangle$ directions in most alkali halides, while in NaI, KBr and KI, the S—S bond of S_2^- is oriented along the $\langle 100 \rangle$ direction. In the case of O_2^-, the p orbitals are parallel to the $\langle 100 \rangle$ direction in sodium salts but are parallel to the $\langle 110 \rangle$ direction in rubidium and potassium salts. Extensive spectroscopic studies have been reported on molecular ions such as NO_2^-, NO_3^-, CrO_4^{2-} and MnO_4^{2-}. Several reviews (Corish *et al.*, 1977; Bridges, 1975; Grimes, 1976) are available on such impurity-doped solids.

5.2.3   Colour centres

In the discussion of defect equilibrium we discussed neutral defects such as V_X; V_X corresponds to an anion vacancy which has captured an electron. Such vacancies give rise to interesting optical phenomena. Defects associated with electrons or holes lead to colouration of the crystals and are known as *colour centres*. The term colour centre also includes impurity centres such as Tl^+, which are responsible for absorption and luminescence in the visible region. Electrons captured by anion vacancies are referred to as F-centres. The absorption maxima of the F-centres in LiCl, NaCl, KCl and RbCl are at 385, 465, 563 and 624 nm respectively, giving colours between yellow and blue. The corresponding emission bands show large Stokes shifts. The oscillator strength of the absorption band as well as the strength of emission of an F-centre are very high. The radiative life-times are $\sim 10^{-6}$ s, almost two orders of magnitude higher than in optical transitions. The spectral features of an F-centre can be understood by treating it as a particle in a box or as a hydrogen-like atom. If the size of the box is taken as the nearest-neighbour distance, the particle in a box model works well in accounting for the energy dependence of the F-band on the nearest-neighbour distance.

Besides the F-centre, many other colour centres have been described in the literature (Fowler, 1968, 1975; Stoneham, 1975). An F^--centre is created by trapping two electrons by an anion vacancy. In an alkaline-earth oxide, a single electron in an anion vacancy gives a charged defect corresponding to a F^+-centre. Two adjacent F-centres form a M-centre; three adjacent F-centres on an equilateral triangle occurring on the (111) plane of alkali halides are referred to collectively as a F_3-centre. A V_K-centre in alkali halides is a hole trapped by two adjacent halide ions in the $\langle 110 \rangle$ direction. Three adjacent centres plus an electron constitute an R-centre. Absorption and emission spectra of many such colour centres have been investigated and they often exhibit zero-phonon emission corresponding to the emission from the ground vibrational state of the electronic excited state to the ground vibrational state of the ground electronic state. In alkaline-earth oxides, both singlet and triplet state F-centres are formed since there are two electrons in a centre. Electron spin resonance spectroscopy is very useful for the investigation of the trapped electrons in colour centres. For example, a V_K-centre in KCl (self-trapped hole between two chloride ions, corresponding to Cl_2^- ion) shows seven equally spaced lines ($2 \times 3 + 1$) in the ESR spectrum due to hyperfine splitting since each chlorine has a nuclear magnetic moment of $\frac{3}{2}$. F-centres give rather broad ESR signals corresponding to an envelope of many hyperfine splittings; ENDOR is employed to reveal the hyperfine interaction.

5.3   Dislocations

Dislocations are linear defects and were first invoked to account for the mechanical properties of solids, particularly the shear strengths. Dislocations play an important role in a variety of solid state phenomena from phase transitions to chemical reactions and the subject has been investigated and reviewed widely (Fine, 1973; Nembach, 1974). The effect of dislocations on the transformations and properties of organic solids has been recognized in recent years (Thomas & Williams, 1971; Jones & Thomas, 1979).

A dislocation marks the boundary between the slipped and unslipped parts of crystal; the simplest kind of dislocation is an *edge dislocation*, involving an extra layer of atoms in a crystal. Atoms in such an edge dislocation do not have their coordination fully satisfied. Two portions of a crystal are slipped past each other by one atomic spacing. A *screw dislocation* transforms successive planes of atoms into the surface of a helix. In Fig. 5.5, the high-resolution electron microscopic image of an edge dislocation in $Ba_2Bi_4Ti_5O_{18}$ is shown. Dislocations in an organic solid are shown in Fig. 5.6. The

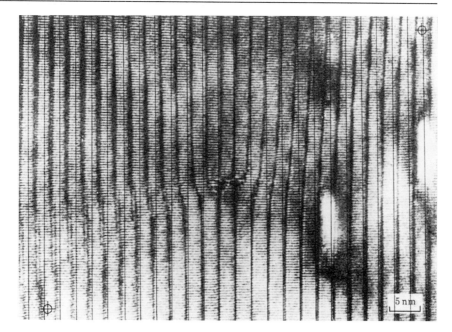

Figure 5.5   Lattice image of a high-strength disolocation in $Ba_2Bi_4Ti_5O_{18}$. Notice the large value of $\vec{b}$ (from Hutchison *et al.*, 1977.)

most important characteristic of a dislocation is its associated slip displacement, or *Burgers vector*, $\vec{b}$, which is readily determined by following a path around the dislocation, called the Burgers circuit (Fig. 5.7). An edge dislocation is normal to its $\vec{b}$, a screw dislocation parallel to it. Strain due to atomic displacements is high around the dislocation. The displacements decrease rapidly beyond the core of the dislocation as can be noticed by an examination of Fig. 5.5. The strain energy of a dislocation, proportional to $|\vec{b}|^2$, is considerably greater near an edge dislocation than a screw dislocation. Generally, dislocations are neither purely edge nor purely screw type and exhibit mixed character. Such a mixed dislocation can be resolved into edge and screw dislocations.

If the Burgers vector of a dislocation corresponds to a lattice translational vector, the dislocation is known as a *perfect dislocation*; otherwise, it is called an *imperfect* or *partial dislocation*. Perfect dislocations are involved in slip (where no change in crystal structure occurs), but imperfect dislocations are involved in phase transitions (e.g. the martensitic transition described in Chapter 4). Perfect dislocations split into *partial dislocations* with Burgers vectors $\vec{b}_1$ and $\vec{b}_2$ (each $< \vec{b}$) in several situations; this process reduces the strain energy and also gives rise to structural changes (e.g. introduction of stacking faults). The necessary condition for such splitting is $|\vec{b}_1|^2 + |\vec{b}_2|^2 < |\vec{b}|^2$.

An edge dislocation is confined to move on its slip plane (conservative motion), and the slip due to the motion of the dislocation is also confined to the slip plane. Movement of a screw dislocation can capture on the plane where it started or else move to any other, parallel to the dislocation line (cross slip). If an edge dislocation were to move

Figure 5.6   Electron micrograph of a network of dislocations in anthracene crystal. (Courtesy of G.M. Parkinson & J.M. Thomas, University of Cambridge.)

normal to the slip plane, it would require an entire row of atoms at the edge of the extra half plane to move away, or the same number of vacancies to migrate into those positions; such a diffusive process is referred to as the *climb* of a dislocation. Motion of a dislocation is usually designated by the *slip plane* and the *slip direction* (= the Burgers vector). For example, in anthracene, the slip plane and slip direction are (010) $\langle 001 \rangle$; (100) $\langle 010 \rangle$ and in naphthalene, they are (001) $\langle 010 \rangle$. Interaction of dislocations

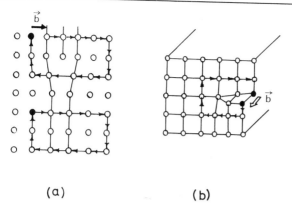

(a)   (b)

Figure 5.7   Burgers circuit for (a) an edge dislocation and (b) a screw dislocation. By moving a certain number of lattice distances to the right and down followed by a corresponding number of distances to the left and up, we get a closed path in the lower portion of (a); in the upper portion containing an edge dislocation, we have the misfit given by $\vec{b}$. In (a), $\vec{b}$ is perpendicular to the dislocation line, while it is parallel in (b).

(belonging to different slip planes) during their motion leads to *jogs* and *kinks* or even vacancies. When a screw dislocation is bent and has an edge component, vacancies migrate towards the bent end of the dislocation, causing a *helical dislocation*. When the splitting of a dislocation on the (111) plane occurs, as expressed by

$$(\tfrac{1}{2})a\langle 01\bar{1}\rangle = (\tfrac{1}{3})a\langle 11\bar{1}\rangle + (\tfrac{1}{6})a\langle \bar{2}1\bar{1}\rangle \tag{5.24}$$

the partial whose Burgers vector is $\tfrac{1}{3}a\langle 11\bar{1}\rangle$ is referred to as a *sessile* dislocation or *Frank partial*. The other partial, $\tfrac{1}{6}a\langle \bar{2}1\bar{1}\rangle$ is referred to as a *glissile* dislocation or a *Shockley partial*.

Dislocations multiply in a facile manner during a plastic deformation process, and several mechanisms for this have been observed by electron miscroscopy. Dislocations are destroyed by the processes of recovery and recrystallization during annealing after plastic deformation. Since dislocations cause low-yield stresses in metals and other solids, solid strengthening is accomplished either by eliminating dislocations or by immobilizing them.

Dislocations are generally investigated by electron microscopy. Decoration (e.g. dislocations in Si by Cu) and chemical etching combined with optical microscopy are also often employed to observe dislocations. Dislocation density (number of dislocation lines intersecting a unit area of a crystal) depends on the method of preparation and the thermomechanical history of the sample. While nearly dislocation-free metal whiskers have been grown, normal carefully grown single crystals have a dislocation density of about 10^2 dislocations cm^{-2}; annealed polycrystalline metals have one of about 10^{-8} dislocations cm^{-2}. In organic solids and molecular crystals, dislocation densities are generally low ($\sim 10^4 \text{cm}^{-2}$) in crystals grown from the vapour phase, but high ($\sim 10^6 cm^{-2}$) in crystals grown from melts. Dislocation energies in organic solids are high since the lattice parameters are an order of magnitude higher than in inorganic solids. On the other hand, formation energies of point defects are much lower ($\sim 1 eV$) in organic solids.

When the dislocation density corresponds to about $10^{10} cm^{-2}$ the number of atoms in the highly modified core region of dislocations will be about $10^{18} cm^{-3}$. These atoms are present in crowded, rarefied or distorted sites. The average extra energy possessed by atoms present in such sites can be as high as 20–40 kJ mol^{-1}. These high energies alter or affect the course of chemical reactions in organic solids. The kinetic barriers are reduced, altering the rates of reactions (wherever the reactions are kinetically controlled). The reactions are further assisted since part of the reaction enthalpy is provided by the dislocations. Catalytic impurities may migrate preferentially to dislocations, thereby enhancing the reaction rates considerably at such locations. Molecules in the cores of dislocations (particularly of a screw dislocation) are oriented differently from those in the dislocation-free region. This very often brings about unusual stereochemical alignments necessary for reactions to occur.

5.4   Planar defects

A surface or an interface of a crystal constitutes a planar, two-dimensional defect. There is considerable difference in the environment of atoms, ions or molecules on a surface and that of atoms in the bulk of a crystal (where they are in a 'perfect' order situation). The change in environment could result in a decrease in the number of interacting neighbours or in the distortion of the polyhedra around the surface atoms. The free energy of surface atoms is higher than those in the bulk. The surface energies of crystalline solids are generally in the range 0.5–2 J m^{-2}. In polycrystalline materials there are *grain boundaries* between various grains. When more than one phase is present in the material then there are necessarily interfaces with substantial interfacial energies.

The interface between two phases may be coherent, incoherent or semicoherent. If there is a perfect matching in the interatomic distances between the planes in contact, the interface is said to be *coherent*. In these interfaces, there will be no disregistry and consequently they possess very low interfacial energies ($\leqslant 0.015 J m^{-2}$). Good examples of coherent interfaces are found in Ni-based super-alloys where the (100) planes of Ni_3Al form coherent interfaces with the (100) faces of the face-centred cubic matrix. $MgFe_2O_4$ precipitates in MgO form coherent interfaces. This is because both in MgO and $MgFe_2O_4$, oxygens are cubic close-packed and hence the precipitate and the matrix are continuous. In the case of cobalt, where the fcc and hcp structures coexist, the coherent interface corresponds to the (0001) plane of the hexagonal phase and the (111) plane of the cubic phase.

When the mismatch in interatomic distances is sufficiently large, *semicoherent* interfaces are formed. Interfaces corresponding to *epitaxial growth* are generally semicoherent and contain arrays of dislocations. Small angle-grain boundaries also constitute semicoherent interfaces. The energy of such boundaries, which also consist of dislocations, depends on the relative orientations of the two grains. A low-angle *tilt*

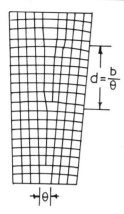

Figure 5.8   Dislocation array in a low-angle tilt boundary.

boundary consists of an array of periodically spaced edge dislocations as shown in Fig. 5.8.

A low-angle *twist boundary* is made up of an array of screw dislocations. Grain-boundary energies are typically low and arrays of dislocations which constitute low-angle boundaries are expected to form during annealing. There are special orientations for which the grain-boundary energies are minimum and the dislocations are such that the Burgers vectors of these misfit dislocations assimilate the atomic mismatch along the boundary. Grids of such misfit dislocations, called *van der Merwe dislocations*, are readily identified by transmission electron microscopy.

The fcc (ABC ABC...) and hcp (ABAB...) arrangements are the two common motifs in solids (Section 1.4). Two situations corresponding to the formation of *stacking faults* can be visualized: (a) by the passage of a Shockley-type partial dislocation (ABC ABC ABC → ABC AB ABC ABC) and (b) by the removal of a partial plane of atoms forming an edge dislocation (Frank partial). Stacking-fault energies vary widely from one solid to another. In α-Al_2O_3, slipping at 1200 K involves three different stacking-fault regions since the Burgers vector splits into four partials; the region between any two partials corresponds to stacking faults.

A layer possessing mirror plane symmetry (with respect to the layers on either side) can be formed by the passage of Shockley partials and such a layer is known as a *twin boundary*. Twinning (reflection) is an important crystallographic operation that relates complex structures to simple ones (see Section 1.6). Twinning is observed in many minerals and is responsible for the banded appearance of some of them (calcite, feldspars). We can also have a planar fault involving the rotation of one part of the crystal, on a specific plane, with respect to another and having a number of lattice sites at the interface common to the structures on either side. Such a planar fault, a *coincidence boundary*, is shown in Fig. 5.9. Antiphase domains constitute another type of two-dimensional defect; across the *antiphase boundary*, the sublattice occupation gets interchanged. For example, if AB is an ordered alloy with A and B atoms occupying α and β sublattices respectively on one side of the antiphase boundary, on the other side A and B occupy β and α sublattices respectively. A domain boundary in

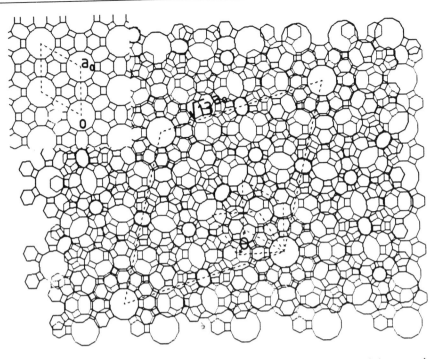

Figure 5.9   A coincidence boundary (001) in Zeolite L. The top part of the crystal is rotated by 32.2° with respect to the bottom part thereby generating a coincidence boundary. (From Terasaki *et al.*, 1984.)

ferromagnetic, ferroelectric and related materials is similar to an antiphase boundary except that the spins or the electric dipoles across the boundary are polarized in different but equivalent low-energy directions.

Planar defects correspond to a state of higher energy relative to the bulk of the solid, and therefore impurities such as aliovalent ions or solute atoms are attracted towards interfaces or grain boundaries in a manner similar to that prevalent in dislocations. For example, Ca^{2+} ions migrate to grain boundaries in polycrystalline sapphire. In a fcc metal, atomic segregation occurs in the region of the stacking-fault ribbon of hcp structures. Such atomic migrations lead to lowering of energies.

5.5   Ordered point defects and superstructures

There is increasing experimental evidence for the superlattice ordering of vacant sites or interstitial atoms as a result of interactions between them. Superlattice ordering of point defects has been found in metal halides, oxides, sulphides, carbides and other systems, and the relation between such ordering and nonstoichiometry has been reviewed extensively (Anderson, 1974, 1984; Anderson & Tilley, 1974). Superlattice ordering of point defects is also found in alloys and in some intermetallic compounds (Gleiter, 1983). We shall examine the features of some typical systems to illustrate this phenomenon, which has minimized the relevance of isolated point defects in many of the chemically interesting solids.

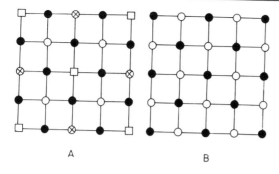

Figure 5.10   Superstructures in Suzuki phases: A, sheet of the superstructure where the open square is the cation vacancy and the crossed circle is the divalent cation impurity, M; B, rocksalt sheet.

5.5.1   Doped alkali halides

The solubility of halides of divalent cations in alkali halides strongly depends on the temperature. For example, the solubility of $CdCl_2$ and $MnCl_2$ in NaCl is around 3×10^{-2} mol% at 650 K and $\sim 2 \times 10^{-5}$ mol% at 450 K. Owing to facile cation diffusion in such doped crystals, precipitation of the dopant can occur at low temperatures; formation of dimers and trimers of the impurity–vacancy pairs has been suggested to explain the rates of such processes. Suzuki (1961) characterized a family of halides of the type $MX_2 \cdot 6AX$ where the divalent cations and cation vacancies are arranged in a $2 \times 2 \times 2$ superstructure of the alkali halide (AX) structure. Sheets of the structure A in Fig. 5.10 alternate with sheets of rocksalt structure B so that the vacant sites and the compensating divalent cation impurities occupy second-neighbour (rather than nearest-neighbour) positions. The ease of formation of *Suzuki phases* results from its topological compatibility with the rocksalt structure. Computer simulation studies (Corish & Quigley, 1982) show that the species formed depends on the relative sizes of dopant and host cations.

5.5.2   Metal chalcogenides and carbides

Chalcogenides of the $3d$-transition metals (except Mn) exhibit wide ranges of stoichiometry and successions of phases with common structural patterns related to CdI_2 and NiAs structures. In the intermediate composition range (between MX_2 and MX), superstructure ordering is found, the occupied and empty octahedral sites adopting a two-dimensional order in the hexagonal array of sites. Many of the superstructure phases exhibit order–disorder transformations. Not all the chalcogenide systems cover the entire composition range. Thus, $Cr_{0.67}S$ (Cr_2S_3) is the upper limit in the Cr–S system while in the telluride system the limit is MTe_2. The $Ni_{2-x}Te_2$ system exists between $Ni_{1.84}Te_2$ and $NiTe_2$; when $x < 0.4$, the structure is defect NiAs type and when $x > 0.4$, the structure is of CdI_2 type. This is true for materials annealed at low temperatures; at high temperatures, however, there is no long-range order.

In the Ti–S system, a number of phases between TiS and TiS_2 have been identified and they all exhibit wide stoichiometric ranges. For example, TiS and TiS_2 have the

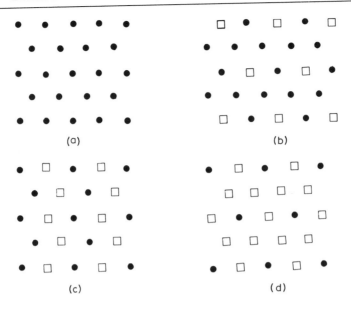

Figure 5.11   Ordered superstructure arrangements of vacant sites in cation layers of chalcogenides of NiAs structure. Fraction of occupied sites: (a) 1; (b) $\frac{3}{4}$; (c) $\frac{1}{2}$; (d) $\frac{1}{4}$.

ranges $Ti_{0.94}S$–$Ti_{1.03}S$ and $Ti_{0.52}S$–$Ti_{0.55}S$, respectively. An irrational fraction of the sites may be occupied in the metal-deficient layers in these sulphides, and the superlattice ordering is incomplete. In Ti_5S_8, for example, the ideal layer sequence would be ... $STi_{1.0}STi_{0.2}STi_{1.0}STi_{0.2}$ .... Although no well-defined superstructure is observed in such sulphides, a high degree of short-range order is present as evidenced from the diffuse scattering between the Bragg reflections from the basic hexagonal subcell. Moret *et al.* (1977) have examined the diffuse streaking and the incommensurate satellite maxima for the compositions in the range $TiS_{1.5}$–$TiS_{1.73}$ and propose a model with two modes of short-range ordering, where the interaction between Ti atoms prohibits the occupation of the nearest-neighbour sites but allows the occupation of the second or third neighbour sites. Microdomains with the two kinds of local order have been identified.

In $Cr_{1-x}X$ ($X = S$, Se or Te) systems, a number of superstructure phases with narrow ranges of stoichiometry occurs. The chalcogenides are within the composition range $0 \leqslant x \leqslant 0.33$, typical of them being Cr_7S_8, Cr_3S_4 and Cr_5S_8 with layer sequences ... $SCr_{1.0}SCr_{0.75}S$ ... , ... $SCr_{1.0}SCr_{0.5}S$ ... and ... $SCr_{1.0}SCr_{0.25}S$ ... respectively; in Cr_2S_3, the sequence is ... $SCr_{1.0}SCr_{0.33}S$ ... Strict alternation of filled and defective layers with good long-range order is found in all these sulphides, only Cr_7S_8 showing some randomization of the empty sites. In Fig. 5.11, typical ordered superstructure arrangements of vacant sites are shown for the purpose of illustration. Similar ordered phases are known in the $V_{1-x}S$ and M_xVS_2 ($M = $ Fe, Co, Ni) systems.

Two-dimensional superstructure ordering has been known to occur in iron sulphides for some time. Thus, in Fe_7S_8 (4C) the ordered structure is defined by the strict alternation of the filled and defective metal layers ... $SFe_{1.0}SFe_{0.75}SFe_{1.0}SFe_{0.75}S$ ... with empty octahedral sites in the incomplete metal layers (Bertaut, 1956). More complex sequences are found in Fe_9S_{10}(5C) and $Fe_{11}S_{12}$(6C); here the symbol nC stands for the number of layers of iron atoms in the repeating stacking sequence. Around 370 K, the vacant sites in the iron layer become statistically distributed in all these sulphides (Nakazawa & Morimoto, 1971). A common basis for the ordered structures and for the irrational superlattice multiplicities in iron sulphides giving rise to regular (not necessarily commensurate) modulation has been proposed (Kato & Kitamura, 1981); a consequence of this type of ordering is to distribute the vacant sites in a spiral fashion with a variable pitch.

Monosulphides of d^1 and d^2 transition metals with the rocksalt structure exhibit cation-deficient nonstoichiometry. For example, in the Zr–S system, compositions between $Zr_{0.63}S$ and ZrS are known, and structural studies on $Zr_{0.77}S$ show that superlattice ordering does not conform to a simple pattern in the distribution of cation vacancies (Conrad & Franzen, 1970). The monoclinic superstructure has 32 sites for 23 cations with only one two-fold position fully occupied and the other equivalent positions showing partial occupancy to different extents. Furthermore, in any given (111) sheet of the cubic subcell (hexagonal sheet of cation sites), the site occupancy is modulated in each row. In the Sc–S system, even Sc_2S_3 shows a superstructure $(2 \times \sqrt{2} \times 3\sqrt{2})$ of cation vacancies in a rocksalt matrix structure, the vacant cation sites in the second neighbour positions being arranged in spirals along the b axis (Dismukes & White, 1964). Scandium sulphide vaporizes congruently above 1970 K around the composition $Sc_{0.8}S$. This sulphide is fully disordered at high temperatures but shows superlattice ordering on annealing (at 970 K), with the vacant sites ordered into alternate (111) sheets of cation sites. Below 570 K, a second type of superlattice ordering (involving the ordering of the vacant sites within a defective cation layer) occurs; this superlattice is incommensurate (Franzen et al., 1983). The ordered superstructures of $Zr_{1-x}S$, $Sc_{1-x}S$ and $Lu_{1-x}S$ have been examined in the context of Landau's theory (Franzen & Folmer, 1984). On the basis of the cluster model and extended Hückel calculations, Burdett & Mitchell (1993) have shown that metal–metal bonding is unimportant in the ordering of vacancies in $Sc_{1-x}S$. Apparently, the formation of cis-divacant $SSc_4\square_2$ units gives rise to the ordering of Sc vacancies into alternating (111) planes. Subtle changes in the electronic structure, arising from a combination of shifts in the Sc $3d$ band and mixing of the Sc $3d$ and the S $3p$ bands, lies at the origin of this type of nonstoichiometry, which is common to the monochalcogenides of early transition metals.

The nonmetal sites in transition-metal carbides are only partly filled, and the carbides generally exhibit nonstoichiometry. Superstructures due to ordering of vacant sites have been found in vanadium carbides, V_6C_5 and V_8C_7 being two such phases. These phases undergo order–disorder transitions at high temperatures and, curiously, exhibit stoichiometric ranges (Emmons & Williams, 1983). Compositions between V_6C_5 and V_8C_7 show two phase transitions and possess a considerable degree of short-range order. The short-range order may not be lost even above the

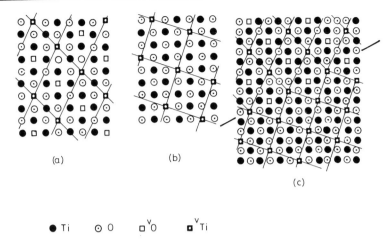

(a)   (b)

(c)

● Ti   ⊙ O   □ $\overset{v}{}$O   ■ $\overset{v}{}$Ti

Figure 5.12   (a) Ordered defects in monoclinic $TiO(Ti_{5/6}O_{5/6})$; (b) ordered defects in (orthorhombic) nonstoichiometric $TiO_{1.2}$; (c) coherent intergrowth of (a) and (b) along the (120) planes of rocksalt structure. Lines indicate unit cell faces of the superstructures. (After Anderson, 1984.)

order–disorder transition temperature. Superstructure ordering has been noticed in a few other carbides as well (e.g. Nb_6C_5).

5.5.3   Metal oxides of rocksalt structure

TiO and VO are two important oxides that deserve special mention because of the high proportion of defects (up to 20% vacancies) in them. In both these oxides, there is a phase transition between the low-temperature and the high-temperature forms. The high-temperature form of TiO with the averaged rocksalt structure spans a wide composition range between $TiO_{0.65}$ and $TiO_{1.25}$ at 1770 K; the low-temperature form (monoclinic) with a narrow composition range is formed below 1270 K (Watanabe et al., 1970). The lattice parameter of the NaCl-type unit cell is 4.182 Å. The superstructure corresponding to the monoclinic unit cell with $a' = 5.855$ Å $= \sqrt{2}a, b' = 9.340$ Å $= \sqrt{5}a, c' = 4.142$ Å $\approx a$, and $\gamma = 107°32' \approx \cos^{-1}(1/\sqrt{10})$, shown in Fig. 5.12(a), can be accommodated in a distorted cubic structure with $a = 4.200$, $b = 4.43$, $c = 4.142$ Å and $\gamma = 89.9$; this structure seems to hold for the oxides $TiO_{0.9}$–$TiO_{1.1}$. With greater deviation from stoichiometry, the superlattice corresponds to an orthorhombic distortion as shown in Fig. 5.12(b). The orthorhombic and monoclinic elements can intergrow in the parent rocksalt structure as shown in Fig. 5.12(c). The high-temperature disordered form of TiO shows diffuse scattering in the diffraction pattern suggesting the presence of some short-range order in the distribution of vacancies (Terauchi & Cohen, 1979). Terauchi et al. (1978) find that there may not be complete long-range order in the low-temperature form, at least for annealing temperatures close to the phase transition.

VO also possesses a high proportion of vacancies just like TiO, but the defect ordering at low temperature is probably quite different from that in low-temperature

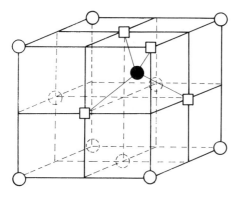

Figure 5.13   The 4:1 cluster of one tetrahedral cation and four vacant octahedral sites in one octant of the rocksalt structure.

TiO. Accordingly, VO is virtually immiscible with TiO (Loehman *et al.*, 1968). Superstructure in VO is formed only at the oxygen-rich end, where the oxygen sublattice is nearly filled and some vanadium atoms occupy tetrahedral sites; typical compositions studied are $V_{52}O_{64}$ and $V_{244}O_{320}$. Diffraction studies show the presence of vanadium atoms in tetrahedral sites. Each tetrahedral cation has four vacant octahedral sites as nearest neighbours (Fig. 5.13) and the cluster so formed, being topologically similar to the rocksalt structure, gives rise to a $2\sqrt{2} \times 2\sqrt{2} \times 2$ superstructure. In $V_{52}O_{64}$, all lattice sites (other than those of the defect clusters) are occupied as in the rocksalt structure, but in $VO_{1.30}$ some of the octahedral sites are partly occupied; in the $V_{52}O_{64}$ superlattice, all the vacant octahedral sites are in the clusters and there are no free vacancies (Andersson & Gjonnes, 1970; Morinaga & Cohen, 1979). VO_x preparations quenched from high temperatures show diffuse X-ray scattering and satellites in electron-diffraction patterns (Morinaga & Cohen, 1979; Rao *et al.*, 1976) due to short-range order, probably involving tetrahedral vanadium atoms. It is noteworthy that the superstructure of the $VO_{1.30}$ phase arises from the ordering of clusters with characteristic structure formed by a local rearrangement of the matrix structure, rather than from ordering of simple defects (Anderson, 1984). Such clusters are responsible for the nonstoichiometry in monoxides of transition metals (after vanadium) such as $Fe_{1-x}O$ and in nonstoichiometric compounds related to the fluorite structure. Another point to note is that, TiO and VO being both metallic, the itinerant electrons can screen long-range interactions and isolated point defects can occur. In the more ionic oxides such as $Fe_{1-x}O$, however, coulombic effects will be strong and the surroundings of point defects are considerably modified, favouring the formation of well-defined vacancy clusters.

Wüstite is always cation-deficient and has a composition in the range $Fe_{0.85}O$–$Fe_{0.95}O$. The reaction for the formation of wüstite is written as

$$2Fe^{2+} + \tfrac{1}{2}O_2 \rightarrow 2Fe^{3+}_{oct} + V_{Fe,oct} + O^{2-} \qquad (5.25)$$

Since wüstite has the NaCl structure, the oxidized Fe^{3+} and the newly created

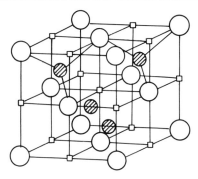

Figure 5.14   Koch–Cohen cluster of defects in $Fe_{1-x}O$ (wüstite.) Hatched circles are tetrahedral Fe atoms and open circles are anions; squares are octahedral Fe vacancies.

vacancies are expected to remain in octahedral sites. From neutron-diffraction studies, Roth (1960) concluded that a substantial proportion of the Fe^{3+} ions occupy tetrahedral sites giving the reaction

$$Fe^{3+}_{oct} + V_i \rightarrow Fe^{3+}_{tet} + V_{Fe,oct}$$

Roth therefore suggested a model of $(V_{Fe}-Fe^{3+}_{tet}-V_{Fe})$ complexes in which Fe^{3+} is tetrahedrally surrounded by oxygens. However, on the basis of X-ray crystallographic studies on quenched $Fe_{0.9}O$, Koch & Cohen (1969) suggested that the defect complex is as large as the unit cell of FeO, consisting of four tetrahedral Fe^{3+} ions in alternate tetrahedral sites (Fig. 5.14) and thirteen vacancies on Fe^{2+} sites. Such a Koch–Cohen cluster corresponds to the chemical formula, FeO, but considering the net charge of the complex, including the virtual charges on the vacancies, it is a negatively charged unit with an antisphalerite structure. Since the oxygen ion packing remains undisturbed, these defect clusters can exist with coherent interfaces with the matrix. The ratio of vacancies to tetrahedral Fe^{3+} ions in the Roth cluster is $2/1 = 2$ whereas in the Koch–Cohen cluster it is $13/4 = 3.25$. On the basis of neutron-diffraction experiments it was established by Cheetham et al. (1971a) that over the temperature range 1073 K to 1473 K, and in the nonstoichiometric range $Fe_{0.9}O-Fe_{0.95}O$, the ratio of octahedral vacancies to tetrahedral Fe^{3+} was nearly 3.25 rather than 2. Gavarri (1978) has found clusters where this ratio is around 2.5.

Theoretical calculations by Catlow and coworkers (Catlow & Mackrodt, 1982) show that such clusters are formed with a net lowering of free energy. Binding energies of defect clusters vary between 1.98 eV (4 : 1 cluster) and 2.52 eV (8 : 3 cluster) with 2.1 eV for the 13 : 4 cluster of Koch and Cohen. The 16 : 5 cluster, which is an element of the spinel structure of Fe_3O_4, is also quite stable (binding energy 2.38 eV), indicating that it may be a precursor in the disproportionation of wüstite. Starting from the perfect rocksalt structure, oxidation through the nonstoichiometric range may well involve the following stages (Anderson, 1984): Isolated vacancies → dipolar associates → 4 : 1 clusters → 6 : 2, 8 : 3, 13 : 4 and similar complex defect clusters → corner-shared 16 : 5

clusters $\rightarrow Fe_3O_4$ (inverse spinel). Such clusters could be present in the $Mn_{1-x}O$ system as well; the cluster binding energies (calculated by Catlow and coworkers) are lower in $Mn_{1-x}O$ than in $Fe_{1-x}O$. Free point defects, if at all, may be present in $Mn_{1-x}O$, but not in $Fe_{1-x}O$. The constitution of wüstite at high temperatures is still not clear, and Bauer *et al.* (1980) have investigated this problem by a study of the incommensurable satellite peaks in the diffraction patterns. It appears that no single cluster species dispersed in the matrix of rocksalt structure can explain the structure at high temperatures. A function that can define the probability of finding Fe_{tet}^{3+} and $V_{Fe,oct}$ along with the displacement of adjacent atoms seems best suited to describe the structure. The incommensurable modulation gives rise to an irrational superstructure, the ratio of $V_{Fe,oct}/Fe_{tet}^{3+}$ as well as the periodicity of the modulation being averaged over the whole crystal.

5.5.4   *Fluorite-related solids*

The fluorite structure generates a variety of derivative structures with anion-excess or anion deficiency; the cation sublattice remains essentially perfect. The unit cell of a compound MX_2 having the fluorite structure corresponds to a face-centred cubic close-packing of M^{2+} ions, all the eight available tetrahedral sites being occupied by X^- ions. An interstitial position of high symmetry is present at the centre. In these sructures, it is now recognized that disorder arises primarily from either anion vacancies or interstitial anions. Extensively investigated among these are the dioxides, which are either anion-deficient, MO_{2-x}, or anion-excess, MO_{2+x}. The nature of the defect complexes in these fluorites seems to be essentially independent of the chemical system studied. In MO_{2-x} compounds, the defect clusters correspond to tightly bound vacancies along a body diagonal, $\langle 111 \rangle$, as shown in Fig. 5.15; the central cation is six-coordinated and is surrounded by six seven-coordinated cations (Thornber & Bevan, 1970). This defect cluster is often referred to as the Bevan cluster. Several oxides

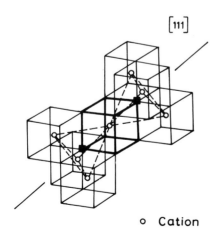

o   Cation

■   Empty anion site

Figure 5.15   Bevan cluster in MO_{2-x} oxides of fluorite structure.

of the formula M_7O_{12} (e.g. Pr_7O_{12}) and intermediate phases in the system Y_2O_3–YOF and Y_2O_3-stabilized ZrO_2, seem to contain Bevan clusters. Since the geometry of these clusters permits four different equivalent orientations, it is easily accommodated in the fluorite structure and accounts for the wide range of stability of the so-called α-phases ($MO_{1.71}$–MO_2). Accordingly, linear strings of the clusters along $\langle 111 \rangle$ of the parent fluorite structure form the motif in the homologous series Pr_nO_{2n-2} and it intergrows with increasing proportion of the fluorite structure in the higher oxides.

Oxides of the formula MO_{2+x} containing excess oxygens seem to accommodate the excess O^{2-} in the high-symmetry interstitial positions when the level of non-stoichiometry is low. But when the nonstoichiometry is high, considerable distortions occur and the interstitial oxygen ions no longer occupy $\frac{1}{2} \frac{1}{2} \frac{1}{2}$ positions in the fluorite structure. Thus, from neutron-diffraction studies on U_4O_9, it has been established that the additional oxygen ions in a unit cell of UO_2 are displaced considerably from the $\frac{1}{2} \frac{1}{2} \frac{1}{2}$ position in the $\langle 110 \rangle$ direction; furthermore, two fluorite oxygens nearest to the interstitial oxygen are also displaced in the $\langle 111 \rangle$ direction, creating two vacancies. The cluster of two vacancies, one interstitial of one kind and two interstitials of another kind is known as the 2:1:2 Willis cluster (Willis, 1964). A more common 2:2:2 cluster involves another interstitial in a normally unoccupied fluorite octahedral site, so that there are two vacancies, two interstitials each of two kinds. Extended Willis clusters such as 3:4:2 (which represents number of vacancies: type I interstitials: type II interstitials) are also considered to be present in MO_{2+x} type of nonstoichiometric materials. Different types of Willis clusters in UO_{2+x} along with a sketch of a fluorite supercell are shown in Fig. 5.16. Analogous defect clusters are found in the rare-earth fluoride–alkaline-earth fluoride solid solutions such as the CaF_2–YF_3 system (Cheetham et al., 1971b). Since the cations in the face-centred cubic arrangement are unaltered, the defect clusters intergrow coherently with the parent structure. Gettmann & Greis (1978) have found superstructure phases of compositions $Ca_{10}Y_5F_{35}$, $Ca_9Y_5F_{33}$ and $Ca_{7+y}Y_{6-y}F_{31-y}$ belonging to the homologous series M_nF_{2n+5} with $n = 15$, 14 and 13 respectively. In the YbF_2–YbF_3 system, Greis (1977) has identified superstructure phases Yb_3F_7, $Yb_{14}F_{33}$ and $Yb_{13}F_{32-y}$. In the mineral $Ca_{14}Y_5F_{43}$ which is a member of the homologous series, the anions are systematically displaced to give rise to a new type of cluster $M_{16}X_{37}$ and the cations are close to the positions of the fluorite structure (Bevan et al., 1982); the cluster $M_{16}X_{37}$ could provide an alternative interpretation of anion-excess fluorites, but does not explain the known structural features and spectroscopic information (Laval & Frit, 1983). Defect energy calculations also do not lend support to this cluster (Catlow et al., 1983). It appears that the nature of defect clusters in anion-excess fluorites is not completely clear.

In systems such as CaO–ZrO_2 and Y_2O_3–ZrO_2, the observed conductivity behaviour does not support the simple isolated vacancy model. There are strong local ordering effects as evidenced from X-ray, neutron and electron diffraction studies, but the nature of the short-range order is not clear. The possible formation of microdomains of ordered superstructures has also been suggested. In anion-excess structures of the type YF_3–YOF and ZrO_2–Nb_2O_5, the excess interstitial anions begin to

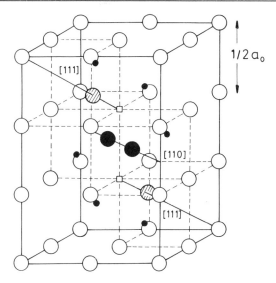

Figure 5.16   Willis clusters in UO_{2+x}. Open circles, oxygens; small filled circles, uranium atoms in the UO_2 structure. A 2:2:2 defect cluster consists of two interstitial oxygen atoms (large filled circles) and two normal oxygen atoms displaced from their ideal positions (open squares) to new interstitial positions (hatched circles). A 2:1:2 defect cluster would consist of one interstitial oxygen and two normal oxygens displaced from their ideal positions. (After Anderson, 1972*b*.)

aggregate on specific crystallographic planes, the boundary climbing outwards until the singularity extends across the entire cross-section of the crystal.

5.6   Crystallographic shear

In the preceding section, we examined how defects in nonstoichiometric compounds are assimilated into the host lattice as structural elements so that the defects are no longer point defects in the ordinary sense and no longer contribute to diffusion and conductivity. An important means by which anion-deficient nonstoichiometry is accommodated in certain transition-metal oxides is to eliminate point defects by the so-called *crystallographic shear* designated by the symbol cs (Anderson, 1972*a*; Anderson & Tilley, 1974). Crystallographic shear was first applied by Wadsley to explain the structures of homologous series of intermediate oxides of tungsten, molybdenum, titanium and vanadium discovered by Magnéli and coworkers. Since point defects are eliminated by cs, the process is 'nonconservative' with respect to the lattice sites; as compared with the parent structure, there is a decrease in the number of anion sites as a consequence of cs.

The principle of cs is best understood with reference to the ReO_3 structure. The ReO_3 structure is built up of ReO_6 octahedra sharing all its corners. A projection of the ReO_3 structure along one of its crystallographic axes is shown in Fig. 5.17(a). When anion deficiency is introduced in the structure, the anion vacancies are eliminated by shearing one part of the structure relative to the other along a certain crystallographic direction.

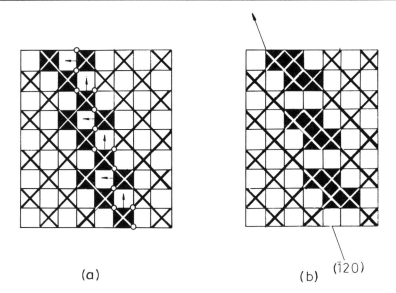

(a)   (b) $(\bar{1}20)$

Figure 5.17   (a) ReO$_3$ structure showing oxygen vacancies which are eliminated by crystallographic shear; (b) the structure after crystallographic shear has occurred.

In the $(\bar{1}20)$ cs illustrated in Fig. 5.17, a plane of oxygen sites in the parent structure is eliminated by shear, resulting in groups of octahedra which share edges in the cs plane. The plane in which the octahedra are differently linked constitutes the shear plane (Fig. 5.17(b)). The direction and magnitude of the displacement of the parent structure along the shear plane are defined by a shear vector. Thus the shear plane shown in Fig. 5.17 is designated $\frac{1}{2}\langle110\rangle$ $(\bar{1}20)$. Only when the shear vector has a component normal to the shear plane can there be an elimination of anion sites. The presence of such a planar defect in a crystal gives good contrast in high-resolution electron micrographs so that HREM becomes invaluable in the study of cs structures. In Fig. 5.18, an HREM lattice image of a $(\bar{1}20)$ cs in a WO$_{3-x}$ crystal is shown for the purpose of visualization. cs provides a means of accommodating anion-deficient nonstoichiometry without introducing point defects and without change of cation coordination; anion coordination, however, changes in the cs plane. In the ReO$_3$ structure, cs can occur not only on $(\bar{1}20)$, but also on $(\bar{1}30)$, $(\bar{1}40)$, etc. planes as well as on (001) planes, although $(\bar{1}20)$ and $(\bar{1}30)$ cs are by far the most common. In the rutile structure, cs generally occurs on $(\bar{1}21)$ and $(\bar{1}32)$ planes, involving face-sharing of octahedra.

   An isolated cs plane or a random array of cs planes, known as the *Wadsley defect*, still gives rise to nonstoichiometry. Regularly recurring cs planes result in a homologous series of stoichiometric intermediate phases. Occurrence of such equidistant cs planes in a crystal indicates a cooperative mechanism. The formula of a cs phase depends on the cs plane index as well as the width of the parent slab between cs planes. If MO$_a$ is the formula of the parent line phase, the formula of the homologous series of oxides resulting from cs may be represented as M_nO_{an-m}, where n is the width (number of

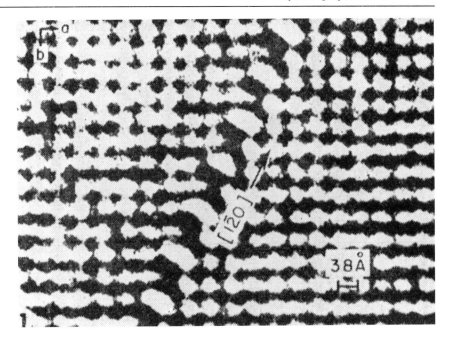

Figure 5.18   HREM lattice image of ($\bar{1}$20)cs in a WO_{3-x} crystal. (From Iijima, 1975.). 3.8 Å separation between WO_6 octahedra is seen.

octahedra) of the parent structure between cs planes and m is the number of anion sites eliminated. In the ReO_3 family of cs phases, we can show that one oxygen is lost for every group of four edge-sharing octahedra in ($\bar{1}$20) cs plane, two oxygens are lost for every group of six edge-sharing octahedra in ($\bar{1}$30) cs plane and, in general, $(m-1)$ oxygens are lost in a ($\bar{1}m0$) cs plane. If $\{1m0\}$ planes form an ordered array in MO_{3-x}, the general formula of the homologous series would be $M_nO_{3n-(m-1)}$. W_nO_{3n-1} and W_nO_{3n-2} are the two homologous series of oxides derived from the parent ReO_3 structure, based on ($\bar{1}$20) and ($\bar{1}$30) cs respectively. Typical homologous series of oxides involving cs are listed in Table 5.2.

Crystallographic shear phases can be regarded as translation modulations of the parent structure, the translation boundaries being cs planes (Section 1.6). Although a number of solids with cs planes has been described in the literature, it is only recently that an understanding of the mechanism of formation and ordering of cs planes has begun to emerge (Tilley, 1980; Catlow & James, 1980). Elastic strain appears to play a crucial role in the ordering of shear planes, and both continuum and atomistic theories have been valuable in providing an insight into this interesting problem. Extensive relaxation around the shear planes seems to be essential for stabilizing these structures.

An interesting feature to note with regard to crystallographic shear is that even very slightly reduced rutile (e.g. $TiO_{1.997}$) shows the presence of cs planes. In slowly cooled samples of TiO_{2-x} with $0.0 < x \leqslant 0.01$, pairs of cs planes precipitate and separate subsequently (Blanchin et al., 1984). When $0 < x \leqslant 0.0035$, novel $\{100\}$ platelet defects

Table 5.2. *Typical homologous series of oxides involving cs*[a]

Parent structure	cs planes	Series formula	Approximate composition range
WO_3	{T20} ReO_3	W_nO_{3n-1}	$WO_3 - WO_{2.93}$
	{T30} ReO_3	W_nO_{3n-2} $(n = 20, 24, 25, 40)$	$WO_{2.93}-WO_{2.87}$
MoO_3	{T20} ReO_3	Mo_nO_{3n-1} $(n = 8, 9)$	—
	{311} MoO_3	Mo_nO_{3n-2} $(n = 18)$	—
TiO_2	{T21} rutile	Ti_nO_{2n-1} $(4 < n < 10)$	$TiO_{1.75}-TiO_{1.90}$
	{T32} rutile	Ti_nO_{2n-1} $(16 < n < \sim 37)$	$TiO_{1.90}-TiO_{1.9375}$

[a]Crystallographic shear with one set of nonintersecting shear planes.

occur along with cs planes; this can be understood in terms of cationic interstitial defects (Bursill *et al.*, 1984). In WO_{3-x}, besides cs planes, hexagonal tunnels and pentagonal bipyramidal columns have been observed. Bursill & Smith (1984) have analysed the extrinsic and/or intrinsic nature of extended defects in TiO_{2-x} and WO_{3-x} by drawing Burgers circuits directly on to the high-resolution electron micrographs; this study has been of value in determining the nature of small defects in these oxides. These authors have also obtained HREM images showing a partial wall of anion vacancy defects.

5.7   Block structures

Binary and certain ternary oxides of niobium adopt a new structural principle to accommodate a variable metal: oxygen ratio. Structures of these oxides can be derived from ReO_3 by the operation of two sets of nearly orthogonal crystallographic shear (Wadsley & Andersson, 1970; Anderson & Tilley, 1974). The two sets of recurrent shear planes, which run through the entire crystal, splice the ReO_3 structure into rectangular blocks (or pillars) of $m \times n$ octahedra in cross-section and infinite length in one dimension (*b*-direction). The blocks of ReO_3 structure share edges with identical or similar blocks. Metal atoms in the resulting structure retain octahedral coordination but they are at two levels ($z = 0$ and $z = \frac{1}{2}$), being displaced by one-half the octahedron diagonal in adjacent blocks. The oxygen sublattice of the resulting structures is the same as in ReO_3 but the coordination number of oxygen is different: 1 at unshared block corners, 2 in the interior of a block, 3 at the edges and 4 at shared corner sites. The blocks can be interconnected in three different ways: (i) the corners can be unshared, creating tetrahedral sites for cations, (ii) corner octahedron edges can be shared at the same level or (iii) the blocks can form ribbons by overlapping edge-sharing between blocks at the same level. Corner linkages of types (ii) and (iii) join the blocks at any one

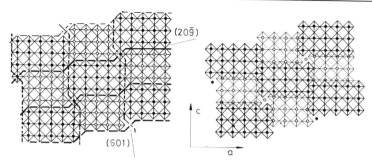

Figure 5.19   Generation of the block structure of $H-Nb_2O_5$ from the ReO_3 structure through cs on $(20\bar{9})$ and (601) planes. The $(3 \times 4)_1 + (3 \times 5)_\infty$ blocks in the $H-Nb_2O_5$ structure are illustrated on the right. (After Anderson & Tilley, 1974.)

level into groups of p blocks ($p = 1, 2$ or ∞). A given set of blocks can be represented by the symbol $(m \times n)_p$. If $p = 1$ or 2, tetrahedral cation sites are created at the free corners. The composition of a $(m \times n)_p$ block is given by $M_{mnp+1} O_{3mnp-p(m+n)+4}$. For a structure involving two or more distinct blocks of type $i, j \ldots$, the chemical formula would be

$$\sum_i M_{mnp+1} O_{3mnp-p(m+n)+4}$$

The fact that the geometrical parameters m, n and p can be varied enables exceedingly small variations in the metal:oxygen ratio to be accommodated by a rearrangement of the block structure, retaining perfect order. More interestingly, it is possible to have different block structures for one and the same composition. Nb_2O_5 for instance exists in fourteen different polymorphic modifications, ten of which appear to possess block structures. The structure of $H-Nb_2O_5$, the most stable high-temperature polymorph, is derived from the ReO_3 structure by recurrent crystallographic shear in two nearly orthogonal directions as shown in Fig. 5.19; the cs planes involved are $(20\bar{9})$ and (601). The resulting block structure is designated $(3 \times 4)_1 + (3 \times 5)_\infty$. A lattice image of $H-Nb_2O_5$ is shown in Fig. 5.20. Between $NbO_{2.4167}$ and $NbO_{2.5000}$ a number of block structures consisting of different types of blocks have been identified: e.g. $(3 \times 4)_\infty$ in $Nb_{12}O_{29}$, $(3 \times 4)_\infty + (3 \times 3)_1$ in $Nb_{22}O_{54}$ and $Nb_{47}O_{116}$ and $(3 \times 4)_2$ in $Nb_{25}O_{62}$. Besides binary niobium oxides, block structures are also well-characterized in the $TiO_2-Nb_2O_5$ and $WO_3-Nb_2O_5$ systems in the composition region $MO_{2.654}-MO_{2.333}$. Descriptions of block structures are available in the literature (Wadsley & Andersson, 1970; McConnell et al., 1976) and they have been investigated adequately by high-resolution electron microscopy (Gruehn & Mertin, 1980).

It must be pointed out that defects in block structures giving rise to non-stoichiometry are possible. Both Wadsley defects and point defects are known in block structure phases. Nonstoichiometry is introduced by the insertion of rows or columns

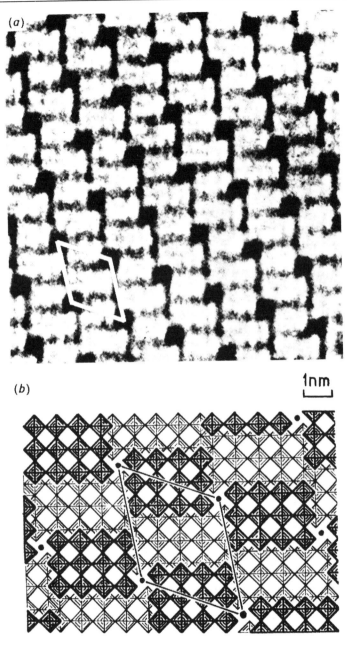

Figure 5.20   (a) Lattice image of H–Nb$_2$O$_5$ showing the two differently sized blocks. (From Allpress & Sanders, 1973.) In (b) the idealized structure is shown.

of blocks of different size which are geometrically compatible with the matrix structure. Thus in the range NbO$_{2.48}$–NbO$_{2.50}$, units of Nb$_{25}$O$_{62}$, $(3 \times 4)_\infty$ structure intergrows with the H–Nb$_2$O$_5$, $(3 \times 4)_1 + (3 \times 5)_\infty$ structure. More interestingly, point defects have been invoked to account for nonstoichiometry in certain block structures.

Germanium niobium oxide, $GeO_2 \cdot 9Nb_2O_5$, is a typical example. The phase is apparently isostructural with PNb_9O_{25}, a $(3 \times 3)_1$ block structure. The composition is, however, incompatible with the structure, unless there are oxygen vacancies or cation interstitials or both. Crystal structure determination has revealed that the oxygen sites are fully occupied and all the octahedral cation sites within the blocks are also fully occupied. The excess cations are accommodated in the channels formed by the corners of the columns. Within these channels, both tetrahedral and octahedral sites are available. The sites are only partly occupied. The relative number of cations in tetrahedral and octahedral sites determines the stoichiometry. Within any one channel, tetrahedral and octahedral cations may be ordered, but there is little correlation between the ordering in the neighbouring channels. Since there is no evidence for superlattice ordering in the germanium niobium oxide, a statistical model with $14/m$ symmetry, as for PNb_9O_{25}, seems appropriate. The Raman spectrum of $GeO_2 \cdot 9Nb_2O_5$ is consistent with the random occupation model (McConnell *et al.*, 1976).

5.8   Infinitely adaptive structures

In the homologous series Ti_nO_{2n-1}, cs occurs on $(\overline{1}21)$ for $n < 10$ and on $(\overline{1}32)$ for $n > 16$ (Table 5.2). A remarkable structural feature of the composition for which $n = 10-14$ is that the $(\overline{1}21)$ plane swings continuously from $(\overline{1}21)$ to $(\overline{1}32)$ through all possible orientations such as (253), (374), (495), etc. As a consequence, almost any composition in the range $TiO_{1.900}$ to $TiO_{1.937}$ can remain structurally ordered. Such continuous series of ordered structures may have apparently irrational cs planes. These structures have been called *infinitely adaptive structures*, a concept put forward by Anderson (1973). In the relevant composition range, the crystallographic shear plane swings continuously, being pivoted in its own zone, from one stable direction to another.

An example of infinitely adaptive structures not involving cs is found in L–Ta_2O_5 and its solid solutions with WO_3 up to the composition $11Ta_2O_5 \cdot 4WO_3$. The structure of L–Ta_2O_5 is related to that of U_3O_8 and consists of edge-shared pentagonal bipyramids and distorted octahedra. Addition of WO_3 to L–Ta_2O_5 results in unique superstructures for every composition (Roth & Stephenson, 1970). The resulting homologous series corresponds to $M_{2m}O_{(16m-2)/3}$ ($M_{10}O_{26}$, $M_{22}O_{58}$, etc.). An infinite series of structures can be built by an ordered stacking of these units in varying proportions. Any composition between $MO_{2.625}$ and $MO_{2.636}$ in this system can be built by using a units of $M_{10}O_{26}$ and b units of $M_{22}O_{58}$ so that the formula $(M_{10a+22b}O_{26a+58b})$ may be made to correspond to any composition in the range by an appropriate choice of a and b. Another system showing infinitely adaptive structures is $Ba_p(Fe_2S_4)_q$; several such systems (Rao, 1985) have been found in recent years. Kittel (1978) points out that the elastic energy associated with the forced coherent registration of two types of lattice planes gives the long-range repulsive interaction necessary for the formation of these long-period structures. Making use of the Saint-Venant theorem of the theory of elasticity, the average elastic energy per unit area per plane in forming a composition A_nB (by the coherent registration of planes of A and B) is given by

$$U = \frac{Kd}{2(N+1)}(N\varepsilon_A^2 + \varepsilon_B^2) = \frac{1}{2}Kd\left(\frac{\Delta a}{a}\right)^2 \frac{N}{(N+1)^2} \tag{5.26}$$

where $\Delta a = a_A - a_B = (\varepsilon_A - \varepsilon_B)a$, a_A and a_B are the lattice constants of pure A and B, εs are the strains, K is the bulk modulus, d is the interplanar spacing and ε_B is taken as equal to $- N\varepsilon_A$.

5.9   Intergrowths

Epitaxy and polytypism are examples of intergrowth across solid–solid interfaces. *Epitaxy* is of great interest since it deals with ways of laying down thin, single-crystal films (of semiconducting materials) on an appropriate substrate, for use in integrated circuits, night-vision devices, etc. The twin plane and the coincidence boundary are planar faults (see Section 5.4) or interfaces which are kinetically stabilized nonequilibrium defects and these do not give rise to changes in stoichiometry of the material. Polytypism similarly does not introduce compositional changes. Many other compositionally invariant intergrowths are also known (Rao & Thomas, 1985), an unusual example being the enantiomeric intergrowth in hexahelicenes. Intergrowth of the zeolite ZSM-5 with ZSM-11 is another interesting example. Intergrowth of amphiboles with triple, quadruple and wider strips of connected pyroxene chains has been encountered. In most such systems, intergrowth is nonrecurrent or disordered.

Nonrecurrent intergrowth with composition changes is found in a variety of systems. All cases of epitaxy are in fact examples of solid–solid interfaces involving changes in composition. Isolated cs planes are good examples of such solid–solid interfaces. Nephrite jade (amphibole) accommodates triple chains of a different composition. Other systems showing such nonrecurrent intergrowth are β and β'' alumina, neighbouring members of the oxides of the formula $Bi_2A_{n-1}B_nO_{3n+3}$ as well as of V_nO_{2n-1}. Fig. 5.21 shows the lattice image of $Sr_3Ti_2O_7$ in region A and also lamellae of different members of the $Sr_{n+1}Ti_nO_{3n+1}$ family with $n = 3, 4, 5, 7$ and 8. Rao & Thomas (1985) have recently discussed nonrecurrent and recurrent intergrowths in the general context of the chemistry of the solid–solid interface. One could, in principle, consider intergrowth as a class of modulated structures where the intergrowth boundary is the periodic perturbation.

Several nonrecurrent intergrowths show a certain degree of order and it is often difficult to decide how many repeats of a particular sequence constitute recurrent intergrowth. There are, however, many systems where recurrent intergrowth occurs over relatively large distances, (Rao, 1985), giving rise to homologous series of structures with large unit cells (Table 5.3).

An example of recurrent intergrowth is provided by the family of hexagonal barium ferrites, M_pY_q, formed by the intergrowth of $BaFe_{12}O_{19}(M)$ and $Ba_2Me_2Fe_{12}O_{22}(Y)$ with $Me = Zn$, Ni or Mg (Anderson & Hutchison, 1975). In Fig. 5.22 we show an electron micrograph of a typical ferrite. These barium ferrites, which may be considered to be intergrowth of perovskite and spinel layers, exhibit a wide range of unit cell dimensions; for example, in MY, $c = 26$ Å and in M_8Y_{27}, $c = 1455$ Å. The complex-

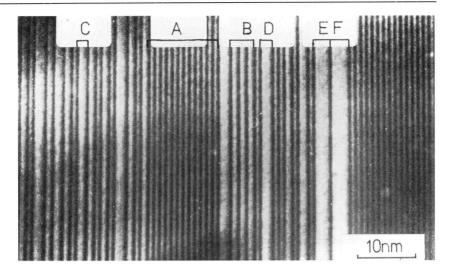

Figure 5.21   Lattice image of $Sr_3Ti_2O_7$ showing nonrecurrent intergrowth of other members of the $Sr_{n+1}Ti_nO_{3n+1}$ family. (From Tilley, 1977.) A, $n = 2$; B, $n = 3$; C, $n = 4$; D, $n = 5$; E, $n = 7$ and F, $n = 8$.

Table 5.3. *Typical homologous series resulting from recurrent intergrowth*

$Ba_{2n+p}Me_{2n}Fe_{12(n+p)}O_{22n+19p}$(Me = Zn, Ni, etc.)[a] with $n = 1$ to 47
and $p = 2$ to 12
$Bi_4A_{m+n-2}B_{m+n}O_{3(m+n)+6}$ with A = Ba or Bi, B = Ti, Nb, Cr, etc.
and $m, n = 1, 2, 3, 4$
A_xWO_3 with A = alkali or alkaline earth metal, Bi, etc.
$(0.0 < x \leqslant 0.1)$
$A_xP_4O_8(WO_3)_{2m}$ with $m = 4$ to 10
$(Na, Ca)_nNb_nO_{3n+2}$ with $n = 4$ to 4.5[b]
$(A_3Nb_6Si_4O_{26})_n(A_3Nb_{8-x}M_xO_{21})^c$ with A = Ba or Sr, M = Ti, Ni, Zn,
etc. and $n = 1$ to 15.

[a]Oxygen-to-metal ratio in the range 1.38–1.43 is accommodated by change in stacking sequence.
[b]In these oxygen excess perovskites ABO_{3+x}, x varies between 0.444 and 0.50 but there is no evidence of point defects.
[c]Oxygen-to-metal ratio varies between 1.9 and 2.0 in these systems with no apparent point defects.

ity of these intergrowths is illustrated by $M_{12}Y_{47}$ with the composition $Ba_{106}Ni_{94}Fe_{708}O_{1268}$ containing 12M units in the primitive cell. The M_pY_q intergrowths cannot be formed when Fe^{3+} is replaced by Al^{3+} ions, indicating the importance of magnetic interactions in stabilizing the long-period structures. An interesting class of recurrent intergrowths is that formed when slabs of WO_3 intergrow with strips of a *hexagonal tungsten oxide bronze* (HTB); these *intergrowth tungsten oxide*

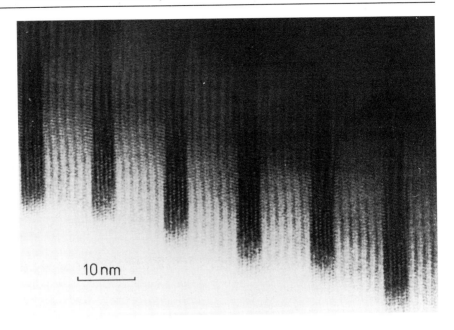

Figure 5.22   Lattice image of $MYMY_6$ intergrowth in the barium ferrite system. (From Anderson & Hutchison, 1975.)

bronzes (ITB) show complex ordered sequences with large periodicities over fairly large distances (Kihlborg, 1978). In Figs 2.8 and 2.10 we showed the image of an ITB formed by bismuth where the WO_3 slabs intergrow with a HTB strip, one tunnel wide. In Fig. 5.23 we show the lattice image of a highly ordered Cs_xWO_3 ITB. The formation of intergrowth bronzes is possibly linked to the growth conditions and may well represent a case where the impurity-rejection model holds (Anderson *et al.*, 1972). Long-period intergrowths of barium siliconiobates (Studer & Raveau, 1978) and $(Na, Ca)_nNb_nO_{3n+2}$ (Portier *et al.*, 1975) seem to resemble the continuous series of ordered structures or the infinitely adaptive structures described by Anderson (1973). Recurrent intergrowths formed by the Aurivillius family of oxides of the formula $(Bi_2O_2)^{2+} (A_{m-1}B_mO_{3m+1})^{2-}$ have been investigated extensively (Gopalakrishnan *et al.*, 1984; Jefferson *et al.*, 1984). This system shows remarkable order even in the absence of magnetic or charge-ordering interactions and forms the homologous series $Bi_4A_{m+n-2}B_{m+n}O_{3(m+n)+6}$. Figure 5.24 shows the high-resolution image of $Bi_9Ti_6CrO_{26}$ formed by the intergrowth of $Bi_4Ti_3O_{12}$ ($m = 3$) and $Bi_5Ti_3CrO_{15}$ ($n = 4$), along with the computer-simulated image and the structural model, to illustrate the ordered arrangement of the $m = 3$ and $n = 4$ lamellae. These systems represent cases where it seems elastic forces alone are responsible for the recurrent intergrowth.

In the various intergrowth systems examined (see Table 5.3) there is no evidence for the presence of point defects. The origin of long-range periodicity in the complex recurrent intergrowth systems is, however, intriguing. The importance of elastic forces in the formation of polytypes, shear structures and infinitely adaptive structures was

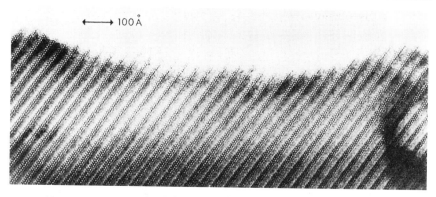

Figure 5.23   Lattice image of Cs_xWO_3. (From Kihlborg, 1978.) The ITB phase has HTB strips (two tunnels wide) separated by seven WO_6 octahedra.

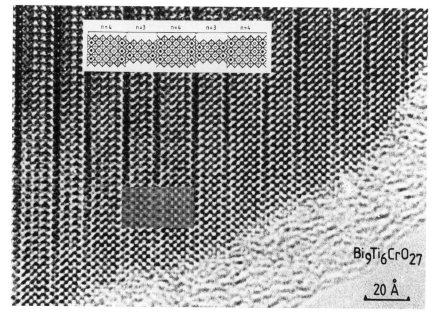

Figure 5.24   High-resolution image (500 kV) of $Bi_9Ti_6CrO_{27}$. (From Gopalakrishnan *et al.*, 1984.)

pointed out earlier. It is likely that elastic forces are responsible for recurrent intergrowths as well (Rao, 1985). An operational criterion for the formation of recurrent intergrowths can be derived by expressing the elastic strain energy in terms of the volume change accompanying the formation of the intergrowths (Gopalakrishnan *et al.*, 1984). Jefferson *et al.* (1984) have recently examined the nature of atomic displacements arising from the elastic strain by comparing the observed high-resolution images with those computed on the basis of different models. Such a structure refinement throws light on the solid–solid interface. Mass density waves (solitons) could be used to understand intergrowth structures.

5.10   Defect perovskite oxides: a case study

Perovskites constitute an important class of inorganic solids and it would be instructive to survey the variety of defect structures exhibited by oxides of this family. Non-stoichiometry in perovskite oxides can arise from cation deficiency (in A or B site), oxygen deficiency or oxygen excess. Some intergrowth structures formed by oxides of perovskite and related structures were mentioned in the previous section and in this section we shall be mainly concerned with defect ordering and superstructures exhibited by these oxides.

A-site vacancies.   Since the BO_3 array in the perovskite structure forms a stable network, the large A cations at twelve-coordinated sites can be missing either partly or wholly. The tungsten bronzes, A_xWO_3, which constitute an important family of nonstoichiometric oxides, may be regarded as A-site deficient perovskites (Ekström & Tilley, 1980). Of the three different structures known for tungsten bronzes, hexagonal, tetragonal and cubic, only the cubic one is related to the perovskite structure consisting of A-site vacancies. The cubic bronzes are referred to as *perovskite tungsten bronzes* (PTB) and two types of PTBs can be distinguished: those with small values of x ($x \ll 1.0$) and others with large values of $x \leqslant 1.0$. PTBs with large x show true bronze characteristics, viz. metallic behaviour and chemical inertness. An important question with regard to the nonstoichiometry of these phases is whether the A-site atoms and vacancies are ordered or not. An early neutron-diffraction study (Atoji & Rundle, 1960) of $Na_{0.75}WO_3$ showed that Na ions are ordered in six out of eight A sites in a doubled perovskite cell. The nature of ordering of A-site vacancies in metallic PTBs is, however, not clear. Random isolated defects or weakly interacting defects can occur in metallic systems unlike in the nonmetallic counterparts.

Besides PTBs, A-site defective perovskite oxides are known to be formed when B = Ti, Nb, Ta and so on. Such compounds exhibit metallic properties and perovskite structures when the B atom occurs in a low oxidation state. Compositions such as $A_{0.5}NbO_3(A = Ba, Pb$ etc.), where niobium is in the highest oxidation state, adopt nonperovskite network structures. An interesting example of a A-site defective perovskite is $Cu_{0.5}TaO_3$, which crystallizes in a pseudocubic perovskite structure. The unit cell is orthorhombic with $a = 7.523, b = 7.525$ and $c = 7.520$ Å and eight formula units per cell. Tantalum atoms form a TaO_3 framework as in a cubic perovskite, while the copper atoms are ordered at the A sites (Longo & Sleight, 1975; Vincent *et al.*, 1978). Besides the perovskite, tetragonal and hexagonal tungsten bronzes, a new variety of tungsten bronzes (ITBs) formed by the intergrowth of WO_3 slabs (*n* octahedra wide) with strips of hexagonal bronze (HTB), one to three tunnels wide, is noteworthy (Kihlborg, 1978). The intergrowth in ITBs can be recurrent or nonrecurrent. The HTB strip in Fig. 5.23 is two tunnels wide and this is the most stable configuration of alkali-metal intergrowth tungsten bronzes. Recently, intergrowth tungsten bronzes of bismuth where the HTB strip is always one tunnel wide have been investigated (Ramanan *et al.*, 1984). What is interesting in the Bi_xWO_3 system is that when $x \leqslant 0.02$, a PTB phase seems to be formed, the ITBs being found only when $x > 0.02$ (see Fig. 2.8 and Fig. 2.10). WO_3 and ReO_3 may be regarded as the limiting cases of A-site vacancy perovskites. Both the oxides possess a corner-linked framework of the octahedra, but unlike ReO_3, WO_3 is never cubic. It shows several polymorphic transitions starting from the low-temperature triclinic structure to more symmetric forms with increasing temperature.

B-site vacancies.   B-site vacancies in perovskite oxides are energetically not favoured because of the large formal charge and the small size of B-site cations. If vacancies are to occur at the B-sites, there must be other compensating factors such as B—O covalency and B—B interaction. While the covalency of the B—O bond would increase with increasing charge and decreasing size of the B cation, the B—B interaction would be favoured by hexagonal rather than cubic stacking of AO_3 layers. Thus, one would expect B-site vacancies to occur more frequently in perovskite oxides of highly charged B cations possessing hexagonal polytypic structures. B-site vacancy perovskites are indeed far more common with hexagonal or mixed hexagonal–cubic AO_3 stackings than with exclusive cubic stacking of AO_3 layers. Rauser & Kemmler-Sack (1980) have listed cubic perovskites exhibiting B-site vacancy. Some of the examples of ordered B-site vacancy perovskites are: $Ba_2Sm_{2/3}UO_6(5)$, $Ba_2Ce^{4+}_{3/4}Sb^{5+}O_6(7)$, $Ba_2Ce^{4+}Sb^{5+}_{4/5}O_6(9)$ and $Ba_2Sm^{3+}U^{6+}_{5/6}O_6(11)$, where the numbers in parentheses represent the number of B-cations per vacancy. Ordering of these vacancies is revealed by the formation of superstructures.

Many well-characterized B-site vacancy hexagonal perovskites are known and the B-site vacancies in these oxides are ordered between h–h layers where the BO_6 octahedra share faces. Accordingly, $Ba_5Ta_4O_{15}$ adopts a five-layer (hhccc) sequence (Shannon & Katz, 1970) where the octahedral site between h–h layers is vacant. The arrangement gives rise to isolated clusters of four octahedra that share each two opposite corners, $[Ta_4O_{15}]^{10-}$. The structure of $Ba_3Re_2O_9$ is similar (Calvo et al., 1978) consisting of the $(hhc)_3$ layer sequence where the ReO_6 octahedra share corners, and the vacancy occurs at octahedral sites between h–h layers. Somewhat exceptional to this general trend are $Ba_3W_2O_9$ and $Ba_3Te_2O_9$ (see, for example, Jacobson et al., 1981). Both these compounds adopt an all-hexagonal BaO_3 layer sequence. In $Ba_3W_2O_9$, two-thirds of the octahedral sites in every layer are filled by tungsten, while in $Ba_3Te_2O_9$ octahedral sites in two adjacent layers are fully occupied, every third layer being completely vacant. Both the orderings give rise to $[B_2O_9]^{6-}$ (B = W,Te) groups comprising two BO_6 octahedra sharing a common face (confacial bioctahedra). In both the structures, B atoms are displaced considerably towards the vacancies, resulting in a minimization of B–B repulsion.

Oxide perovskites consisting of ordered B-site vacancies exhibit novel luminescence properties. $Ba_3W_2O_9$, for instance, shows an efficient blue (460 nm) photoluminescence below 150 K, unlike $BaWO_4$ and Ba_3WO_6 which consist of isolated WO_4 tetrahedra and WO_6 octahedra respectively (Blasse & Dirksen, 1981). It has been suggested that clustering of WO_6 octahedra in $Ba_3W_2O_9$ is responsible for the efficient *luminescence*. More interestingly, B-site vacancy ordered perovskites show different colour emissions, when they are doped with different rare-earth activators. For instance, when $Ba_{3-x}Sr_xLaScW_2\square O_{12}$, a 12-layered $(hhcc)_3$ perovskite, is doped with Eu^{3+} and

Tb^{3+} simultaneously, a green terbium emission $(Tb^{3+} : {}^5D_4 \rightarrow {}^7F_5)$ at 547 nm and a red europium emission $(Eu^{3+} : {}^5D_0 \rightarrow {}^7F_2)$ at 615 nm are seen. The former shows up when the phosphor is excited in the WO_6 charge-transfer band at 300 nm and the latter emission is seen when excitation occurs at 340 nm corresponding to charge-transfer from oxygen to Eu^{3+}. Similarly, $Sr_3La_2W_2\square O_{12}$ doped with Eu^{3+} and Tm^{3+} shows red and blue emissions when excited suitably, and the same host doped with Er^{3+} and Tm^{3+} shows green and blue emissions (Brown & Kemmler-Sack, 1983). The family of bismuth oxides of the general formula $Bi_mB_{m-1}O_{3m}(B = Ti, Nb$ and W), first described by Aurivillius (1950), may also be regarded as B cation-deficient perovskites. Typical members of this family are Bi_2WO_6, Bi_3TiNbO_9 and $Bi_4Ti_3O_{12}$.

Anion-deficient perovskites and vacancy-ordered structures.

Anion-vacancy nonstoichiometry in perovskite oxides is more common than that involving vacancies in the cation sublattice. Anion-vacancy nonstoichiometry is known with both ABO_3 and BO_3 frameworks. Nonstoichiometric tungsten trioxide, WO_{3-x}, is a typical example of the latter. Anion-deficient nonstoichiometry in ABO_{3-x} is not generally accommodated by the cs mechanism. Defect ordering in ABO_{3-x} oxides involves a conservative mechanism in the sense that the vacancies are assimilated into the structure, resulting in large supercells of the basic perovskite structure. The type of superstructure formed depends, however, on the identity of the B cation.

One of the best-characterized perovskite oxides with ordering of anion vacancies is the *brownmillerite* structure exhibited by $Ca_2Fe_2O_5$ and Ca_2FeAlO_5 (Grenier et al., 1981). The compositions could be considered as anion-deficient perovskites with one-sixth of anion sites being vacant. The orthorhombic unit cell of the brownmillerite structure $(a = 5.425, b = 5.598$ and $c = 14.768$ Å for $Ca_2Fe_2O_5)$ arises because of vacancy-ordering and is related to the cubic perovskite as $a \simeq \sqrt{2}a_c, b \simeq \sqrt{2}a_c, c \simeq 4a_c$. Oxygen vacancies are ordered in alternate $(001)BO_2$ planes of the cubic perovskite structure such that alternate [110] rows of oxygens are missing (Fig. 5.25). The layer sequence along the c-axis is therefore $CaO-FeO_2-CaO-FeO$ .... A slight shift of the iron atoms in the FeO layers leads to a tetrahedral coordination of iron in this layer resulting in alternate sheets of FeO_6 octahedra and FeO_4 tetrahedra (OTOT'...) in the c-direction (Fig. 5.25). A new oxide, Ca_2FeCoO_5, possessing the brownmillerite structure, in which Fe^{3+} and Co^{3+} occupy tetrahedral and octahedral sites respectively, has been reported (Vidyasagar et al., 1984).

$Ca_2Mn_2O_5$ and $Ca_2Co_2O_5$, which have compositions similar to brownmillerite, have been reported (Poeppelmeier et al., 1982; Vidyasagar et al., 1984), but their structures involve different modes of vacancy-ordering. $Ca_2Mn_2O_5$ adopts an orthorhombic structure with $a = 5.432, b = 10.242$ and $c = 3.742$ Å, the relationship with the cubic perovskite structure being $a \simeq \sqrt{2}a_c, b \simeq 2\sqrt{2}a_c$ and $c \simeq a_c$. Oxygen vacancies in $Ca_2Mn_2O_5$ are ordered in every (001) BO_2 plane of the cubic perovskite such that one-half of the oxygen atoms in alternate [110] rows are removed (Fig. 5.26). The composition of the plane is now $MnO_{1.5}$. If these are stacked in the sequence $CaO-MnO_{1.5}-CaO-MnO_{1.5}$ ... the structure of $Ca_2Mn_2O_5$ is obtained. An important feature of the structure, which distinguishes it from brownmillerite, is that

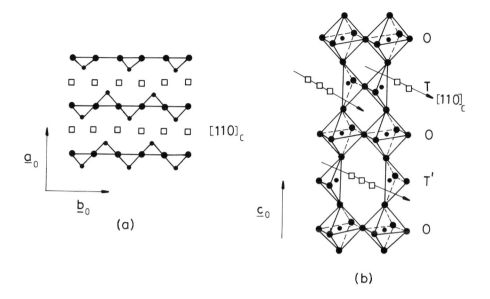

Figure 5.25   (a) Vacancy ordering in the *ab* plane of $Ca_2Fe_2O_5$ of brownmillerite structure; (b) alternating sequence of octahedral and tetrahedral layers in the *c* direction. Small filled circles, iron; large filled circles, oxygen; open squares, oxygen vacancy.

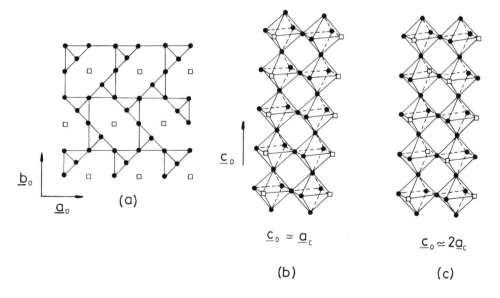

Figure 5.26   (a) Vacancy ordering in the *ab* plane of $Ca_2Mn_2O_5$. Vacancy ordering parallel to the *c* direction in $Ca_2Mn_2O_5$ and $Ca_2Co_2O_5$ is shown in (b) and (c) respectively. Filled circles, oxygen; open squares, oxygen vacancy.

manganese is square-pyramidally coordinated. The $3d^4$ configuration of Mn^{3+} favours square-pyramidal coordination rather than tetrahedral coordination, thereby determining the anion-vacancy-ordering in $Ca_2Mn_2O_5$. Members of the $Ca_2Fe_{2-x}Mn_xO_5$ series with three different coordinations (octahedral, tetrahedral and square-pyramidal) around the transition metal ions have been characterized (Vidyasagar, et al., 1986). $Ca_2Co_2O_5$ also crystallizes in an orthorhombic structure with $a = 11.12$, $b = 10.74$ and $c = 7.48$ Å. Comparison of these parameters with those of $Ca_2Mn_2O_5$ reveals a doubling of the a and c axes in $Ca_2Co_2O_5$. The vacancy-ordering in $Ca_2Co_2O_5$ is similar to that in $Ca_2Mn_2O_5$, but the doubling of the c axis is likely to be due to the ordering scheme shown in Fig. 5.26. $Sr_2Mn_2O_5$ is found to have the same defect ordering as in $Ca_2Mn_2O_5$ (Caignaert et al., 1985). $Sr_2Co_2O_5$, on the other hand, crystallizes in an anion-deficient 2H perovskite structure at low temperatures and transforms to the brownmillerite structure abo ve 1170 K (Grenier et al., 1979). $SrCoO_{3-x}(0 < x < 0.5)$ prepared under high oxygen pressures (50–2600 bars) adopt a cubic perovskite structure.

$SrTiO_{2.5}$, $SrVO_{2.5}$ and $KTiO_{2.5}$ are a few other anion-vacancy perovskites with the composition $ABO_{2.5}$. $KTiO_{2.5}$ shows little similarity to the perovskites, each titanium having a trigonal bipyramid oxygen coordination. X-ray diffraction studies of $SrTiO_{2.5}$ and $SrVO_{2.5}$ show cubic perovskite unit cells with $a \simeq 3.9$ Å, implying that anion vacancies are random. There is no evidence for superlattice ordering in $SrTiO_{2.5}$ (Tofield, 1978; see also Alario–Franco & Regi, 1977). It is noteworthy that both $SrTiO_{3-x}$ and $SrVO_{3-x}$ exhibit metallic properties, indicating that d-electrons associated with B cations are itinerant.

In many of the ABO_{3-x} perovskites, nonstoichiometry spans the composition range $0 < x \leqslant 0.5$ and we have so far discussed only the structures of the end members with $x = 0.5$. It is of interest to see what structures prevail at intermediate compositions. In $SrFeO_{3-x}$, MacChesney et al. (1965) reported a cubic structure when $0 < x < 0.12$, a tetragonally distorted perovskite structure when $0.15 < x < 0.28$, and a mixture of the perovskite and brownmillerite phases when $0.28 < x < 0.5$. $SrFeO_{2.75}$ was investigated by Tofield et al. (1975), who found a supercell related to the perovskite as $a \simeq 2\sqrt{2}a_c$, $b \simeq 2a_c$ and $c \simeq 2\sqrt{2}a_c$ in the electron-diffraction patterns. The vacancy-ordering suggested for this phase is related to that of brownmillerite; every other oxygen atom is lost in alternate $[110]_c$ strings to give five-coordination to half the iron atoms, the other half retaining octahedral coordination as in perovskite. In $Sr_xNd_{1-x}FeO_{3-y}$, superstructures involving random doubling of one of the perovskite cell axes (in a, b or c directions) are found (Alario-Franco et al., 1982) and this has been considered to be due to the existence of a microdomain texture, where the domains intergrow randomly in three dimensions.

Reller et al. (1984) have investigated the anion-deficient $CaMnO_{3-x}$ system over the range $0 < x \leqslant 0.5$ in great detail. Five distinct compositions, $CaMnO_{2.5}$, $CaMnO_{2.556}$, $CaMnO_{2.667}$, $CaMnO_{2.75}$ and $CaMnO_{2.80}$, with ordered oxygen vacancies have been identified and, in all these, structural features of the parent perovskite are preserved. The structure of $CaMnO_{2.5}$, possessing square-pyramidal coordination of Mn^{3+}, was discussed earlier. In the other compositions, the proportion of square-pyramids relative to octahedra decreases as we go from $CaMnO_{2.5}$ to $CaMnO_{3.0}$. These structures may be pictured as superlattice repeats of the parent undistorted perovskite, but with the superlattice mesh rotated by an angle R in the (001) plane. The superlattice in $CaMnO_{2.8}$ is $\sqrt{5} \times \sqrt{5}$, but rotated by 26.5° and this structure gives a tetragonal unit cell with $a = b = 8.34$ Å and $c = 7.46$ Å. In order to check this structure, lattice images were computed and compared with the HREM image; the observed agreement tends to support the structural model. The structure of $CaMnO_{2.75}$ can be represented as $\sqrt{2} \times 2\sqrt{2}R\,45°$ or $\sqrt{2} \times 4\sqrt{2}R\,45°$, while that of $CaMnO_{2.667}$ is $\sqrt{2} \times 3\sqrt{2}R\,45°$. Four ordered structures of $CaMnO_{2.5}$ have been observed, but the most common one is the $\sqrt{2} \times 2\sqrt{2}R\,45°$ phase described by Poeppelmeier et al. (1982). The relation between the defect structures of the members of the $CaMnO_{3-x}$ system and their catalytic activity has been commented upon by Reller et al. (1984).

Vacancy-ordering in anion-deficient $LaNiO_3$ was investigated some time ago by Gai & Rao (1975a). Thermogravimetry in air and oxygen provided evidence for the formation of a series of phases of the general formula $La_nNi_nO_{3n-1}$ with $n = 7, 9, 13$ and 30. An electron diffraction study of the anion-deficient samples suggested the formation of superstructures due to vacancy-ordering. An orthorhombic or body-centred tetragonal cell with $a \simeq c \simeq 2\sqrt{2}a_c$ and $b \simeq 2a_c$ was identified by electron diffraction. Vidyasagar et al. (1985) have examined vacancy-ordering in $La_2Ni_2O_5$ and $La_2Co_2O_5$ by electron diffraction. While $La_2Co_2O_5$ belongs to the brownmillerite structure type, $La_2Ni_2O_5$ adopts a new vacancy-ordered tetragonal superstructure of the perovskite type with $a \simeq c \simeq 2a_c$ involving Ni^{2+} in both octahedral and square-planar coordination. There are indications (from HREM) for the presence of cs planes in slightly reduced $LaNiO_3$.

Stability of $LaBO_3$ (B = V, Cr, Mn, Fe, Co or Ni) perovskites at 1270 K in a reducing atmosphere has been investigated by Nakamura et al. (1979) by thermogravimetry. The results reveal that the stability depends on the B atom, varying in the order $LaVO_3 \sim LaCrO_3 > LaFeO_3 > LaMnO_3 > LaCoO_3 > LaNiO_3$. Under the experimental conditions employed in this study, the $LaBO_{3-x}$ type of oxides is not obtained.

Grenier et al. (1981) have investigated $Ca_xLa_{1-x}FeO_{3-y}$ (compositions intermediate between $Ca_2Fe_2O_5$ and $LaFeO_3$) and $CaTi_{1-x}Fe_xO_{3-y}$ (compositions intermediate between $Ca_2Fe_2O_5$ and $CaTiO_3$) systems in an attempt to characterize anion-vacancy ordering at compositions intermediate between brownmillerite and perovskite. Their work has revealed the existence of a homologous series of phases of the general formula $A_nB_nO_{3n-1}$ (A = La,Ca;B = Ti,Fe). The structure of these phases consists of a succession of $(n-1)$ perovskite-like BO_6 sheets (O layers) and a sheet of MO_4 tetrahedra (T layers) in the c-direction. When $n = 2$ we have the brownmillerite phase, where the sequence of octahedral and tetrahedral layers ... OTOT′ OTOT′ ... with T and T′ standing for two different orientations of tetrahedra in the structure (Fig. 5.25). When $n = \infty$, there will be no tetrahedral layers and the sequence is ... OOOO ... corresponding to the perovskite structure. Other members of this series are $Ca_2LaFe_3O_8$, $Ca_3Fe_2TiO_8$ ($n = 3$) and $Ca_4Fe_2Ti_2O_{11}$ ($n = 4$). The layer sequences in the $n = 3$ and $n = 4$ members are ... OOT OOT ... and ... OOOT OOOT ... respectively. An interesting member of this series is $Ca_4YFe_5O_{13}$, which corresponds to

$n = 2.5$ in the series. The structure of this oxide, determined by high-resolution (1 MeV) electron microscopy (Bando *et al.*, 1981), consists of a regular intergrowth of $n = 2$ and $n = 3$ units of the $A_nB_nO_{3n-1}$ series, the polyhedral layer sequence being ... OTOOTOTOOT .... Other ordered intergrowth phases reported are $Ca_7Fe_6TiO_{18}$ ($n = 2.33$) and $Ca_5Fe_4TiO_{13}$ ($n = 2.5$).

Studies (Alario-Franco *et al.*, 1983; Calbet *et al.*, 1983) have shown the non-stoichiometric behaviour in such Fe containing systems to be highly complex, the type of long-range ordering (or the absence of it) being dependent on sample preparation conditions. Intermediate compositions in the $Ca_xLa_{1-x}FeO_{3-y}(\frac{2}{3} < x < 1)$ system exhibit two different types of behaviour. When the preparations are carried out in a nonoxidizing atmosphere at relatively low temperatures (1370 K), the material exhibits a disordered intergrowth structure comprising fragments of $Ca_2LaFe_3O_8$ ($n = 3$) and $Ca_2Fe_2O_5$. When the samples are prepared at 1670 K in air, they become oxygen-rich and exhibit a totally different structural behaviour. Power X-ray diffraction of the latter samples shows a deceptively simple cubic perovskite structure with $a \sim 3.85$ Å. Electron diffraction and HREM reveal the structure to be more complex, consisting of microdomains of brownmillerite or $Ca_2LaFe_3O_8$, which are intergrown three-dimensionally to give an arrangement that appears cubic perovskite-like on the average. The dimension of individual domains is of the order $\sim 10^6$ Å^3 and decreases with increasing oxygen content. It is significant that, even at the highest temperatures of preparation, the system does not exhibit classical nonstoichiometric behaviour consisting of randomly distributed point defects; apparent point defects are always structurally ordered, giving microdomain texture of the material.

Anion-deficient hexagonal perovskites exhibit interesting structural behaviour involving polytypism. $BaMO_{3-x}(M = Mn, Fe, Co$ etc.) are prototype examples of this type of nonstoichiometric system. The nonstoichiometry is accommodated in these systems by adjusting the relative proportion of cubic and hexagonal stacking of AO_{3-x} layers. HREM provides direct verification of the stacking arrangement of the BaO_3 layers in $BaMO_3$ polytypes (Gai & Rao, 1975*b*).

The results of Negas & Roth (1971) on the $BaMnO_{3-x}$ system clearly illustrate the structural complexity of the system. With decreasing oxygen content, cubic stacking increases in the structure. The structural changes accompanying oxygen non-stoichiometry are, however, not true polytypic transformations. The $BaFeO_{3-x}$ system (Mori, 1970; Zanne & Gleitzer, 1971) is equally complex and shows a different structural behaviour: $BaFeO_{2.5}$ (triclinic, brownmillerite type), $BaFeO_{2.62-2.64}$(rhombohedral), $BaFeO_{2.67}$(triclinic), $BaFeO_{2.75-2.81}$(tetragonal) and $BaFeO_{2.69-2.95}$(6H hexagonal). A 12H phase has also been described; its composition is not known definitely. A structural investigation (Jacobson, 1976) of 6H $BaFeO_{2.79}$ shows that oxygen vacancies are unevenly distributed between the cubic and hexagonal layers, which are arranged in the sequence $(cch)_2$. The hexagonal layers have the composition $BaO_{2.5}$ and the cubic layers $BaO_{2.835}$. The oxygen vacancy-ordering in $BaO_{2.5}$ layers appears to be similar to the ordering in $SrO_{2.5}$ layers of $SrFeO_{2.5}$ where one [110] string of oxygen in every four is removed, converting part of the face-shared FeO_6 oxtahedra into corner-shared FeO_4 tetrahedra. The structure of 6H$BaFeO_{2.79}$ is to be contrasted with that of the manganese system. In 4H $Ba_{0.5}Sr_{0.5}MnO_{2.85}$, for instance, all the vacancies are found in the hexagonal layers, converting pairs of face-sharing octahedra in the parent 4H structure into edge-sharing trigonal bipyramids (Hutchison & Jacobson, 1977). The difference in structural behaviour between the iron and manganese systems has been attributed to the difference in coordination tendencies of Fe^{3+} and Mn^{3+}; high-spin Mn^{3+} (d^4) prefers a trigonal bipyramidal or square-pyramidal environment, while high-spin Fe^{3+} (d^5) accepts both octahedral and tetrahedral coordinations.

Structures of the members of the $BaCoO_{3-x}$ systems are interesting. Zanne et al. (1972) found the following phases in this system at 1178 K: $2HBaCoO_{3.0-2.85}$, $7HBaCoO_{2.575-2.520}$, $12HBaCoO_{2.49-2.43}$, $15HBaCoO_{2.23-2.10}$ and an orthorhombic $BaCoO_{2.07}$. Among the various phases reported, a 12H phase is well established in this system. The structure of $12HBaCoO_{2.6}$ has been determined by HREM and powder neutron diffraction (Jacobson & Hutchison, 1980). The layer sequence of this phase is $(ccchhh)_2$. Oxygen vacancies are nonrandom in this structure, being accommodated by replacing one-sixth of BaO_3 layers by BaO_2 layers. As a consequence, one-third of the cobalt atoms are tetrahedrally coordinated by oxygen. The CoO_4 tetrahedra are linked through corners to strings of four face-shared octahedra that contain the remaining cobalt. Assuming that tetrahedral-site cobalts are $4+$ and octahedral-site cobalts are $3+$, one arrives at the ideal composition of the 12H phase as $BaCoO_{2.67}$. Replacement of BaO_3 by BaO_2 layers leading to tetrahedral coordination of B metal atoms is a novel way of accommodating oxygen deficiency in ABO_3 perovskites, which may be prevalent in other systems as well.

An interesting case (Er-Rakho et al., 1981) of a ABO_{3-x} perovskite with high oxygen deficiency ($x > 0.5$), where the B metal is present in three different coordination environments, is $La_3Ba_3Cu_6O_{18-y}$. Phases with varying y adopt a tetragonal cell related to the perovskite structure $a_T \simeq \sqrt{2}a_c$ and $c_T \simeq 3a_c$, where a_c is the cubic perovskite parameter. The detailed structure of $La_3Ba_3Cu_6O_{14+x}$ has been determined (Torardi et al., 1987; David et al., 1987). The structure is closely related to that of the superconducting copper oxide $YBa_2Cu_3O_7$, consisting of square-pyramidally coordinated copper (Cu–O sheets) and square-planar copper (Cu–O chains). Excess oxygen in the lanthanum compound is located interstitially converting partially the chain copper to square-pyramidal and/or octahedral coordination.

The discovery of superconductivity in cuprates has led to the synthesis of several new anion-deficient perovskite oxides (Anderson et al., 1993). These are $LaSrCuGaO_5$, $LaSrCuAlO_5$, $YBaCuFeO_5$, $LaBa_2Cu_2TaO_8$, $LnSr_2Cu_2GaO_7$ (Ln = rare earth), and $Ba_2La_2Cu_2Sn_2O_{11}$, besides $YBa_2Cu_3O_7$. Among these, $LaSrCuGaO_5$ is isostructural with brownmillerite, while $LaSrCuAlO_5$ has a slightly different orthorhombic structure $(2a_p \times 2\sqrt{2}a_p \times \sqrt{2}a_p$, a_p = the cell edge of cubic perovskite), where the CuAl-CuAl coordination polyhedral sequence is STST parallel to the a axis (S = square pyramidal). $YBaCuFeO_5$ and its cobalt analogue adopt a tetragonal ($a_p \times a_p \times 2a_p$) perovskite superstructure where both Cu(II) and Fe(III) have a square-pyramidal oxygen coordination. $LaBa_2Cu_2TaO_8$ crystallizing in a tetragonal ($a_p \times a_p \times 3a_p$) cell is related to the $YBa_2Cu_3O_7$ structure, which has an orthorhombic $a_p \times a_p \times 3a_p$ cell. The difference between the two structures is that octahdrally coordinated Ta(V) replaces the planar coordinated Cu. Accordingly, the CuTaCu polyhedral sequence in $LaBa_2Cu_2TaO_8$ is SOS, whereas the CuCuCu polyhedral sequence in $YBa_2Cu_3O_7$ is SPS, parallel to the c-axis (P = planar coordination). The structure of $LnSr_2Cu_2GaO_7$ is different, having an orthorhombic $6a_p \times \sqrt{2}a_p \times \sqrt{2}a_p$ cell where the CuGaCu-CuGaCu polyhdra are STSSTS parallel to the a-axis. $Ba_2La_2Cu_2Sn_2O_{11}$ and its titanium analogue crystallize in a $\sqrt{2}a_p \times \sqrt{2}a_p \times 4a_p$ tetragonal structure where CuSnSnCu coordination polyhedra are SOOS parallel to the c-axis. The structure of the tin compound is in fact more complex due to a structural modulation. The superstructure is different from that of $Ca_4Fe_2Ti_2O_{11}$ where the TiFeTiFe polyhedra are OTOO, parallel to the unique axis. Oxygen-vacancy ordering is also possible in layered perovskites. For example, a brownmillerite-type ordering is reported for $CsCa_2Nb_2AlO_9$ (Uma & Gopalakrishnan, 1994).

Anion-excess nonstoichiometry.   Anion-excess nonstoichiometry in perovskite oxides is not as common as anion-deficiency nonstoichiometry, probably because introduction of interstitial oxygen in perovskite structures is energetically unfavourable. There are a few systems which show apparent oxygen-excess nonstoichiometry: $LaMnO_{3+x}$, $Ba_{1-x}La_xTiO_{3+x/2}$ and $EuTiO_{3+x}$. Tofield & Scott (1974) investigated oxidative nonstoichiometry in $LaMnO_{3+x}$ by neutron diffraction. These workers suggest three possible models to accommodate oxygen-excess in the perovskite structure: (i) interstitial oxygen at $(\frac{1}{2}00)$ or $(\frac{111}{444})$ positions of the cubic perovskite structure, (ii) cation vacancies at A and/or B sites, leaving a perfect oxygen sublattice and (iii) formation of new oxidized phases. In $LaMnO_{3.12}$, diffraction results show that oxygen-excess is accommodated by vacancies at A and B sites with partial elimination of La(as La_2O_3), the composition of the perovskite being $La_{0.94}\square_{0.06}Mn_{0.98}\square_{0.02}O_3$. Recent studies (Hervieu *et al.*, 1995) have shown that Mn^{4+} in $LaMnO_3$ is created by cation vacancies in both the A and B sites, present randomly.

Where there is a large oxygen-excess, new phases possessing perovskite-related structures are formed. One of the ways of accommodating oxygen-excess, while retaining the features of the perovskite structure, is to slice the cubic perovskite parallel to (110) to give slabs of composition $(A_{n-1}B_nO_{3n+2})_\infty$ and stack them one over the other by adding a layer of A atoms in between. The process gives rise to a homologous series of the general formula $A_nB_nO_{3n+2}$. Important examples of this family are $Ca_2Nb_2O_7$ and $La_2Ti_2O_7$, which are $n = 4$ members of the series.

Nonstoichiometry in oxides of structures related to perovskites.   The first family of oxides that we consider here will be that related to the K_2NiF_4 structure. These oxides of the general formula A_2BO_4 contain alternating ABO_3 perovskite layers and AO rocksalt layers. Structural chemistry of these oxides has been reviewed recently (Ganguly & Rao, 1984). Oxygen-excess nonstoichiometry is found in La_2NiO_4. Oxygen-deficient nonstoichiometry is found in Ca_2MnO_4, which can be reduced topotactically to $Ca_2MnO_{3.5}$ (Poeppelmeier *et al.*, 1982). The unit cell dimensions are $a = 5.30$, $b = 10.05$ and $c = 12.24\,Å$, the relationship with the tetragonal K_2NiF_4 structure being $a \simeq \sqrt{a_T}$, $b \simeq 2\sqrt{2}a_T$ and $c \simeq c_T$. The anion vacancies are ordered along alternate $[110]_c$ rows as in $CaMnO_{2.5}$ to give sheets of (MnO_5) square-pyramids. Vidyasagar *et al.* (1984) have characterized an oxide of the formula $Ca_2FeO_{3.5}$ with $a = 14.79$, $b = 13.71$ and $c = 12.19\,Å$. The structure of this oxide is different from that of $Ca_2MnO_{3.5}$. It is likely that the ordering of anion vacancies in $Ca_2FeO_{3.5}$ is similar to that in the brownmillerite structure, consisting of alternating rows of tetrahedra and octahedra in the perovskite-like layers of the K_2NiF_4 structure.

Oxide pyrochlores, $A_2B_2O_6O'$, can tolerate vacancies at the A and O' sites, giving phases $A_2B_2O_6$ $\square(ABO_3)$ and A $\square B_2O_6$ $\square(AB_2O_6)$; a small number of $A_2B_2O_6$ compounds adopt the pyrochlore structure in preference to the perovskite structure when the A and B ions are highly polarizable but not too electropositive (Subramanian *et al.*, 1983). Typical of these oxides are $Tl_2B_2O_6(B = Nb, Ta\ or\ U)$ and $A_2Sb_2O_6$. In the A–Sb–O system (A = Na, K or Ag), a series of nonstoichiometric compounds, ASb_yO_z, the maximum ratio of A:Sb depending on A, is known. Some of these compounds are superionic conductors.

References

Alario-Franco, M. A. & Regi, M. V. (1977) *Nature* **270**, 706.

Alario-Franco, M. A., Joubert, J. C. & Levy, J. P. (1982) *Mater. Res. Bull.* **17**, 733.

Alario-Franco, M. A., Calbet, J. M. G., Regi, M. V. & Grenier, J. C. (1983) *J. Solid State Chem.* **49**, 219.

Allpress, J. G. & Sanders, J. V. (1973) *J. Appl. Cryst.* **6**, 165.

Anderson, J. S. (1972a) in *Surface and Defect Properties of Solids* (eds Roberts, M. W. & Thomas, J. M.) Vol. 1, The Chemical Society, London.

Anderson, J. S. (1972b) in *Solid State Chemistry*: NBS Special Publication 364 (eds Roth, R. S. & Schneider, S. J.) US Department of Commerce, Washington D.C.

Anderson, J. S. (1973) *J. Chem. Soc. Dalton Trans.* 1107.

Anderson, J. S. (1974) in *Defect and Transport in Oxides* (eds Seltzer, M. S. & Jaffe, R. I.) Plenum Press, New York.

Anderson, J. S. (1984) *Proc. Indian Acad. Sci. (Chem. Sci.)* **93**, 861.

Anderson, J. S., Browne, J. M. & Hutchison, J. L. (1972) *Nature* **237**, 5351.

Anderson, J. S. & Tilley, R. J. D. (1974) in *Surface and Defect Properties of Solid* (eds Roberts, M. W. & Thomas, J. M.) Vol. 3, The Chemical Society, London.

Anderson, J. S. & Hutchison, J. L. (1975) *Contemp. Phys.* **16**, 443.

Anderson, M. T., Vaughey, J. T. & Poeppelmeier, K. R. (1993) *Chem. Mater.* **5**, 151.

Andersson, B. & Gjonnes, J. (1970) *Acta Chem. Scand.* **24**, 2250.

Atoji, M. & Rundle, R. E. (1960) *J. Chem. Phys.* **32**, 627.

Aurivillius, B. (1950) *Ark. Kemi* **2**, 519.

Bando, Y., Sekikawa, Y., Nakamura, H. & Matsui, Y. (1981) *Acta Crystallogr.* A**37**, 723.

Bauer, E., Pianelli, A., Aubry, A. & Jeannot, F. (1980) *Mater. Res. Bull.* **15**, 323.

Bertaut, E. F. (1956) *Acta Crystallogr.* **6**, 357.

Bevan, D. J. M., Strahl, J. & Greis, O. (1982) *J. Solid State Chem.* **44**, 75.

Blanchin, M. G., Bursill, L. A. & Smith, D. J. (1984) *Proc. Roy. Soc. London* **A391**, 351.

Blasse, G. & Dirksen, G. J. (1981) *J. Solid State Chem.* **36**, 124.

Bridges, F. (1975) *CRC Crit. Rev. Solid State Science* **5**, 1.

Brown, R. & Kemmler-Sack, S. (1983) *Naturwissenschaften* **70**, 463.

Burdett, J. K. & Mitchell, J. F. (1993) *Chem. Mater.* **5**, 1465.

Bursill, L. A., Blanchin, M. G. & Smith, D. J. (1984) *Proc. Roy. Soc. London* **A391**, 373.

Bursill, L. A. & Smith, D. J. (1984) *Nature* **309**, 319.

Caignaert, V., Nguyen, N., Hervieu, M. & Raveau, B. (1985) *Mater. Res. Bull.* **20**, 479.

Calbet, J. M. G., Regi, M. V., Alario-Franco, M. A. & Grenier, J. C. (1983) *Mater. Res. Bull.* **18**, 285.

Calvo, C., Ng, N. N. & Chamberland, B. L. (1978) *Inorg. Chem.* **17**, 699.

Catlow, C. R. A. & James, R. (1980) in *Chemical Physics of Solids and Surfaces* (Sr. Reporters Roberts, M. W. & Thomas, J. M.) Vol. 8, The Chemical Society, London.

Catlow, C. R. A. & Mackrodt, W. C. (eds) (1982) *Computer Simulation of Solids*, Lecture Notes in Physics, Springer-Verlag, Berlin.

Catlow, C. R. A., Chadwick, A. V. & Corish, J. (1983) *J. Solid State Chem.* **48**, 65.

Catlow, C. R. A., Chadwick, A. V., Greaves, G. N. & Moroney, L. M. (1984) *Nature* **312**, 601.

Cheetham, A. K., Fender, B. E. F. & Taylor, R. I. (1971a) *J. Phys. C* **4**, 2160.

Cheetham, A. K., Fender, B. E. F. & Cooper, M. J. (1971b) *J. Phys. C* **4**, 3107.

Conrad, B. R. & Franzen, H. F. (1970) in *The Chemistry of Extended Defects in Nonmetallic Solids* (eds O'Keeffe, M. & Eyring, L.) North-Holland, Amsterdam.

Corish, J., Jacobs, P. W. M. & Radhakrishna, S. (1977) in *Surface and Defect Properties of Solids* (Sr. Reporters Roberts, M. W. & Thomas, J. M.) Vol. 6, The Chemical Society, London.

Corish, J. & Quigley, J. M. (1982) in *Computer Simulation of Solids* (eds Catlow, C. R. A & Mackrodt, W. C.) Springer-Verlag, Berlin.

David, W. I. F., Harrison, W. T. A., Ibberson, R. M., Weller, M. T., Grasmeder, J. R. & Lanchester, P. (1987) *Nature* **328**, 328.

Dismukes, J. P. & White, J. G. (1964) *Inorg. Chem.* **3**, 1220.

Ekström, T. & Tilley, R. J. D. (1980) *Chem. Scripta* **16**, 1.

Emmons, G. H. & Williams, W. S. (1983) *J. Mater. Sci.* **18**, 2589.

Er-Rakho, L., Michel, C., Provost, J. & Raveau, B. (1981) *J. Solid State Chem.* **37**, 151.

Fine, M. E. (1973) in *Treatise on Solid State Chemistry* (ed. Hannay, N. B.) Vol. 1, Plenum Press, New York.

Fowler, W. B. (1968) *Physics of Colour Centres*, Academic Press, New York.

Fowler, W. B. (1975) in *Treatise on Solid State Chemistry* (ed. Hannay, N. B.) Vol. 2, Plenum Press, New York.

Franzen, H. F., Tuenge, R. T. & Eyring, L. (1983) *J. Solid State Chem.* **49**, 206.

Franzen, H. F. & Folmer, J. C. W. (1984) *J. Solid State Chem.* **51**, 396.

Gai, P. L. & Rao, C. N. R. (1975a) *Z. Naturforsch.* **30A**, 1092.

Gai, P. L. & Rao, C. N. R. (1975b) *Pramana* **55**, 274.

Ganguly, P. & Rao, C. N. R. (1984) *J. Solid State Chem.* **53**, 193.

Gavarri, J. (1978) Thesis, Univ. Paris VI.

Gettmann, W. & Greis, O. (1978) *J. Solid State Chem.* **26**, 255,

Gillan, M. J. & Jacobs, P. W. M. (1983) *Phys. Rev.* **B28**, 758.

Gleiter, H. (1983) in *Physical Metallurgy*, 3rd edn (eds Cahn, R. W. & Haasen, P.) North-Holland, Amsterdam.

Gopalakrishnan, J., Ramanan, A., Rao, C. N. R., Jefferson, D. A. & Smith, D. J. (1984) *J. Solid State Chem.* **55**, 101.

Greis, O. (1977) *Z. Anorg. Allgem. Chem.* **430**, 175.

Grenier, J. C., Ghodbane, S. Demazeau, G., Pouchard, M. & Hagenmuller, P. (1979) *Mater. Res. Bull.* **14**, 831.

Grenier, J. C., Pouchard, M. & Hagenmuller, P. (1981) *Structure and Bonding* **47**, 1.

Grimes, N. W. (1976) *Contemp. Phys.* **17**, 71.

Gruehn, R. & Mertin, M. (1980) *Angew. Chem. Int. Edn. (Engl.)* **19**, 505.

Hervieu, M., Mahesh, R., Rangavittal, N. & Rao, C. N. R. (1995), *Euro. J. Solid State Inorg. Chem.* **32**, 79.

Hutchison, J. L. & Jacobson, A. J. (1977) *J. Solid State Chem.* **20**, 417.

Hutchison, J. L., Anderson, J. S. & Rao, C. N. R. (1977) *Proc. Roy. Soc. London* **A355**, 301.

Iijima, S. (1975) *J. Solid State Chem.* **14**, 52.

Jacobson, A. J. (1976) *Acta Crystallogr.* **B32**, 1087.

Jacobson, A. J. & Hutchison, J. L. (1980) *J. Solid State Chem.* **35**, 334.

Jacobson, A. J., Scanlon, J. C., Poeppelmeier, K. R., Longo, J. M. & Cox, D. E. (1981) *Mater. Res. Bull.* **16**, 359.

Jefferson, D. A., Uppal, M. K., Rao, C. N. R. & Smith, D. J. (1984) *Mater. Res. Bull.* **19**, 1403.

Jones, W. & Thomas, J. M. (1979) *Prog. Solid State Chem.* **12**, 101.

Kato, K. & Kitamura, M. (1981) *Acta Crystallogr.* **A37**, 307.

Kihlborg, L. (1978) *Chem. Scripta* **14**, 187.

Kittel, C. (1978) *Solid State Comm.* **25**, 519.

Koch, F. & Cohen, J. B. (1969) *Acta Crystallogr.* **B25**, 275.

Kröger, F. A. (1974) *The Chemistry of Imperfect Crystals*, Vol. 2, North-Holland, Amsterdam.

Laval, P. L. & Frit, B. (1983) *J. Solid State Chem.* **49**, 237.

Libowitz, G. G. (1973) in *Treatise on Solid State Chemistry* (ed. Hannay, N. B.), Vol. 1, Plenum Press, New York.

Loehman, R., Rao, C. N. R. & Honig, J. M. (1968) *J. Phys. Chem.* **73**, 1781.

Longo, J. M. & Sleight, A. W. (1975) *Mater. Res. Bull.* **10**, 1273.

MacChesney, J. B., Sherwood, R. C. & Potter, J. F. (1965) *J. Chem. Phys.* **43**, 1907.

McConnell, A., Anderson, J. S. & Rao, C. N. R. (1976) *Spectrochim. Acta* **32A**, 1067.

Moret, R., Huber, M. & Comes, R. (1977) *J. Phys. Colloq.* **C7**, 202.

Mori, S. (1970) *J. Phys. Soc. Japan* **28**, 44.

Morinaga, M. & Cohen, J. B. (1979) *Acta Crystallogr.* **A35**, 745, 975.

Mott, N. F. & Littleton, M. J. (1938) *Trans. Faraday Soc.* **34**, 485.

Nakamura, T., Petzow, G. & Gauckler, L. J. (1979) *Mater. Res. Bull.* **14**, 649.

Nakazawa, H. & Morimoto, N. (1971) *Mater. Res. Bull.* **6**, 345.

Negas, T. & Roth, R. S. (1971) *J. Solid State Chem.* **3**, 323.

Nembach, E. (1974) in *Treatise on Solid State Chemistry* (ed. Hannay, N. B.), Vol. 2, Plenum Press, New York.

Poeppelmeier, K. R., Leonowicz, M. E., Scanlon, J. C., Longo, J. M. & Yelon, W. B. (1982) *J. Solid State Chem.* **45**, 71.

Portier, R., Carpy, A., Fayard, M. & Galy, J. (1975) *Phys. Stat. Solidi* **A30**, 683.

Ramanan, A., Gopalakrishnan, J., Uppal, M. K., Jefferson, D. A. & Rao, C. N. R. (1984) *Proc. Roy. Soc. London.* **A395**, 127.

Rao, C. N. R. (1985), *Bull. Mat. Sci.* **7**, 155.

Rao, C. N. R., Gai, P. L. & Ramasesha, S. (1976) *Phil. Mag.* **33**, 387.

Rao, C. N. R. & Thomas, J. M. (1985) *Acc. Chem. Res.* **18**, 113.

Rauser, G. & Kemmler-Sack, S. (1980) *J. Solid State Chem.* **33**, 135.

Reller, A., Thomas, J. M., Jefferson, D. A. & Uppal, M. K. (1984) *Proc. Roy. Soc. London* **A394**, 223.

Roth, R. S. & Stephenson, N. C. (1970) in *The Chemistry of Extended Defects in Nonmetallic Solids* (eds Eyring, L. & O'Keeffe, M.) North-Holland, Amsterdam.

Roth, W. L. (1960) *Acta Crystallogr.* **13**, 140.

Shannon, J. & Katz, L. (1970) *Acta Crystallogr.* **B26**, 102.

Sørensen, O. T. (1981) *Nonstoichiometric Oxides*, Academic Press, New York.

Stoneham, A. M. (1971) *J. Phys.* **F1**, 778.

Stoneham, A. M. (1975) *Theory of Defects in Solids*, Clarendon Press, Oxford.

Studer, F. & Raveau, B. (1978) *Phys. Stat. Solidi* **A48**, 301.

Subramanian, M. A., Aravamudan, G. & Subba Rao, G. V. (1983) *Prog. Solid State Chem.* **15**, 55.

Suzuki, K. (1961) *J. Phys. Soc. Japan* **16**, 67.

Taylor, A. & Doyle, N. J. (1972) *J. Appl. Cryst.* **5**, 201.

Terasaki, O., Thomas, J. M. & Ramdas, S. (1984) *J. Chem. Soc. Chem. Comm.* 216.

Terauchi, H., Cohen, J. B. & Reed, T. B. (1978) *Acta Crystallogr.* **A34**, 556.

Terauchi, H. & Cohen, J. B. (1979) *Acta Crystallogr.* **A35**, 646.

Tewary, U. K. & Bullough, R. (1972) *J. Phys.* **F2**, 269.

Thomas, J. M. & Williams, J. O. (1971) *Prog. Solid State Chem.* **6**, 271.

Thornber, M. R. & Bevan, D. J. M. (1970) *J. Solid State Chem.* **1**, 536.

Tilley, R. J. D. (1977) *J. Solid. State Chem.* **21**, 293.

Tilley, R. J. D. (1980) in *Chemical Physics of Solids and Surfaces* (Sr. Reporters Roberts, M. W. & Thomas, J. M.), Vol. 8, The Chemical Society, London.

Tofield, B. C. (1978) *Nature* **272**, 713.

Tofield, B. C. & Scott, W. R. (1974) *J. Solid State Chem.* **10**, 183.

Tofield, B. C., Greaves, C. & Fender, B. E. F. (1975) *Mater. Res. Bull.* **10**, 737.

Torardi, C. C., McCarron, E. M., Subramanian, M. A., Horowitz, H. S., Michel, J. B., Sleight, A. W. & Cox, D. E. (1987) in *Chemistry of High Temperature Superconductors* (eds Nelson, D. L., Whittingham, M. S. & George, T. F.), American Chemical Society, Washington, D.C.

Uma, S. & Gopalakrishnan, J. (1994) *Chem. Mater* **6**, 907.

Uppal, M. K., Rao, C. N. R. & Sangster, M. J. L. (1978) *Phil. Mag.* **38**, 341.

Vidyasagar, K., Gopalakrishnan, J. & Rao, C. N. R. (1984) *Inorg. Chem.* **23**, 1206.

Vidyasagar, K., Reller, A., Gopalakrishnan, J. & Rao, C. N. R. (1985) *J. Chem. Soc. Chem. Comm.* 7.

Vidyasagar, K., Ganapathi, L., Gopalakrishnan, J. & Rao, C. N. R. (1986) *J. Chem. Soc. Chem. Comm.* 449.

Vincent, H., Bochu, B., Aubert, J. J., Joubert, C. C. & Marezio, M. (1978) *J. Solid State Chem.* **24**, 245.

Wadsley, A. D. & Andersson, S. (1970) in *Perspectives in Structural Chemistry* (eds Dunitz, J. D. & Ibers, J. A.), Vol. 3, John Wiley, New York.

Watanabe, D., Terasaki, O., Jostsons, A. & Castles, J. R. (1970) in *The Chemistry of Extended Defects in Nonmetallic Solids* (eds Eyring, L. & O'Keeffe, M.), North-Holland, Amsterdam.

Willis, B. T. M. (1964) *J. Physique* **25**, 431.

Wollenberger, H. J. (1983) in *Physical Metallurgy*, 3rd edn (eds Cahn, R. W. & Haasen, P.), North-Holland, Amsterdam.

Zanne, M. & Gleitzer, C. (1971) *Bull. Soc. Chim. Fr.* 1567.

Zanne, M., Courtois, A. & Gleitzer, C. (1972) *Bull. Soc. Chim. Fr.* 4470.

6 Structure–property relations

6.1 Introduction

Relating properties of substances to their structures has been a major objective of modern chemistry and this also happens to be a prime concern of solid state chemists. Some of the aspects of importance in solid state chemistry are electronic, magnetic, superconducting, dielectric and optical properties. We shall briefly discuss these properties and present highlights in the solid state chemistry of some interesting classes of materials. An important class of materials is that of *ferroics*, which possess several orientation states that can be switched from one to another by the application of an appropriate force; ferroelectric materials form a subgroup of this class of materials. Other classes of materials discussed are amorphous solids, mixed-valence compounds, low-dimensional solids and liquid crystals which are of considerable importance. We have also devoted attention to different types of metal–nonmetal transitions and have briefly examined the question, 'what makes a metal?'. While a detailed discussion of the theory of electronic structure would be outside the scope of the book, we have presented the necessary background material at an elementary level and discussed some of the typical results obtained from empirical theory as well as electron spectroscopy.

6.2 Electrons in solids

In order to correlate the structure and physical properties of solids, it is essential to have a description of valence electrons that bind the atoms in the solid state. Two limiting descriptions of atomic outer electrons in solids are available: the band theory and the localized-electron theory or the *ligand-field theory*. When there is appreciable overlap between orbitals of neighbouring atoms, the *band theory* of Bloch and Wilson is applicable. The theory assumes that valence electrons are shared equally by all the like atoms in the solid. In this model, the energy U required to transfer a valence electron from one site to an occupied orbital on another equivalent site is small as compared to the band-width, $W(U \ll W)$; in the extreme case, $U = 0$. The localized-electron theory is applicable when interatomic interactions are weak and electrons are tightly bound to the atomic core. Localized electrons are characterized by a large U (of the order of the atomic ionization potential minus the electron affinity in the zero-order approximation) and small $W(U \gg W)$. When $U \sim W$, we have the third possibility of *strongly correlated electrons* in solids. Outer s and p electrons which are weakly bound to the atomic core and interacting strongly with the neighbouring atoms are well described by the band model. 4*f* electrons of the rare earths which are screened from the neighbouring atoms by the $5s^2 5p^6$ electrons are always localized in solids and are well described by the ligand-field theory. It is to be pointed out that the 4*f* electrons are localized even in rare-earth metals, indicating that electrical conduction is not always an indication of electron delocalization. Outer *d* electrons of transition-metal compounds are intermediate; in the inorganic complexes of transition metals, the *d* electrons are localized, being described by ligand-field theory. In transition-metal oxides and related materials, we come across both localized and itinerant electron behaviour. In transition metals, the *d* electrons are itinerant, being in narrow *d* bands that overlap the broad *s–p* bands.

6.2.1   *Band model*

Band theory is a one-electron, independent particle theory, which assumes that the electrons are distributed amongst a set of available stationary states following the Fermi–Dirac statistics. The states are given by solutions of the Schrödinger equation

$$\mathbf{H}\psi_{nk} = \mathbf{E}\psi_{nk} \tag{6.1}$$

where the Hamiltonian operator **H** includes not only the kinetic energy term, $(p^2/2m)$, but also a crystal potential $V(x)$, which accounts for the interaction of an electron with other particles in the crystal lattice. $V(x)$ has the translational periodicity of the crystal structure, viz. $V(x + a) = V(x)$ where *a* is the lattice constant. The wave functions (*Bloch functions*) in one dimension are therefore of the form $\exp(i\,k\,x)\,u_{nk}(x)$, where u_{nk} is periodic with the same periodicity as the crystal potential,

$$u_{nk}(x + a) = u_{nk}(x) \tag{6.2}$$

The wave function is modulated by $u_{nk}(x)$.

Each wave function is characterized by an integer, the band index *n* and the wave vector **k**. Unlike in the simple free-electron model, where the quadratic dependence of energy on wave number *k* is continuous, energy dependence on *k* according to the band model is discontinuous, showing breaks at *k* equal to π/a, $2\pi/a$, $3\pi/a$ etc. The dependence of energy eigenvalue E_{nk} on the crystal momentum $(\hbar k)$ for different values of *n* describes the electronic band structure of a crystalline solid. The discontinuities in energy that give rise to allowed (energy bands) and forbidden (band gaps) regions of energy arise from the periodic potential: at values of $k = \pm n\pi/a$, the electron wave is Bragg-reflected, since $k = \pm n\pi/a$ becomes equivalent to the Bragg condition, $2a \sin \theta = n\lambda$, because $k = 2\pi/\lambda$ and $\sin \theta = 1$. Thus at $k = \pi/a$, for instance, the solutions to equation (6.1) are $\exp[i(\pi/a)x] \pm \exp[-i(\pi/a)x]$, which correspond to two different allowed energies, giving rise to the energy gap. This range of *k* is known as the first *Brillouin zone*. At values of *k* far from $\pm n\pi/a$, the energy shows free-electron quadratic dependence on *k*. Simple band structures of inorganic solids are calculated by employing a linear combination of metal and anion orbitals (*tight binding scheme*) just as in MO calculations.

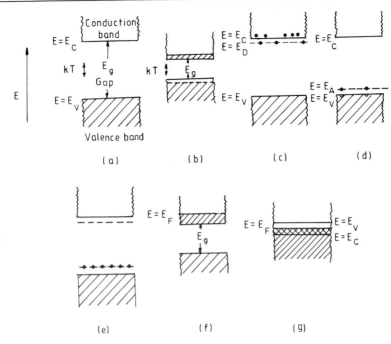

Figure 6.1   Schematic band structures of solids: (a) insulator ($kT \ll E_g$); (b) intrinsic semiconductor ($kT \sim E_g$); (c) and (d) extrinsic semiconductors; donor and acceptor levels in n-type and p-type semiconductors respectively are shown. (e) compensated semiconductor; (f) metal; (g) semimetal; top of the valence band lies above the bottom of the conduction band.

Application of the Fermi–Dirac distribution shows that all states lying lower in energy than the *Fermi energy*, E_F, are occupied by electrons, while those above E_F are unoccupied at 0K. In a metal E_F lies within a band (the highest occupied band is partly filled). In an *insulator* or a *semiconductor*, E_F lies between bands with an energy gap separating the highest lying-filled band (*valence band*) and the lowest lying-empty band (*conduction band*). Thus Bloch–Wilson band theory classifies crystalline solids into metals and insulators according to whether or not their electronic structures involve partly filled bands. Physical properties of solids are determined by the electron states close to the *Fermi surface*, which is the locus of k values corresponding to the boundary between the occupied and the empty levels. Energy band schemes for different types of solids are shown in Fig. 6.1; examples of electronic materials are given in Table 6.1.

The above simple picture of solids is not universally true because we have a class of crystalline solids, known as Mott insulators, whose electronic properties radically contradict the elementary band theory. Typical examples of Mott insulators are MnO, CoO and NiO, possessing the rocksalt structure. Here the only states in the vicinity of the Fermi level would be the $3d$ states. The cation d orbitals in the rocksalt structure would be split into t_{2g} and e_g sets by the octahedral crystal field of the anions. In the transition-metal monoxides, TiO–NiO ($3d^2$–$3d^8$), the d levels would be partly filled and hence the simple band theory predicts them to be metallic. The prediction is true in TiO

Table 6.1. *Different types of electronic materials*

Metallic materials	
3d compounds	TiO, CrO$_2$, TiS, CoS$_2$, CuS$_2$
4d compounds	NbO, MoO$_2$, RuO$_2$
5d compounds	ReO$_3$, A$_x$WO$_3$
Others	Doped polyacetylene, TTF–TCNQ
Semiconductors or insulators	
3d compounds	MnO, CoO, NiO, Fe$_2$O$_3$, Cr$_2$O$_3$
	MnS, MnS$_2$, FeS$_2$
4d and 5d compounds	MoO$_3$, WO$_3$, Nb$_2$O$_5$, Rh$_2$O$_3$
Others	Organic crystals e.g. anthracene
Materials showing nonmetal–metal transition[a]	
3d compounds	V$_2$O$_3$, VO$_2$, V$_3$O$_5$, V$_4$O$_7$, V$_6$O$_{13}$,
	Ti$_2$O$_3$, Ti$_3$O$_5$, NiS, CrS, FeS, NiS$_2$
4d compounds	NbO$_2$
4f compounds	EuO, SmS, SmTe
Others[b]	Doped semiconductors (e.g. P:Si), expanded metals (e.g. Hg, Cs at $T > 2200$ K), Na–NH$_3$, Xe–Na films
Materials showing superconducting transition	
	TiO, NbO, LiTi$_2$O$_4$, BaPb$_{0.8}$Bi$_{0.2}$O$_3$,
	NbS$_2$, Pb$_{0.9}$Mo$_6$S$_8$, LuRh$_4$B$_4$, YBa$_2$Cu$_3$O$_7$

[a] These are thermal transitions in most cases, excepting 4f compounds (where the transition occurs by application of pressure).
[b] Doped semiconductors, expanded metals, metal–ammonia solutions and rare gas–metal films where the transition occurs because of change of donor concentration or density.

and to some extent in VO. But stoichiometric MnO, CoO and NiO are all good insulators, showing antiferromagnetic ordering. The insulating nature of FeO may be understood if we assume that the t_{2g} sub-band is completely filled for the $3d^6$ configuration (assuming Fe^{2+} is in the low-spin state), but the insulating nature of MnO, CoO and NiO cannot be understood in terms of simple band theory. Similar problems arise with transition-metal oxides of other structures. Kleiner (1967) has shown by a symmetry analysis that the sesquioxides of corundum structure with 1, 3 and 5 d electrons per cation can in principle be insulators: the electrical properties of Ti$_2$O$_3$($3d^1$), Cr$_2$O$_3$ ($3d^3$) and Fe$_2$O$_3$ ($3d^5$) do conform to this prediction; Ti$_2$O$_3$, however, shows metallic conductivity above ~ 400 K. Also, according to this model, V$_2$O$_3$ ($3d^2$), Mn$_2$O$_3$ ($3d^4$) and Co$_2$O$_3$ ($3d^6$), if they existed in the corundum structure, would most likely be metallic. In reality, only V$_2$O$_3$ occurs in corundum structure above ~ 150 K and this phase is metallic.

By invoking antiferromagnetic ordering, it is possible to account for the insulating behaviour of some oxides at $T = 0K$; magnetic order leads to a doubling of the size of the primitive cell and consequently an exchange splitting of all the bands. This may be valid for NiO and MnO but not for CoO; to explain the insulating behaviour of this last oxide we have to assume that the crystallographic distortion to a low symmetry below T_N introduces an energy gap. The concept of *spin-density waves* is employed to describe magnetically ordered insulators. The behaviour of such monoxides at higher temperatures is not explained, since the compounds are predicted to show metallic conductivity at temperatures higher than T_N (MnO, CoO and NiO are semiconducting at all temperatures). Elementary band theory does not account for the properties of these oxides even qualitatively. The reason for this failure is the neglect of electron correlation arising from large U and small W (Hubbard, 1963). Band theory modified by considering electron correlations and electron–phonon interactions would predict a low-temperature insulating state of some of these oxides. Although simple LCAO approximation provides a simple method to calculate band structures of oxides and other solids, it becomes necessary to employ more sophisticated methods to fully understand their properties. Thus, the electronic structure of certain insulators has been treated within the framework of the Hartree–Fock approach (Brandow, 1977). Energy band calculations based on the Hartree–Fock formulation, Xα methods and density functional theories have been reviewed by Koelling (1981); for semiempirical methods see Messmer (1977).

Besides magnetic perturbations and electron–lattice interactions, there are other instabilities in solids which have to be considered. For example, one-dimensional solids cannot be metallic since a periodic lattice distortion (*Peierls distortion*) destroys the Fermi surface in such a system. The perturbation of the electron states results in charge-density waves (CDW), involving a periodicity in electron density in phase with the lattice distortion. Blue molybdenum bronzes, $K_{0.3}MoO_3$, show such features (see Section 4.9 for details). In two- or three-dimensional solids, however, one observes *Fermi surface nesting* due to the presence of parallel Fermi surface planes perturbed by periodic lattice distortions. Certain molybdenum bronzes exhibit this behaviour.

Electron correlation plays an important role in determining the electronic structures of many solids. Hubbard (1963) treated the correlation problem in terms of the parameter, U. Figure 6.2 shows how U varies with the band-width W, resulting in the overlap of the upper and lower Hubbard states (or in the disappearance of the band gap). In NiO, there is a splitting between the upper and lower Hubbard bands since $W < U$. Thus, the relative values of U and W determine the electronic structure of transition-metal compounds. Unfortunately, it is difficult to obtain reliable values of U. The Hubbard model takes into account only the d orbitals of the transition metal (single band model). One has to include the mixing of the oxygen p and metal d orbitals in a more realistic treatment. It would also be necessary to take into account the presence of mixed-valence of a metal (e.g. Cu^{2+}, Cu^{3+}).

The important feature of magnetic insulators is that, being nonmetallic, they have a band gap and possess unpaired electrons. They show crystal-field transitions due to the presence of open-shell (d^n) ions. Mott proposed that electron repulsion can be responsible for the breakdown of the normal band properties of transition-metal

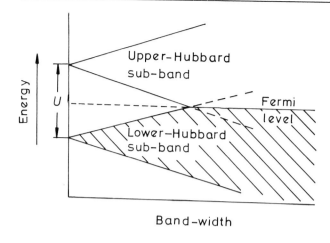

Figure 6.2   Variation of Hubbard U with band-width.

oxides and related materials. This model gives a simple interpretation of the properties of magnetic insulators (e.g. $3d$-transition-metal oxides with small band-widths) and of metallic systems (e.g. some of the metal sulphides). These ideas are incorporated in the Hubbard model. In the so-called *Mott-insulators*, the band gap involves the split Hubbard states. The realization that the top-filled band in many of the insulating transition-metal oxides has considerably more oxygen p character than the metal d character has given rise to the concept of *charge-transfer insulators*. Whether one should consider oxides like NiO to be Mott or charge-transfer insulators is a problem of interest. A transition between a magnetically insulating state to a metallic state (which occurs with increase in the W/U ratio) is called the *Mott transition* (Mott, 1956).

6.2.2   *Localized-electron model*

The localized-electron model or the ligand-field approach is essentially the same as the Heitler–London theory for the hydrogen molecule. The model assumes that a crystal is composed of an assembly of independent ions fixed at their lattice sites and that overlap of atomic orbitals is small. When interatomic interactions are weak, intraatomic exchange (Hund's rule splitting) and electron–phonon interactions favour the localized behaviour of electrons. This increases the relaxation time of a charge carrier from about 10^{-15} s in an ordinary metal to $\sim 10^{-12}$ s, which is the order of time required for a lattice vibration in a polar crystal.

Localized d or f electrons retain their one-atom manifolds, except that states arising from different d^n or f^n are split by crystal field and spin-orbit coupling. Multiplet splittings due to spin-orbit coupling are larger than crystal-field splittings of $4f^n$ levels; the converse is the case for $3d^n$ levels. The difference in energy between $d^n(f^n)$ and $d^{n+1}(f^{n+1})$ manifolds corresponds to the amount of free atom U that is decreased due to interatomic interaction in the solid.

The localized-electron model has the advantage that it can predict the insulating ground state of solids. At higher temperatures, however, electron–phonon and electron–electron interactions become important, especially when the bands are narrow, as in the case of d-bands. Much work has been carried out on the effect of electron–phonon interactions on the transport properties of oxide materials. The strength of the electron–phonon interaction (*polaron*) determines the *effective mass* of the electrons. If the interaction is sufficiently large, the charge carriers move along with the associated polarization (small polaron); the mobility would then be small and thermally activated. The problem of charge transport is treated in terms of the classical diffusion theory. In the presence of an electric field, preferential diffusion of electrons occurs through the crystal, giving rise to a net current. Such uncorrelated *hopping* of polarons is encountered in compounds having the same cation in more than one valence state (e.g. Fe_3O_4, Pr_7O_{12}, $Na_xV_2O_5$). If the electron–lattice interaction is small, the electron movement is only affected slightly and we then have large polarons.

6.2.3   Chemical bond approach

A semiempirical approach based on crystal chemistry has been developed to explain the electronic properties of transition-metal oxides and related materials (Goodenough, 1971, 1974). With empirically derived criteria for the overlap of cation–cation and cation–anion–cation orbitals, Goodenough has attempted to rationalize the nature of d-electrons in transition-metal compounds and provide a unified picture of the magnetic and electrical properties of a variety of transition-metal oxides, sulphides and related solids. This approach, although qualitative, appeals to chemical intuition and enables an understanding of the conditions that give rise to localized and collective d-electrons. Conceptual phase diagrams are constructed in terms of the transfer energy b_{ij}, defined as

$$b_{ij} = (\psi_i H \psi_j) \simeq \varepsilon_{ij}(\psi_i \psi_j) \tag{6.3}$$

where H is the interaction term and ε_{ij} is one-electron energy; b_{ij} measures the strength of interaction between localized orbitals ψ_i and ψ_j on neighbouring like atoms i and j and is proportional to the overlap integral $(\psi_i \psi_j)$. Although it is not possible to measure or calculate b_{ij} exactly, it is possible to predict its variation in a series of isostructural solids, since the relative magnitudes of the overlap integral and orbital energies can be estimated.

In solids where cation–cation interaction is significant, b_{ij} can be related to R^{-1}, where R is the cation–cation separation. In cases where cation–anion–cation interaction is important, b_{ij} is related to the covalent mixing parameter, λ, of the cation–anion orbitals. For octahedrally coordinated cations, as in rocksalt and perovskite structures, the relevant mixing parameters are λ_σ and λ_π in the following 'molecular' wave functions

$$\psi_e = N_e(f_e - \lambda_\sigma \phi_\sigma) \tag{6.4}$$
$$\psi_{t_2} = N_t(f_{t_2} - \lambda_\pi \phi_\pi) \tag{6.5}$$

where N_e and N_t are normalization constants, f_e and f_{t_2} are the cation d orbitals of e_g

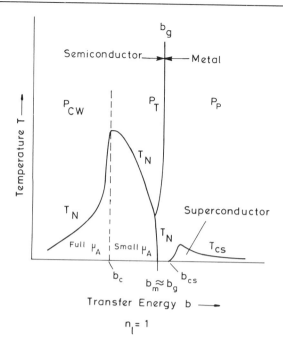

Figure 6.3   Conceptual T–b phase diagram for $n_l = 1$. (Reprinted from Goodenough, 1974 by courtesy of Marcel Dekker, Inc.)

and t_{2g} symmetry respectively, and ϕ_σ and ϕ_π are the anion sp_σ and p_π orbitals of appropriate symmetry. In general $\lambda_\sigma > \lambda_\pi$ and $b_{ij} \sim \lambda^2$. For small values of b, the outer d electrons are localized and for large values, they are itinerant. There is a critical value of transfer energy, b_c, in a series of isostructural solids that separates the collective electron regime from the localized electron regime. Since b is related to R and λ, expression for the critical values R_c and λ_c are given in terms of the position of the metal in the periodic table, the principal quantum number of the d orbital, the oxidation state and the total spin of the cation, as well as the electronegativity of the anion. Calculated values of R_c in oxides of rocksalt, rutile and corundum structures do indeed separate members showing localized-electron behaviour from those exhibiting collective electron properties. Critical transfer energies b_m, b_g and b_{cs} relevant to change of various physical properties are also defined. For $b > b_m \sim b_g$, spontaneous magnetism disappears and metallic conductivity sets in. Other important electronic properties that can be related to b are the antiferromagnetic ordering temperature, $T_N \sim b^2/U$, for $b < b_c$ and d band-width, $W = 2zb$, where z is the number of nearest neighbours.

A typical conceptual (T–b) phase diagram for the case of a single d electron per interacting orbital ($n_l = 1$) is shown in Fig. 6.3. The physical significance of the phase diagram is as follows: The magnetic interaction in a system with $n_l = 1$ is antiferromagnetic. Since T_N is proportional to b^2, T_N increases up to the critical value, b_c. For $b < b_c$, the localized electrons exhibit Curie–Weiss paramagnetism above T_N. For $b > b_c$, the states are band-like, and at b_m, spontaneous magnetism disappears ($T_N = 0$ at $b = b_m$)

and a first-order semiconductor-to-metal transition is predicted. For $b > b_m$, the d electrons are truly collective, exhibiting Fermi surface-dependent electronic properties and Pauli paramagnetism. A superconducting state may be stabilized for $b > b_{cs}$ at low temperatures. Wilson (1972) considers that the ratio W/U is equivalent to b_{ij} and makes use of Goodenough's conceptual phase diagrams to understand Mott insulation (see Section 6.5) in transition-metal compounds.

In order to apply the conceptual phase diagrams to interpret electronic properties of isostructural series of transition-metal compounds, Goodenough constructs one-electron energy-level diagrams assuming most probable hybridization of cationic and anionic orbitals. These diagrams are similar to the one-electron molecular orbital-energy-level diagrams of isolated molecules (Cotton, 1971), except that discrete energy levels are allowed to become band states where necessary, depending on cation–cation and cation–anion–cation interactions in solids. A typical one-electron energy-level diagram applicable to ReO_3 is shown in Fig. 6.4. We see the close similarity of the diagram to that of a discrete octahedral complex ion such as $Ti(H_2O)_6^{3+}$. In the Goodenough diagram for ReO_3, the filled bonding states (which are primarily anionic $2s$, $2p$) and the empty antibonding states (which are mainly cationic $6s$, $6p$) are split by a large gap ($\sim 5\,eV$) due to the difference in electronegativity between the constituent atoms. The $5d$ states of the cation, which are antibonding with respect to the anionic $2s$ and $2p$ states, lie in the gap. Electronic properties of the material are determined by the nature of the d-states, viz. localized or itinerant. In ReO_3, both t_{2g} and e_g orbitals are itinerant because of large λ_π and λ_σ; the t_{2g} band is partly occupied, making the material metallic. A similar diagram is applicable to ABO_3 perovskites; whether the t_{2g} and e_g states are localized or itinerant in a perovskite depends on the magnitude of covalent mixing, λ_π and λ_σ, of B and O orbitals (Section 6.4.1). Similar one-electron energy-level diagrams can be constructed for different series of solids including rocksalt, rutile and pyrite structures. We shall discuss the use of some of these diagrams in Section 6.4.

6.2.4   Cluster models

Between the localized-electron model and the band model, we have cluster models which in their simplest form take into account the interaction of a metal atom with the surrounding ligand atoms (oxygens in the case of oxides). Basically, cluster calculations are molecular orbital calculations carried out on a cluster. In Fig. 6.5 we show the molecular orbital diagram of a transition-metal ion octahedrally surrounded by six oxygens (O_h point group). The different orbitals arising from combinations of the metal orbitals and oxygen p orbitals are readily worked out. In principle, one can deal with larger clusters instead of the smallest cluster corresponding to the first coordination sphere. Molecular orbital calculations of varying degrees of sophistication have been carried out on a variety of transition metal–oxygen and other clusters. Besides *ab initio* methods, approximate methods such as the $X\alpha$ method (which employs an approximation for the electron repulsion term) have been employed. Configuration interaction (CI) (involving mixtures of different electron configurations of the cluster) is included in many of the recent calculations on clusters. Specially significant are the so-called charge-transfer impurity models (the impurity hole or electron is taken to be localized

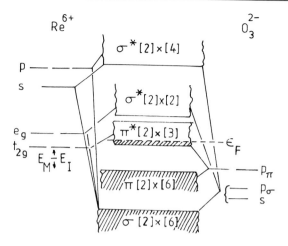

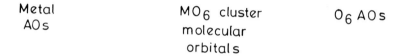

Figure 6.4   Schematic energy level diagram for ReO$_3$. (From Goodenough, 1971.)

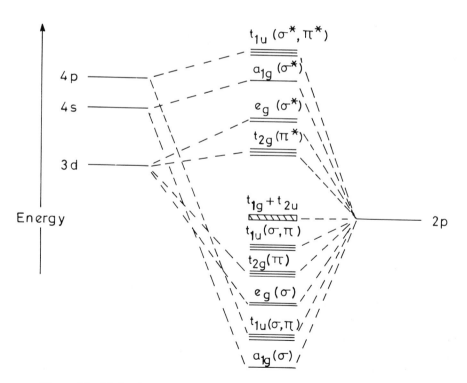

Figure 6.5   Molecular orbital diagram of MO$_6$ cluster formed by a transition metal with oxygen.

on a specific ion) which take into account states corresponding to $d^n, d^{n+1}L, d^{n+2}L^2$ etc. where L is the hole in the oxygen $2p$ band states.

6.3   Properties

Physical properties of solids arise as responses to external stimuli (physical forces). The important physical forces are mechanical stress, electric field, magnetic field and temperature. Stress produces strain, a magnetic field causes magnetization and an electric field gradient produces a current flow. Physical properties can be classified into four categories: equilibrium, steady-state, hysteretic and irreversible (Koerber, 1962; Newnham, 1975). Equilibrium properties are measured when the system is in thermal equilibrium with its surroundings. During the measurement of a steady-state property, on the other hand, the system is not in thermal equilibrium with the surroundings but it does not change with time. An example of a steady-state property is electrical resistivity, which relates potential gradient to charge flow. Hysteretic properties are those for which it is difficult to define a functional relationship between the extensive and intensive variables (the relation between the two variables is irreversible and depends on whether the intensive variable is increased or decreased); a typical example would be the magnetization–magnetic field relationship of a ferromagnetic material. Measurement of irreversible properties leaves permanent change in the material. For example, the Moh's hardness of a solid is obtained by making a scratch. Equilibrium and steady-state properties of crystals are formulated mathematically by employing tensors (Nye, 1957). We give here a brief description of magnetic, electrical, dielectric and optical properties of solids.

6.3.1   *Magnetic properties*

When a substance is placed in a magnetic field H, it develops a certain amount of magnetization (magnetic moment per unit volume) M given by $M = \chi H$, where χ is the magnetic susceptibility. The magnetic induction B is defined as

$$B = H + 4\pi M = \mu H \tag{6.6}$$

where μ is the permeability; we see that $\mu = 1 + 4\pi\chi$. Substances with weak negative magnetic susceptibility are called *diamagnetic*, while those with positive susceptibility are known as *paramagnetic*. Ordered arrays of atomic magnetic moments in solids give rise to other types of magnetic behaviour which are discussed later in this section.

Magnetic properties are due to the orbital and spin motions of electrons in atoms. The relation between the magnetic dipole moment $\boldsymbol{\mu}$ and the angular momentum $\mathbf{J}$ of an electron of charge e and mass m can be expressed as

$$\boldsymbol{\mu} = -(e/2m)\,\mathbf{J} \tag{6.7}$$

The angular momentum is quantized in units of $h/2\pi$ and the lowest nonzero value of $\boldsymbol{\mu}$ is the *Bohr magneton, β*

$$\beta = eh/4\pi m \tag{6.8}$$

Its value is 9.2732×10^{-24} JT^{-1}(SI) or 9.2732×10^{-21} erg Oe^{-1} (CGS).

The orbital motion of the electron about the nucleus in an atom gives rise to a magnetic moment μ_L which is related to the magnitude of the orbital angular momentum L by the expression

$$\mu_L = \frac{eL}{2mc} = \beta[L(L+1)]^{1/2} \tag{6.9}$$

Similarly, the spin magnetic moment is given by

$$\mu_S = \frac{eS}{mc} = 2\beta[S(S+1)]^{1/2} \tag{6.10}$$

where S is the magnitude of the spin angular momentum. L and S are respectively the orbital and the spin quantum numbers. The effective magnetic moment μ is related to the magnitude of the total angular momentum J by

$$\mu = g\frac{eJ}{2mc} = g\beta[J(J+1)]^{1/2} \tag{6.11}$$

The g (Landé) factor is a measure of the relative contribution of the spin and orbital angular momenta to the effective magnetic moment. $g = 0$ when $S = 0$ and $g = 2$ when $L = 0$. Equation 6.11 gives the magnetic moment of a single atom. An atom with angular momentum J has $2J + 1$ energy levels in a magnetic field and their energies are

$$E_M = -g\beta M_j H \tag{6.12}$$

where M_j ranges from $+ J$ to $- J$ in integral steps. For a collection of N noninteracting atoms, population of the various energy levels is governed by the Boltzmann distribution. The relative population of a given sublevel is

$$P(E_M) = \frac{\exp(-E_M/kT)}{\Sigma \exp(-E_M/kT)} \tag{6.13}$$

The total magnetic moment can be added up if the populations of the various sublevels are known. Assuming that the spacing of the multiplet levels is large compared to the splitting of the levels of a given J, the mean magnetization can be expressed as

$$M = Ng\beta \, JB_J(Y) \tag{6.14}$$

where $B_J(Y)$ is known as the *Brillouin function* and $Y = g\beta H/kT$.

$$B_J(Y) = \frac{2J+1}{J} \coth[(J+\tfrac{1}{2})Y] - \frac{1}{2J}\coth(Y/2) \tag{6.15}$$

For paramagnetic substances the condition $Y \ll 1$ is applicable, in which case

$$B_J(Y) = (J + 1)Y/3 \qquad (6.16)$$

The molar susceptibility is then

$$\chi_M = \frac{M}{H} \simeq Ng^2\beta^2 J(J + 1)/3kT = \frac{C}{T} \qquad (6.17)$$

This is the *Curie law* in which the Curie constant

$$C = Ng^2\beta^2 J(J + 1)/3k = P_{eff}^2 N\beta^2/3kT \qquad (6.18)$$

where $P_{eff} = g[J(J + 1)]^{1/2}$. Equation (6.17) is derived by assuming that each para-magnetic atom is independent. In solids there is often interaction between atomic moments providing an internal field. Hence the paramagnetic susceptibility is described more often by the *Curie–Weiss law* (6.19) than the Curie law.

$$\chi_M = \frac{C}{T - \theta} \qquad (6.19)$$

where θ is the Weiss constant. In the derivation of equation (6.17) another assumption made is that all the atoms have a unique value of J. In most cases this assumption is valid because the excited J states have energies much higher than kT. The effect of admixing higher energy states is to add a small temperature-independent *Van Vleck paramagnetism* to the much larger Curie–Weiss component.

Rare-earth metals and their salts are typical paramagnets at ordinary temperatures, following the Curie–Weiss law. Experimental P_{eff} values generally agree well with those calculated from the ground state J values, the behaviour being in accordance with the localized nature of $4f$ electrons. Sm and Eu show discrepancies with theory, arising from the presence of higher multiplets within energies comparable to kT. Eu^{3+} deserves special mention; $J = 0$ ground state ($^7F_0$) of this ion would predict zero P_{eff}. The observed nonzero moment is therefore due to population of higher multiplets.

Paramagnetism of transition metals and their salts is different from that of rare earths. This is because of the involvement of d-electrons in crystal binding (Section 6.2.3). In compounds of the first row transition metals, where definite electronic configurations can be assigned to the cations, agreement between calculated and experimental P_{eff} is poor. If, however, it is assumed that orbital moment plays no part in the magnetism and that the moment is wholly due to spin, good agreement is found between experiment and theory in many cases. The reason for this loss (quenching) of orbital moment lies in the inhomogeneous crystal field of the surroundings, which interacts strongly with the cation d-orbitals.

Itinerant electrons of simple metals (Li, Na, etc.) show a weak temperature-independent paramagnetism (*Pauli paramagnetism*), whose origin can be understood in terms of the free-electron theory (Kittel, 1976). Most electrons in the conduction band do not respond to the applied field because most states are fully occupied. Only those electrons near the Fermi level E_F within the energy range $\sim kT$ contribute to the susceptibility. This fraction is roughly T/T_F where $T_F = E_F/k$. Since $T_F \gg T$ for nearly all accessible temperatures the susceptibility is small and is given by

$$\chi \simeq N\beta^2/kT_F \tag{6.20}$$

Magnetism in solids is more complex than that of isolated atoms because of the possibility of interaction (coupling) between atomic moments. The coupling between moments which is responsible for cooperative magnetism of solids has its origin in the Pauli exclusion principle: electrons of parallel spin tend to stay away from each other. A pair of electrons of like spin localized on an atom is lower in energy than a pair with opposite spin by an amount called intra-atomic exchange energy. Consequently, there is a statistical correlation for electrons of like spin with each surrounded by a void due to the local depletion of parallel spin electrons. This is called *exchange coupling* of which we can distinguish two classes. *Direct exchange* occurs between moments on atoms that are close enough to have significant overlap of their wave functions. Exchange coupling is strong but decreases rapidly with increasing interatomic distance. *Indirect exchange*, on the other hand, couples moments over relatively large distances. It can act through an intermediary nonmagnetic ion (superexchange) or through itinerant electrons (RKKY, Ruderman, Kittel, Kasuya and Yosida). Superexchange occurs in insulators while RKKY coupling is important in metals.

The exchange energy H_{ex} of atoms i and j in a solid separated by a distance r_{ij}, having spins S_i and S_j respectively, can be expressed as

$$H_{ex} = -\sum_{ij} J(r_{ij})\mathbf{S}_i \cdot \mathbf{S}_j \tag{6.21}$$

where J is called the exchange parameter. For direct intraatomic exchange, as between two electrons belonging to the same atom, J is positive, leading to Hund's rule. For direct interatomic exchange, J can be positive or negative. If J is positive, a parallel alignment of spins is favoured. For negative J, an antiparallel alignment of spins results in low energy. For indirect exchange, J can be positive or negative (superexchange) or oscillatory (RKKY).

Superexchange describes interaction between localized moments of ions in insulators that are too far apart to interact by direct exchange. It operates through the intermediary of a nonmagnetic ion. Superexchange arises from the fact that localized-electron states as described by the formal valences are stabilized by an admixture of excited states involving electron transfer between the cation and the anion. A typical example is the 180° cation–anion–cation interaction in oxides of rocksalt structure, where antiparallel orientation of spins on neighbouring cations is favoured by covalent mixing of the anion p orbital with the cation d orbitals on each side. A variety of superexchange interactions are possible, depending on the structure of the solid and the electronic configuration of cations. Two important forms of superexchange are correlation superexchange and delocalization superexchange. *Delocalization superexchange* involves transfer from one cation to another, be it due to cation–cation or cation–anion–cation interaction. *Correlation superexchange* is restricted to cation–anion–cation interaction only, since it requires spin correlation between the coupled cations; Anderson (1959), Kanamori (1959) and Goodenough (1963) have given rules for determining the sign and magnitude of superexchange. For example, 180° cation–anion–cation interaction between half-filled orbitals is antiferromagnetic. A 90° cation–anion–cation interaction between half-filled orbitals is ferromagnetic provided the orbitals are bonded to orthogonal anion orbitals. Superexchange involving σ-bonds is stronger than that involving π-bonds. Thus, the Néel temperature, T_N (see p. 299) of $LaCrO_3$ is less than that of $LaFeO_3$. In the $3d$-transition-metal monoxides, T_N increases in the order $MnO < FeO < CoO < NiO$ because the σ-interaction increases in that order. For cations of the same electronic configuration, superchange is stronger for the higher valency cation (e.g. $Fe^{3+} > Mn^{2+}$).

The RKKY interaction provides a mechanism for the coupling of localized magnetic moments in itinerant electron systems. It is unique because J oscillates from positive to negative as the separation between magnetic ions changes. The mechanism can be understood as follows: a magnetic ion induces an oscillatory spin polarization in the conduction electron in its neighbourhood. The oscillatory spin polarization dies off rather slowly (as the inverse cube) with distance from the local spin. This modulated spin polarization of the itinerant electrons is felt by the moments of other magnetic ions within the range, leading to an oscillatory indirect coupling (Fig. 6.6). *RKKY interaction* is responsible for, among others, magnetic ordering in rare-earth metals and for a number of phenomena involving nuclear magnetic resonance in metals. In disordered metallic systems, where the separation between magnetic ions is random, RKKY interaction leads to positive and negative coupling. This leads to so-called *frustration*. A frustrated system has no unique microscopic arrangement for its ground state; an infinite number of equivalent states is available to the system. The concept of frustration has wider implications in other branches of science, where it is called 'structural disequilibrium'.

We shall now briefly describe the various types of magnetism in solids. Magnetism in solids can be cooperative or noncooperative. In the cooperative type, mutual interaction between moments is important, while in the noncooperative type, the individual moments behave independent of one another. There are five basic types of magnetism in solids (Fig. 6.7). *Diamagnetism*, which is a property of closed electron shells, is a noncooperative magnetism, characterized by a weak negative, temperature-independent, magnetic susceptibility. *Ideal paramagnetism* is another noncooperative magnetism that arises from atomic moments that are identical and located in isotropic surroundings, sufficiently separated from one another. The temperature-dependent susceptibility follows the Curie law (eqn. 6.17). Ideal paramagnetism is the exception rather than the rule because the susceptibility of real systems follows the Curie–Weiss

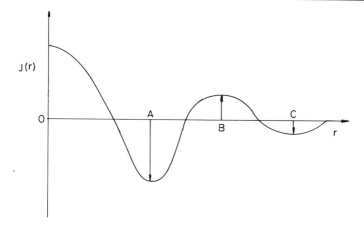

Figure 6.6   Interaction between magnetic impurities through RKKY mechanism. Interaction between the magnetic impurity at the origin O and that at a distance $r = OA$ is antiferromagnetic; interaction with a second impurity at $r = OB$ is ferromagnetic. It is seen that the interaction is oscillatory and dies away slowly as the inverse cube of r.

law. Curie–Weiss law behaviour implies a cooperative magnetism that can manifest itself if the thermal energy of the moments in the paramagnet is reduced sufficiently; at temperatures $T < T_o$, an ordered magnetic state exists.

Ferromagnetism is a cooperative phenomenon in which there is a long-range collinear order of all the moments in the solid ($J > 0$). A ferromagnetic solid is spontaneously magnetized even in the absence of the field. To maximize its magnetostatic energy, a crystalline ferromagnet divides into domains which are spontaneously magnetized nearly to saturation, but the moment of each domain is oriented so as to produce a zero net moment. An external field changes the size of the domains, enlarging those of favourable orientation at the expense of others. Thus in ferromagnetism external field is just an agent to make evident on a macroscopic scale the ordering that exists microscopically. Typically, magnetization rises sharply at lower fields, as the domains with more favourable alignments expand at the expense of others, and saturates when the maximum domain alignment is reached. With increasing temperature, thermal energy increases, becomes comparable to and eventually exceeds the exchange energy. The spontaneous magnetization thus decreases with temperature to disappear at the *Curie temperature* T_c. Above T_c an ideal ferromagnet becomes a paramagnet obeying the Curie–Weiss law. Both crystalline and amorphous ferromagnets are known; typical examples of the latter are the various compositions of Fe, Co, Ni with B (metallic glasses) (Cahn, 1980).

Antiferromagnetism, like ferromagnetism, is characterized by long-range ordering of identical moments. But since the exchange parameter J is negative, the moments of neighbouring atoms are exactly opposed; there is no overall spontaneous magnetization. An antiferromagnet below the *Néel* (ordering) *temperature*, T_N, consists of two identical interpenetrating sublattices in which the spins of one sublattice are

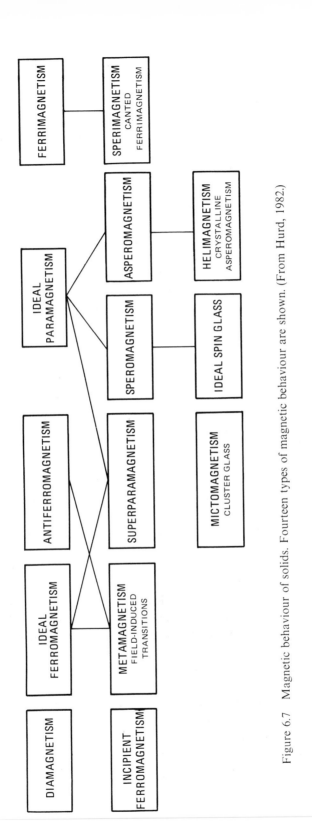

Figure 6.7   Magnetic behaviour of solids. Fourteen types of magnetic behaviour are shown. (From Hurd, 1982.)

opposed to those of the other. Simple antiferromagnetism can exist only in crystalline solids since it is not possible to divide an amorphous system into two identical sublattices. Most antiferromagnets are insulating solids (e.g. MnO, NiO, MnS, etc.) but antiferromagnetism is known among metals and alloys as well (e.g. Cr, α-Mn, Pt_3Fe). Itinerant d-electron antiferromagnets do not follow the Curie–Weiss law above T_N.

Ferrimagnetism requires two or more magnetic species that are chemically different. They occupy two kinds of lattice sites in ferrimagnets, producing two sublattices A and B as in spinels. The moments of ions in each sublattice are ferromagnetically coupled, but the coupling between the moments of A and B is antiferromagnetic. Since the net moments of A and B are different, there is a resultant spontaneous magnetization. The temperature dependence of ferrimagnetism is similar to that of ferromagnetism, except that the spontaneous magnetization decreases more rapidly with increasing temperature. In the paramagnetic state, there is deviation from the Curie–Weiss law, particularly close to T_c. Ferrimagnetism can exist in the amorphous state also, but the A and B sublattices are random. Examples of amorphous ferrimagnets are $LnFe_2$ (Ln = Gd or Tb) where Fe–Fe and Ln–Ln couplings are ferromagnetic but Fe–Ln interaction is antiferromagnetic.

Magnetic order in ferro-, antiferro- and ferrimagnets involves a simple collinear arrangement of atomic moments, either parallel or antiparallel. These are common but special cases occur of a more general situation where noncollinear arrangements are possible. A magnetic state in which localized moments are ordered in a crystal but locked up in different orientations below some ordering temperature, with some orientations more likely than the rest, is called *helimagnetism* (Fig. 6.7). $MnAu_2$ is a prototypical example where the moments show a helical structure. In the tetragonal structure, the magnetic moments lie in the *ab* plane normal to the *c*-direction but their direction rotates about 50° from plane to plane. Rare-earth metals show more complicated forms of helimagnetism where the moments rotate along the surface of a cone rather than in a plane. Certain solids comprising two (or more) antiferromagnetic lattices that are canted at an angle leave a net magnetization. They are called *weak ferromagnets* or *canted antiferromagnets*. Weak ferromagnetism can arise from differences in single-ion anisotropy, or from Dzyaloshinsky–Moriya interaction, or from both. Rutile type NiF_2 is an example of weak ferromagnetism arising from single-ion anisotropy. Dzyaloshinsky–Moriya interaction, which results in asymmetric arrangement of anion moments, arises from a strong spin-orbit coupling of the anion, which upsets superexchange interaction in certain systems. Typical examples of weak ferromagnets arising from this interaction are α-Fe_2O_3, $CoCO_3$ and CrF_3.

A number of other types of magnetic order have come to be recognized, many of them being relevant to amorphous and disordered solids (for a summary see Hurd, 1982). These new types of magnetic order include metamagnetism, spero-, aspero- and sperimagnetism, spin-glass and mictomagnetism (see Fig. 6.7). In amorphous solids, where the crystal field varies from point to point, equation (6.21) is modified as follows to include local anisotropy

$$H = -\sum_i \mathbf{D}_i(S_z)_i^2 - \sum_{ij} J(r_{ij})\mathbf{S}_i\cdot\mathbf{S}_j \qquad\qquad (6.22)$$

where $\mathbf{D}$ is the axial crystal field strength and S_z is the total spin of the ion along the local easy direction z. Equation (6.22) expresses the energy of interaction arising from both the local anisotropy and exchange effects. In an amorphous solid $\mathbf{D}$ is fixed and positive but the easy axes z_i are randomly distributed.

In *metamagnetism*, there is a field-induced magnetic transition from a state of low magnetization to one of relatively high magnetization (Stryjewski & Giordano, 1977). A typical metamagnet is an antiferromagnet below its Néel temperature, but, with increasing applied field strength, crystal anisotropy forces are overcome, changing the magnetic structure abruptly. A *superparamagnet* is a ferromagnet whose particle size is too small to sustain the multidomain structure. The small particle size removes the constraints on magnetic moments, permitting them to fluctuate as in an ideal paramagnet. In superparamagnetism, there is no hysteresis in the field dependence of the magnetization (M versus H is a single-valued curve at a given temperature). Superparamagnetism is found when fine ferromagnetic particles are distributed in amorphous gels or in a non-magnetic matrix. A *speromagnet* is one in which localized moments of a given species are locked into random orientations with neither net magnetization nor a regular pattern of local ordering beyond nearest neighbours. Speromagnetism is believed to occur mainly in systems where the J_{ij} are random or negative but strongly frustrated.

A crystalline alloy comprising magnetic atoms of a given species incorporated into a nonmagnetic host can order speromagnetically when cooled below a critical tempera-ture. This is called 'spin-glass freezing', characterized by a cusp in the low-field a.c. susceptibility at a temperature called T_{SG}. The randomness of the orientations of the moments is reminiscent of the randomness of the atomic positions in an ordinary glass; hence the name *spin glass*. The orientations of the frozen moments are fixed indefinitely and do not fluctuate with time, but above T_{SG} the moments fluctuate to give a paramagnetic state. What is significant in a spin glass is that, as $T \to T_{SG}$, the various spins begin to interact with each over long range and the system seeks its ground-state configuration for the particular distribution of spins. The existence of such a ground state for randomly coupled spins without any short-range order was pointed out by Edwards & Anderson (1975). Typical examples of spin glasses are iron or manganese dissolved in copper or gold. The problem of spin glass continues to attract attention, see Ford (1982) and Mydosh (1982). Spin glass behaviour has been found in Mn-doped LaNiO$_3$ (Vasanthacharya *et al.*, 1984) and in insulating systems. e.g. Eu$_{1-x}$Sr$_x$S and Mn$_{1-x}$O (see, e.g., Hauser & Waszczak, 1984).

A spin-glass state exists in crystalline alloys only within a limited range of concentration of magnetic solute. The concentration must be high enough to give mutual interactions but low enough to avoid cluster formation. For concentrations below the dilute limit, the Kondo regime obtains and magnetic atoms are screened from interaction with others. When the concentration of magnetic atoms is large

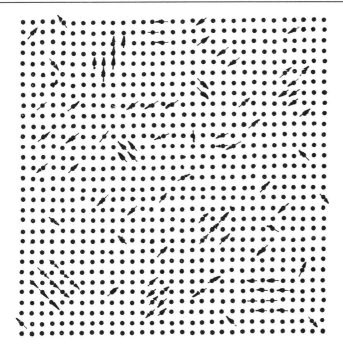

Figure 6.8   Representation of a spin glass with about 10% of impurities. Mictomagnetic clusters are seen in certain regions. (After Mydosh, 1975.)

enough, clusters are formed. A *mictomagnet* is similar to a spin glass except that local correlations of magnetic atoms are dominant because of clustering. There is direct exchange in clusters but the moments of individual clusters are coupled indirectly by RKKY interaction. A 10% spin glass showing mictomagnetic clusters is represented schematically in Fig. 6.8.

The problem of dilute magnetic impurities in itinerant electron systems in general is particularly important both in magnetism and electrical transport. The interaction between an isolated *d*-transition-metal atom and the delocalized electrons of a metal is described by *s–d* mixing (Bell & Caplin, 1975). Because of the mixing, an itinerant *s*-electron temporarily resides in the *d*-atomic state before tunnelling back into the delocalized states. In other words, the *d* level hybridizes with the conduction band states to form a *virtual bound state* (VBS). The concept of VBS is able to explain the magnetic properties of 3*d* metal impurities in metallic solvents such as Al, Cu, Ag, Au and Zn. The VBS model thus explains why in some solvents the transition-metal impurity is magnetic and in others not. For example, Fe is magnetic in Au, Ag and Cu and is not in Zn and Al. The eigenstates (VBS) of the impurity are of two kinds, local and band. If the mixing is weak, local effects dominate and magnetism survives. A strong admixture of band states destroys the magnetism. The strength of mixing is roughly proportional to the E_F of the solvent metal. When E_F is high, as in Al ($\sim 12\,eV$), all the 3*d* metal atoms lose their magnetism. E_F of noble metals is smaller ($\sim 4\,eV$) and many transition-metal impurities retain their magnetism in noble-metal solvents.

The presence of a magnetic impurity in a metallic solvent spin-polarizes the s-electron population. This cloud of antiferromagnetic spin-polarization resonating about the magnetic ion is called *Kondo binding*. Below a characteristic temperature called the Kondo temperature, T_K, the antiparallel spin cloud neutralises the ion's moment and reduces it to zero in the theoretical limit. The ideal Kondo situation prevails at absolute zero and when the magnetic impurity is isolated; an approximate Kondo regime obtains in a very dilute magnetic alloy at low temperatures.

Besides the varied ground-state ordering of moments in solids discussed, various *magnetic excitations* can occur in the different magnetic states and the subject has been discussed well (Mattis, 1981). One of the elementary excitations is a quantized *spin wave* (*magnon*) in a classical ferromagnet. In the ground state of a ferromagnet all the spins are oriented parallel. We can form an excited state by reversing one of the spins. Instead of this, we can let all the spins share the reversal, in which case the spins take a wave-like arrangement. The elementary excitations of a spin system having a wave-like form are called *magnons*. Magnons are analogous to phonons and they can be similarly treated to derive magnon dispersion relations. The magnon spectrum of a solid can be determined by inelastic neutron scattering.

6.3.2   *Electrical properties*

D.c. electrical conductivity, thermal conductivity, Seebeck effect and Hall effect are some of the common electron-transport properties of solids that characterize the nature of charge carriers. On the basis of electrical properties, solid materials may be classified into metals, semiconductors, and insulators where the charge carriers move in band states (Fig. 6.1); there are other semiconductors and insulators where charge carriers are localized and their motion involves a diffusive process (Honig, 1981). We shall briefly present the important relations involved in interpreting the transport phenomena in solids.

An expression for the *electrical conductivity* of a metal can be derived in terms of the free-electron theory. When an electric field **E** is applied, the free carriers in a solid are accelerated but the acceleration is interrupted because of scattering by lattice vibrations (phonons) and other imperfections. The net result is that the charge carriers acquire a drift velocity $\langle v_d \rangle$ given by

$$\langle v_d \rangle = \frac{-e\mathbf{E}\tau}{m} = -u_d\mathbf{E} \tag{6.23}$$

where e is the electron charge, m its mass, τ the relaxation time (average time between scattering events) and u_d drift mobility (drift velocity per unit electric field).

The current density **J** can be expressed as

$$\mathbf{J} = ne\langle v_d \rangle = \frac{ne^2\tau}{m}\mathbf{E} \tag{6.24}$$

Equation (6.24) expresses Ohm's law, where the proportionality constant between **J** and $\mathbf{E}$ defines electrical conductivity σ. Thus

$$\sigma = ne^2\tau/m = ne^2l/mv \tag{6.25}$$

since mean free path $l = v\tau$, where v is the total (not the drift) velocity. For a metal, v, τ and l in equation (6.25) refer to electrons with energy E_F, since only those carriers near E_F can be scattered. Thus the expression for electrical conductivity of a metal is

$$\sigma = \left(\frac{8\pi}{3n}\right)^{1/3} ne^2l/m$$

or

$$\sigma = \left(\frac{8\pi}{3}\right)\left(\frac{2m}{h^2}\right)^{3/2} E_F^{3/2} e\frac{u_r}{T^r} \tag{6.26}$$

where $-3 < r < 3$ is a scattering index. From (6.26) onwards m represents the *effective mass* of the charge carriers. The temperature dependence of conductivity in a metal arises from the variation of l (and u) with temperature; l is inversely proportional to the number of phonons. Above the Debye temperature, the density of phonons varies as T and hence $\sigma \propto T^{-1}$, while below it density of phonons goes as T^3. Since the scattering is weak, resulting in memory effect, $\sigma \propto T^{-5}$ at low temperatures.

In semiconductors, where the number of carriers is small, all the carriers respond to the applied field. The average kinetic energy $\frac{1}{2}mv^2$ can be equated to $\frac{3}{2}kT$ and thus

$$\sigma = ne^2l/(3mkT)^{1/2} \tag{6.27}$$

For intrinsic semiconductors, in which both electrons and holes are present

$$\sigma = n_i e(u_n + u_p) \tag{6.28}$$

where

$$n_i = (N_c N_v)^{1/2}\exp(-E_g/2kT) \tag{6.29}$$

Here N_c and N_v are the effective densities of states in the conduction and valence bands respectively, and us are the mobilities. The final expression for the electrical conductivity of an intrinsic semiconductor is given by

$$\sigma = 2eu_p(b + 1)[2\pi(m_n m_p)^{1/2}kT/h^2]^{3/2}\exp(-E_g/2kT) \tag{6.30}$$

where $b \equiv u_n/u_p$. Since the pre-exponential factor is not very sensitive to temperature, a plot of $\ln \sigma$ versus $1/T$ should yield a straight line with slope proportional to E_g.

For the case of semiconductors doped with donors (*n*-type) and acceptors (*p*-type), three regions can be distinguished. In the low-temperature extrinsic region ($kT \ll E_g$), the conductivity is given by

$$\sigma = \frac{4\sqrt{2}g(N_d - N_a)}{3h^{3/2}}[le^2(2\pi m_n kT)^{1/4}\exp(-E_d/2kT)] \tag{6.31}$$

when the number of donors N_d > the number of acceptors N_a per unit volume. In equation 6.31, g is the statistical weight factor; $g = \frac{1}{2}$ for electrons and 2 for holes. Thus $\sigma \sim \exp(-E_d/2kT)$ since $T^{1/4}$ in the preexponential term has little influence. In the high-temperature (intrinsic) region, equation 6.30 is the expression for conductivity. At intermediate temperatures (exhaustion region), extrinsic conductivity gradually gives way to intrinsic conductivity, and the temperature dependence of conductivity is essentially due to the temperature dependence of mobility, $u \sim T^{-n}(n \simeq \frac{3}{2})$, since the number of carriers remains constant (all the dopant levels are ionized).

Semiconductors characterized by localized charge carriers (*hoppers*) are different from itinerant semiconductors discussed above in that the mobility is activated in the case of hoppers, $u = u_o(T)\exp(-E_u/kT)$, and the expression for conductivity is

$$\sigma = N_t\frac{g_i}{g}\left[\frac{N_d - N_a}{N_a}\right]eu_o(T)\exp(-(E_t + E_u)/kT) \tag{6.32}$$

where g_i and g are degeneracy factors, N_t is the density of localized sites providing charge carriers, E_t is the energy required to ionize charge carriers and E_u is the activation energy associated with mobility.

The *thermoelectric effect* is due to the gradient in electrochemical potential caused by a temperature gradient in a conducting material. The Seebeck coefficient α is the constant of proportionality between the voltage and the temperature gradient which causes it when there is no current flow, and is defined as ($\Delta V/\Delta T$) as $\Delta T \to 0$ where ΔV is the thermo-emf caused by the temperature gradient ΔT; it is related to the entropy transported per charge carrier ($\alpha = -S^*/e$). The Peltier coefficient π is the proportionality constant between the heat flux transported by electrons and the current density; α and π are related as $\alpha = \pi/T$.

The expression for the Seebeck coefficient of metals is

$$\alpha = -\frac{2\pi^2 mkT}{3h^2 e}[8\pi/3n]^{2/3} = -\frac{k}{e}\left[\frac{\pi^2}{3}\right]\left[\frac{kT}{E_F}\right] \tag{6.33}$$

Thus, metals are characterized by a small $|\alpha|$ that increases linearly with temperature.

The Seebeck coefficient of an intrinsic semiconductor is given by

$$\alpha = \pm\frac{k}{e}\left[\frac{S^*}{k} - \frac{E_F}{kT}\right] \tag{6.34}$$

where $S*$ is the entropy transported by charge carriers. $S*/k < 10\,\mu V/K$ in ionic solids like metal oxides. E_F is the Fermi energy relative to the appropriate band edge. The sign of α is positive for hole conduction and negative for electron conduction. Since

$$\frac{E_F}{kT} = \ln[c/(1-c)] \text{ or } = \ln\left[\frac{n}{N}\right] \tag{6.35}$$

where $(1-c)$ (or N) represents the density of unoccupied states in the relevant band and c (or n) the density of charge carriers in the band. For extrinsic semiconductors $|\alpha|$ is generally large, in the range of 0.1 to $1\,mV/K$, and decreases with increasing temperature. A maximum value of α occurs for small c and there is a change of sign for α as c varies from 0 to 1. For localized extrinsic semiconductors (small polaron hopping), c is generally constant and hence α is nearly temperature-independent.

Hall effect.   The Hall coefficient R, of a solid with a single type of charge carrier is

$$R = \pm\frac{C_r}{ne} \tag{6.36}$$

where $C_r \sim 1(3\pi/8 < C_r < 315\pi/512)$. If R and σ are known, then the product

$$\sigma R = neu_d\left(\frac{C_r}{ne}\right) = C_r u_d \tag{6.37}$$

is a measure of the drift mobility.

Electrical transport properties provide useful criteria for distinguishing localized and collective electrons in solids. Thus, the temperature dependence of drift mobility u for collective electrons $(b > b_c)$ is different from the behaviour for small polarons $(b < b_c)$. For collective electrons, u goes as $T^{-3/2}$ and when the bands are narrow, mobility becomes thermally activated; $u \sim \exp(-E_a/kT)$, where E_a is the activation energy for hopping. Mobility is small $(< 0.1\,cm^2/Vs)$ for localized semiconductors exhibiting hopping conduction and large $(> 1\,cm^2/Vs)$ in the band limit. Experimentally, the product of Hall coefficient and electrical conductivity gives a measure of the drift mobility. *Quantum Hall effects* (Klitzing et al., 1980) will find many applications; the Hall resistance is quantized corresponding to $R = h/ie^2$ ($i = $ integer).

The motion of ions through solids results in both charge as well as mass transport. Whereas charge transport manifests itself as ionic conductivity in the presence of an applied electric field, macroscopic mass transport (diffusion) occurs in a concentration gradient. Both ionic conductivity and diffusion arise from the presence of point defects in solids (Section 5.2). For a solid showing exclusive *ionic conduction*, conductivity is written as

$$\sigma = \sum_i n_i q_i e u_i \qquad (6.38)$$

where the summation is taken over all the charged species i; n_i denotes the concentration of the ith type ion bearing a net charge $q_i e$ and possessing mobility u_i. In alkali halides where Schottky defects give rise to electrical conductivity, n refers to the concentration of vacancies in the cation and anion sublattices, n_c and n_a. For a pure crystal, $n_c = n_a$ but their mobilities are not equal, u_c being much larger than u_a. The expression for the electrical conductivity of a Schottky defect solid where conductivity is dominated by the motion of cations is given by

$$\sigma = \left(\frac{\sigma_o}{T}\right) \exp[-E_A/kT] \qquad (6.39)$$

Here $\sigma_0 = Ne^2 a^2 v/k$ is a constant for a given crystal and $E_A = \frac{1}{2}E_s + E_m$. N is the number of lattice sites of one kind per unit volume, a the jump distance and v is the appropriate vibration frequency. E_s is the Schottky defect formation energy and E_m is the activation energy required for ion migration.

The expression for the *diffusion coefficient* is

$$D = va^2 \exp[-(\tfrac{1}{2}E_s + E_m)/kT] \qquad (6.40)$$

Comparing the expressions for ionic conductivity and diffusion coefficient, we see that

$$\frac{\sigma}{D} = \frac{Ne^2}{kT} \quad \text{or} \quad \frac{u}{D} = \frac{e}{kT} \qquad (6.41)$$

Equation (6.41) is known as the *Nernst–Einstein relation*, originally deduced for the mobility of colloid particles in a liquid, but also valid for ionic solids.

In the early work on ionic conductivity, the data were presented as $\log \sigma$ vs. T^{-1} plots which showed two linear regions joined by a 'knee', the position of the 'knee' being dependent on the purity of the sample. But, since the mobility of ions due to point defects not only depends exponentially on T^{-1} but also inversely on T (cf. equation 6.39), a plot of $\log (\sigma T)$ vs. T^{-1} is more appropriate. Earlier, in Chapter 5, we showed a typical plot for 'pure' NaCl (Fig. 5.2) and explained the origin of the various regions. From the slopes of the linear regions, energies associated with various processes can be readily estimated. Thus in the intrinsic region, $E_{I,II} = E_m^c + E_s/2$, $E_{III} = E_m^c$ and $E_{IV} = E_m^c + E_a/2$, where E_s is the energy of formation of a Schottky pair, E_m^c is the cation migration energy and E_a is the association energy between an impurity and an oppositely charged vacancy. Temperature dependence of cation diffusion coefficient in alkali halides also shows a similar behaviour. Diffusion coefficients determined by tracer technique D_T and those obtained from electrical conductivity using

Nernst–Einstein relation D_σ show subtle differences because ionic motion in tracer experiments is correlated. The *Haven ratio*, D_T/D_σ, is proportional to the correlation factor. Electrically neutral defects contribute to D_T but not to D_σ and therefore the Haven ratio will be abnormally large in the presence of neutral defects. On the other hand, the presence of electronic conduction in the solid results in abnormally small Haven ratios. Mechanisms of diffusion and ionic conductivity in terms of point defects in ionic solids have been discussed widely in the literature (Haven, 1978; Corish *et al.*, 1977). In certain solids, conductivity of one of the ions is unusually large, being comparable to that of aqueous electrolytes; such fast ion conduction is discussed in Section 7.2. Ionic transport in glasses has been treated in terms of weak electrolyte theory (Ingram *et al.*, 1980). The mixed alkali effect in ionic glasses wherein two isovalent ions retard each other is intriguing.

6.3.3   *Superconductivity*

When the electrical resistance of a material vanishes below a critical temperature $T_c > 0\,K$, we refer to it as a superconducting transition. The transition was first observed in mercury by Kamerlingh Onnes in 1911. Fig. 6.9(a) shows schematically two possible types of transition; a sharp discontinuity in resistivity found in pure, homogeneous materials and a broad transition seen in inhomogeneous materials. Typical resistivities of materials in the superconducting state are of the order of 10^{-23} ohm cm compared to the lowest resistivity of $\sim 10^{-13}$ ohm cm in normally conducting metals. The temperature interval ΔT_c over which the transition between the normal and the superconducting state occurs, may be as small as $10^{-4}\,K$ or several degrees in width, depending on the material. In the superconducting state, materials exhibit perfect diamagnetism, excluding magnetic field below a critical field H_c (Meissner–Ochsenfeld effect). When $H > H_c$, superconductivity is destroyed and the material reverts to the normal state. The Meissner–Ochsenfeld effect is a fundamental property of the superconducting state and is often used as a means of detecting superconducting transitions. The entropy of the superconducting state is lower than the entropy of the normal state; the difference in entropy near $0\,K$ between the superconducting and normal states of a material relates to the electronic specific heat γ:

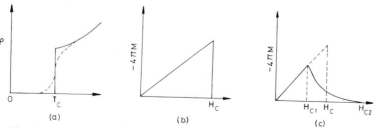

Figure 6.9   Properties of superconductors: (a) resistivity–temperature curve for a pure (solid) and an impure (broken) superconductor; (b) magnetization as a function of external field for type I superconductor; (c) magnetization curve for a type II superconductor.

$(S_S - S_N)_{T\to 0} = -\gamma T$. The behaviour of γ in the superconducting state is different from that of the normal state; γ is a linear function of temperature in the normal state but its temperature dependence is exponential in the superconducting state. The superconducting transition at zero magnetic field is a second-order phase transition since there is discontinuity in specific heat but no latent heat change.

Two kinds of superconductors are distinguished, depending on their magnetic behaviour. Materials in which superconductivity is destroyed abruptly at a critical field H_c are known as Type I superconductors (Fig. 6.9(b)). H_c is a function of temperature. Pure materials show this behaviour. In Type II superconductors, the magnetic flux starts to penetrate the material at a field H_{c1} lower than the thermodynamic critical field H_c. The magnetization then gradually decreases with increasing field strength until at $H = H_{c2}$ the superconducting state is completely destroyed (Fig. 6.9(c)). Between H_{c1} and H_{c2}, the material is said to be in a *vortex state* consisting of rod-shaped regions of normal conductivity within the superconducting bulk substance.

Because of the possibility of exciting technical applications, superconductivity has been widely investigated and the number of materials found showing superconductivity is continually growing. These include pure metals, alloys, intermetallic compounds and semiconductors. Some of the important structure types of superconducting materials are A-15 (β-W), rocksalt, tungsten bronze, perovskite, spinel and Chevrel phases; A-15 compounds (e.g. Nb_3Ge, Nb_3Sn) exhibit high T_c around 23 K. Ternary phases exhibiting superconductivity are known (Subba Rao & Balakrishnan, 1984). Among the ternary superconductors, Chevrel phases are unusual because not only do they exhibit highest critical magnetic fields but, more interestingly, magnetic order and superconductivity coexist in some of them. Superconductivity is also known in the glassy state, typical examples being Be and Ga, whose transition temperatures in the glassy state (8 K) are higher than in the crystalline state (Duwez & Johnson, 1978). Some semiconductors become superconducting at high pressures; high-pressure phases of silicon and germanium are superconducting. Although most nonmagnetic metals exhibit superconductivity, it is significant that alkali metals and coinage metals do not become superconducting down to very low temperatures. Certain organic charge-transfer salts of the type bis(ethylenedithiolo) tetrathiafulvalene triiodide (BEDT-TTF)$_2$I$_3$ exhibit superconductivity at ambient pressures (Jerome, 1985). The highest T_c in organic superconductors is around 13 K (Williams *et al.*, 1990).

Attempts have been made continually to understand the origin of superconductivity and predict its occurrence in materials. Bardeen, Cooper and Schrieffer (BCS) explained the nature of the superconducting state as resulting from the formation of bound electron pairs (*Cooper pairs*) whose energy is lower than the free-electron energy in a normal metal. The pairing occurs through electron–lattice interaction in such a way that an electron of wave vector $+ k$ and spin-up is paired with an electron of wave vector of $- k$ and spin-down. BCS theory does not provide an answer to the question of which substances can become superconducting and which cannot. Matthias (1955) suggested that the occurrence of superconductivity strongly depends on the crystal structure and on the average number of valence electrons per atom. For instance, in alloys of Nb_3Ge and BCC structures, T_c is maximum when the electron concentration is between 4 and 5 or 6 and 7. Krebs (1975) has explained superconductivity in terms of structure and chemical bonding. According to him, superconductivity is possible only if the structure permits formation of extended (delocalized) states, which at least in one spatial direction are not intersected by planar or conical nodal surfaces and if the corresponding energy band is not fully occupied. Allen & Mitrovic (1982) have discussed various aspects of attainable maximum T_c. It seems that any limit set by mode softening is wrong; lattice instability seems to be crucial.

The entire situation changed when Bednorz and Müller (1986) discovered 30 K superconductivity in a cuprate of the La–Ba–Cu–O system. This broke the apparent 23 K barrier for the T_c. Since 1986, a variety of cuprates have been synthesized and characterized with the maximum T_c reaching ~ 150 K. The cuprates are unusual in many ways and do not seem to follow the BCS model. We shall discuss the cuprates at length later in Chapter 7. Alkali and alkaline-earth derivatives of C_{60} (buckminsterfullerene) exhibit superconductivity with a maximum T_c of 35 K. We shall examine these fullerides along with other organic superconductors in Chapter 7.

6.3.4 *Dielectric and optical properties*

The dielectric constant of a medium is the constant of proportionality in the relation between the dielectric displacement and the strength of the electric field. The polarization of the medium is given by

$$P = \varepsilon_v \chi_e E = \varepsilon_v (\varepsilon_r - 1)E \tag{6.42}$$

where ε_v is the dielectric constant of vacuum, ε_r the dielectric constant of the material relative to vacuum and χ_e the electric susceptibility. The molar polarization in the case of a solid is expressed in terms of the *Clausius–Mossotti equation*:

$$\frac{\varepsilon_r - 1}{\varepsilon_r + 2} \cdot \frac{M}{\rho} = \frac{N_o \alpha}{3\varepsilon_v} \tag{6.43}$$

where M is the molecular weight, ρ the density, N_o the Avogadro number and α the polarizability.

Reorientation of dipoles in the direction of the field is characterized by a relaxation time, τ. When τ is small compared to the frequency of the applied voltage, instantaneous polarization occurs; when τ is large, the resultant polarization will be free from orientational contribution. A dispersion of polarization and hence of ε_r occurs when the frequency and τ^{-1} are close to each other and ε_r in this region is a complex quantity, $\varepsilon_r = \varepsilon_r' - i\varepsilon_r''$. In the region of dispersion there is a phase lag between the applied field and the instantaneous polarization; when the phase angle δ is small, $\tan \delta = \varepsilon_r''/\varepsilon_r'$. Furthermore,

$$\varepsilon_r' = \varepsilon_\infty - \frac{\varepsilon_s - \varepsilon_\infty}{1 + \omega^2 \tau^2} \tag{6.44}$$

$$\varepsilon_r'' = \frac{\omega\tau(\varepsilon_s - \varepsilon_\infty)}{1 + \omega^2\tau^2} \qquad (6.45)$$

Here, ε_s and ε_∞ are the static (low-frequency) and optical (high-frequency) dielectric constants respectively.

The susceptibility can be written as

$$\chi_e = \frac{P}{\varepsilon_v E} = \varepsilon_r - 1 = \frac{N\alpha/\varepsilon_v}{1 - (N\alpha/3\varepsilon_v)} \qquad (6.46)$$

When $(N\alpha/3\varepsilon_v) = 1$, both polarization and susceptibility go to infinity. At a critical temperature, T_c, the randomizing effect of temperature is balanced by the orienting effect of the internal field. Under such conditions, χ_e is given by the *Curie–Weiss law*

$$\chi_e = \frac{3T_c}{T - T_c} \qquad (6.47)$$

where T_c is the *Curie temperature*. This law is generally written in the form

$$\varepsilon_r = \varepsilon_\infty + \frac{C}{T - T_c} \qquad (6.48)$$

where the Curie constant $C = Np^2/3\varepsilon_v k$. This law is of importance in discussing dielectric properties of ferroelectrics and related materials.

The behaviour of dielectrics in alternating electric fields may be treated in the framework of forced harmonic oscillation. The displacement is then given by

$$x(t) = \frac{eE_o\exp(i\omega t/m)}{\omega_o^2 - \omega^2 + i\gamma\omega} \qquad (6.49)$$

where γ is the damping constant, ω_o is the natural frequency of vibration and e and m are the charge and mass of the vibrating particle. The complex susceptibility, $\chi = \chi' - i\chi''$, can be obtained from this equation. A maximum in χ'' at a frequency ω means that the material absorbs energy from the field (resonance absorption), and χ' varies rapidly with frequency in this region (anomalous dispersion). The dielectric catastrophe, which occurs when the ratio between molar refractivity R of atoms in the gaseous state and the molar volume in the condensed state becomes unity, marks the onset of metallization of a solid (Edwards & Sienko, 1983).

The effect of electromagnetic radiation on matter is to induce a dipole. In a transparent dielectric medium, only the velocity of electromagnetic radiation is reduced, depending on the refractive index of the medium, which is determined by its density. The propagation constant of electromagnetic waves is given by

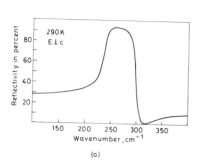
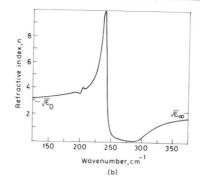

Figure 6.10   Optical properties of CdS: (a) experimental reflectance spectrum of single crystals of CdS; (b) refractive index n of CdS obtained from data given in (a) through the Kramers–Kronig relation.

$$K = \frac{n\omega}{c} + \frac{ik\omega}{c} \tag{6.50}$$

where n and k are real and imaginary parts of the refractive index. The absorption coefficient is equal to $2k\omega/c$ while the reflection coefficient is

$$R = \frac{(n-1)^2 + k^2}{(n+1)^2 + k^2} \tag{6.51}$$

The two refractive indices are related to the dielectric function by

$$\varepsilon_r(\omega) = \varepsilon'_r + i\varepsilon''_r = [n(\omega) + ik(\omega)]^2 \tag{6.52}$$

and we therefore have $\varepsilon'_r = n^2 - k^2$ and $\varepsilon''_r = 2nk$. The quantities $(n^2 - k^2)$ and $2nk$ are related to each other through *Kramers–Kronig relations* (Hodgson, 1971). If nk is known over the entire range of frequencies, $n^2 - k^2$ may be determined and vice versa. Experimentally one measures R and employs the Kramers–Kronig relations for understanding the behaviour of the systems. In Fig. 6.10 we show the reflectance spectrum of CdS crystal and the refractive index n obtained from the reflectance measurements through Kramers–Kronig analysis (Balkanski, 1971).

The response of electrons in a solid to an applied electromagnetic field is understood by

$$\varepsilon_r = 1 + \frac{\omega_p^2}{\omega_o^2 + \omega^2 + i\gamma\omega} \tag{6.53}$$

where ω_p, known as the *plasma frequency*, is given by $[(4\pi ne^2)/m]^{1/2}$; ω_o is natural frequency, m the electron mass, and γ the damping constant. For free electrons, as in a metal, $\omega_o = 0$ and for insulators ω_o has a finite value. The imaginary and the real parts of the dielectric constant become equal at the plasma frequency.

The dielectric constant of a transparent solid varies as a function of the frequency of the oscillating electric field in the nonabsorbing region of the electromagnetic spectrum. The refractive index similarly varies with the frequency (it generally increases with ω). Such dispersion behaviour of materials is of importance in the choice of materials for prisms, etc. Solids absorb electromagnetic radiation by different modes depending on the nature of bonding present.

Tightly bound electrons and ions give rise to narrow resonance absorptions, electrons in the ultraviolet region and ions in the infrared region. Resonance absorption in the infrared is also called *restrahlen absorption*. Absorption of radiation by loosely bound electrons is generally due to *interband transitions*; such absorption bands in semiconductors are broad and featureless, but with sharp absorption edge. The absorption edge of ZnS($E_g = 3.7\,\text{eV}$) is at 0.34 μm, while in InSb ($E_g = 0.2\,\text{eV}$) it is at 6.2 μm. Direct interband transitions occur when the difference between conduction-band and valence-band energies corresponds exactly to the energy of the transition, $hv = E_c(\vec{k}) - E_v(\vec{k})$. If the minimum of the conduction-band and the maximum of the valence-band energy states do not occur at the same k, indirect transitions occur.

Absorption of radiation occurs due to the excitation of electrons from impurities in solids. Transition-metal ions give absorption bands due to d–d transitions that are Laporte forbidden. Both donor and acceptor impurities in semiconductors give absorption bands due to photoionization at energies lower than the gap energies. *Excitons* (electron-hole pairs) in insulators give characteristic bands, the energy depending on how tightly bound the electron-hole pair is.

Electrons in metals and semiconductors give rise to *free-carrier absorption*, the absorption coefficient being proportional to the square of the incident wavelength (hence high in the infrared region for most metals). The reflectivity of metals is related to the plasma frequency, ω_p, by the relation

$$R \simeq 1 - 2\left(\frac{\omega}{\omega_p^2 \tau}\right)^{1/2} \tag{6.54}$$

Since the quantity in parenthesis is very small in metals, reflectivity is almost total up to high frequencies.

Inelastic scattering of radiation in solids is typified by the *Raman effect*, which involves the creation or annihilation of phonons or magnons. If a single phonon is involved, the scattering event is referred to as the first-order Raman effect; in second-order Raman effect two phonons are involved. The polarizability associated with a phonon mode can be represented as a power series of the phonon amplitude, u, as follows:

$$\alpha = \alpha_0 + \alpha_1 u + \alpha_2 u^2 + \alpha_3 u^3 + \dots \tag{6.55}$$

In the first-order effect $\alpha_1 u$ is always finite; the second-order effect arises from the $\alpha_2 u^2$

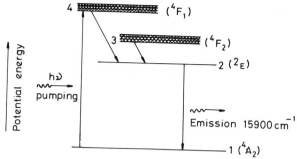

Figure 6.11   Schematic energy levels of Cr^{3+} ion in ruby used in laser operation.

term. *Brillouin scattering* is a special case of Raman scattering where the phonons involved are from the acoustic branch.

The electron-hole pair produced in the photo-excitation process reverts back to the original state by releasing energy to the lattice through creation of phonons or radiative recombination, giving rise to *luminescence*. Such a de-excitation can be induced at impurity sites called activators (e.g. Tl^+ in alkali halides). Absorption and emission processes in *photoluminescence* need not occur at the same impurity sites. For example, in $Ca_3(PO_4)_2$ doped with Ce and Mn, absorption occurs at Ce sites (sensitizer) and recombination at Mn sites (activator). This process involves energy transfer between two centres and the luminescence intensity therefore depends on the average distance between impurity centres. The frequency of luminescence depends on the activator ion; thus ZnS doped with Ag^+, Cu^{2+} and Mn^{2+} has emission bands in blue, green and yellow respectively. Electron-hole pairs can be introduced into semiconductors by applying high voltages, and radiative recombination gives rise to electro-luminescence. In the commonly used *light-emitting diode* (LED) visual displays, GaP is used to get electro-luminescence in the red.

Laser (*light amplification through stimulated emission of radiation*) action in materials such as ruby (0.01 at % Cr^{3+} in Al_2O_3) can be understood in terms of Fig. 6.11. Population inversion is achieved by optical pumping to energy levels 3 and 4 from which de-excitation occurs through a nonradiative process to level 2. Lasing occurs from level 2 to level 1 (giving 6943 Å radiation). The stimulated light has the special property of coherence; it has the same phase and direction of propagation as the incident photon. *Laser action* has been produced in semiconductor diodes (heavily doped III–V semiconductor p–n junctions), by applying a bias voltage.

We have assumed hitherto that the polarization produced in a solid is proportional to the electric field of the radiation (linear behaviour). The electric fields produced can be very high in laser beams and nonlinear effects therefore become significant. Polarization would then contain multiples of frequencies, as given by

$$P = \varepsilon_v\chi_1 E_0 exp(i\omega t) + \varepsilon_v\chi_2 E_0^2 exp(i2\omega t) + \varepsilon_v\chi_3 E_0^3 exp(i3\omega t) + \dots \quad (6.56)$$

In the linear case, only the first term contributes. The number of photons with

Table 6.2. *Examples of different types of transition-metal oxides*

d^0 metal oxides Sc_2O_3, TiO_2, V_2O_5, CrO_3, ZrO_2, Nb_2O_5, MoO_3, HfO_2, Ta_2O_5, WO_3	Diamagnetic semiconductors or insulators when pure, but exhibit n-type extrinsic conduction when doped or slightly reduced.
d^n metal oxides TiO, NbO, CrO_2, ReO_2, RuO_2, OsO_2, MoO_2, RhO_2, WO_2, IrO_2 and ReO_3	Metallic and Pauli paramagnetic (CrO_2 is ferromagnetic)
Ti_2O_3, Ti_3O_5, Ti_4O_7, Ti_5O_9, V_2O_3, V_3O_5, V_4O_7, VO_2, NbO_2 and Fe_3O_4	Exhibit temperature-induced nonmetal–metal transition
MnO, FeO, CoO, NiO, Cr_2O_3, Fe_2O_3 and Mn_3O_4	Mott insulators
f^n metal oxides PrO_2, Ln_2O_3 (Ln = rare earth), Pr_nO_{2n-2}, Tb_nO_{2n-2} and EuO	Insulators or hopping semiconductors. Paramagnetism characteristic of f^n configuration. EuO shows nonmetal–metal transition.

frequencies 2ω, 3ω ... will be small. Harmonic generation at 2ω is possible only if $\chi_2 \neq 0$; noncentrosymmetric crystals such as KH_2PO_4 (KDP) can do this.

6.4   Case studies
Relations between the structure and properties have been investigated in a variety of solids such as metal oxides, chalcogenides, pnictides and halides. In addition to studying model systems for testing theoretical predictions, solid state chemists have been preparing new classes of solids as well as novel members of known types of solids. In this section, we have chosen three classes of solids, viz. metal oxides, metal sulphides and metal fluorides, to discuss structure–property relations; we shall concentrate especially on their electrical and magnetic properties.

6.4.1   *Metal oxides*
It is convenient to discuss the oxides of transition metals and nontransition metals separately. Typical oxides of nontransition metals are Na_2O, MgO, Al_2O_3 and SiO_2. Their electronic structures consist of a filled valence band (derived mainly from oxygen $2p$) and an empty conduction band (derived from the outer shells of metal atoms) separated by a large gap (~ 10 eV). They are therefore diamagnetic insulators under ordinary conditions. Since the intrinsic activation energy for electronic conduction is higher than the energy required for the creation and migration of point defects, ionic conduction predominates over electronic conduction in such oxides at moderately high temperatures (Adler, 1975).

Transition-metal oxides constitute an enormous group of solids exhibiting a wide variety of electrical and magnetic properties (Goodenough, 1971; Rao & Subba Rao, 1974). Two classes of transition-metal oxides can be distinguished: those in which the metal ion has a d^0 electronic configuration and those where the d shell is partly filled (Table 6.2). The former class of oxides has a filled oxygen $2p$ valence band and an empty metal d conduction band. The energy gaps are around 3–5 eV; high-purity materials of this class exhibit intrinsic electronic conduction only at high temperatures. Metal oxides with d^0 cations at octahedral sites exhibit spontaneous ferroelectric and antiferroelectric distortions. Many of them lose oxygen at high temperatures, becoming nonstoichiometric. Oxygen loss or insertion of electropositive metal atoms into these oxides places electrons in the conduction band. The nature of electronic conduction in such materials depends on the strength of electron–phonon coupling and the width of the conduction band derived from metal d states. When the coupling is large and the band is narrow, small polarons are formed and such materials exhibit hopping conduction, e.g. $Na_xV_2O_5$. On the other hand, when the conduction band is broad, the material exhibits metallic properties, e.g. Na_xWO_3.

Transition-metal oxides containing partly filled d states can be metallic or semiconducting. A small group of them shows temperature-induced nonmetal–metal transitions (Table 6.2). Magnetic properties are also quite varied, ranging from Curie–Weiss paramagnetism through spontaneous magnetism to Pauli paramagnetism. In Table 6.2 we have also included typical rare-earth metal oxides containing localized $4f^n$ electrons (Eyring, 1974, 1979).

Metal oxides with d^n configuration exhibit metallic properties when the overlap between orbitals of the valence shells of constituent atoms is large. Two kinds of metallic behaviour can be distinguished: one due to strong cation–cation interaction arising from a small cation–cation separation and the other due to strong cation–anion–cation interaction arising from a large covalent mixing of oxygen $2p$ orbitals with cation d orbitals. Several isostructural series of transition-metal oxides, which include rocksalt, corundum, rutile and perovskite structures among others, exhibit systematic changes in electronic properties, where at least one member of the series shows properties characteristic of itinerant electrons, and the others properties of localized electrons. In these solids, the Fermi energy E_F that coincides with energy of the d^n manifold, lies within the energy gap between anionic bonding states and cationic antibonding states and determines the electronic properties. We shall discuss the electronic properties of a few typical isostructural series of transition-metal oxides in the light of concepts presented in Section 6.2.

Monoxides of 3d transition metals, TiO to NiO, possess the rocksalt structure and exhibit properties shown in Table 6.3. While TiO and VO exhibit properties characteristic of itinerant (or nearly itinerant) d electrons, MnO, FeO, CoO and NiO show localized electron properties. The properties can be understood in terms of the possible cation–cation and cation–anion–cation interactions in the rocksalt structure (Fig. 6.12(a)). Direct cation–cation interaction can occur through the overlap of cationic t_{2g} orbitals across the face diagonal of the cubic structure. When this interaction is strong ($R < R_c$ and $b > b_c$), cationic t_{2g} orbitals are transformed into a cation sublattice t_{2g}^* band; if this band is partly occupied, the material would be

Table 6.3. *Properties of 3d metal monoxides with rocksalt structure*

Oxide	$R(Å)$	$R_c(Å)^a$	Properties
TiO	2.94	3.02	metallic, Pauli–paramagnetic
VO	2.89	2.92	semimetal, weak temperature dependence of χ
MnO	3.14	2.66	semiconductor, Curie–Weiss, $T_N = 122\,K$
FeO	3.03	2.95	semiconductor, Curie–Weiss, $T_N = 198\,K$
CoO	3.01	2.87	semiconductor, Curie–Weiss, $T_N = 293\,K$
NiO	2.95	2.77	semiconductor, Curie–Weiss, $T_N = 523\,K$

^{a}After Goodenough (1971).

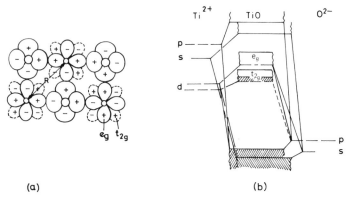

(a)   (b)

Figure 6.12   (a) Orbitals in the (100) plane of rocksalt structure showing cation–cation and cation–anion–cation interaction; (b) schematic energy band diagram of TiO.

metallic. The properties of TiO conform to this picture (Fig.6.12(b)). VO is less metallic and its magnetic susceptibility becomes temperature-dependent at low temperatures (Fig. 6.13) because the increased nuclear charge contracts the radial extension of $3d$ orbitals of vanadium and therefore decreases the overlap of the t_{2g} orbitals. Increase of nuclear charge across the TiO–NiO series has the effect of lowering the $3d$ manifold energy relative to the top of the anionic $2p$ states. This has two consequences. (i) The radial extension of the $3d$ wave functions decreases, resulting in a sharp decrease in the cation–cation transfer energy b_{cc}, rendering the t_{2g} electrons localized beyond VO. (ii) The cation–anion–cation transfer energy b_{ca} involving the e_g orbitals of cations increases across the series, since the covalent mixing parameter, λ, between the cation and anion orbitals is inversely proportional to the energy difference between the cationic d states and the anionic p states; this energy difference decreases across the series, being a minimum at NiO. The rocksalt structure permits $180°$ cation–anion–cation superexchange interaction through the metal e_g orbitals (Fig. 6.12(a)), which explains the antiferromagnetic behaviour of MnO, FeO, CoO and NiO. The increase of T_N in the series is explained in terms of the increase in the cation–anion transfer energy

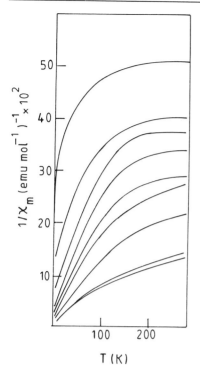

Figure 6.13   Temperature dependence of reciprocal molar magnetic susceptibility of VO_x. x varies from 0.79 for the curve at the top to 1.32 for the curve at the bottom. $x = 1.02$ and 0.99 for the curves 5 and 6 from the bottom. (After Banus & Reed, 1970.)

b_σ across the series. These four oxides are typical Mott insulators, remaining semiconducting up to the highest temperatures studied (1400 K).

A number of ternary oxides, $LiMO_2$ (M = V, Cr, Mn, Fe, Co and Ni) crystallize in ordered rocksalt structures. In M = V, Cr, Co and Ni compounds, the M^{3+} and Li^+ ions are ordered in alternate (111) cation planes, introducing a unique [111] axis and rhombohedral symmetry. In $LiVO_2$, the V–V distance is 2.84 Å $< R_c(V^{3+}) = 2.95$ Å. The susceptibility is Curie–Weiss-like at high temperatures but shows a sharp drop below 460 K, probably indicating cation–cation bonding in the (111) planes (Bongers, 1975). $NaVO_2$ which is isostructural with $LiVO_2$ does not show a similar magnetic transition since V–V distance is larger than $R_c(V^{3+})$ in this compound. $ACrO_2$ (A = Li, Na, K) compounds, which are isostructural with $LiVO_2$, exhibit two-dimensional antiferromagnetism (Le Flem et al., 1977). In $LiCoO_2$ and $LiNiO_2$, the trivalent transition-metal ions appear to be in the low-spin state. Lithium can be reversibly removed from some of these oxides by chemical or electrochemical means, rendering them useful electrode materials in solid state batteries (Chapter 3).

EuO is an interesting oxide in the rocksalt family (Honig & Van Zandt, 1975). It is a ferromagnetic ($T_c = 69$ K) insulator when pure and stoichiometric. Unlike the $3d$ oxides, superexchange interaction in EuO is ferromagnetic because the charge transfer

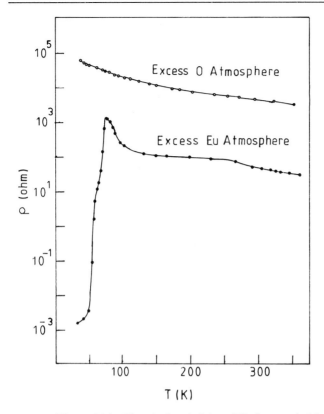

Figure 6.14   Electrical resistivity of EuO annealed in a europium-rich atmosphere (EuO_{1-x}) and an oxygen-rich atmosphere ($Eu_{1-x}O$). (Following Honig & Van Zandt, 1975.)

is from $4f$ to an empty $5d$ orbital. The electronic structure can be described as consisting of localized Eu $4f^7$ states separated from the empty conduction bands of Eu $5d$ (t_{2g}) and Eu $6s$ by about 1.4 eV. Below the localized $4f^7$ states lies a filled band of oxygen $2p$ states. Stoichiometric EuO shows an insulator–metal transition at 300 kbar. The mechanism of this transition involves promotion of a $4f^7$ electron into the $5d$ band with an accompanying fluctuation of the valence state of europium. EuO shows non-stoichiometry on both sides of the stoichiometric composition, $Eu_{1-x}O$ and EuO_{1-x}. Transport properties of these two defect oxides are different. In $Eu_{1-x}O$ holes are introduced into the $4f^7$ level and conduction is due to small polaron hopping; there is no anomaly in the resistivity at T_c. In EuO_{1-x}, electrons are introduced into the $5d$ band. The samples show a sharp semiconductor–metal transition at $50 K < T_c$ and a pronounced maximum in the resistivity above T_c (Fig. 6.14). This transition is remarkable because it is gigantic (the resistivity changes by about 13 orders in some samples) and the metallic state occurs on the low-temperature side. Although the electronic properties of this interesting oxide are not fully understood, it is clear that the transition in EuO_{1-x} is associated with an abrupt change in the number of carriers. The change is probably brought about by a splitting of the conduction band below T_c. The

Table 6.4. *Properties of transition-metal dioxides with rutile-type structure*

Oxide	$R(\text{Å})$	$R_c(\text{Å})^a$	Properties
TiO_2	2.96	3.0	Diamagnetic semiconductor, $E_g \sim 3\,eV$
VO_2	2.88	2.94	Semiconductor–metal transition at $T_t = 340\,K$
	$(2.65, 3.12)^b$		
CrO_2	2.92	2.86	Ferromagnetic ($T_c = 398\,K$) metal
$\beta\text{-}MnO_2$	2.87	2.76	Antiferromagnetic ($T_N = 94\,K$) semiconductor. Anomaly in resistivity at T_N.
MoO_2	2.52, 3.10	—	Diamagnetic metal. Structure distorted to mono clinic symmetry due to M–M bonding.
WO_2	2.49, 3.08	—	Diamagnetic metal. Structure distorted to mono clinic symmetry due to M–M bonding.
$RuO_2{}^c$	3.14	—	Pauli–paramagnetic metal

^{a}After Goodenough (1971).
^{b}Monoclinic structure.
^{c}OsO_2, RhO_2, IrO_2 and PtO_2 show similar properties.

stabilized band edge crosses the donor level associated with anion vacancies so that, at low temperatures, the donor levels give their electrons to the $5d$ band. In another model (Torrance *et al.*, 1972), the charge carriers trapped at vacancy sites are assumed to interact with the magnetic moments of nearby Eu^{2+} ions through exchange. In the low-temperature ferromagnetic state, all spins are aligned and the interaction does not hinder hopping between europium sites. The bound state wave function tends to be large and overlaps for sufficiently large defect concentration so that band formation and metallic conduction occur. As the temperature is increased, the Eu spins progressively disorder, and hopping from the vacancy becomes increasingly less likely. The electron tends to be more localized, forming the so-called magnetic polarons. In the magnetically disordered state, conduction occurs via hopping process.

Dioxides of several transition metals crystallize in the rutile structure (Fig. 1.7(b)). Their electronic properties are summarized in Table 6.4. The rutile structure provides the possibility of $135°$ cation–anion–cation interaction between corner-shared octahedra and $90°$ cation–anion–cation interaction between edge-shared octahedra. In addition, a direct cation–cation interaction is possible along the c-direction. Besides the octahedral crystal field at the cations, the tetragonal structure provides an axial field, which has the effect of splitting the triply degenerate metal t_{2g} orbitals into a nondegenerate $t_{\|}$ and a doubly degenerate $t_{\perp}$ orbitals. While $t_{\perp}$ can form π^* bands with anion p_{π} orbitals, $t_{\|}$ forms cation sublattice d bands if the cation–cation distance along the c axis $R < R_c$. Thus the rutile structure is unique, providing an opportunity for the material to become metallic through cation–cation interaction and/or through cation–anion–cation interaction. Electronic properties of rutile-related oxides have been discussed on the basis of these considerations by Rogers *et al.* (1969) and Goodenough (1971).

Some of the dioxides of transition metals exhibit a distorted (monoclinic) rutile

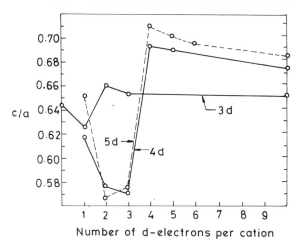

Figure 6.15   Variation of c/a ratio in rutile-type oxides with d electron configuration. (After Rogers *et al.*, 1969.)

structure where metal–metal distances alternate along the chain, indicating formation of metal–metal bonds. There is a direct correlation between the c/a ratio (Fig. 6.15) and the formation of metal–metal bonds in the rutile family of oxides. The c/a ratio is minimum for $3d^1$, $4d^2$, $5d^2$, $4d^3$ and $5d^3$ configurations and the corresponding dioxides exhibit monoclinic distortion due to metal–metal bonding. Among the first-row transition-metal dioxides, TiO_2 and VO_2 are the only rutile oxides where the intercationic distance is less than R_c. Ti^{4+} has no d electrons and therefore the material is an insulator. TiO_2 is easily reduced and the members Ti_nO_{2n-1} so formed exhibit interesting electronic properties (Rao & Subba Rao, 1974; Honig & Van Zandt, 1975).

   VO_2 is monoclinic at room temperature and transforms to the tetragonal structure at 340 K. The structural transition is accompanied by a semiconductor–metal transition. In the tetragonal form, since $R < R_c$, $t_{\parallel}$ orbitals overlap to produce a cation sublattice d band; this band overlaps the π^* band formed by $t_{\perp}$ orbitals. Since the highest band is partly filled, tetragonal VO_2 is metallic (Fig. 6.16(a)). In the monoclinic form, the cations are displaced to produce cation–cation pairs along the c axis, the intercation distances being alternately 2.65 and 3.12 Å instead of the uniform 2.88 Å distance in the tetragonal phase. There is also a ferroelectric component to the monoclinic distortion. The distorted structure traps vanadium d electrons in homopolar metal–metal bonding and the material is nonmagnetic and semiconducting (Fig. 6.16(b)). There is no magnetic ordering in the low-temperature phase and the sharp drop in magnetic susceptibility at the transition temperature is just an indication of the disappearance of the Fermi surface. We discuss the metal–insulator transition in VO_2 at some length in Section 6.5. In solid–solutions of VO_2 with other rutile type oxides, the transition temperature is lowered by substitutional impurities that expand the lattice and raised by ionic substitution that contracts the lattice. The effect is not a simple chemical simulation of hydrostatic pressure; whether the substituted ion suppresses or enhances the ferroelectric component of the transition is important.

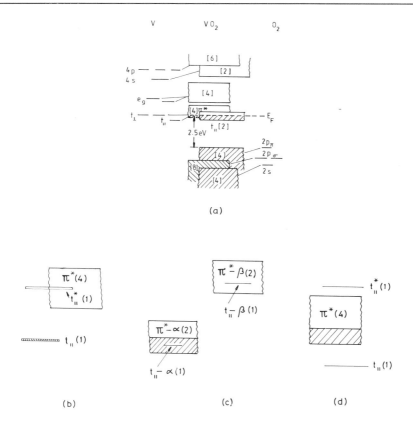

Figure 6.16   Schematic energy-band diagrams for rutile-related oxides: (a) tetragonal VO_2; (b) monoclinic VO_2 showing formation of metal–metal bonds through $t_\parallel$ electrons; (c) ferromagnetic CrO_2 showing the simultaneous presence of localized $t_\parallel$ electrons and collective π^* electrons; (d) MoO_2 and WO_2 showing the presence of metal–metal bonding and collective π^* electrons. (After Rogers et al., 1969 and Goodenough, 1971.)

Rationalizing the properties of CrO_2, MoO_2 and WO_2 is not as straightforward as in the case of VO_2. Tetragonal CrO_2 with the largest c/a ratio (among the $3d$ metal dioxides) is ferromagnetic and metallic while both MoO_2 and WO_2, with a distorted structure, are nonmagnetic and metallic. With two d electrons per transition-metal atom, the properties critically depend on whether $R < R_c$ or $R > R_c$. In CrO_2, one of the d electrons is in $t_\parallel$ and the other in $t_\perp$; the $t_\parallel$ electron would be localized since $R > R_c$ (large c/a ratio). But the $t_\perp$ electron finds itself in the π^* band formed through cation–anion–cation interaction. Intra-atomic exchange splits the states into α spin and β spin components (Fig. 6.16(c)). The α spin states are singly occupied and a ferromagnetic cation–anion–cation interaction between localized $t_\parallel$ electrons and collective π^* electrons renders the material simultaneously ferromagnetic and metallic.

In MoO_2 and WO_2, $R < R_c$ and one d electron per metal atom is involved in cation–cation homopolar bonding through $t_\parallel$ orbitals. The extra d electron partly fills the π^* band, rendering the oxides metallic (Fig. 6.16(d)). This simplified picture does not, however, account for the fact that the M—M bond order in these oxides is more than one. It appears that the true situation is more complex, involving simultaneous participation of $t_\perp$ electrons in M—M bonding as well as M—O π bonding. In MnO_2, with three d electrons per cation, both t and π^* electrons are localized ($R > R_c$ and $b_\pi < b_c$) and the oxide exhibits spin-only magnetism and is a semiconductor. There is an anomaly in electrical resistivity and specific heat at the magnetic ordering temperature, $T_N = 94$ K, in this oxide.

Sesquioxides of first-row transition metals crystallizing in the corundum structure (Ti_2O_3, V_2O_3, Cr_2O_3 and α-Fe_2O_3) show interesting properties. Ti_2O_3 and V_2O_3 exhibit semiconductor–metal transitions at 410 K and 150 K respectively. In Ti_2O_3, the transition is broad, occurring over a large temperature interval. In V_2O_3, the transition is accompanied by an antiferromagnetic ordering. Cr_2O_3 and α-Fe_2O_3 are both antiferromagnetic insulators. These properties can be rationalized qualitatively in terms of cation–cation and cation–anion–cation interactions in the corundum structure (Goodenough, 1971, 1974). In this structure, two different cation–anion–cation interactions (135° and 90°) and cation–cation interactions (in the basal plane and along the hexagonal c axis) are important. The structure provides a trigonal component to the octahedral crystal field, which splits the t_{2g} orbitals into a_{1g} (directed along c_H) and $e_g(\pi)$ directed in the basal plane. In Ti_2O_3, with one d electron per atom and a small hexagonal c/a ratio, both the cation–cation distances are smaller than the critical value ($R_{cc} < R_{bb} < R_c$) and the a_{1g} band is therefore filled and separated from the empty $e_g(\pi)$ band by a finite gap, accounting for the semiconducting behaviour at low temperatures. In V_2O_3, with two d electrons per vanadium and a large c/a ratio, $R_{cc} < R_{bb} < R_c$ and the $e_g(\pi)$ band is more stable than the a_{1g} band. The metallic nature of rhombohedral V_2O_3 indicates that both the bands overlap to some extent (Section 6.5).

In Cr_2O_3 and Fe_2O_3, the a_{1g} and $e_g(\pi)$ orbitals are half-filled. In Cr_2O_3, the e_g electrons are localized but the a_{1g} electrons are likely to be intermediate because the intercationic distance in the c direction is less than R_c, causing variation of the atomic moment with temperature. In Cr_2O_3, with no $e_g(\sigma)$ electrons, the cation–anion–cation interaction is weaker than in Fe_2O_3, with two $e_g(\sigma)$ electrons. Accordingly, T_N in Fe_2O_3 is much higher (953 K) than in Cr_2O_3 (307 K); Fe_2O_3 exhibits weak, parasitic ferromagnetism in the range $253 < T < 953$ K. In this temperature range, the atomic moments lie nearly in the basal plane. Antisymmetric spin coupling is parallel to the c axis and the anisotropic superexchange cants spins in the basal plane to produce a net moment. At $T < 260$ K, the spins are aligned parallel to the c axis, and the parasitic ferromagnetism disappears. On application of a strong magnetic field in the c direction, a first-order spin-flip transition is observed around 260 K (*Morin transition.*).

There are a number of isostructural ternary oxide families whose electrical and magnetic properties have been extensively studied. In ferrimagnetic spinels (see pp. 30–31, 299) strong A–B interaction between the tetrahedral and octahedral site cations favours antiparallel alignment of the spins. Electrical properties of oxide spinels are largely determined by interaction among B site cations. In garnets (e.g. $Y_3Fe_5O_{12}$) antiparallel coupling of magnetic moments of the octahedral and tetrahedral site Fe^{3+} results in ferrimagnetism.

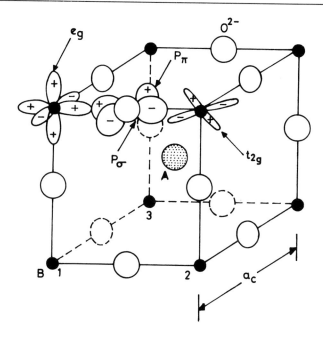

Figure 6.17   Perovskite structure, showing the possibility of cation–anion–cation interaction along the cube edge.

Ferrimagnetic $CoCr_2O_4$ has a conical spiral configuration. The cooperative Jahn–Teller effect shown by some of the spinels (e.g. $FeCr_2O_4$) is of considerable interest (Engelman, 1972). Other oxides showing this effect are rare-earth zircons (e.g. $TbVO_4, DyVO_4$) and $PrAlO_3$ (Gehring & Gehring, 1975). In vandate spinels, $AV_2^{3+}O_4$, the d-electrons are localized when $2.88\,\text{Å} < R_{V-V} < 2.97\,\text{Å}$. Fe_3O_4, which is an inverse spinel, has been of much interest in the past several decades, and we shall discuss some aspects of this oxide later in this chapter. It is noteworthy that the spinels $Li_{1-x}M_x^{2+}Ti_2O_4$ (M = Mg, Mn) and $Li_{1+x}Ti_{2-x}O_4$ show superconductivity (Johnston et al., 1973). In certain spinels such as $Ga_{0.8}Fe_{0.2}NiCrO_4$, spin-glass behaviour with randomly frozen clusters has been observed (Muraleedharan et al., 1985).

The perovskite structure is ideally suited for the study of 180° cation–anion–cation interaction of octahedral site cations (Fig. 6.17); cation–cation interaction is remote because of the large intercation distance along the cube-face diagonal. *Perovskite oxides* exhibit a variety of electronic properties. Thus, $BaTiO_3$ is ferroelectric, $SrRuO_3$ is ferromagnetic, $LaFeO_3$ is weakly ferromagnetic and $BaPb_{1-x}Bi_xO_3$ is superconducting. Goodenough & Longo (1970) and Nomura (1978) have compiled the properties of known perovskites. Several perovskite oxides exhibit metallic conductivity, typical examples being ReO_3, A_xWO_3, $LaTiO_3$, $AMoO_3$(A = Ca, Sr, Ba), $SrVO_3$ and $LaNiO_3$. Metallic conductivity in perovskite oxides is exclusively due to strong cation–anion–cation interaction.

We have listed important perovskite oxides containing B-site transition-metal atoms

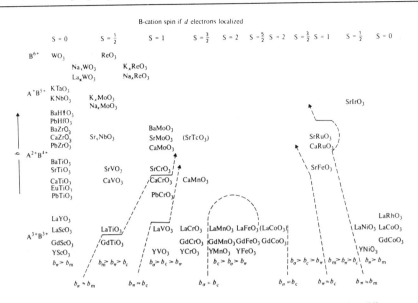

Figure 6.18 Perovskite oxides containing transition-metal ions in different spin configurations. Oxides are grouped into various regions on the basis of transfer energy b. (Following Goodenough, 1971.)

in Fig. 6.18; following Goodenough (1971, 1974), oxides having the same d electron configuration are grouped together in the columns of this figure. The entries in each column are arranged in the decreasing order of B cation–anion transfer energy b (B–O covalency) from top to bottom. Covalent mixing parameters λ_σ and λ_π (and hence b_σ and b_π) increase with increasing valence state of the B cation; for the same valence state mixing varies as $5d > 4d > 3d$. The influence of A cations on B–O covalency is indirect; acidic A cations decrease B–O covalency. $\lambda_\sigma > \lambda_\pi$ in all compounds.

The dotted lines in Fig. 6.18 representing $b_\pi = b_m$ (b_m is the critical value for spontaneous magnetism), $b_\pi = b_c$ and $b_\sigma = b_c$, separate oxides exhibiting localized electron behaviour from those with collective electron properties. Compounds in column 1 are insulators because the B cations are of d° electron configuration. Most of the compounds in column $2(S = \frac{1}{2})$ are metallic and Pauli paramagnetic; the line $b_\pi = b_m$ separates $LaTiO_3$ from $GdTiO_3$ because $GdTiO_3$ is a semiconductor with a ferromagnetic $T_c = 21\,K$. There is reason to believe, however, that stoichiometric $LaTiO_3$ is a semiconductor. $AMoO_3$ (A = Ca, Sr, Ba) and $SrCrO_3$ in the third column $(S = 1)$ are metallic and Pauli paramagnetic. Other compounds in this column are semiconducting and antiferromagnetic. The line $b_\pi = b_m$ separates metallic and Pauli paramagnetic $SrCrO_3$ from the antiferromagnetic semimetal $CaCrO_3$. The line $b_\pi = b_c$ separates $PbCrO_3$ from $LaVO_3$ because the latter exhibits a crystallographic transition at $T < T_N$ characteristic of localized electrons. The region $b_m > b_\pi > b_c$ appears to be quite narrow as revealed by electrical, magnetic and associated properties. Pressure experiments are valuable in the study of this region. Thus $dT_N/dP < 0$ in $CaCrO_3$ while

$dT_N/dP > 0$ in $YCrO_3$ and $CaMnO_3$. Since increasing pressure increases b_π (by decreasing lattice dimensions), $dT_N/dP > 0$ for $b_\pi < b_c$ (localized behaviour) and $dT_N/dP < 0$ for $b_m > b_\pi > b_c$ (collective behaviour). Compounds in columns 4, 5 and 6 are antiferromagnetic insulators. Since the intra-atomic exchange $\sim S(S+1)$ decreases the covalent mixing, it is natural that maxima in the curves $b_\pi = b_c$ and $b_\sigma = b_c$ corresponding to smallest values of b_π and b_σ occur in the middle of the columns with $S = \frac{5}{2}$. Rare-earth orthoferrites with $S = \frac{5}{2}$ are antiferromagnetic insulators and exhibit parasitic ferromagnetism. The important contributions here are: (a) Fe^{3+} spins canted in a common direction either by cooperative buckling of oxygen octahedra or by anisotropic superexchange, and (b) canting of antiferromagnetic rare-earth sublattice because of interaction between two sublattices.

$LaCoO_3$ is shown twice in Fig. 6.18 in both the $S = 2$ column and the $S = 0$ column at the end, since Co^{3+} can occur in the low- or the high-spin configuration. The compound exhibits a transition from a localized-electron state to a collective-electron state (Section 6.5). In the ninth column of Fig. 6.18, perovskites containing d^4 cations are placed. Of the three compounds in this column, $SrRuO_3$ is a ferromagnetic metal ($T_c = 160$ K); $CaRuO_3$ is antiferromagnetic ($T_N = 110$ K) with a weak ferromagnetism. Since both the compounds have the same RuO_3 array, the change from ferromagnetic to antiferromagnetic coupling is significant. Although $SrFeO_3$ is placed in the same column on the assumption that $Fe^{4+}:3d^4$ is in the low-spin state, recent work (Takano & Takeda, 1983) shows that Fe^{4+} in this oxide is in the high-spin state down to 4 K. $CaFeO_3$, on the other hand, shows disproportionation of Fe^{4+} to Fe^{3+} and Fe^{5+} below 290 K. In the last but one column containing $S = \frac{1}{2}$ cations, metallic and Pauli paramagnetic $LaNiO_3$ should be separated from antiferromagnetic $YNiO_3$ and $LuNiO_3$, indicating that in $LaNiO_3$ $b_\sigma > b_m$ and in $YNiO_3$ $b_\sigma < b_m$. Similarly, $LaCoO_3$ should be separated from $LaRhO_3$ because the latter is a narrow gap semiconductor with a filled $t_{2g}(\pi^*)$ band and an empty $e_g(\sigma^*)$ band.

Oxide pyrochlores of the general formula $A_2B_2O_7$ show interesting electronic properties (Subramanian et al., 1983). Ferromagnetic pyrochlores of rare earths have also been described (Subramanian et al., 1988). A composition-dependent metal–semiconductor transition has been found in $A_2(Ru_{2-x}A_x)O_{7-y}$ where $A =$ Bi or Pb (Beyerlein et al., 1988).

Hexagonal, cubic, and intergrowth bronzes formed by WO_3 with alkali, hydrogen, and other metals were discussed in earlier chapters. Electrical transport and other properties of WO_3 bronzes have been reviewed in the literature (Doumerc et al., 1985; Greenblatt, 1994). MoO_3 forms different varieties of bronzes (Greenblatt, 1988, 1994): blue bronzes of the type $A_{0.3}MoO_3$ (A = K,Tl,Rb), which are quasi-one-dimensional metals with charge-density wave (CDW) instability; purple bronzes, $A_{0.9}Mo_6O_{17}$ (A = Na,K), which are quasi-two-dimensional metals with CDW instability; $Li_{0.9}Mo_6O_{17}$ which is one-dimensional and superconducting ($T_c \approx 2$ K); red bronzes $A_{0.33}MoO_3$ (A = K,Tl,Rb), which are semiconducting; and $Li_{0.33}MoO_3$, which is violet and three-dimensional with low resistivity (Tsai et al., 1987; Collins et al., 1988). Hydrogen molybdenum bronzes, H_xMoO_3, of different compositions ($0 < x \leq 2.0$) with structures related to MoO_3, have been characterized (Greenblatt, 1988). Conductivity measurements have been made on some of these hydrogen bronzes (Barbara *et al.*, 1988). Anisotropic electronic properties of $CsP_8W_8O_{40}$, which has a unique structure, have been measured (Wang & Greenblatt, 1988).

6.4.2   *Metal sulphides*

Structures and properties of transition-metal chalcogenides are quite different from those of the corresponding oxides because of the large covalency of the metal–chalcogen bonds. Covalency reduces the formal charge on the metal atom and favours metal–metal bonding. The factors that contribute to differences between metal oxides and metal sulphides are the following: (a) large polarizability of anions, favouring the formation of layered structures with van der Waals bonding between layers, (b) sulphur–sulphur bonding, giving rise to molecular anions and (c) participation of anion $3d$ orbitals in bonding, stabilizing trigonal prismatic coordination of anions. Electrical and magnetic properties of transition-metal sulphides have been investigated extensively (Hulliger, 1968; Jellinek, 1972: Rao & Pisharody, 1976). While the electrical properties vary from semiconducting to metallic (superconducting in some cases), magnetic properties vary from weak diamagnetism (through localized magnetism) to temperature-independent Pauli paramagnetism. Jellinek (1968) has discussed these variations in general terms on the basis of a qualitative band model. Goodenough (1974) has made a comparative study of the properties of fluorides, oxides and sulphides of divalent transition metals in terms of the conceptual phase diagrams (Section 6.2.3).

The general features of transition-metal monosulphides and disulphides can be rationalized in terms of one-electron energy diagrams similar to those of oxides; the $(n + 1)s$ and $(n + 1)p$ orbitals of the transition metal interact with the $3s$ and $3p$ orbitals of sulphur to form a valence band (mainly due to sulphur) and a conduction band (mainly due to metal), the energy gap between them being a few electron volts. The nd orbitals of the transition metal, which occur in the gap and are partly filled, determine the electronic properties. They can remain localized on the cation or form band states, depending on the crystal structure and internuclear distances. In some instances, the d bands overlap with either the conduction band or the valence band. The d states are usually split up into sublevels due to ligand fields, Jahn–Teller distortion and intra-atomic exchange interaction. Besides these, metal–metal interactions can also split d bands into narrow sub-bands.

The d^n manifolds in $3d$ metal monosulphides become more stable with increasing atomic number until the shell is half-filled, just as in the monoxides. The larger covalency in monosulphides leads, however, to a smaller U and hence a smaller transfer energy or the formation of itinerant states. A smaller transfer energy would also mean a larger critical distance R_c for the formation of cation sublattice d bands through cation–cation interaction. Such cation sublattice d bands are formed in TiS and VS. Stoichiometric VS is unstable at atmospheric pressure and seems to disproportionate to V_7S_8 and V_9S_8. The magnetic susceptibility of nonstoichiometric $V_{1-x}S$ phases (Fig. 6.19) reveals the presence of both localized and itinerant d-electrons. In CrS and MnS, the d electrons are localized. The localized moments found in CrS and MnS as well as

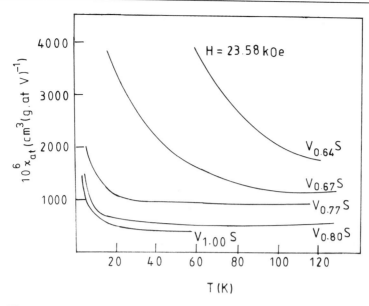

Figure 6.19   Magnetic susceptibility of vanadium sulphides as a function of temperature and composition. (After De Vries & Haas, 1973.)

the Jahn–Teller distortion in CrS reveal that $b_\sigma < b_c$. In CoS, NiS and CuS, we find definite evidence for itinerant σ^* electrons due to $b_\sigma > b_c$. In CuS, the value of U is very small since $b_\sigma > b_m$, bringing the d^{10} manifold of copper close to the top of the valence band, thus rendering the formal valence of copper ambiguous. Indeed, the structure of CuS (covellite) is different from that of the other monosulphides containing two types of copper atoms and S^{2-} and S_2^{2-} ions. Electronic properties of FeS (Fig. 6.20) reveal that $b_\sigma \approx b_c$ and both itinerant and collective d electrons coexist in this sulphide. The high-temperature (NiAs) structure of FeS distorts at 413 K to form triangular metal clusters. The spontaneous magnetism ($T_N = 600$ K), the spin-flip transition at 433 K and the high-spin state of iron indicate that the α-spin electrons are localized but the β-spin electrons are delocalized in FeS. NiS with the NiAs structure is metastable and exhibits a first-order phase transition at $T_N = 264$ K without change in crystal symmetry; the change in electronic properties (Fig. 6.20) at T_N indicates that the transition is from a strongly correlated antiferromagnetic phase to a weakly correlated Pauli-paramagnetic phase and that $b_\sigma = b_m$. Photoemission studies of $3d$ metal monosulphides in the valence-band region are consistent with their electronic properties, showing definite metallic character for TiS, VS and CuS and insulating character for CrS and MnS (Gopalakrishnan et al., 1979).

Electronic properties of the disulphides of the fourth, fifth and sixth group transiton metals crystallizing in the CdI_2(1T), MoS_2(2H, 3R) and related polytypic structures have been rationalized by similar considerations. Covalent interaction between quadrivalent cations and sulphide ions is so strong that collective d states are formed in all the cases. The properties crucially depend on the electronic configuration and the

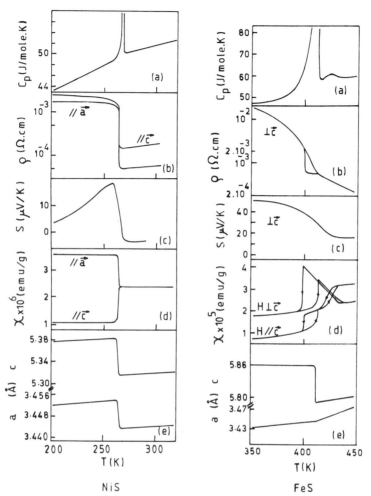

Figure 6.20   Changes in physical properties of NiS and FeS at the phase transition: (a) specific heat; (b) resistivity; (c) thermoelectric power; (d) magnetic susceptibility and (e) lattice parameters. (After Coey *et al.*, 1976.)

coordination of the transition metal. Thus, disulphides of d° cations (Ti, Zr, Hf) are expected to be diamagnetic semiconductors, those of d^1 cations (V, Nb, Ta) metals, and those of d^2 cations (Mo, W) in trigonal prismatic coordination semiconductors (Fig. 6.21). Electrical, magnetic and optical properties of layered disulphides (Hulliger, 1977; Subba Rao & Shafer, 1979) are broadly consistent with this picture, with the exception of TiS_2, which shows metallic behaviour. TiS_2 appears to acquire metallic character through a slight overlap of the empty $3d$ band with the top of the filled valence band (Fig. 6.21). This conclusion is supported by photoemission studies (Williams, 1976), which show a finite density of states near E_F, unlike other Group IVb disulphides. The first peak in the TiS_2 spectrum is unmistakably d-like and shows the expected increase in intensity between HeI and HeII. The results support presence of definite $Ti(3d)/S(3p)$ overlap in this compound, unlike in the other Group IVb disulphides.

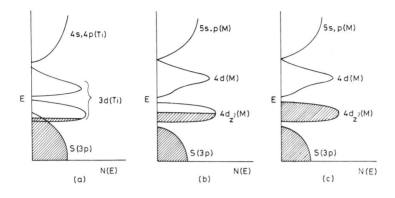

Figure 6.21   Schematic band structures of (a) $TiS_2(1T)$; (b) NbS_2 (2H) and (c) MoS_2.

Wilson & Yoffe (1969), as well as Huisman *et al.* (1971), have proposed qualitative energy band schemes for Groups Vb and VIb dichalcogenides containing transition metals in the trigonal prismatic coordination. The trigonal prismatic ligand field splits the d orbitals into three sets: d_{z^2} (lower in energy), $d_{x^2-y^2}$, d_{xy} and d_{xz}, d_{yz}, which form narrow bands. The d bands may or may not overlap with the valence or conduction bands. In the model of Huisman *et al.*, there is considerable overlap of the d bands with the conduction and valence bands, while in the Wilson–Yoffe model, the d_{z^2} band does not overlap with the valence band. Both the models account for the metallic and the semiconducting properties of d^1 and d^2 dichalcogenides respectively. Band structure calculations and photoemission studies support many of the features of the simple models. Group Vb transition-metal disulphides and diselenides exhibit intrapolytypic transitions at low temperatures due to the formation of charge-density waves (Section 4.9). The transitions are accompanied by anomalies in magnetic susceptibility and electrical resistivity as well.

   The pyrite disulphides, $MS_2(M = Mn–Cu)$, consisting of S_2^{2-} ions and octahedral-site divalent cations, display a wide range of electronic properties (Bongers, 1975). MnS_2 is an antiferromagnetic ($T_N = 78$ K) insulator; susceptibility above T_N is characteristic of high-spin $Mn^{2+}(3d^5)$. FeS_2 is a nonmagnetic semiconductor indicating a low-spin $Fe^{2+}(3d^6)$. CoS_2 is a ferromagnetic metal ($T_c = 115$ K); the saturation magnetization corresponds to 0.8 unpaired electrons per cobalt. NiS_2 is, surprisingly, an antiferromagnetic semiconductor ($T_N \sim 50$ K and $\theta = -1800$ K); CuS_2 is metallic and superconducting ($T_c < 1.5$ K). The changes in properties of MS_2 pyrites from localized d electrons in MnS_2 to strongly correlated itinerant electrons in CoS_2 and NiS_2 to weakly correlated itinerant electrons in CuS_2 can be understood in terms of the expected variation of $b_\sigma \sim \varepsilon_\sigma \lambda_\sigma^2$ on passing from one compound to another. In MnS_2, the d electron manifold is localized as in MnO and MnS. In the other pyrites, covalent mixing of metal e_g orbitals through σ bonding is sufficiently strong ($b_\sigma > b_c$) to create not only a narrow σ^* band of itinerant d electrons but also a low-spin state for cations. Thus FeS_2 has a filled $t_{2g}(e_\pi$ and $a_1)$ that is separated from $e_g(\sigma^*)$ by a definite band gap

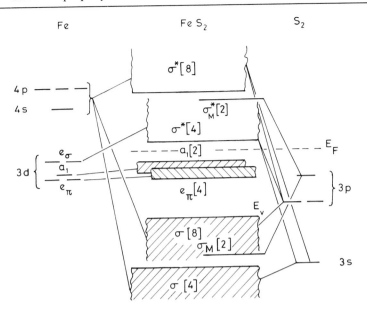

Figure 6.22   Schematic band structure of pyrite, FeS$_2$. (Reprinted from Goodenough, 1974 by courtesy of Marcel Dekker, Inc.)

of ~ 0.9 eV (Fig. 6.22). The d bands of CoS$_2$ and NiS$_2$ are essentially similar except that the σ^* band is split by correlation energy. The quarter-filled e_g^* band of CoS$_2$ is spin-polarized to give α and β spin components. If the two components did not overlap, CoS$_2$ would be a ferromagnetic semiconductor and the ferromagnetic moment would correspond to one electron. Metallic conductivity and reduced moment of CoS$_2$ indicate a small overlap of α and β spin components. NiS$_2$ with a half-filled σ^* band is a Mott insulator; the e_g band is split into spin-up and spin-down sub-bands due to electron repulsion. In contrast to NiS$_2$, NiSe$_2$ is a metal and the NiS$_{2-x}$Se$_x$ solid solutions (Bouchard et al., 1973; Wilson, 1985) show a semiconductor–metal transition (Fig. 6.23). Under high pressures (30–40 kbar), NiS$_2$ shows a transition to the metallic state (Wilson, 1972). In CuS$_2$, $b_\sigma > b_m$ and the Cu d band probably drops below the top of the anion $3p$ band creating holes as indicated by the p-type metallic conduction. X-ray photoelectron spectra of pyrite disulphides in the core and valence regions are consistent with the electronic structures discussed here (van der Heide et al., 1980).

The trisulphides (and triselenides) of Ti, Zr, Hf, Nb and Ta crystallize in one-dimensional structures formed by MS$_6$ trigonal prisms that share opposite faces. Metal atoms in these sulphides are formally in the quadrivalent state, and part of the sulphur exists as molecular anions, M^{4+}S$_2^{2-}$S^{2-}. TaS$_3$ shows a metal–insulator transition of the Peierls type at low temperatures (Section 4.9). NbS$_3$ adopts a Peierls distorted insulating structure suggesting the possibility of a transformation to a metallic phase at high temperatures, but does not transform completely to the undistorted structure. Electronic properties and structural transitions of these sulphides have been reviewed (Rouxel et al., 1982; Meerschaut, 1982; Rouxel, 1992).

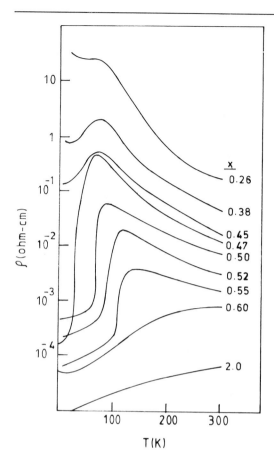

Figure 6.23   Electrical resistivity of $NiS_{2-x}Se_x$ showing semiconductor–metal transition. (After Bouchard et al., 1973.)

A large number of binary metal-rich sulphides (S/M ratio < 1.0) possessing metal–metal bonds are known (Franzen, 1978). Typical of them are V_3S, V_5S_4, Zr_2S, Hf_2S, Ta_6S and $Nb_{21}S_8$. Stoichiometry of such solids cannot be understood in terms of the traditional concepts of valence, owing to the presence of extensive metal–metal bonding.

Ternary sulphides containing two different metal atoms are legion. Both the metal atoms in such solids may be transition metals, or one metal is alkali or alkaline earth and the other transition metal. Examples for the first category are the thiospinels and the MM'_2S_4 phases belonging to the Cr_3S_4 type. Their properties have been adequately reviewed in the early literature (Goodenough, 1967). Ternary sulphides containing alkali metals and transition metals that exhibit chain and layered structures have been discussed by Bronger (1981). Among the large number of ternary sulphides, $BaVS_3$ and $Ba_{1+x}Fe_2S_4$ are noteworthy. $BaVS_3$, crystallizing in a chain ($CsNiCl_3$) structure, exhibits structural and electronic transitions below 240 K arising from coupling of V^{4+} :$3d^1$ electrons in the V–V chain (Sayetat et al., 1982). $Ba_{1+x}Fe_2S_4$ is a unique ternary sulphide system exhibiting infinitely adaptive chain structures. Electrical and magnetic properties show drastic changes with barium nonstoichiometry. Stoichiometric $BaFe_2S_4$, like $KFeS_2$, is a linear chain antiferromagnet showing semiconducting behaviour. Other members of the $Ba_{1+x}Fe_2S_4$ family, on the other hand, are mixed-valence iron sulphides exhibiting Curie–Weiss paramagnetism arising from extra electrons associated with Fe^{2+} (Swinnea & Steinfink, 1982). Ternary rare-earth sulphides $LnMS_3$ (Ln = La, Nd, Gd and M = Ti, V or Cr), crystallizing in a layer structure consisting of alternating LnS and MS_2 layers, are known. The titanium and vanadium compounds show itinerant d electron properties whereas the chromium compounds exhibit localized Cr^{3+} d electrons (Murugesan et al., 1981). The change in electronic properties in this series on passing from titanium and vanadium to chromium is reminiscent of $LaMO_3$ perovskites and $NaMS_2$ sulphides. Ternary chalcogenides which have attracted considerable attention because of their unusual electronic properties are the Chevrel phases, $M_xMo_6X_8$ (X = S, Se, Te). Structures and electronic properties of these solids have been extensively discussed in the literature (Yvon, 1979; Maple & Fisher, 1982; Schöllhorn, 1983).

Among the binary sulphides of rare earths (Holtzberg et al., 1970), the monosulphides crystallizing in the rocksalt structure and Ln_3S_4 sulphides adopting the Th_3P_4 structure are noteworthy. The monosulphides, with the exception of Sm, Eu and Yb, are metallic; magnetic moments indicate trivalency for the rare earths. The metallic nature of these sulphides arises from one of the electrons of the rare earth getting delocalized in the $5d$ state. Monosulphides of Sm, Eu and Yb are insulating and their magnetic properties indicate divalent f^6, f^7 and f^{14} electronic configurations respectively for the rare-earth atom. The divalent monosulphides exhibit pressure-induced transitions to the metallic state due to *valence fluctuation* in the rare earth (Jayaraman et al., 1973, 1975). The transition in SmS has been extensively investigated. The isostructural transition to the metallic phase in SmS occurs at 6.5 kbar and it can be induced by substitution of tervalent rare earths such as Gd^{3+}. The change to the metallic state arises from delocalization of a $4f$ electron into the $5d$ band $(4f^6 5d^0 \rightarrow 4f^5 d^1)$, which is accompanied by a fluctuation of samarium valence from 2^+ toward 3^+.

Ln_3S_4 (Ln = La, Ce, Pr, Nd, Sm, Eu, Gd), crystallizing in the Th_3P_4 structure, constitutes another isostructural series of sulphides. Eu_3S_4 and Sm_3S_4 are mixed (intermediate) valence compounds containing trivalent and divalent cations on equivalent sites. Both of them exhibit hopping semiconduction with an activation energy of 0.1–0.2 eV. Magnetic susceptibility of these solids is consistent with the presence of divalent and trivalent rare-earth ions. In Eu_3S_4, there is an abrupt change in resistivity around 185 K. While there is indication of some kind of charge ordering in Eu_3S_4 at low temperatures, there is no ordering in Sm_3S_4 down to very low temperatures. The activation energy for electron hopping in Eu_3S_4 and Sm_3S_4 decreases with pressure, $dE_a/dP = -1.8 \pm 0.3$ meV/kbar for Eu_3S_4 (Röhler & Kaindl, 1980). It is surmised from a linear extrapolation of the variation of activation energy with pressure that the activation energy would vanish at around 110 kbar and the material would transform to a homogenous mixed-valent compound with metallic conductivity. A neutron λ-diffraction study of Eu_3S_4 (Wickelhaus *et al.*, 1982) has shown a small rhombohedral distortion below the transition temperature, but it is too small to indicate a separation of rare-earth ions into Eu^{2+} and Eu^{3+}. It is more probable that the rare-earth ion sites split into two or more sets of intermediate valency, each having its own activation energy for electrical conduction. Mixed valency of Sm_3S_4 and Eu_3S_4 has not yet been completely understood (Holtzberg, 1980; Wachter, 1980). Other Ln_3S_4 compounds containing tervalent rare earths exhibit metallic behaviour. They exhibit metal-deficient nonstoichiometry, the limiting composition being Ln_2S_3. In the series Ln_3S_4–Ln_2S_3 there is a continuous change from metallic to insulating behaviour (the structure remaining the same); some of the members exhibit superconductivity (Ikeda *et al.*, 1980). The thermoelectric efficiency of some of the materials has been investigated (Taher & Gruber, 1981).

6.4.3   *Metal tellurides*
Transition-metal tellurides occupy a special place amongst the chalcogenides. In the dichalcogenides, MX_2 (X = S, Se, Te), S and Se exist as discrete anions, X^{2-} or X_2^{2-}, with the oxidation state of the chalcogen as -2 or -1, forming layered (CdI_2 or MoS_2) or three-dimensional (pyrite, marcasite etc.) structures. On the other hand, this clear-cut distinction between layered and three-dimensional structures gets blurred in the transition-metal ditellurides, MTe_2 (Jobic *et al.*, 1992), since the loss of bond directionality and decrease in electronegativity give rise to anionic *sp* valence states of Te which are higher in energy than those of S or Se. This results in polymeric anionic networks with fractional oxidation states in many of the tellurides. The formation of anionic associations in the solid state and the cationic–anionic redox-competition in chalcogenides has been extensively examined by Rouxel (1993).

IrTe$_2$ is an illustrative case to examine. While IrS_2 and $IrSe_2$ crystallize in a marcasite-related structure where half of the anions form pairs and the cation is trivalent, $Ir^{3+}X^{2-}(X_2^{2-})_{1/2}$ (X = S, Se) (Hulliger, 1968), $IrTe_2$ has traditionally been assigned a layered CdI_2 structure. Jobic *et al.* (1991) have shown that the structure of this solid is three-dimensional forming fractional Te–Te bonds with an average anion oxidation state of -1.5. The structure preserves the trivalency of iridium, but the tellurium anions are polymeric instead of being discrete. A similar behaviour obtains in other MTe_2 phases where M = Co, Ni, Cu, Zn, Rh, Pd and Pt (Jobic *et al.*, 1992). The disulphides and diselenides of these metals crystallize in pyrite or marcasite structures, but the ditellurides are isostructural with $IrTe_2$ adopting a three-dimensional CdI_2-like structure; the polymerized anions possess a fractional oxidation state between -1.0 and -1.5, as reflected in short c/a ratios and short Te–Te distances. The effect of short interlayer Te–Te distances in CdI_2-type MTe_2 compounds has been examined by electronic structure calculations (Canadell *et al.*, 1992). The results reveal that the top of Te *p*-states overlaps significantly with the bottom of the metal *d*-states, causing a substantial electron transfer from the *p*- to the *d*-states. The $p \rightarrow d$ electron transfer depends crucially on the interlayer Te–Te distance.

Another interesting example of the special behaviour of tellurides is provided by WTe_2 and its relatives (Mar *et al.*, 1992). WS_2 and WSe_2 crystallize in the layered MoS_2

structure where W(IV) is trigonal-prismatic coordinated, giving rise to insulating and diamagnetic properties. WTe_2 (and the high-temperature form of $MoTe_2$) crystallize in a distorted CdI_2-like structure and are metallic; the Mo compound is superconducting. The presence of metal–metal bonding with almost no Te–Te bonding in a CdI_2-like structure accounts for the properties of WTe_2 and $MoTe_2$. In the isotypic analogues, $MM'Te_4$ (M = Nb, Ta; M' = Ru, Os, Rh, Ir), metal–metal bonding decreases and Te–Te bonding increases as the d-electron count increases preserving the WTe_2 structure. Similar considerations seem to apply for $TaPtTe_5$ and its analogues containing Nb, Ni or Pd (Mar & Ibers, 1991) where Te–Te interaction exists across the van der Waals gap between adjacent layers. Solid state chemistry of transition-metal tellurides appears to be a potentially fruitful area which deserves greater attention in the years to come.

6.4.4   Metal nitrides

Since the discovery of high T_c superconductivity in copper oxides, there has been considerable effort to discover other superconductors containing transition metals. Among the various possibilities, transition-metal nitrides have been considered to be good candidates because nitrogen lies to the left of oxygen in the periodic table forming a trinegative anion (N^{3-}) whose anionic p-states are higher in energy ($\sim 1.5\,eV$) than those of oxygen (DiSalvo, 1990). Accordingly, nitrides are expected to produce extensive mixing of cation/anion states at the Fermi level – a situation similar to that in superconducting cuprates – with many transition metals left of copper in the periodic table. Several ternary nitrides have been synthesized in recent years. Among them, mention must be made of Ca_3MN (M = group IV or group V element) which crystallize in an antiperovskite structure (Chern et al., 1992), MNiN (M = Ca, Sr, Ba) which possess a layered structure with linear Ni–N–Ni chains (Chern & DiSalvo, 1990), $MTaN_2$ (M = Na, K, Rb, Cs), $LiMN_2$ (M = Mo, W) and MWN_2 (M = Mn, Fe, Co) – all possessing layered structures (Jacobs & von Pinkowski, 1989; Elder et al., 1992; Bem et al., 1994). Among these, CaNiN, where nickel exists in a formal oxidation state of $+1:3d^9$, is metallic and paramagnetic ($0.39\,\mu_B$). $LiMoN_2$, which has a layered structure where Mo(V):$4d^1$ exists in trigonal prismatic coordination and Li in octahedral coordination, is also metallic but Pauli paramegnetic. $MnWN_2$ occurs in two different modifications, α and β. The α-form is isostructural with $LiMoN_2$ and has the AA'BB'CC' sequence of metal layers. The β-form adopts the ABAC sequence similar to $FeTa_3N_4$. In spite of many efforts, it has not been possible to prepare ternary nitrides with mixed valency of the transition metal showing superconducting properties.

6.4.5   Metal fluorides

Unlike metal oxides and sulphides, the fluorides are highly ionic solids exhibiting localized electron properties because of the high electronegativity of fluorine. Covalency of the metal–fluorine bond is never large enough to create itinerant electrons (excepting probably in PdF_3 at high pressures). The small covalency also decreases cation–cation interaction in fluorides. While the electronic properties of fluorides are determined by the nature of metal–fluorine bonds, the crystal structures

Table 6.5. *Structural analogies between fluorides and oxides*

Formula	Structure	Examples
AX	NaCl	Alkali fluorides and alkaline-earth oxides
AX_2	Rutile	Transition-metal difluorides and dioxides
AX_3	ReO_3	VF_3 and ReO_3
ABX_3	Perovskite	$KNiF_3$ and $SrTiO_3$
A_2BX_4	K_2NiF_4	K_2NiF_4 and La_2NiO_4
AB_2X_4	Spinel	Li_2NiF_4 and $MgFe_2O_4$
$A_3B_3C_2X_{12}$	Garnet	$Na_3Li_3Fe_2F_{12}$ and $Y_3Fe_5O_{12}$

are dictated by ionic size and Madelung energy. The radius of fluorine ion (1.33 Å) is comparable to the oxygen ion radius (1.40 Å) and we therefore find structural similarities between several families of metal oxides and metal fluorides (Table 6.5). Size similarity also enables substitution of the oxide ion by the fluoride ion in some instances (e.g. $VO_{2-x}F_x$ and $MoO_{3-x}F_x$). Since fluorine can form only a uninegative ion, the formal valence of a cation in a fluoride is lower than in an isostructural oxide, but the extreme electronegativity of fluorine favours higher oxidation states in the case of some metals. The electrical, magnetic and optical properties of metal fluorides are quite fascinating and have been increasingly investigated in recent years (Hagenmuller 1983a, 1985; Portier, 1976; Tressaud & Dance, 1977, 1982). Electrical properties of oxides range from those of good metals to good insulators; transition-metal fluorides, on the other hand, are all good insulators.

Fluorides possessing tungsten bronze structures $A_xMF_3(M = Fe$ or $V)$ are known. In the K_xFeF_3 system, for example, the hexagonal tungsten bronze structure is obtained for the compositions $0.18 < x < 0.25$, the tetragonal bronze structure for $0.40 < x < 0.50$ and the cubic perovskite bronze structure for $0.95 < x \leqslant 1.0$. Similar bronze structures are known in the K_xVF_3 system. Unlike oxide bronzes, fluoride bronzes are antiferromagnetic (or ferrimagnetic) insulators.

Metal fluorides are better ionic conductors than metal oxides (Portier *et al.*, 1983). Several metal fluorides exhibit fast fluoride-ion conduction with conductivities comparable to those of the best ionic conductors (Fig. 6.24) (Section 7.2); typical examples are $PbSnF_4$, $Pb_{1-x}Bi_xF_{2+x}$ and $NH_4Sn_2F_5$. Many of the fluoride ion conductors possess fluorite-related structures. The presence of cations with lone pairs as well as a stoichiometric excess of fluoride ions seem to generally favour high ionic conduction as typically illustrated by the $Pb_{1-x}Bi_xF_{2+x}$ system (Hagenmuller *et al.*, 1981). The extra fluoride ions occupying interstitial sites in the fluorite structure displace neighbouring fluoride ions from their ideal positions. The resulting disorder increases the conductivity up to $x = 0.25$. For $x > 0.25$, conductivity decreases owing to increasing order; at $x = 0.5$, a new ordered phase $PbBiF_5$ is formed. As a rule, the maximum in conductivity and minimum in activation energy correspond approximately to the middle of the solid solution range. $PbSnF_4$ which has a conductivity of

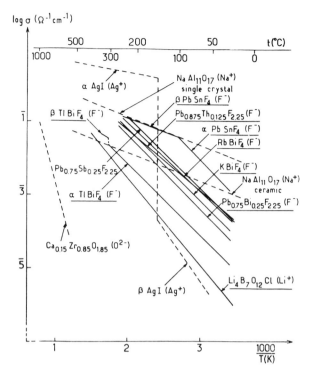

Figure 6.24   Ionic conductivity of fluoride-ion conductors. For comparison, data for typical Li^+, Na^+, Ag^+ and O^{2-} ion conductors are also shown. (After Hagenmuller, 1983a.)

10^{-1} ohm^{-1} cm^{-1} at 370 K with an activation energy of 0.14 eV appears to be the best known fluoride-ion conductor.

Fluoride-ion conductors are potentially useful in many electrochemical devices which include solid state batteries, electrochromic displays and gas sensors. $PbSnF_4$ can be used as solid electrolyte in cells of the type $Ag|PbSnF_4|Pt$ to measure O_2 pressures as well as pressures of reducing gases like H_2 or HF. Fluorine-intercalated graphite can be used as an electrode in batteries of the type $Li|LiClO_4$ in propylene carbonate$|CF_x$. Insulating fluorides could, in principle, be used as capacitors but for the fact that they are poor dielectrics; their permittivity is not generally very high, but when it is, the material shows high dielectric loss due to high ionic conductivity.

Transition-metal fluorides provide ideal model systems for checking theoretical magnetic models (Section 6.8). Since they are generally good insulators, magnetic properties can be studied without the complication of electron delocalization. In addition, transition-metal ions are invariably found in an octahedral environment and the MF_6 octahedra can link themselves to form one-, two- or three-dimensional (1D, 2D, 3D) fluoride structures or exist as isolated units. An example of the latter is A_3NiF_6 (A = alkali metal) containing isolated NiF_6 octahedra, which exhibits a low-spin–high-spin transition of Ni^{3+}.

When MF_6 octahedra are shared, magnetic interaction between localized spins on neighbouring atoms is described by the Heisenberg exchange term $H_{ij} = \sum_{ij} J_{ij} S_i \cdot S_j$, where J_{ij} is the exchange integral. The dominant contribution to J_{ij} comes from superexchange(Section 6.3.1). When the interaction is through σ bonding, $J = -2b^2/U$ and the sign of J can be predicted from the rules of Anderson, Kanamori and Goodenough. If superexchange is between half-filled cation orbitals as in $KNiF_3$, the interaction is antiferromagnetic. If the electron transfer is from a half-filled cation orbital to an unoccupied orbital the interaction will be ferromagnetic as between Ni^{2+} and Mn^{4+}. While the same rules govern magnetic interactions in both oxides and fluorides, the interaction is much weaker in fluorides, as revealed by the ordering temperatures. Thus $T_N = 738$ K for $LaFeO_3$, while in FeF_3, with a similar antiferromagnetic structure, $T_N = 394$ K. FeF_3 and CoF_3 show weak ferromagnetism below T_N. Owing to the transparency in the visible region, FeF_3 finds use as a magneto-optic material. It can also be used as a magnetic 'bubble' material in memory devices in place of $LaFeO_3$ (see Section 7.4).

Ferrimagnetic fluorides constitute a particularly important class of magnetic materials (Tressaud & Dance, 1977). Ferrimagnetism is known in a variety of fluorides of chiolite, $6H$-$BaTiO_3$, weberite and other structures. These structures are different from the structures of ferrimagnetic oxides. In oxides, ferrimagnetism is common in spinel, garnet and magnetoplumbite structures where ferrimagnetism is due to the occupation of tetrahedral, octahedral or dodecahedral sites by magnetic ions. In fluorides, however, transition-metal cations invariably occupy octahedral sites; ferrimagnetism results from the way in which the octahedra are linked. In the fluorides of hexagonal $BaTiO_3$ structure, for instance, one of the sublattices consists of octahedra linked by corners and the other octahedra linked by faces; one-third of the metal ions are present in corner-shared octahedra and two-thirds in face-shared octahedra. Both the sublattices have different magnetic moments, which are coupled antiferromagnetically, leading to ferrimagnetism; $CsFeF_3$ is a typical material possessing this structure. Fe–Fe interaction between face-shared octahedra is ferromagnetic and that between corner-shared octahedra is antiferromagnetic, and the net result is ferrimagnetism with $T_c = 60$ K. Isostructural $RbMnF_3$ is, however, antiferromagnetic since all interactions between the neighbouring octahedra are antiferromagnetic in this compound. Chiolites, $Na_5M_3F_{14}$(M = V, Cr, Fe, Co) are an important family of ferrimagnetic fluorides. MF_6 octahedra in this structure are corner-linked, forming M_3F_{14} layers; one-third of the octahedra share four corners and two-thirds share two corners. The sodium ions occupy vacant sites in and between the layers (Fig. 6.25(a)). Within each layer magnetic interactions are antiferromagnetic(nearly $180°$ superexchange). The net moments of each layer are aligned parallel to the c axis, resulting in ferrimagnetism (Fig. 6.25(b)).

Another interesting class of magnetic fluorides is the $Na_2M^{2+}M^{3+}F_7$ weberites, where chains of corner-sharing $M^{2+}F_6$ octahedra are linked by isolated $M^{3+}F_6$ octahedra. The structure provides different possibilities of magnetic behaviour. When both the cations are magnetic, the solid is generally ferrimagnetic due to $M^{2\pm}F$–M^{3+} superexchange. When M^{3+} ions are nonmagnetic, as in Na_2NiAlF_7, the structure provides magnetically isolated $(M^{2+}F_5)_n$ chains which show one-dimensional magnetic

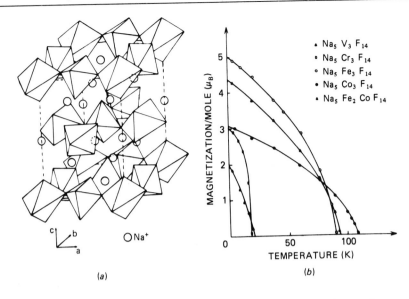

Figure 6.25   (a) The chiolite structure; (b) temperature dependence of magnetization of $Na_5M_3F_{14}$ chiolites. (After Tressaud & Dance, 1977.)

behaviour. When M^{2+} ions are diamagnetic, isolated $M^{3+}F_6$ octahedra give rise to paramagnetic behaviour in this structure.

Ferromagnetism, which requires indirect or direct magnetic interaction between unequally occupied cation orbitals, is shown by fluorides of the type $M^{2+}Mn^{4+}F_6$(M = Ni, Zn or Cd) crystallizing in ordered VF_3 structure. $NiMnF_6$ has the highest $T_c = 39$ K so far reported for ferromagnetic fluorides. PdF_3 is an interesting fluoride with an ordered $NiMnF_6$ type structure (which corresponds to $Pd^{2+}Pd^{4+}F_6$). It is a ferromagnetic insulator at ordinary pressures, but under high pressures the material seems to undergo a valence transition, $Pd^{2+} + Pd^{4+} \rightarrow 2Pd^{3+}$, accompanied by a transition to the metallic state (Demazeau *et al.*, 1980).

A number of transition-metal fluorides exhibit low-dimensional magnetism (Tressaud & Dance, 1982). Low-dimensional magnetism (see Section 6.8) can arise even in crystallographically three-dimensional structures owing to the *cooperative Jahn–Teller effect* (CJTE) (Section 4.10). Fluorides containing d^4 (high-spin), d^7 (low-spin) and d^9 cations show distortion of the octahedron due to CJTE. Two possible orderings are distinguished (Reinen, 1979): ferrodistortive (all the long metal–ligand bonds are parallel) and antiferrodistortive (long metal–ligand bonds are directed alternately perpendicular to each other). Magnetic couplings in ferro- and antiferrodistortive orderings are different. For d^4 cations elongated octahedra of antiferrodistortive order show ferromagnetism, while compressed octahedra of ferrodistortive order exhibit antiferromagnetic interaction. The perovskite $KCrF_3$ is an antiferromagnetic solid adopting an antiferrodistortive structure. The magnetic structure is similar to that of $LaMnO_3$ (A type). Superexchange interaction $d^0_{x^2-y^2} - p - d^1_{z^2}$ leads to ferromagnetic layers which are antiferromagnetically coupled through the empty $d_{x^2-y^2}$ orbitals.

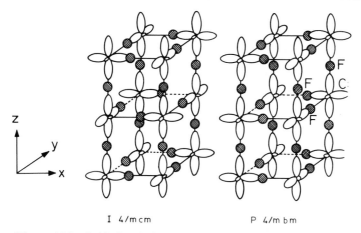

I 4/m cm   P 4/m b m

Figure 6.26   Orbital ordering in the two polymorphs of KCuF₃.

KCuF$_3$ is a 1D antiferromagnet (A type) with the spins lying in the ab plane. The magnetic behaviour is again a consequence of antiferrodistortive ordering of distorted octahedra. Interaction between two half-filled $d_{x^2-y^2}$ orbitals occurs along the c axis while interchain coupling is through filled d_{z^2}–half-filled $d_{x^2-y^2}$ interaction in the ab plane. There are two forms of tetragonal KCuF$_3$, one with $I4/mcm$ symmetry and the other $P4/mbm$ symmetry; 1D character is more pronounced in the latter, as can be seen from the orbital ordering (Fig. 6.26). CsNiF$_3$ is the only fluoride crystallizing in the hexagonal 2H perovskite structure where infinite chains of face-sharing NiF$_6$ octahedra exist parallel to the c axis. It exhibits one-dimensional ferromagnetism at high temperatures ($70 < T < 300$ K) and three-dimensional antiferromagnetism at low temperatures (Fig. 6.27). Neutron diffraction and specific heat measurements show that the three-dimensional transition occurs at 2.65 K. The three-dimensional magnetic structure consists of ferromagnetic planes parallel to the c axis, which are coupled antiferromagnetically.

Among the fluorides with layered structures, the K$_2$NiF$_4$ family has been widely investigated. K$_2$NiF$_4$ itself is a two-dimensional Heisenberg antiferromagnet (see Section 6.8) with $T_N = 97$ K and $J/k = -50$ K. The isostructural K$_2$CoF$_4$ behaves as a two-dimensional, $S = \frac{1}{2}$, Ising system ($T_N = 107.8$K, $J/k = -97$K) (Fig. 6.28). K$_2$CuF$_4$ (and its rubidium and caesium analogues) crystallizing in an orthorhombic–distorted K$_2$NiF$_4$ structure are two-dimensional Heisenberg ferromagnets. The distortion of the structure and ferromagnetic properties arise from ordering of the elongation axis of CuF$_6$ octahedra alternately in the a and b directions (Friebel & Reinen, 1974). The intra-layer exchange constant J/k is ~ 11 K and the value for interlayer coupling is ~ 0.03 K.

Ionic fluorides with large optical gaps exhibit high transparency to electromagnetic radiation. MgF$_2$, for instance, is transparent from $\sim 10^6$ cm^{-1} (corresponding to the energy threshold for the electronic transition from the valence band to the conduction band) to $\sim 10^3$ cm^{-1} (maximum frequency of lattice vibrations). The transparency of metal fluorides has led to their use as windows and prisms in optical instruments (see

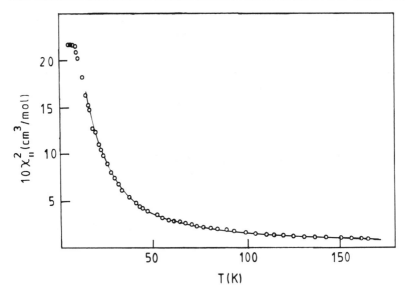

Figure 6.27   Single-crystal magnetic susceptibility data of CsNiF$_3$. (After Tressaud & Dance, 1982.) The solid line is the theoretical curve for $J/k = 10.2$ K.

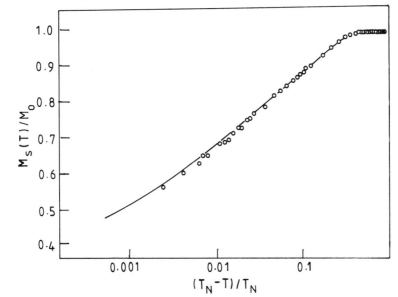

Figure 6.28   Reduced sublattice magnetization as a function of reduced temperature for K$_2$CoF$_4$. The solid line is the theoretical fit to the two-dimensional Ising model. (After Tressaud & Dance, 1982.)

Section 7.6). The property also renders them attractive as host materials for luminescent ions, Eu^{2+} emission in certain fluoride hosts being a good example. In oxide and chloride hosts, Eu^{2+} gives a broad emission from states of $4f^6 5d^1$ configuration, but in fluoride hosts where the bonding between Eu^{2+} and the host is weak and hence provides a weaker ligand field, the emission occurs from $^6P_{7/2}$, which lies lower in energy than the states derived from $4f^65d^1$. Monochromatic $(27\,800\,\text{cm}^{-1})^6P_{7/2} \to {}^8S_{7/2}$ emission is seen in Eu^{2+}-doped $BaAlF_5$, $BaSiF_6$ and BaY_2F_8.

Tervalent rare earths (with the exception of Ce^{3+} and Eu^{3+}) cannot easily be excited by UV irradiation for luminescence purposes, but many of them have excited levels close to the $^6P_{7/2}$ level of Eu^{2+}. Thus, it is possible to have resonant and phonon-assisted energy transfer from excited Eu^{2+} to tervalent rare-earth ions if they can be doped in a suitable host lattice. BaY_2F_8 can be simultaneously doped with divalent Eu and tervalent Tb, Ho or Er, resulting in strong green emission from the trivalent ions due to energy transfer from Eu^{2+}; emission does not occur in the absence of Eu^{2+} (Hagenmuller, 1983b). Fluoride hosts have also been useful in upconversion phosphors which convert infrared light to visible light. The process involves transfer of two photons from the sensitizer to the activator doped in the same lattice; Yb^{3+} and Er^{3+} function as sensitizer and activator respectively.

6.5   Metal–nonmetal transitions

Metal–nonmetal (M–NM) transitions can be broadly classified into three categories: (i) transitions in crystalline solids that occur between extended states with a change in structure; (ii) the *Mott transition* occurring between extended and localized states and (iii) the *Anderson transition* occurring between extended and localized states, of particular relevance to noncrystalline solids. The first class of transitions is described within the framework of the band theory of electrons in solids. The band structure of a crystalline solid made up of atoms with an even number of electrons (filling an integral number of bands) can be made to change over to a situation where the empty and filled bands cross or overlap due to a change in pressure or temperature or by suitable doping. Such band-overlap or band-crossover M–NM transitions are found in many materials (e.g. Yb, oxides of Ti and V) and are generally accompanied by a change in the crystal structure and in some instances by a change in magnetic ordering as well. The Mott transition from a metallic to a nonmetallic state occurs when the band-width decreases sufficiently to become smaller than the intrasite electron–electron energy, because of localization induced by electron correlation. In the Anderson transition, localization occurs because of disorder (as in amorphous materials); that is, a M–NM transition occurs when the electron band-width becomes less than the width of the distribution of random site energies. In this section, we shall present a rather simplistic account of the different M–NM transitions with illustrative examples; for more detailed treatments, the reader is referred to specialized monographs on the subject (Mott, 1974; Honig & Van Zandt, 1975; Elliott, 1984; Edwards & Rao, 1995). Edwards, Ramakrishnan & Rao (1995) have recently reviewed the present status of the subject.

Transitions between extended states.   Several oxides of Ti and V exhibit M–NM transitions due to changes in band structure caused by changes in crystal structure and symmetry as well as magnetic ordering. We shall consider the three oxides V_2O_3, VO_2 and Ti_2O_3 which typify M–NM transitions with different magnitudes of changes in conductivity (Fig. 6.29) and of varying complexity.

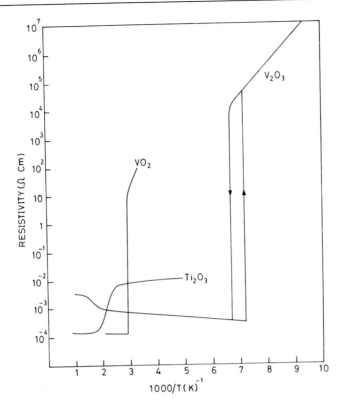

Figure 6.29   Temperature dependence of electrical resistivity of Ti_2O_3, VO_2 and V_2O_3 through the metal–nonmetal transition.

V_2O_3 undergoes a *first-order transition* (monoclinic–corundum) accompanied by a 10 *million-fold* jump in electrical conductivity at 150 K and by a *magnetic* transition (antiferromagnetic–paramagnetic). The c/a ratio and the volume change abruptly at the transition. The high-temperature metallic phase has considerable resistivity. Application of pressure makes V_2O_3 increasingly metallic suggesting that this oxide is near a critical region; accordingly doping with Cr or Ti has marked effect on the transition, the latter having the effect of positive pressure and the former that of negative pressure. V_2O_3 also shows a second-order transition around 400 K with a small conductivity anomaly (Fig. 6.29).

VO$_2$ undergoes a *first-order transition* (monoclinic–rutile) accompanied by a 10 *thousand-fold* jump in conductivity at 340 K. There is no magnetic ordering in the low-temperature phase and the material remains paramagnetic throughout. In pure VO_2 there is an intermediate monoclinic phase with a small temperature range of stability and this phase gets stabilized on doping with Al or Cr.

Ti$_2$O$_3$ undergoes a *second- or higher-order transition* (no change in crystal symmetry) with a 100-*fold* jump in conductivity and accompanied by a gradual change in c/a ratio and volume around 410 K. The material is paramagnetic throughout. Replacement of

Ti by V up to 10% makes the system metallic, the c/a ratio of the solid solution corresponding to that of the high-temperature metallic phase of Ti_2O_3.

Several explanations have been proposed to account for the appearance of the energy gap at a temperature leading to the M–NM transition (Van Zandt & Honig, 1974). A simple mechanism involving band-crossing (or band-broadening) with increasing temperature can account for the gradual transition in Ti_2O_3 associated with a small change in conductivity. A mechanism based on crystal distortion (the low-temperature low-symmetry structure being associated with an energy gap) can explain the essential features of the transition in VO_2. However, the large conductivity jump in V_2O_3 is difficult to explain on the basis of crystal distortion even when the antiferromagnetism of the low-temperature phase (doubling the periodicity and hence opening a gap) is taken into account. Models taking electron–lattice interaction, exciton binding energy and other factors into account are available in the literature. According to a model, the density of states curve for the d band in V_2O_3 has a set of high peaks and deep valleys in alternation (Kuwamoto et al., 1980). The Fermi level in close proximity to one of the minima may be replaced by a band gap which can be opened by Cr or Al doping. Change in oxygen stoichiometry or Ti doping shifts the Fermi level in such a way as to render the material metallic.

Besides the three oxides mentioned above, many of the *Magnéli phases* (Ti_nO_{2n-1} and V_nO_{2n-1}), Ti_3O_5, NbO_2 and a few other oxides also exhibit M–NM transitions. M–NM transitions are also found in NiS, CrS, FeS and other chalcogenides; the transition in NiS can be accounted for based on the accompanying magnetic transition (antiferromagnetic–paramagnetic) (Section 6.4.2).

The alloy system $(V_{1-x}M_x)_2O_3$ deserves special mention because of its spectacular M–NM transitions. The present situation with regard to these transitions may be summarized as follows (Honig, 1982). M–NM transitions in this system are shown schematically in Fig. 6.30 where the roman numerals correspond to the different regimes.

I $0.0 < x < 0.005$ (M = Cr or Al): At low temperatures, the oxide is an antiferromagnetic insulator (AFI) and shows the major M–NM transition around 167 K. Above 350 K the resistivity increases by a factor of 3 and becomes constant above 800 K.

II $0.005 < x < 0.018$: These compositions show three transformations. The first is from the AFI phase to a metallic (M) phase around 170 K while the second is from the M phase to a paramagnetic insulating (I) phase at $T = T_o$ in the range 190–385 K. T_o decreases with increasing x and at the same time, the temperature of the AFI–M transition increases. Thus, the stability range of the metallic M phase decreases with increasing Cr or Al content. The resistivity of the I phase gradually decreases giving way to a second metallic phase. M′.

III $0.018 < x < 0.10$: In this regime there is a AFI–I transition around 170 K and the M′ is reached beyond 800 K; the M phase is missing.

IV   The $(V_{1-x}Ti_x)_2O_3$ system is similar to I, except that the temperature of the AFI–M transition is reduced; for $x \geqslant 0.055$, the M–AFI discontinuity disappears.

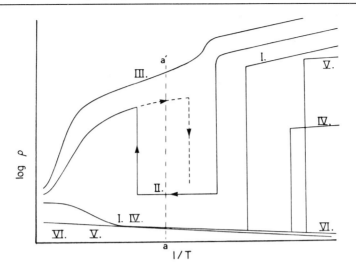

Figure 6.30   Metal–nonmetal transition in $(V_{1-x}M_x)_2O_3$ (M = Cr, Al or Ti) and $V_{2(1-y)}O_3$ systems. (After Honig, 1982.) See text for explanation of the various curves.

V $V_{2(1-y)}O_3$: AFI–M transition temperature decreases with increase in y and for $y > 0.009$ there is no transition.

VI $0.055 < x < 0.1$ of Ti or $0.009 < y < 0.035$: The material is a high-resistivity metal possibly with antiferromagnetic ordering around 10 K.

All the above transitions are accompanied by changes in Seebeck coefficient, structural parameters, heat capacity and other characteristics. The V_2O_3 system has been explained in terms of a thermodynamic model which uses different free energy expressions for electrons in the itinerant and localized regimes (Honig & Spalek, 1986).

Localized electron to itinerant electron (band) state transitions.   Rare-earth cobaltates, $LnCoO_3$(Ln = La or rare earth) have been reported to undergo first-order phase transitions around 1200 K which seem to be essentially governed by the change in the electronic entropy (Raccah & Goodenough, 1967; Bhide *et al.*, 1972; Jadhao *et al.*, 1976). The cobaltates become metallic above the transition, but there is no marked jump in conductivity. The temperature-evolution of the electronic (and spin) configurations of cobalt has been investigated by employing Mössbauer spectroscopy and other techniques. At low temperatures, cobalt ions are in the diamagnetic low-spin state (t_{2g}^6) and transform to the high-spin state ($t_{2g}^4 e_g^2$) with increase in temperature, the two spin states being clearly distinguished in Mössbauer spectra. Electron hopping between the two spin states gives rise to charge-transfer states ($Co^{2+} + Co^{4+}$) and associated increase in electrical conductivity in the 0–650 K range. As the temperature is increased further, the e_g elecrons tend to form a σ^* band; accordingly, the central shift in Mössbauer spectra shows a decrease in this temperature region (700–1000 K) due to progressive increase in the cation–anion orbital overlap. Since no change in crystal

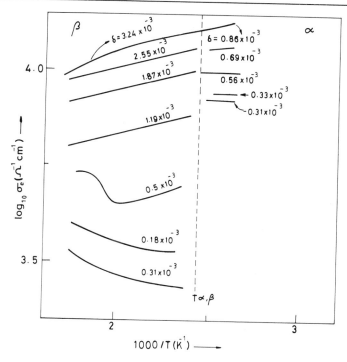

Figure 6.31   Composition-controlled metal–nonmetal transition in $Ag_{2+\delta}Se$ system. (After Shukla *et al.*, 1981.)

symmetry is noticed at the transition, it was considered that the entire entropy change was electronic in origin. Thornton *et al.* (1986) have shown that the $e_g \rightarrow \sigma^*$ transition occurs continuously without an associated structural change. The metallic state of $LaCoO_3$ occurs above 650 K and grows with increasing temperature in the 350–650 K range.

Mixed conductors.   Silver chalcogenides, $Ag_{2+\delta}X$ (X = S, Se or Te) transform to a high-symmetry phase (around 410 K) where the Ag^+ ions are randomly distributed, giving rise to super-ionic conductivity. These materials are also good electronic conductors (close to metallic) and show interesting electronic behaviour as a function of temperature as well as of composition (Shukla *et al.*, 1981). Thus, in the high-temperature phase, $Ag_{2+\delta}Se$ shows metallic behaviour of electronic conductivity for high values of δ. With decrease in δ, the electronic conductivity shows evidence for an interesting transition (Fig. 6.31), the ionic conductivity of the high-temperature phase being essentially independent of δ. The magnitude of change in electronic conductivity at the phase transition is also determined by stoichiometry. In the low-temperature phase, the material conducts like a semimetal around 400 K and the conductivity decreases substantially at low temperature, the magnitude of the decrease depending on the value of δ.

Mott transition.   Wigner (1938) introduced the idea of electron–electron interactions and suggested that at low densities, a free-electron gas should 'crystallize' to a nonconductive state. Mott (1949) suggested that an insulating state can be obtained in materials where the bands in the vicinity of the Fermi level are narrow. If the total reduction in kinetic energy does not overcome the total increase in potential energy due to the additional coulomb repulsion between electrons in the ionized states of a partly filled band, then the ground state of the system should be nonconducting. In his later papers, Mott (1956, 1967) proposed that if an electron is removed from the vicinity of one atom and placed on another atom in the above type of insulator, the free hole and the free electron would attract each other by a coulomb interaction and form a bound state or exciton, allowing neither hole nor electron to participate in electrical conduction. However, if many carriers are present, the electron and the hole will attract through a screened coulomb interaction with a screening constant. At higher values of the screening constant, the interaction becomes very weak and a sharp transition can result from a state with no free carriers to one with a large number. According to Mott, this transition should occur at a critical lattice constant and does not necessarily imply a phase transition since the change need be sharp only at $T = 0$ K. Such insulator–metal transitions are referred to as Mott transitions.

What makes a solid metallic?   Hubbard (1963) showed that the energy gap in a solid vanishes at a critical lattice constant, giving rise to a metal. According to Hubbard, the criterion for a M–NM transition to occur is $W \lesssim U$. The M–NM transition occurs at a critical value of $W/U \sim 1.2$ (see Fig. 6.2).

According to Mott, a M–NM transition in a solid may be induced by application of pressure as the lattice parameter passes through the critical value. Since the lattice constants of real crystals can be varied only through a limited range even by the application of superpressures, the problem of Mott transition is better investigated in doped semiconductors and other chemical systems. Mott predicted that the M–NM transition should occur at a critical charge-carrier concentration, n_c, given by $n_c^{1/3} a_H \simeq 0.25$, where a_H is the Bohr radius of the atom in the nonmetallic regime. Edwards & Sienko (1983) have employed the Mott criterion to describe the onset of metallic character in a wide range of systems. Experimentally, it is found that the onset of metallic character in a large number of systems is described by the criterion, $n_c^{1/3} a_H^* = 0.26 \pm 0.05$, where a_H^* is now the radius of the realistic wave function for the isolated species, obtained directly from measurements in the localized-electron regime.

The M–NM transition has been a topic of interest from the days of Sir Humphry Davy when sodium and potassium were discovered; till then only high-density elements such as Au, Ag and Cu with lustre and other related properties were known to be metallic. A variety of materials exhibit a transition from the nonmetallic to the metallic state because of a change in crystal structure, composition, temperature or pressure. While the majority of elements in nature are metallic, some of the elements which are ordinarily nonmetals become metallic on application of pressure or on melting; accordingly, silicon is metallic in the liquid state and nonmetallic in the solid state. Metals such as Cs and Hg become nonmetallic when expanded to low densities at high temperatures. Solutions of alkali metals in liquid ammonia become metallic when the concentration of the alkali metal is sufficiently high. Alkali metal tungsten bronzes (e.g. Na_xWO_3) become metallic at a critical electron (Na) concentration. There are many other systems which show concentration-dependent M–NM transitions, typical examples being metal–noble gas thin films (e.g. Na:Ar), doped semiconductors (e.g. P:Si), alloys (e.g. Cu:Au) and several complex metal oxides.

The criterion propounded by Herzfeld in 1927 is instructive in understanding the metallic state. According to the *Herzfeld criterion*, a material is metallic when $R/V \geqslant 1$ and insulating when $R/V < 1$. Here R is the molar refractivity of the atomic species in the gaseous state and V is the molar volume in the condensed state. This criterion explains the metallic character of the elements in nature (Edwards & Sienko, 1983). Rao & Ganguly (1986) have shown that the latent heat of vaporization provides a satisfactory criterion to delineate metallic elements from nonmetallic ones.

Anderson transition.   In disordered (or noncrystalline) materials, the reciprocal lattice vector becomes redundant and k is no longer a good quantum number. The electronic structure of these materials cannot be described in terms of band structure (E versus k). Unlike in crystals where the insulating side of a M–NM transition corresponds to a filled band, in disordered materials the electron states are localized. The wave function is concentrated near a centre composed of a few atoms and has negligible amplitude elsewhere in the solid. The amplitude spatially decays as exp-αR with distance R, where α represents the inverse localization length. An important concept introduced in connection with the electron states of disordered materials is that of the *mobility edge* (Mott & Davis, 1979). According to this concept, there exists a critical energy below which electrons are spatially localized in the region of a single atomic site and above which they are spatially extended (as in a crystal), although phases of the wave function may be random from site to site. The result is shown in Fig. 6.32, where the states in the tails of the band are localized. Anderson (1958) in his classic paper showed how localization appears as a consequence of disorder in a purely independent-electron picture. The disorder in this model occurs in the form of random site energies (Fig. 6.33). The criterion for localization is that the energy difference between localized sites is greater than the band-width. The Anderson transition is one where a localization–delocalization transition occurs because the Fermi level is pushed through the mobility edge by a change in composition, temperature or some other variable.

Electrical conduction in such disordered materials proceeds by a process of phonon-assisted tunnelling in which the probability of transfer between two sites is given by

$$P = v \exp(-2\alpha R)\exp(-W/kT) \tag{6.57}$$

where the first exponential factor gives the tunnelling probability between two sites having localized wave functions associated with them and the second exponential factor represents the probability of a phonon existing with energy W (energy difference between the two sites); v is the phonon frequency. At low temperatures, the probability of electron transfer can be optimized by tunnelling not to the nearest site but to a more

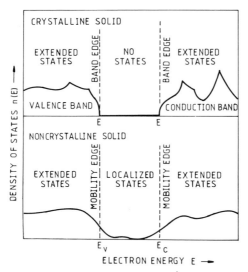

Figure 6.32 Schematic density-of-states diagram for crystalline and amorphous semiconductors in the vicinity of highest occupied (E_v) and lowest empty (E_c) states.

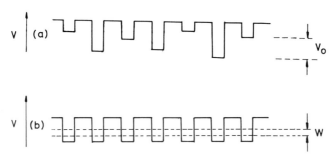

Figure 6.33 (a) Anderson localization due to disorder in site potentials. For comparison, potentials in a regular lattice are also shown in (b), W is one-electron band-width in the absence of the random potential V_0. Localization is determined by (W/V) ratio.

distant one which has a smaller energy difference, W. Such optimization led Mott to propose the law for variable-range hopping as

$$\sigma = \sigma_o \exp(-AT^{-1/4}) \tag{6.58}$$

where $A = 2.1[\alpha^3/kN(E_F)]^{1/4}$ and $N(E_F)$ denotes the density of states at the Fermi level. Such behaviour has been observed in amorphous semiconductors as well as oxides of the type $La_{1-x}Sr_xCoO_3$. The picture of a well-defined mobility edge with electrical conduction at high temperatures taking place by means of electron transport in bands of extended states beyond the mobility edge has been questioned. A theory based on polaron transport has been proposed by Emin (1977). The concept of mobility edge has

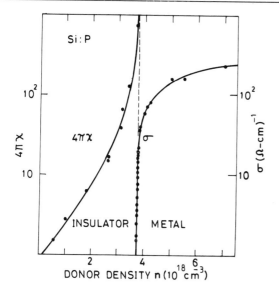

Figure 6.34 Electrical conductivity and dielectric susceptibility of phosphorus-doped silicon at $T \sim 10^{-2}$ K. The dielectric susceptibility shows a divergence as the transition is approached from the insulator side. Notice the sharp, but continuous threshold in $\sigma(n)$ on the metallic side. (After Hess *et al.*, 1982.)

been questioned by Abrahams *et al.* (1979) who use scaling methods (similar to those in phase transitions) to show that the mobility and the conductivity change continuously through the localization transition. This has been supported by studies of P-doped Si (Fig. 6.34); conductivities below the value of the *minimum metallic conductivity* $(\sigma_{min} \approx 500\,\mathrm{ohm}^{-1}\,\mathrm{cm}^{-1}$ at 0 K) proposed by Mott (1972, 1983) as the limiting conductivity of a disordered material on approaching the localization transition $(E = E_c)$ from the metallic side, have been measured. The sharp divergence in dielectric susceptibility is reminiscent of the Herzfeld criterion.

The gap states in amorphous materials are known to result in charged defects, transport occurring through the hopping of *bipolarons*. In chalcogenide glasses, the bipolarons correspond to over-coordinated (C_3^+) and under-coordinated (C_1^-) centres.

Composition-controlled M–NM transitions in oxides.   There are many systems which change from nonmetallic to metallic behaviour based on composition. For example, in the system $(V_{1-x}Ti_x)_2O_3$, when $x = 0.90$ or 0.10, the material is metallic at all temperatures; the transition from NM to M with change in x is quite abrupt. What we shall refer to here are oxide systems where there is no M–NM transition in the parent material, but the replacement of one of the cations by another progressively renders the oxide metallic (Edwards & Rao, 1985; Rao & Ganguly, 1985). $Ln_{1-x}Sr_xCoO_3$ (Ln = La or rare earth) is a good example of such a system (Rao *et al.*, 1975; Rao *et al.*, 1977). $LnCoO_3$ ($x = 0$) is an insulator at ordinary temperatures, but the resistivity decreases progressively with increase in x until, at $x = 0.4$–0.5, the oxide

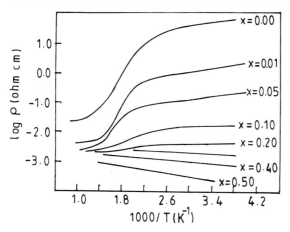

Figure 6.35   Resistivity behaviour of $Nd_{1-x}Sr_xCoO_3$. (From Rao *et al.*, 1975.)

is metallic (Fig. 6.35). Somewhere at an intermediate value of x the temperature-coefficient of resistivity changes sign and the value of the conductivity at this composition (10^2–10^3 ohm^{-1} cm^{-1}) is close to the '*minimum metallic conductivity*' defined by Mott (1972) in connection with his work on disordered materials. What is interesting is that the σ_{min} value is exhibited by oxide systems that show a change in the sign of the temperature coefficient of resistivity, at least at relatively high temperatures (> 50–100 K). The situation is, however, more complex since many of these oxides show lower conductivities than σ_{min} at low temperatures (Raychaudhuri, 1991).

Another example of such behaviour is found in $LaNi_{1-x}M_xO_3$(M = Mn, Cr etc.) where the temperature coefficient of resistivity changes sign at a critical value of x (Fig. 6.36); $LaNiO_3$(x = 0) is a d-band metal and the resistivity increases with increase in x. $La_{1-x}Sr_xVO_3$ also shows similar σ_{min} behaviour (Sayer *et al.*, 1975).

The situation with $La_{1-x}A_xMnO_3$ (A = Ca or Sr) is somewhat different. The material becomes ferromagnetic when x is relatively high ($\gtrsim$ 0.25 or so) because of the Mn^{3+}–O–Mn^{4+} interaction (*Zener double exchange*). Fast hopping of electrons from Mn^{3+} to Mn^{4+} makes the d-electrons itinerant and the material shows metal-like behaviour at temperatures slightly below the ferromagnetic Curie temperature (T_c). One therefore observes a maximum in the resistivity–temperature plot around a temperature T_p close to T_c (Fig. 6.37). It is possible to increase the Mn^{4+} content in parent $LaMnO_3$ by chemical means (see Chapter 3) and make it an itinerant electron ferromagnet with an insulator–metal transition. It has been found recently (Ju *et al.*, 1994, McCormack *et al.*, 1995, Mahesh *et al.*, 1995, Urushibara *et al.*, 1995) that these manganates show a large change (decrease) in resistivity on the application of a magnetic field (*giant magnetoresistance*). The decrease can be close to 100% relative to the resistivity in the absence of the field (see Fig. 6.37).

Giant magnetoresistance (GMR) in rare-earth manganates $Ln_{1-x}A_xMnO_3$ (Ln = La, Pr, Nd; A = Ca, Sr, Ba, Pb) is essentially governed by the Mn^{4+} content, around 30% being optimal. Thus, cubic $LaMnO_3$ with $\sim$ 30% Mn^{4+} (in the absence of aliovalent cation substitution) exhibits GMR. GMR is favoured by low T_c and high

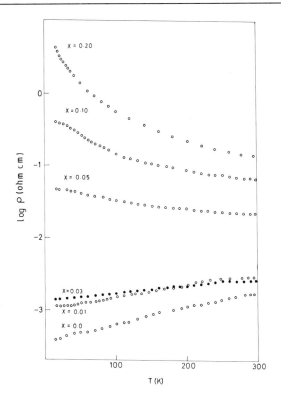

Figure 6.36   Resistivity behaviour of $LaNi_{1-x}Mn_xO_3$.

resistivity at T_p. The resistivity in the 'metallic regime' at $T < T_c$ is rather high compared to Mott's maximum metallic resistivity. The $T_c(T_p)$ increases with the weighted average radius of the A-site cations (Fig. 6.38) while the magnitudes of GMR and peak resistivity decrease (Fig. 6.39).

Composition-controlled M–NM transitions occur in the bronzes, A_xWO_3; but $A_{0.3}MoO_3$ (A = K, Rb) shows a M–NM transition associated with charge density waves, nonohmic transport and quasi one-dimensional character (Schlenker *et al.*, 1985). Similar nonohmic transport is found in $NbSe_3$ and TaS_3.

6.6   Metal clusters

As mentioned in Section 3.4, clusters of metal atoms of varying sizes can be prepared. The presence of alkali atom clusters in the vapour phase is well documented. Such clusters have a much lower ionization energy than that of an isolated atom and also have a high electron affinity. The probability of electron transfer is therefore considerably greater in a metal cluster. It is indeed known in the case of caesium that as the density of caesium increases (from isolated atoms in a low-density gas to a liquid), larger clusters form and charge-transfer becomes increasingly favoured; as the density

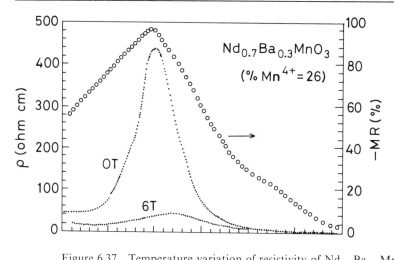

Figure 6.37   Temperature variation of resistivity of $Nd_{0.7}Ba_{0.3}MnO_3$ at 0 and 6 Tesla. Variation of the % magnetoresistance is also shown (After Mahesh *et al.*, 1995).

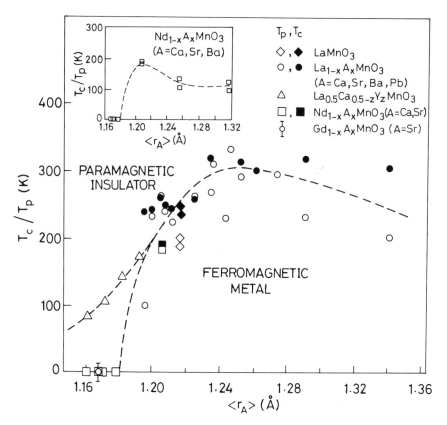

Figure 6.38   Variation of T_p (T_c) of various rare-earth manganates with weighted average radius of the A-site cations, $\langle r_A \rangle$ (After Mahesh *et al.*, 1995). The area in the bottom left hand corner represents the ferromagnetic insulator regime.

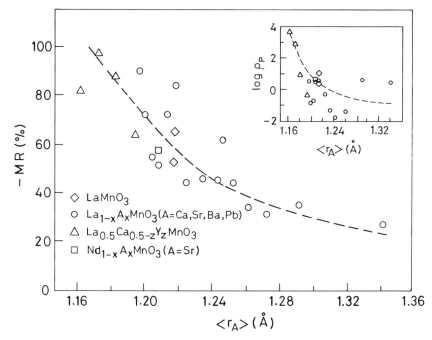

Figure 6.39   Variation of the % magnetoresistance with $\langle r_A \rangle$. Inset shows the variation of the peak resistivity, ρ_p, at T_p (After Mahesh et al., 1995).

approaches the critical value, the energies of the cluster bands overlap, resulting in good conduction. When one looks at such metal clusters, there is always the question as to how many atoms are needed in a cluster to imitate or reflect the properties of the bulk metal (Edwards & Sienko, 1983). If one is interested in transport properties such as electrical conduction, complete delocalization of the electron wave function would be required to make the electrons itinerant; this would require a large number of atoms in the aggregate. Spin delocalization, however, requires a smaller number of atoms since this occurs without the formation of free carriers. The width of the 'd' band or the cut-off at the Fermi edge in photoelectron spectra could also be employed to distinguish metal clusters as distinct from bulk metals. Metal particulates (similar to those on the surfaces of supported metal catalysts (Chapter 8)) constitute a type of metal aggregate somewhere between high nuclearity metal clusters and bulk metals. Recent studies on well-characterized gold clusters (Rao et al., 1993) have shown nonmetallicity to occur when the cluster diameter is $\lesssim 1$ nm.

There are families of metal cluster compounds (Fig. 6.40) containing metal clusters surrounded by ligands (Lewis & Green, 1982). In small cluster compounds, the electrons are paired, but in large clusters there will be closely spaced electronic levels, as in metal particles. In such clusters, quantum size effects would be expected. Benfield et al. (1982) have found intrinsic paramagnetism in $H_2Os_{10}(CO)_{24}$ below 70 K as expected of an osmium particle of approximate diameter of 10 Å; the excess paramagnetism increases with cluster size in osmium compounds (Johnson et al., 1985).

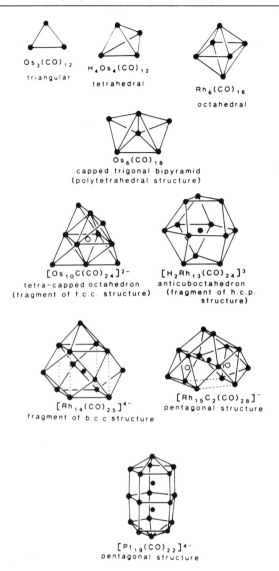

Figure 6.40   Typical metal-atom clusters found in transition-metal carbonyls. For clarity, the carbonyl groups are not shown. Open circles denote carbon atoms in the cluster compound. (After Edwards & Sienko, 1983.)

Metal cluster compounds simulate surface species produced by the interaction of molecules with metal surfaces (Muetterties *et al.*, 1979) and this is of value in understanding heterogeneous catalysis. The development of selective catalysts for the C_1 chemical industry employing CO (and possibly CO_2) as the raw material has resulted in major efforts in metal cluster research. Criteria have been developed to distinguish between cluster catalysis and mononuclear catalysis. Typical of the catalysts investigated hitherto are $[Ir_4(CO)_{12-x}(PPh_3)_x]$ where Ph = phenyl and $x = 1, 2$ or 3.

Table 6.6. *Typical mixed-valence solids*

Compound	Classification in the Robin–Day scheme	Importance
Pb_3O_4	Class I	Red lead
Sb_2O_4	Class I	Mineral cervantite
$Fe_4[Fe(CN)_6]_3 \cdot 14H_2O$	Class II	Dye and pigment (Prussian blue)
V_nO_{2n-1}	Class II	Semiconductor–metal transitions
$Li_xNi_{1-x}O$	Class II	Hopping semiconductor
$La_{1-x}Sr_xMnO_3$	Class III	Ferromagnetism
$BaBi_{1-x}Pb_xO_3$	Class III	Superconductivity
$LiTi_2O_4$	Class III	Superconductivity
$K_2Pt(CN)_4Br_{0.30} \cdot 3H_2O$	Class III	Molecular metal; Peierls instability
Na_xWO_3	Class III	Bronze lustre and metallic at high x
$M_xMo_6S_8$	Class III	Superconductivity
Fe_4S_4 Ferredoxins	Class III	Electron transfer (enzymes)

6.7   Mixed-valence compounds

Chemical compounds consisting of an element (usually a metal) in two different oxidation states are said to exhibit mixed valency. Mixed-valence chemistry is as old as chemistry itself, some of the well-known mixed-valence compounds being Prussian blue ($Fe_4[Fe(CN)_6]_3 \cdot 14H_2O$), magnetite ($Fe_3O_4$), and heteropoly tungsten and molybdenum blues. Mixed-valence chemistry, however, encompasses a large variety of solids with fascinating properties (Table 6.6) formed by nearly a third of the elements in the periodic table, and there has been a recent upsurge of interest in the subject (Day, 1981). Since variable valency is a prerequisite for mixed valency, it is quite common among the compounds of transition metals, Ce, Eu and Tb, as well as some of the post-transition elements with stable ns^2 and ns^o electronic configurations such as Ga, Sn, Sb, Tl, Pb and Bi. Most mixed-valency compounds contain electronegative counterpart anions such as halides, oxide, sulphide or molecular ligands containing electronegative atoms. In order for a solid to be called a mixed-valence compound, we should be able to assign definite oxidation states that differ by integral numbers (one or at the most two units) to the element showing mixed valency. Fe_3O_4 consisting of iron in the 2^+ and 3^+ oxidation states, is a mixed-valence compound, whereas the alloy Nb_3Ge, where we cannot specify the oxidation states of the constituents, is not. We should also make a distinction between the mixed-valence compounds of the Fe_3O_4 type and valence-fluctuating systems such as Ce and SmS, where a fluctuation in the electron configuration between $4f^n$ and $4f^{n-1}d^1$ occurs (Falicov *et al.*, 1981). (see Section 2.2.7). Electronic properties associated with f-electrons receiving much attention recently are those arising from *heavy fermion behaviour* (e.g. $CeCu_2Si_2$, UPt_3) wherein the carriers exhibit large effective masses (Stewart, 1984).

In certain mixed-valence compounds, the presence of more than one oxidation state can be recognized from the formula, as for example Pb_3O_4 and V_nO_{2n-1}, while in some others the formula indicates an apparently integral oxidation state although the oxidation state is rather unusual for the element in question. Typical examples of the latter category are Sb_2O_4, $BaBiO_3$ and $Pt(NH_3)_2Cl_3$; experimental evidence shows that we are not dealing with Sb(IV), Bi(IV) and Pt(III) states in these compounds but with Sb(III, V), Bi(III, V) and Pt(II, IV), and these solids should indeed be formulated as $Sb^{3+}Sb^{5+}O_4$, $BaBi_{0.5}^{3+}Bi_{0.5}^{5+}O_3$ and $[Pt(NH_3)_4]^{2+}[PtCl_6]^{2-}$. In all such systems, X-ray photoelectron spectroscopy can readily identify the presence of mixed valency (Rao *et al.*, 1979). $CsAuX_3$ (X = Cl,Br,I), possessing properties similar to those of $BaBiO_3$ undergoes a pressure-induced metal–nonmetal transition driven by the Au(I)(III) → Au(II) valence change (Kojima *et al.*, 1994).

Robin & Day (1967) have proposed a classification of mixed-valence compounds based on the valence delocalization coefficient, α, the magnitude of which depends on the energy difference between the two states $M_A^{n+}M_B^{(n+1)+}$ and $M_A^{(n+1)+}M_B^{n+}$, where A and B are two different sites. When ΔE is large as in Pb_3O_4, α is small; such compounds belong to class I. If the two sites are similar but crystallographically distinguishable, the compounds are considered to belong to class II. In class III, α becomes large and the two sites occupied by the mixed-valent cations are identical. Properties associated with the different classes would be different (Table 6.6). For example, in class I compounds, electron hopping between the sites is not favoured since ΔE is large. In class III compounds, on the other hand, the electrons would be delocalized. The ligands which bridge the cations play a role in determining the intervalence transfer; the greater the metal–ligand overlap, the higher the probability of electron transfer (Mayoh & Day, 1972). In order to describe the electron transfer in mixed-valence compounds properly, one would have to consider the coupling between the electronic and vibrational motions. Experimentally, the frequency of optical intervalence transition gives an estimate of the energy required for thermally activated electron transfer. The intensity of the optical intervalence transition gives information on α. One of the most characteristic features of mixed-valent class II compounds is the structureless broad intervalence absorption band in the visible and infrared. A vibrational coupling model has been developed to calculate the absorption profiles (Piepho *et al.*, 1978); a good example of analysis of such absorption profiles is the study of $(CH_3NH_3)_2Sb_xSn_{1-x}Cl_6$ by Prassides & Day (1984). When the electrons are not completely delocalized and they hop from site to site in marginal semiconductors, the strength of interaction between the electrons and the lattice (polarons) becomes an important factor.

Mixed valency occurs in minerals (e.g. Fe_3O_4), metal–chain compounds, dimers and oligomers and metal complexes, and even in organic and biological systems (Brown, 1980; Day, 1981). Among the dimeric and oligomeric metal complexes exhibiting mixed valency, the pyrazine-bridged Ru (II, III) ammine complex,

$$[(NH_3)_5Ru—N\bigcirc N—Ru(NH_3)_5]^{5+}$$

synthesized by Creutz & Taube (1973), has received much attention. The important question with regard to this family of complexes is whether they belong to class II or III. With identical ligands around each metal ion, the first impression is to consider the *Creutz–Taube complex* as belonging to class III. Optical absorption shows an intense band in the visible (550 nm), which is characteristic of Ru(II). It certainly supports the idea that a distinct Ru(II) can be identified on the time scale of optical transition (10^{-14} s). However, the intervalence band centred at $\sim$ 1550 nm is insensitive to solvent effects and a bit too narrow to be called a class II behaviour. XPS shows doublets in the core-level Ru($3d$) and ($3p$) spectra, indicating that the individual oxidation states can be distinguished on this time scale (10^{-16} s) as well. Hush (1975) has argued that the creation of core-hole by photoionization would relax the system into a localized state even if it were originally delocalized; core–shell photoelectron spectroscopy therefore does not appear to provide the means to determine the extent of localization or delocalization in the valence shell. Infrared spectroscopy has shown that the NH_3 rocking mode (800 cm^{-1}), Ru–NH_3 stretching mode (449 cm^{-1}) and Ru–pyrazine stretching mode (316 cm^{-1}) of the Creutz–Taube complex; all occur at values intermediate between those of the corresponding Ru(II) and Ru(III) complexes. This has been taken as evidence that the valence electron is delocalized in the time scale of infrared spectroscopy (10^{-12}–10^{-13}s). Results of the various measurements on this complex are somewhat conflicting because of the different time scales; it appears that the electron-transfer rate is somewhere between 10^5–10^{12}s^{-1}. Mixed valency in compounds like $La_{1-x}Sr_xCoO_3$ (Rao *et al.*, 1975, 1977) is determined by rate of electron transfer and so by composition; that in $MoFe_2O_4$ and other solids resulting from fast electron transfer is discussed (Ramdani *et al.*, 1985).

In the metal–chain compounds, we can distinguish two types of mixed-valent systems, one where the chain is entirely composed of metal atoms (class III) and the other in which the metal and the bridging ligand alternate (class II). Wolfram's red salt, $[Pt(C_2H_5NH_2)_4]\,[Pt(C_2H_5NH_2)_4Cl_2]\cdot4H_2O$ is an example of the former, consisting of Pt(II) and Pt(IV) ions bridged by chloride ions. In KCP type of mixed-valent compounds, mixed valency is achieved by partly oxidizing the Pt ions to an average oxidation state $(2 + x)$ with $x \approx 0.3$. Partial oxidation is accomplished by removing some of the cations as in $K_{1.75}Pt(CN)_4\cdot1.5H_2O$ or by introducing extra anions $K_2Pt(CN)_4Br_{0.3}\cdot3H_2O$. Even the Hg chain compound $Hg_{3-x}AsF_6$ (see Section 4.9) is mixed-valent.

Fe_3O_4 has the inverse spinel structure, with all the Fe^{2+} ions and half of the Fe^{3+} ions located in octahedral sites (B sites) in the oxygen network and the remaining half of the Fe^{3+} ions located on tetrahedral sites (A sites). It undergoes a ferrimagnetic–paramagnetic transition around 850 K and another transition around $T_V = 123$ K (*Verwey transition*). The material is a semiconductor both above and below the Verwey transition. Some changes in properties have also been observed near 200 K and 12 K, but these are not very significant. The properties of Fe_3O_4 in the region of the Verwey transition and above have been a subject of great interest, and an entire issue of the *Philosophical Magazine* (*B*42, No. 10, 1980) was devoted to the transition.

Detailed structural investigations employing neutron diffraction and other techniques suggest that charge ordering of Fe^{2+} and Fe^{3+} ions (and therefore the long-range order) is established below T_V. Cation strings, a and b, run along the $[\bar{1}10]$ and $[110]$ directions respectively (on alternate (001) planes). While three Fe^{2+} ions in succession are followed by a Fe^{3+} ion along one a chain, on an adjacent a chain three Fe^{3+} ions are followed by one Fe^{2+}. In the b chain, cation ordering occurs with a pair of Fe^{3+} ions followed by a pair of Fe^{2+} ions in alternation. Successive $-b$–a–b– planes are stacked, perpendicular to the c axis, in such a way that proximate cations in three successive planes are in groups, forming hexagonal rings of alternate Fe^{2+} and Fe^{3+} ions. All the cations along b strings are members of rings and only a quarter of the a-string cations are involved in ring formation. The synchronous displacement of three electrons to their nearest-neighbour position inside any ring produces an interchange of Fe^{2+} with Fe^{3+}. A significant fraction of hexagonal rings always exists in the 'inverted' charge configuration, thereby randomizing charge along the b strings, but leaving three-quarters of the a-chain constituents in an ordered arrangement below T_V. This rationalizes earlier experimental findings that it is the a-plane cations which order at low temperatures.

The existence of superstructure lines just above T_V in critical neutron scattering and the detailed investigation of elastic and inelastic neutron scattering show the existence of a soft mode with wave vector $k = (00\xi)$ that 'condenses' at $k = (00\frac{1}{2})$. Proceeding from one Fe^{2+} (or from one Fe^{3+}) ion in the a plane at $z = \frac{1}{8}$ to the corresponding position in the a plane at $z = \frac{9}{8}$, one arrives at the complementary charge (namely, Fe^{3+} or Fe^{2+}) respectively. One has to advance by twice the unit lattice distance along c to duplicate the same ionic configuration at $z = \frac{17}{8}$ that prevails at $z = \frac{1}{8}$. Formally, this corresponds to the existence of a charge-density wave (with wavelength $\lambda = 2c$) which couples strongly to the corresponding phonon mode with the same wave vector. At the transition, the ordering of the charges (leading to the establishment of the CDW) occurs simultaneously with a net atomic displacement that lowers the symmetry.

Although the structural characterization of Fe_3O_4 near the Verwey transition is fairly satisfactory, many finer details are not yet understood, (Honig, 1982, 1986), including the actual structure of the low-temperature phase (rhombohedral, orthorhombic, monoclinic or triclinic). Electrical properties around T_V are also not fully characterized. There is uncertainty regarding the nature of variation of conductivity with temperature. It is not clear whether the itinerant character of charge carriers assumed by some workers is valid. Most of the data on transport properties seem to suggest a small polaron model. What is rather puzzling is that the resistivity decreases by two orders of magnitude at T_V, accompanied by the loss of long-range order. Recent studies have shown that the Verwey transition and the associated changes in conductivity and heat capacity are very sensitive to oxygen stoichiometry (Honig, 1986).

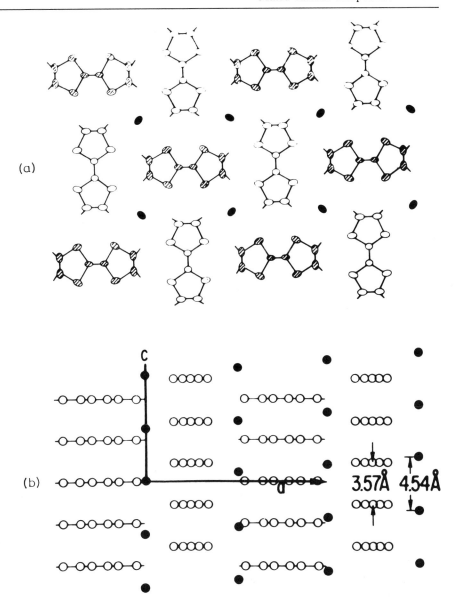

Figure 6.41   Crystal structure of TTF-$Br_{0.79}$. Projections (a) down the a-axis and (b) down the b-axis. (After Torrance & Silverman, 1977.)

There are several interesting families of inorganic mixed-valence compounds that we have not discussed here (see Yvon, 1979; McCarley, 1982). For example, there are metal-cluster compounds such as the Chevrel phases, $M_xMo_6X_8(X = S$ or Se) and condensed metal-cluster chain compounds such as $TlMo_3Se_3$, Ti_5Te_4, $NaMo_4O_6$ and $M_xPt_3O_4$. TTF halides and TTF–TCNQ complexes (Section 1.9) constitute molecular mixed-valent systems in which the mixed valency is associated with an entire molecule; the charge on TTF in such compounds is nonintegral. The structure of TTF–$Br_{0.79}$ and

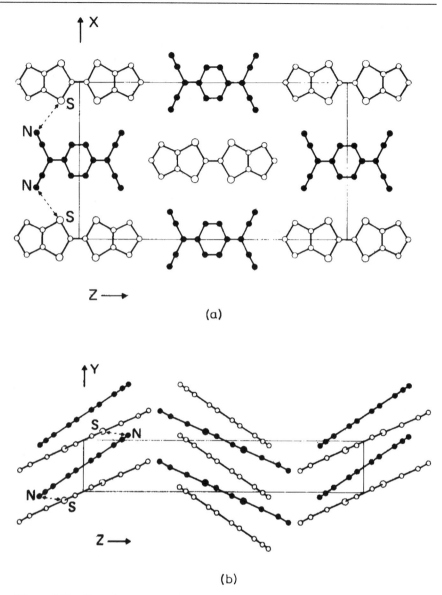

Figure 6.42   Crystal structure of HMTTF–TCNQ. (a) Projection on the plane perpendicular to the stacking axis and (b) projection on a plane containing the stacking axis. (After Greene *et al.*, 1976.)

such solids consists of stacks of TTF molecules parallel to the *c*-axis. The Br⁻ ions are also arranged in columns parallel to the *c*-axis. However, the repeat distances of the cation (3·57 Å) and anion (4.54 Å) columns are different (Fig. 6.41). The structure is incommensurate along the *c*-axis because the lattice periodicities of the two subunits are not simple multiples (or fractions) of one another. It is important to note that the periodicity of the TTF sublattice is independent of stoichiometry, whereas that of the bromide sublattice is stoichiometry dependent. Hence the charge transfer in the salt can be expressed as $f = 3.57/c(\text{Br})$. Nonstoichiometric compositions in the TTF–halide systems seem to be stabilized because of electrostatic (Madelung) factors. Calculations show that the Madelung energy is maximum around a halogen content of 0.7 to 0.8. Beyond this composition, repulsion between like charges along the stacking axis begins to dominate, decreasing the net binding energy. Optical absorption shows a new peak around 0.7 eV in TTF–$\text{Br}_{0.79}$. The peak, which is not present in the spectra of stoichiometric salts, has been assigned to a mixed-valence intrastack charge-transfer transition between a neutral TTF and adjacent TTF^+.

The structure of HMTTF–TCNQ, a typical TTF–TCNQ-type complex, is shown in Fig. 6.42. The segregated stacking in this structure is a characteristic feature of the highly conducting organic charge-transfer system of the TTF–TCNQ family. In HMTTF–TCNQ (HMTTF-hexamethylenetetrathiafulvalene), the separation between donor molecules is 3.57 Å while that between TCNQ molecules is 3.23 Å along the stacks. To make the anion and cation sublattices commensurate with each other, the molecules are stacked in a staggered configuration such that the normals to the molecular planes are not parallel to the stacking axis but make an angle with it. The angles are different for the donor and acceptor stacks; in the HMTTF–TCNQ structure, the values are respectively 23.8° and 34.2°.

What makes the TTF–TCNQ family distinct from the other salts of TCNQ with cations, such as alkali metals and tetramethylammonium, is that the charge transfer, f, in the TTF–TCNQ family is incomplete ($f < 1$). TTF–TCNQ members are also different from the TTF–halides; in the TTF–halides, where the charge on each halide atom is unity, partial charge transfer (mixed valency) is realized by the formation of nonstoichiometric materials, while in the TTF–TCNQ family, the composition is stoichiometric (1:1), but mixed valence arises because of partial electron-transfer.

Evidence for incomplete electron-transfer (mixed valency) has come from a number of physical studies. Optical absorption studies show a low-energy peak at 0.3 eV, which is not present in insulating salts such as $\text{K}^+(\text{TCNQ})^-$. Moreover, the absorption is polarized parallel to the stack axis. The absorption is therefore clearly due to mixed-valence intrastack electronic transition. TTF–TCNQ undergoes a transition from conducting to insulating state at 59 K. This transition is characterized by a subtle periodic modulation of the lattice due to a coupling of the conduction electrons with the lattice (CDW). This shows up in the form of satellite reflections surrounding the Bragg peaks in the diffraction experiment. Because the structure is sinusoidally modulated, the Bragg peaks caused by the average structure remain essentially unchanged. Since the satellite peaks are the result of interaction between conduction electrons and the lattice, their positions are determined by the extent of charge transfer f; a value of $f = 0.59$ has been obtained for TTF–TCNQ from the diffraction satellites. A comparison of charge transfer in a variety of TCNQ salts with the reduction potential of cations shows that only those cations with a reduction potential $E_r = 0.0$ to 0.5 V vs. SCE lead to mixed-valence or incomplete transfer (Torrance, 1979). When the potential is too high (perylene, pyrene, anthracene etc.), there is no charge transfer and when it is too low, the transfer is complete.

The effect of f on the conductivity of TCNQ salts can be visualized as follows: for electrical conduction to occur, electrons must move from one TCNQ to another. When the charge transfer is complete, the process can be represented as

$$TCNQ^- + TCNQ^- \rightarrow TCNQ^\circ + TCNQ^{2-}$$

which involves creation of a dianion. Understandably the energy involved would be prohibitive and hence TCNQ salts with $f = 1$ are insulators. In the mixed-valence salts, the electrons can move easily along the stack by the process, $TCNQ^- + TCNQ^\circ \rightarrow TCNQ^\circ + TCNQ^-$, which does not require creation of dianions. This localized picture is, however, only qualitative. A more accurate description would involve the band model. The relation between mixed-valence and electrical properties is seen in HMTTF–TCNQ and HMTTF–TCNQF$_4$. Both are isostructural but HMTTF–TCNQ is metallic, while the tetrafluoro-substituted TCNQ salt is semiconducting; the conductivity of the latter is about seven orders of magnitude less than the former. This difference arises because TCNQF$_4$ is a much stronger acceptor than TCNQ and hence the charge transfer is complete in HMTTF–TCNQF$_4$.

6.8   Low-dimensional solids

Chemists are by and large preoccupied with three-dimensional structures and most of solid state chemistry deals with three-dimensional solids. However, there has been increasing interest in lower-dimensional solids which show spectacular anisotropy in their properties. One is familiar with graphite that is metallic in two dimensions and a semiconductor in the third dimension; the striking directional differences in the properties of mica(sheet) and asbestos (fibre) are common experience. The platinum chain compound KCP referred to earlier reflects visible light and conducts electricity like a metal only in the chain direction. If one looks at a crystal of KCP with a polaroid oriented so that the electric vector of the light is parallel to the chain axis, it is highly reflecting and copper-coloured; if the polaroid is turned through a right angle, it is pale yellow and transparent like any other ionic crystal. The situation is similar with Wolfram's red salt. Most of the synthetic metals or molecular metals are low-dimensional solids; many of the exotic materials being tried for superconductivity are also low-dimensional (Keller 1975, 1977; Miller & Epstein, 1978; Hatfield, 1978; Alcacer, 1980; Miller, 1982).

It is convenient to classify low-dimensional solids into two categories, chain (essentially one-dimensional) and layer (essentially two-dimensional). Examples of the chain compounds are KCP and other Pt-chain compounds, polymeric (SN)$_x$, polyacetylene, Hg$_{3-x}$AsF$_6$ with Hg chains, $[(CH_3)_4N]MnCl_3$, KCuF$_3$ and RbFeCl$_3$. Examples of layer compounds are graphite-related systems, Ta and Nb chalcogenides, K$_2$NiF$_4$, (RNH$_3$)$_2$MCl$_4$ and CoCl$_2$ (R = CH$_3$ etc., M = Cr, Mn etc.). We shall briefly examine the magnetic, electrical and optical properties as well as phase transitions of typical members of this extraordinary class of materials. (Day, 1983; Subramanyam & Naik, 1985.) also see *Phil. Trans. Roy. Soc. London* 1985, **A314**.

In understanding the magnetic behaviour of solids it is necessary to take into

Table 6.7. *Examples of low-dimensional magnetic systems*

	Dimension		S	Interaction[a]	T_c^b (K)	T_c/θ
	Spin	Lattice				
KCuF$_3$	3	1	$\frac{1}{2}$	AFM	38 (243)	0.20
(CH$_3$)$_4$NMnCl$_3$	3	1	$\frac{5}{2}$	AFM	0.84 (55)	0.011
CsCuCl$_3$	3	1	$\frac{1}{2}$	FM	10.4	
CsCoCl$_3$	1	1	$\frac{1}{2}$	AFM	8	0.08
RbFeCl$_3$	2	1	1	FM	2.55	
CsNiF$_3$	2	1	1	FM	2.61	0.12
K$_2$CuF$_4$	3	2	$\frac{1}{2}$	FM	6.5	0.28
K$_2$NiF$_4$	3	2	1	AFM	97.2	0.36
CoCl$_2$	2	2	$\frac{1}{2}$	FM	10	0.8
CoBr$_2$·6H$_2$O	2	2	$\frac{1}{2}$	AFM	3.14	0.6
K$_2$CoF$_4$	1	2	$\frac{1}{2}$	AFM	107	0.55

[a] FM = Ferromagnetic; AFM = Antiferromagnetic.
[b] Numbers in brackets indicate the temperature at which the susceptibility is a maximum.

account not only the dimensionality of the lattice (1 to 3), but also the dimensionality of the spin or of the order parameter (1 to 3), which give rise to nine possible types of magnetic systems. In addition, the coupling parameter J can be positive (ferromagnetic) or negative (antiferromagnetic) and this makes 18 different types of systems possible (see Table 6.7). Magnetism in various model systems has been discussed by de Jongh & Miedema (1974). One-dimensional magnetic systems in which there is magnetic interaction only along one direction do not show a magnetic phase transition (one-dimensional Ising systems) if there are only short-range interactions. If there are deviations from ideal one-dimensional behaviour, it is possible to observe a phase transition at nonzero temperatures to a three-dimensionally ordered structure. If J is the intrachain interaction strength and J' is the interchain interaction strength, then the ratio J'/J determines the temperature at which the phase transition takes place. The two-dimensional spin-half Ising model on a square lattice shows a phase transition in the absence of an external magnetic field. This is an exact result and was first solved by Onsager. The two-dimensional Heisenberg system cannot show any long-range order, and in such systems the magnetic sublattice cannot have a spontaneous magnetization. However, there is no thermodynamic argument which excludes the possibility of the susceptibility divergence, although the zero-field magnetization may vanish. The temperature at which the susceptibility diverges in the absence of deviation from ideal behaviour is called the *Stanley–Kaplan temperature* T_{SK}. T_{SK} is the lower limit to the transition temperature T_c occurring due to deviations from ideal two-dimensional

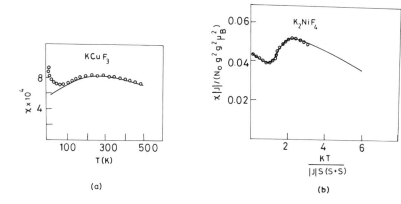

Figure 6.43   (a) χ–T plot of KCuF$_3$ (circles) compared with the theoretical plot (solid line) for a one-dimensional Heisenberg model. (b) χ–T plot of K$_2$NiF$_4$ (circles) and the theoretical fit (solid line) to a two-dimensional Heisenberg model.

behaviour. Only the three-dimensional magnetic system shows long-range order irrespective of the spin dimension.

KCuF$_3$ and K$_2$NiF$_4$ are well-known examples of one-dimensional and two-dimensional magnetic systems respectively. KCuF$_3$ has a tetragonal distortion of the perovskite structure. The ordering of the orbitals shown in Fig. 6.26 is such that there is a strong overlap along the c-axis and almost zero overlap along the ab plane; J'/J in KCuF$_3$ is 2.7×10^{-2}, and neutron-diffraction studies have shown antiferromagnetic ordering at 39.5 K while magnetic susceptibility shows a maximum around 275 K. The susceptibility data agree well with those calculated from a one-dimensional Heisenberg system (Fig. 6.43(a)). In K$_2$NiF$_4$, the interaction between Ni^{2+} ions is antiferromagnetic within a layer and there is little interaction perpendicular to the layer. The magnetic susceptibility to K$_2$NiF$_4$ shows a broad maximum around 230 K (Fig. 6.43(b)), typical of a two-dimensional magnetic system; $d\chi/dT$ is maximum at 100.5 K, while neutron diffraction shows that a three-dimensional ordering occurs at 97.1 K. The latter studies also show that the short-range correlations persist up to 200 K. Thus the broad maximum in the susceptibility is to be associated with short-range interactions in a two-dimensional system. Specific heat measurements show a small anomaly at 98.7 K and a broad maximum around 200 K, which confirm that there is considerable short-range interaction above T_c.

K$_2$CoF$_4$ is a two-dimensional $S = \frac{1}{2}$ Ising antiferromagnet. The Ising anisotropy arises out of the orbital degeneracy of the Co^{2+} ground state and the distortion required to lift the degeneracy. The susceptibility of this compound (Fig. 6.28) compares well with the theoretical curve of the quadratic Ising lattice. T_c coincides with the temperature at which $(d\chi/dT)$ is maximum. The specific heat curve shows a logarithmic singularity at T_c, confirming the two-dimensional Ising behaviour.

The spin and lattice dimensionality of a system can best be determined by studying the thermodynamic behaviour of the system near the transition temperature. In the absence of these studies, one of the quantities of importance in determining the lattice dimensionality of a system is its T_c/θ value, where T_c is the actual critical temperature and θ is the critical temperature predicted by the mean field theory. Table 6.7 gives the relationship between T_c and θ for various lattice dimensionalities and spin dimensionalities, as well as spin values.

Investigations of the magnetic properties of low-dimensional solids continue to be pursued actively and many interesting systems have come to light recently. V_3O_5 exhibits a metal–nonmetal transition at 425 K and a magnetic transition at lower temperatures (Jhans & Honig, 1981). A maximum in the magnetic susceptibility occurs at 12 K, but the actual transition found from heat-capacity measurements is at 76 K. The 76 K transition reflects a change-over from three- to one-dimensional chain ordering in V_3O_5 (Honig, 1982).

Electronic properties of two-dimensional systems have been reviewed by Ando *et al.* (1982). La_2NiO_4 appears to undergo a M–NM transition in the *ab*-plane around 500 K (Rao *et al.*, 1984); stoichiometric samples of this oxide undergo antiferromagnetic ordering. Crossover from 2–d Heisenberg to 2–d Ising type has been reported in several solids of the K_2NiF_4 family. Ganguly & Rao (1984) have reviewed antiferromagnetic (e.g. Ca_2MnO_4) and ferromagnetic (e.g. $La_{0.5}Sr_{1.5}CoO_4$) oxides of K_2NiF_4 structure. In Ln_2CuO_4, the Cu^{2+} ions do not contribute to the susceptibility. High- temperature superconductivity in these cuprates has been a subject of intense study (see Section 7.7). Superconductivity around 1 K has been found in Sr_2RuO_4 (Maeno *et al.*, 1994). Compounds of the A_2CrX_4 formula are transparent ferromagnets (Day, 1979) and detailed structural, magnetooptical and related studies on some of these interesting halides are being pursued (see, for example, Janke *et al.*, 1983; Fyne *et al.*, 1984). $(RNH_3)_2MX_4(R = CH_3$ etc., M = Cr, Mn etc.) show ferromagentism (M = Cr) and antiferromagnetism (M = Mn), and their visible absorption spectra relate to the magnetic ordering (Day, 1985). $FeCl_2$ with a layer structure is metamagnetic; under the influence of a magnetic field the antiferromagnetic phase transforms to a ferromagnetic phase accompanied by increasing opacity (because of the coexistence of the two magnetic phases of different refractive indices (Robbins & Day, 1973)). In low dimensions, competing interactions between ordered chains or layers of spins give rise to canted or incommensurate structures. $NiBr_2$ and $Ni_{1-x}M_xBr_2(M = Fe$ or Mn) show a transition to a helimagnetic phase (Moore & Day, 1985).

Layered materials of K_2NiF_4 structure show interesting phase transitions. For example, compounds of the type $(RNH_3)_2MX_4$ undergo several phase transitions that are related to the ways the NH_3^+ group hydrogen bonds to the halide ions of the layers, the organic chain, R, being in turn affected by the hydrogen bonding (Rao *et al.*, 1981). These compounds have been considered to be inorganic–organic molecular composites by Day (1985). Salts with diacetylene cations which can undergo polymerizations have also been prepared (Ledsham & Day, 1981).

Layered chalcogenides such as TaS_2 can accommodate large molecules such as stearamide. We shall be discussing the intercalation chemistry of these and other layered materials in Chapter 8. TaS_2 is metallic and superconducting ($T_c = 0.8$ K) and the superconducting transition is increased to 3.5 K by incorporation of pyridine; the interlayer separation increases from 3 to 6 Å. Octadecylamine increases the interlayer

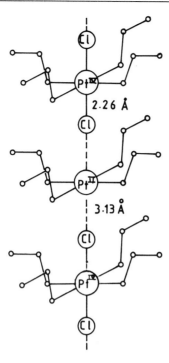

2.26 Å

3.13 Å

Figure 6.44   Structure of Wolfram's red salt.

spacing to $50\,\text{Å}$ increasing T_c to $3\,\text{K}$ (see Sections 3.2.2 and 8.7). A few other low-dimensional chalcogenides were discussed earlier in Chapter 4.

Wolfram's red salt, $[\text{Pt}(\text{C}_2\text{H}_5\text{NH}_2)_4]\,[\text{Pt}(\text{C}_2\text{H}_5\text{NH}_2)_4\text{Cl}_2]\cdot4\text{H}_2\text{O}$, (Fig. 6.44) shows an intervalence absorption band in the visible region due to $\text{Pt}(\text{II})(d_{z^2}) \rightarrow \text{Pt}(\text{IV})(d_{z^2})$ transition, possibly through the participation of orbitals of bridging halide ions. The resonance Raman spectrum shows a large progression of lines in X–Pt–X symmetric stretching mode due to a displacement of the electronically excited state along this vibrational coordinate (Clark, 1977). In tetracyanoplatinites, the energy of the peak varies as the inverse cube of the Pt–Pt distances, indicating thereby that the transition dipoles are all aligned along the Pt chain (Day, 1975).

Following the early work of Krogmann (1969), there has been much interest in the electrical properties of $\text{K}_2\text{Pt}(\text{CN})_4\text{Br}_{0.30}\cdot3\text{H}_2\text{O}$ (KCP) and related compounds. These studies have the purpose of realizing a high-temperature superconductor as predicted by Little (1964); another aspect of interest is the Peierls instability (1955) of one-dimensional metallic conductors (with respect to distortion). KCP is metallic around room temperature, and as the temperature is lowered the conductivity begins to fall and eventually shows the temperature-dependence of a semiconductor (Fig. 6.45). Several other anion-deficient and cation-deficient platinum complexes have been examined with respect to the metal–nonmetal transitions resulting from a Peierls distortion; the transition is affected greatly by the ligands, anions and cations (Underhill, 1985).

The polymer $(\text{SN})_x$ is not only metallic but becomes a superconductor at $0.26\,\text{K}$ (Hatfield, 1978). Another quasi one-dimensional compound showing superconductiv-

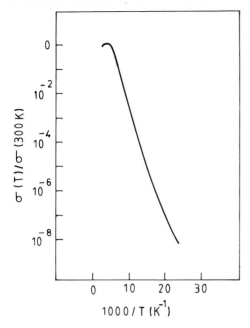

Figure 6.45   Electrical conductivity of KCP as function of temperature.

ity is $Hg_{3-x}AsF_6$ (Fig. 6.46), referred to earlier in Sections 4.9 and 6.7. Other Hg chain compounds of golden lustre of the general formula $Hg_{3-x}MF_6$ (M = Sb, Nb, Ta, As), containing nonintersecting chains of Hg atoms parallel to the a and b axes have been made (Gillespie *et al.*, 1985); these compounds transform to stoichiometric Hg_3MF_6, which have silvery lustre and possess a layer structure of sheets of close-packed Hg atoms. Electrical properties of these compounds have been measured. While the anisotropic conductivity results from a Fermi surface consisting of a set of cylinders (with their axes along the c-axis), the superconducting properties result from mercury trapped in isolated regions (Datars *et al.*, 1985a, b).

Since the early discovery of the large conductivity peak in TTF–TCNQ around 60 K (Coleman *et al.*, 1973) (Fig. 6.47) many studies have been carried out on TTF-related systems (Subramanyam, 1981; Soos & Klein, 1976). Some aspects related to conduction in these systems were mentioned in the previous section. $(TMTSF)_2ClO_4$ and related compounds (TMTSF = tetramethyltetraselenafulvalene) are found to show superconductivity at low temperatures (Jerome, 1985).

Polyacetylene, $(CH)_x$, has probably made the biggest news in this area (Chien, 1984; MacDiarmid & Heeger, 1980). Polyacetylene is doped chemically or electrochemically to produce p-type or n-type materials. The conductivity can be raised to the metallic regime by doping with I_2, Li, $AgClO_4$ and other species (Etemad *et al.*, 1982; MacDiarmid *et al.*, 1985). While polyacetylene itself has a conductivity of 10^{-9} ohm^{-1} cm^{-1}, $[CH\text{-}(AsF_5)_y]_x$ and $(CH\text{-}I_y)_x$ have conductivities in the ranges 10^{-4}–5×10^3 and 10^{-4}–50 ohm^{-1} cm^{-1} respectively (Fig. 6.48). The *cis/trans* content of the polymer can be controlled during synthesis.

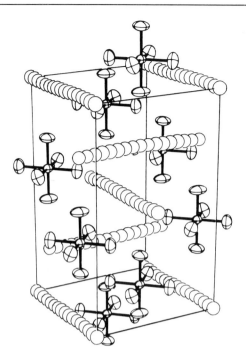

Figure 6.46   Structure of $Hg_{3-x} AsF_6$. (After Schultz *et al.*, 1978.)

Solitons are considered to be important defect states in these conjugated polymers (see Fig. 6.48). It has however been shown that correlation energy is the more important factor in giving rise to the energy gap in $(CH)_x$ (Soos & Ramasesha, 1983). Other polymers related to polyacetylene are polythiophene, polypyrrole, poly-phenylenesulphide, and polyparaphenylene (Section 3.3). Extensive measurements on doped polyacetylenes have been reported in the last five years and these materials, unlike other conducting polymers such as $(SN)_x$, seem to have good technological potential.

cis   *trans*

There are many other low-dimensional solids that have not been included here, polydiacetylene (Bloor, 1985) being one of them. Porphyrinic molecular metals show anisotropic conductivity, where the charge carriers pertain to both the metal and the ligand (Hoffman & Ibers, 1983). The conductivity reaches values greater than 10^3 ohm^{-1} cm^{-1} in some systems and does not appear to come down to the insulator value

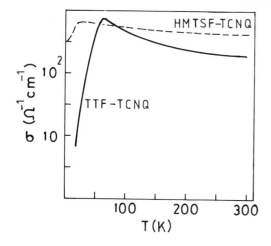

Figure 6.47   Electrical conductivity of HMTSF–TCNQ and TTF–TCNQ as a function of temperature.

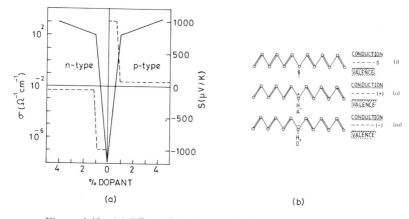

(a)   (b)

Figure 6.48   (a) Effect of doping on the electrical conductivity (solid line) and thermopower (broken line) of polyacetylene. (Following Etemad *et al.*, 1982.); (b) solitons in *trans*-polyacetylene: (i) neutral, (ii) positive and (iii) negative solitons. Arrow marks the boundary between the two symmetric configurations. A, acceptor; D, donor. (Following Subramanyam & Naik, 1985.)

even at low temperatures. Here again, the metal–ligand complex is mixed-valent (partly oxidized), with incomplete charge transfer, and holes created by iodine oxidation seem to hop between the ligand and the metal. The chemical flexibility of porphyrinic metals may indeed offer certain advantages for possible applications and merits further investigation.

Dimensionality reduction in solids (Lee & Holm, 1990) has received some attention. Three-dimensional metal cluster solids of the kind $[M_6Q_8]X_3$ (M = Re, Mo; Q = chalcogen; X = halogen) can be rendered into two- , one- and zero-dimensional phases sequentially by the incorporation of AX type halides (e.g. A = Tl). The stepwise dimensional reduction from 2D to 0D has been realized in the series: $[Re_6Se_8]Cl_2$, $Tl[Re_6Se_8]Cl_3$, $Tl_2[Re_6Se_8]Cl_4$ and $Tl_5[Re_6Se_8]Cl_7$ (Long *et al.*, 1995). Dimensionality reduction also occurs in perovskite oxides of the general formula $ALaSrNb_2M^{II}O_9$; compounds with A = Na adopt the 3D structure, while those with A = Cs possess a 2D structure containing (100) cut perovskite slabs (Gopalakrishnan *et al.*, 1995). Similar dimensionality reduction has been reported in the conducting halide perovskite $CH_3NH_3SnI_3$ (Mitzi *et al.*, 1994, 1995 a, b). While layered compounds of the general formula $R_2(CH_3NH_3)_{n-1}Sn_nI_{3n+1}$ ($n = 1$ to 5 and $R = C_4H_9NH_3$), crystallize with (100) terminated perovskite sheets, compounds of the formula R'_2 $(CH_3NH_3)_mSn_mI_{3m+2}$, where $R' = [NH_2C(I) = NH_2]$, crystallize in a novel structure containing (110) oriented $CH_3NH_3SnI_3$ perovskite sheets separated by the R' cations. The tin environment is fairly regular in these solids indicating that the 5s electrons of Sn(II) are delocalized into the layer conduction-band rather than being localized into lone pairs. The conductivity of these compounds increases with the thickness of the perovskite slab. It is noteworthy that the effect of reducing dimensionality on the electronic properties had been examined sometime ago in oxides of the $(LaO)(LaNiO_3)_n$ or $La_{n+1}Ni_nO_{3n+1}$ and $(La, Sr)_{n+1}Mn_nO_{3n+1}$ (Rao *et al.*, 1988). These oxides tend to become metallic with increase in n. Note that $n = 1$ is two-dimensional with the K_2NiF_4 structure and $n = \infty$ is the three-dimensional perovskite.

6.9   Some results from semi-empirical theoretical investigations and electron spectroscopy

Semi-empirical molecular orbital calculations within the framework of extended Hückel theory (EHT) have been employed to understand the structures of solids (Canadell & Whangbo, 1991; Hoffmann, 1988). Hoffmann (1988) has written a fine introduction to the subject and shown how the frontier orbital and interaction diagram picture can be applied to solids. The importance of such methods lies in the fact that they bridge the results from the ligand-field treatment of molecular complexes with the results of tight-binding (band structure) calculations of crystalline solids. The EHT-tight binding (EHTB) approximation has been exploited widely by Hoffmann, Burdett and Whangbo to throw light on several problems of interest. EHTB would not be the best way to obtain optimized geometries of inorganic solids, but it provides reliable energy dispersions in the valence-band region in known structures, being insensitive to small inaccuracies in atomic parameters. The EHTB method also provides reasonable density of states (if properly parametrized) and reliable Fermi surfaces. The method is therefore convenient to investigate band structures and to visualize the disposition of the metal d-orbitals.

Hoffmann and coworkers have examined Mo chalcogenides and solids of $ThCr_2Si_2$ structure (Hughbanks & Hoffmann, 1983; Hoffmann & Zheng, 1985). Burdett and others (1981, 1982) have applied the ideas of molecular EHT to simple inorganic solids to explain and understand structural relationships. Burdett (1980, 1982) treats a typical unit as the representative fragment of the solid structure in EHT calculations. Using the puckered $A_3X_3H_6$ ring as the fragment, structures of several tetrahedrally coordinated AX solids have been predicted through XX, AX and AA contacts in systems containing

7, 8 and 9 electrons per AX unit. The predictions are borne out in the structures of the Cu^IS part of covellite and 8-electron wurtzites and sphalerites such as ZnS, AlN and $LiGaO_2$. The bond overlap populations calculated for the fragment are of the same order as the bond lengths in these solids. The puckered sheet is energetically favoured compared to the planar analogue as the electronegativity difference between A and X in the A_3X_3 ring increases.

By analogy with molecular systems, Burdett and coworkers (1981, 1982) have rationalized distortions of AX and AX_2 structures as responses to the presence of extra electrons in antibonding orbitals. For example, the symmetric rocksalt structure stable for 8 electrons per AX unit distorts to less symmetric structures by bond-breaking as more electrons are added; in GeP (9 electrons) one bond around every atom is broken and in GeS and GeTe (10 electrons), three mutually perpendicular linkages around every atom are broken. Similarly, the structure of red PbO is regarded as a distorted version of CsCl because of the presence of lone-pair electrons on lead. The structures of AX_2 solids, $CaC_2 \rightarrow FeS_2$ (pyrite and marcasite) $\rightarrow CaF_2$ (fluorite), TiO_2 (rutile) $\rightarrow PbCl_2 \rightarrow XeF_2$, can be understood as responses of adding extra electrons to the symmetric CaC_2 structure which has 10 electrons per CaC_2 unit. With additional electrons, first the X_2 units tilt to give the pyrite and marcasite structures. With 16 electrons per AX_2 unit, the X_2 bond is broken to give structures containing isolated X atoms (fluorite and rutile). Eventually with 22 electrons, the structure of molecular XeF_2 is generated by breaking six of the eight AX linkages of the fluorite structure. EHT calculations also give a diagram similar to that obtained by plotting the principal quantum number against the electronegativity difference, showing resolution of 4-, 6- and 8-coordinated structures of A_mX_n type solids (Burdett & Rosenthal, 1980). The coordination number is determined by the X–X nonbonded repulsions as well as by the stabilizing interactions. The MO method thus provides an alternative to the ionic model to examine inorganic solid structures.

Hückel calculations have been employed extensively in other approaches such as the angular overlap model and the method of moments developed by Burdett and coworkers. Stabilities of crystal structures, pressure- and temperature-induced transitions, dynamical pathways in reactions and other phenomena have been analysed using angular overlap models. Thus, the electronic control of rutile structures and the stability of the defect structure of NbO have been examined (Burdett, 1985; Burdett & Mitchell, 1993). In the case of NbO, the structure is stable at d^3 involving the formation of square-planar Nb with the non-bonding $Nb(4d_{z^2})$ orbital mixing with Nb(5s) orbital and enhancing the Nb–Nb bonding. Using pair potentials, the defect structure is attributed to local repulsive forces. The method of moments has been employed to probe the cause of defect ordering in perovskites such as $Ca_2Mn_2O_5$ (Burdett & Kulkarni, 1988). Distortions in the NiO_6 octahedra in $BaNiLn_2O_5$ have been explained on the basis of a model which combines results from MO theory, bond structure and atom–atom potential arguments (Burdett & Mitchell, 1990). Burdett *et al.* (1994) have recently attempted to answer the fundamental question 'why some solids are so complex while others are so simple' based on three considerations. They first examine some of the topological restrictions imposed by local coordination and stoichiometry. They then make use of the traditional orbital picture to show how the local structure is determined by electron configuration. Finally, they use the method of moments to determine the factors that influence the stability of structures (identity of nearest-neighbour atoms and linkages as a function of electron count).

A few examples from the EHTB band structure calculations would be instructive to examine. The Magneli phase, Mo_8O_{23}, is metallic above 360 K and forms an incommensurate phase between 360 K and 285 K and transforms to a commensurate phase at 285 K. The MoO_6 octahedra in Mo_8O_{23} are all distorted and the structure can be considered to be built up of Mo_4O_{15} slabs and Mo_4O_{14} chains on the one hand and as 3D Mo_8O_{23} on the other. In the first case, the band structure would be a combination of the band structures of the slabs and chains provided there is no appreciable interaction between them. The unit cell corresponds to $(Mo_8O_{23})_2$ with four electrons which fill the two bottommost bands arising from the Mo_4O_{15} slabs. Calculations have revealed the presence of two rather closely spaced bands in addition to a band of even lower energy. Thus Mo_8O_{23} turns out to be a semimetal. As shown by Sato et al. (1982), the electronic instability in Mo_8O_{23} is not caused by the nesting of Fermi surfaces but by a classical mechanism of concerted pairwise rotation of MoO_6 octahedra within the layers (Canadell & Whangbo, 1990).

Red molybdenum bronzes of the formula $A_{0.33}MoO_3$ (A = K, Rb, Cs or Tl) possess 2D structures, with the MoO_3 layers consisting of edge-sharing MoO_6 octahedra. They are present in eclipsed and staggered double chains which are humped and the cations are located between the layers. It was not clear whether these bronzes were semiconductors with a band gap, since semiconducting behaviour can result from partially filled bands when they are Mott insulators. The repeat unit in these bronzes is

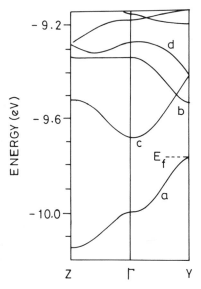

Figure 6.49   Energy dispersion of the t_{2g} block bands in Mo_6O_{18} slabs in $Tl_{0.3}MoO_3$ (after Ganne et al., 1986).

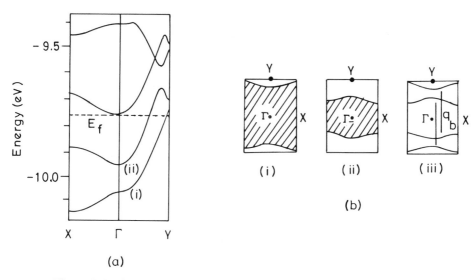

Figure 6.50   (a) Energy dispersion of the t_{2g} bands in $Mo_{10}O_{30}$ blocks; (b) Fermi surfaces of the lowest bands (i) and (ii) and the combined Fermi surface (iii) (after Whangbo & Schneemeyer, 1986).

a Mo_6O_{18} slab with the t_{2g} bands as shown in Fig. 6.49 (Ganne *et al.*, 1986). Since there are ($0.33 \times 6 = 2$) electrons, they fill the *a* band completely and there would be an indirect gap of 0.12 eV to band *c*. The bands are quite dispersive unlike in a Mott insulator. The red bronzes are therefore considered to be semiconductors based on this analysis. Hump octahedra present in these chains do not contribute to the lower t_{2g} bands of the bronzes, but influence the structures through strong intrachain distortions.

Blue bronzes of the formula $A_{0.3}MoO_3$ (with A = K, Rb or Tl) are metallic at room temperature. They contain $Mo_{10}O_{30}$ layers (10 MoO_6 octahedra in a repeat unit) wherein the hump octahedra do not contribute to the low-energy bands. The t_{2g} bands of $Mo_{10}O_{30}$ block are shown in Fig. 6.50(a) where the dashed lines indicate electron filling (3 electrons). In Fig. 6.50(b), Fermi surfaces are shown for the two bands (separately in (i) and (ii) and together in (iii)). The nested portions of the Fermi surfaces are clearly evident in (iii). The Fermi surfaces suggest that blue bronzes are 1D metals in agreement with experiment (Whangbo & Schneemeyer, 1986). An incommensurate CDW transition occurs leading to the formation of a semiconductor around 180 K.

Purple molybdenum bronzes, AMo_6O_{17} (A = Na, K) are 2D metals. Unlike in the blue bronzes, 2D metallic character in these bronzes persists even after CDW formation and the nature of carriers changes from electrons in the normal state (above the CDW transition at 120 K) to holes in the CDW state. About 50% of the carriers are removed below 120 K. The crystal structure of KMo_6O_{17} can be considered as made up of hexagonal Mo_6O_{17} layers built from Mo_4O_{21} units, the latter consisting of an array of 4 MoO_6 octahedra sharing 3 axial corner oxygens. Band structures of KMo_6O_{17} and Mo_2O_9 layers bear a close resemblance (Whangbo *et al.*, 1991).

$Na_{0.25}TiO_2$ is a three-dimensional material, but its structure consists of Ti_2O_8 double octahedral chains from which layers and double chains are built. The pseudo one-dimensional character of $Na_{0.25}TiO_2$ originates from such a structure, giving rise to an interesting metal–insulator transition at 630 K, followed by a transition to an incommensurate phase below 430 K. $Li_{0.3}MoO_3$ is a violet blue coloured bronze and unlike the Na and K bronzes, has a triclinic structure and is semiconducting. Its structure consists of Mo_6O_{24} chains which determine the band structure. The band structure of $Li_{0.3}MoO_3$ suggests that it has to be semiconducting in all directions.

Structures of diphosphate tungsten bronzes (DPTB) with the general formula $A_x(P_2O_4)_4(WO)_{4m}$ (x varying between 1 and 2 and m generally equal to 4) are related to the structure of Mo_8O_{23}. DPTBs are quasi 2D metallic materials and their quasi two-dimensional behaviour is not due to 2D bands but due to 1D bands present in orthogonal directions. The band structure of $Li_{0.9}Mo_6O_{17}$ (the lithium purple bronze) proposed by Whangbo & Canadell (1988) is interesting. It exhibits a pseudo 1D metallic character while the structure is three-dimensional. The low-energy bands originate from Mo_4O_{18} chains possessing two essentially flat bands. They are partially filled in one direction, thereby exhibiting instabilities at temperatures lower than 25 K.

6.9.1   *Results from electron spectroscopy*

The electronic structures of transition-metal and rare-earth compounds have been investigated extensively by employing techniques of electron spectroscopy. These investigations have provided estimates of various interaction strengths that determine the electronic structure of solids with strongly interacting electrons. Let us consider the well-known case of NiO. Several efforts have been made to explain the insulating behaviour of NiO, particularly from band structure calculations within the single-particle approximation. These studies generally assume that the insulating behaviour of NiO is not related to intra-atomic correlation effects. An insulating state of NiO has been obtained within such an approach, but the predicted band gap is considerably lower than that observed experimentally. Mott had pointed out many years ago the need to go beyond the single-particle model and to take the Coulomb correlations into account in understanding the properties of transition-metal compounds. The Hubbard Hamiltonian gives ground states which are metallic or insulating depending on the relative magnitude of the hybridization and the Coulomb interaction strengths. Such a model which takes into account only the transition-metal d^n energy levels and ignores the ligand $2p$ levels altogether is of very limited use.

A very successful approach to understand the electronic structures of transition-metal compounds is due to Zaanen, Sawatzky & Allen (1985) (ZSA). They proposed a phase diagram to explain the insulating and metallic properties as well as the electronic structure of transition-metal compounds in terms of the charge-transfer energy, Δ, between the ligand $2p$ and metal $3d$ levels (i.e. the energy associated with $d^n \rightarrow d^{n+1}L^1$ transition with L^1 representing a hole in the ligand levels) and the on-site Coulomb correlation energy, U_{dd}, and thereby showed how the cationic and anionic energy levels were both equally important. These workers calculated the gap arising in the charge excitation spectrum of a single transition-metal impurity hybridizing with a

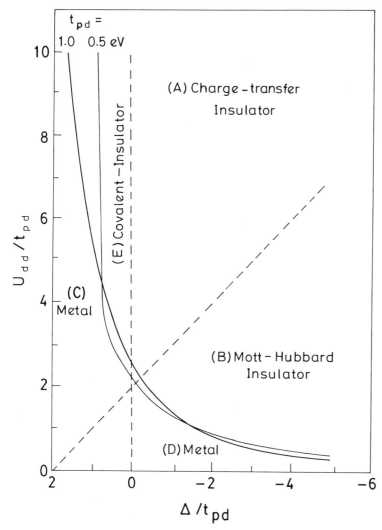

Figure 6.51   Phase diagram for transition-metal compounds in the U/t and Δ/t space showing the parameter ranges for metallic and insulating states (after Sarma *et al.*, 1992).

ligand derived band within the charge-transfer impurity model. The interaction parameters characterizing the system are: (i) the charge-transfer energy, Δ, (ii) the intra-atomic Coulomb interaction strength, U_{dd}, within the $3d$ manifold [$U_{dd} = E(d^{n+1}) + E(d^{n-1}) - 2E(d^{n})$], and (iii) the hybridization strength, t, between the $3d$ and the ligand level. With this model, ZSA calculated the phase line separating metals from insulators in terms of U_{dd}/t and Δ/t. The ZSA diagram could explain differences between compounds of the early transition metals and of the late transition metals as well as the dependence of the conductivity gap, in a related series of compounds of a transition metal on the anion electronegativity. The limitation of the ZSA approach was the neglect of transition-metal translational symmetry implicit in the impurity

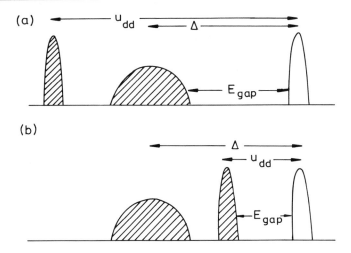

Figure 6.52   Schematic electron addition and removal spectra representing the electronic structure of transition-metal compounds for different regimes of the parameter values: (a) charge-transfer insulator with $U > \Delta$; (b) Mott–Hubbard insulator $\Delta > U$ (From Rao *et al.*, 1992).

model. Such a model describes systems with small hopping interactions reasonably well, but the large hybridization strengths between the transition-metal $3d$ and the oxygen $2p$ states in oxides suggest the need to go beyond the impurity model. Furthermore, such an impurity model cannot satisfactorily describe the magnetic properties.

In Fig. 6.51 we show the ZSA type phase diagram obtained by Sarma *et al.* (1992) which takes into account the translational symmetry as well. This phase diagram is similar to the original ZSA phase diagram and suggests the existence of different regimes of insulating and metallic properties giving rise to (A) charge-transfer insulators, (B) Mott–Hubbard insulators (C) p–d type metals and (D) d-band metals. The phase diagram also identifies a new insulating region (marked E in Fig. 6.51) in addition to the two insulating regions (regions A and B) obtained by ZSA.

The features of the different regions of the phase diagram are as follows: Region A corresponding to the charge-transfer insulators is characterised by $U_{dd} \gg \Delta > \Delta_c (U)$, where $\Delta_c (U)$ is a critical Δ value of the bare charge-transfer energy depending on U_{dd}. The electronic structure of this region is schematically represented by the excitation spectrum shown in Fig. 6.52(a), where the hatched part denotes an electron removal and the other an electron addition spectra. Thus, the essentially oxygen $2p$ derived band appears between the lower and upper Hubbard bands. This region has the lowest energy charge excitation primarily involving a configuration $d^n p^m$ going to a $d^{n+1} p^{m-1}$ configuration and the band gap in this region is mainly dictated by Δ. Late transition-metal compounds are expected to belong to this regime. In contrast to this, the Mott–Hubbard region (region B in Fig. 6.51) is characterised by $\Delta \gg U_{dd} > \Delta_c (U)$, where $\Delta_c(U)$, is the critical U_{dd}-value depending on Δ. In this limit, one obtains the schematic electronic structure shown in Fig. 6.52(b). Here the lowest energy charge excitation involves a $d_i^n d_j^n$ state being excited to $d_i^{n+1} d_j^{n-1}$ state; the band gap is seen to be primarily controlled by the on-site Coulomb correlation energy U_{dd} (see Fig. 6.52(b)). This is the expected situation for the early transition-metal compounds. From Fig. 6.52(a), it seems obvious that for a small Δ in the $U_{dd} \gg \Delta$ regime, the upper Hubbard band overlaps the top of the oxygen $2p$ derived band, thereby giving rise to a metallic ground-state. This is the origin of the metallic region C in Fig. 6.51. In the other limit of $\Delta \gg U_{dd}$ (see Fig. 6.52(b)), a sufficiently small U_{dd} leads to a metallic ground state (region D in Fig. 6.51) due to the overlap of the lower and upper Hubbard bands. These descriptions are somewhat modified in the presence of finite hybridization strength, mixing the transition-metal d and oxygen p states. It is to be noted however that both charge-transfer and Mott–Hubbard insulators retain a finite gap in the charge excitation spectrum in the limit of vanishing hybridization strength. Region E is the covalent insulator regime, characterized as a state which should have been a metal in absence of any hybridization mixing, but is driven to the insulating state due to the presence of large hybridization matrix elements.

The ZSA phase diagram and its variants provide a satisfactory description of the overall electronic structure of stoichiometric and ordered transition-metal compounds. Within the above description, the electronic properties of transition-metal oxides are primarily determined by the values of Δ, U_{dd} and t. There have been several electron spectroscopic (photoemission) investigations in order to estimate the interaction strengths. Valence-band as well as core-level spectra have been analysed for a large number of transition-metal and rare-earth compounds. Calculations of the spectra have been performed at different levels of complexity, but generally within an Anderson impurity Hamiltonian. In the case of metallic systems, the situation is complicated by the presence of a continuum of low-energy electron-hole excitations across the Fermi level. These play an important role in the case of the rare earths and their intermetallics. This effect is particularly important for the valence-band spectra.

Analysis of the valence-band spectrum of NiO helped to understand the electronic structure of transition-metal compounds. It is to be noted that the crystal-field theory cannot explain the features over the entire valence-band region of NiO. It therefore becomes necessary to explicitly take into account the ligand(O2p)–metal (Ni3d) hybridization and the intra-atomic Coulomb interaction, U_{dd}, in order to satisfactorily explain the spectral features. This has been done by approximating bulk NiO by a cluster $(NiO_6)^{10-}$. The ground-state wave function Ψ_g of this cluster is given by,

$$\Psi_g = a|d^2 L^0> + b|d^1 L^1> + c|d^0 L^2>,$$

where $|d^2 L^0>$ (in the hole configuration) denotes the ionic $Ni^{2+} O^{2-}$ (or the $3d^8 2p^6$ electron configuration) state, $|d^1 L^1>$ represents the $Ni^{1+} O^{1-}$ (or the $3d^9 2p^5$) state, and the $|d^0 L^2>$ is the $3d^{10} 2p^4$ state. Hybridization between the Ni $3d$ and O $2p$ states mixes the three states in order to substantially lower the energy of the ground-state compared to the ionic state $|d^2 L^0>$ or the $Ni^{2+} O^{2-}$ state. (Note the importance of covalences). Assuming the contribution of $|d^0 L^2>$ in Ψ_g to be small, ignoring its effects, the ground-state is found to have the symmetry $^3A_{2g}$. This implies that the $3d^8$

configuration ($|d^2L^0>$) with $|t_{2g}^3, t_{2g\uparrow}^3 e_{g\uparrow}^2, {}^3A_{2g}>$ hybridizes with various $|d^1L^1>$ configurations with the same symmetry to form the ground-state wave function. The final state following the photoemission process has one less electron and can have ${}^2E_g$, ${}^2T_{1g}$, and ${}^4T_{1g}$ symmetries due to dipole selection rules. The corresponding $3d^7$ (or $|d^3L^0>$) configurations are given by $|t_{2g}^6 e_g^1, {}^2E_g>$, $|t_{2g}^5 e_g^2, {}^2T_{1g}>$, and $|t_{2g}^5 e_g^2, {}^4T_{1g}>$; the $|d^2L^1>$ and $|d^1L^2>$ states also are chosen with the proper symmetries. Estimating the transfer integrals from band structure calculations and using the various energies (e.g. U_{dd} and Δ) as fitting parameters, the photoemission spectrum for the cluster has been calculated. From the best fit to the experimental spectrum, the charge-transfer energy, Δ, and the Coulomb energy, U_{dd}, are estimated to be about -3.5 and $9\,eV$ respectively. These estimates suggest that NiO is a charge-transfer insulator in terms of the phase diagram shown in Fig. 6.51 (regime A).

Use of core-level spectra to understand the electronic structure of oxides is illustrated by the Cu $2p$ core-level spectrum of insulating cuprates such as La_2CuO_4 (which are the parent compounds for the high T_c superconducting cuprates). These cuprates are modelled by the CuO_2 square-planar sheet. At the simplest level, one Cu $3d_{x^2-y^2}$ orbital and four oxygens (with only p_x and p_y orbitals) around it are taken into account. One then includes the hybridization matrix elements, ts, the bare site energies, εs as well as the Coulomb interaction strength, U_{dd}, within the Cu $3d$ levels. Insulating cuprates have one hole per Cu site as in stoichiometric La_2CuO_4. This implies that U_{dd} plays no role in this case. Since the Cu $3d_{x^2-y^2}$ orbitals belong to CuO_4 cluster, only the linear combination, $1/2\,(p_x^1 - p_y^2 - p_x^e + p_y^4)$, belongs to the b_{1g} representation. The electronic structure of the CuO_4 cluster within the above approximation will be controlled entirely by the interaction between the Cu $3d_{x^2-y^2}$ and the oxygen derived b_{1g} orbitals. The relevant parameters in the model are therefore the energy differences and the hybridization strength between these two levels and the core hole-valence hole repulsion strength, U_{dx}. The value of Δ would be equal to $\varepsilon_p - \varepsilon_d$ in the absence of oxygen–oxygen interaction; inclusion of oxygen–oxygen near-neighbour interactions shift Δ without making any qualitative change in the results. The effective hybridization strength, t, between the two states is $2t_{pd}$ where t_{pd} is the interaction strength between the Cu $3d_{x^2-y^2}$ and the oxygen p_x^1 orbitals.

Experimental Cu $2p$ core-level spectra of cuprates exhibit a two-peak structure with one peak at about $933\,eV$ (main peak) and the other at $942\,eV$ (satellite). Although the model described in the previous paragraph is simplistic, it provides an analytical expression for the Cu $2p$ core-level spectrum. The spectrum is calculated as a function of Δ, t and U_{dx} to obtain the energy difference between the main Cu $2p$ peak and the satellite, ΔE, and the ratio of their intensities, I_s/I_m. In the specific case of La_2CuO_4, I_s/I_m is 0.33 and ΔE is $8.5\,eV$. In order to obtain a suitable range of parameters to describe the core-level spectrum of La_2CuO_4 within the cluster approximation, one constrains the hybridization strength, t, within the reasonably wide bounds of 1.5 and $3.0\,eV$, while freely varying Δ and U_{dx}. In order to fit the experimental spectrum within the experimental uncertainties, Sarma finds that the value of U_{dx} has to be between 7.0 and $8.0\,eV$. The experimental core-level spectrum thus provides a reasonably narrow range of estimate of U_{dx} although the Δ and t values can have a wide range.

The calculation of the core levels of La_2CuO_4 discussed above was performed within the cluster approximation wherein the dependence of the spectral shape on the oxygen band-width is missed out. In order to include the effect of the oxygen $2p$ derived band on the spectral shape, the core-level photoemission calculation has to be performed within the impurity model which includes the O $2p$ band-width (instead of a single O $2p$ level). For reasonable values of the hybridization strength, t, the range of values of Δ is such that the Cu $3d_{x^2-y^2}$ appears close to the bottom of the oxygen $2p$ derived band-width or lies within it. This is the regime of the covalent insulator in terms of the diagram in Fig. 6.51 (region E). The ground-state of La_2CuO_4 is associated with a strongly mixed-valent character from photoemission evidence.

We discussed above the analysis of core-level spectra from cuprates which contain only a single hole in the d-band per Cu ion. This makes U_{dd} an irrelevant parameter within an impurity model. However, analysis can also be carried out for systems where U_{dd} plays a significant role. This is illustrated by the analysis of the core-level spectrum of $LaCoO_3$ carried out within the impurity model (Chainani et al., 1992). This oxide is modelled by the $(CoO_6)^{9-}$ cluster with the transition-metal ion being formally in the $3+$ oxidation state and in an octahedral crystal field. One has to therefore take into account interactions between various configurations such as $|d^6>, |d^7L^1>, |d^8L^2>$. Moreover the $3d$ manifold is split into t_{2g} and e_g sub-bands due to crystal-field splitting and these are further split into up-spin and down-spin levels due to exchange interaction. The competition between the crystal-field splitting that stabilizes the low-spin configuration and the exchange interaction that favours the high-spin configuration gives rise to the spin-state transition in $LaCoO_3$. The low-spin configuration is the ground-state of this oxide. Further complication arises from the fact that in these cases two different hybridization interactions characterize the system. Taking into account all these interactions, calculations have been carried out for the low- as well as the high-spin states of $LaCoO_3$. The analysis shows that the results of the calculation within the high-spin state is incompatible with the experimental spectrum, whereas the calculation based on the low-spin configuration provides a good description. The values of the parameters are $\Delta = 4.0$, $U_{dd} = 3.4$ and $t = 3.8\,eV$. The ground-state with these parameters turns out to be a very mixed state with 38.5% d^6, 45.4% d^7L^1 and 14.5% d^8L^2 characters. Such a ground-state places $LaCoO_3$ in the regime between the Mott–Hubbard, charge-transfer and covalent insulators in terms of the phase diagram shown in Fig. 6.51.

We have hitherto examined the method of obtaining various electronic interaction parameters from the analysis of electron spectroscopic results in order to provide a description of the electronic structure of correlated systems. Electron spectroscopies and related methods have been used in different ways to provide important information concerning a wide variety of systems. Thus, extensive investigations have helped in understanding different aspects of rare-earth and actinide based mixed-valent as well as heavy Fermion systems. High-resolution photoemission studies have revealed the opening of the superconducting gap in superconducting cuprates. The nature of the doped hole states at the Fermi energy in these cuprates has been established by carrying

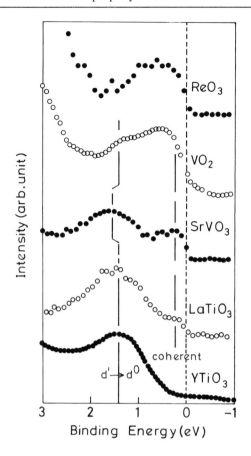

Figure 6.53   Valence-band spectra of YTiO$_3$, LaTiO$_3$, SrVO$_3$, VO$_2$ and ReO$_3$ (from Fujimori *et al.*, 1992 as adopted by Sarma, 1995).

out detailed site and symmetry projected electron spectroscopic measurements on single crystals. The studies have shown that these states have strongly admixed Cu ($3d_{x^2-y^2}$) and O($2p_{xy}$) character.

The use of electron spectroscopies in understanding insulator–metal transitions in transition-metal oxides is noteworthy (see Sarma, 1995, for a detailed review). In Fig. 6.53, we show the UV photoelectron spectra (valence region) of a few d^1 oxides arranged in the order of decreasing U/W ratio. YTiO$_3$ is an insulator and hence there is no spectral intensity at E_F; instead, there is a localized feature due to the occupied lower Hubbard band around 1.5 eV below E_F. LaTiO$_3$, on the other hand, is close to the insulator–metal boundary and shows a weak intensity feature at E_F (probably arising from excess oxygen). As we progressively go towards metallic oxides with small U/W (SrVO$_3$, VO$_2$, ReO$_3$), there is a transfer of spectral weight from the lower Hubbard band to the states at E_F. The feature due to the lower Hubbard band however remains even in metallic samples. Actually, the changes in Fig. 6.53 reflect a systematic increase in W and hence a decrease in U/W. In the La$_{1-x}$Sr$_x$CoO$_3$ system, where the isulator–metal transition is brought about with increase in x, the valence-band spectra show a gap at E_F for $x = 0.0$ (parent $LaCoO_3$) and a finite density of states of E_F with increase in x. The doped hole states with a substantial oxygen p-character populate the top of the valence band (Sarma, 1995).

6.10   Ferroics

An important characteristic of ferromagnetic materials is the hysteresis loop found in the relationship between magnetization and magnetic field. The electrical analogue of a ferromagnet is a ferroelectric that shows a hysteresis loop in the relationship between polarization and electric field and exhibits spontaneous polarization in the absence of an external electric field. *Ferroelectric materials* possess permanent dipole moments, the dipoles arising from the absence of a centre of symmetry. If energy considerations favour an antiparallel arrangement of the permanent dipoles on adjacent planes instead of a parallel arrangement, such a crystal is called an *antiferroelectric*. Both ferroelectric and antiferroelectric materials show a dielectric constant anomaly at a critical temperature, but the latter do not show a polarization–electric field hysteresis loop. In this section we shall discuss the behaviour of a general class of materials called *ferroics* (Newnham & Cross, 1981), which includes ferroelectrics and antiferroelectrics.

A ferroic may be defined as a material possessing two or more orientation states or domains which can be switched from one to another through the application of one or more appropriate forces (Newnham, 1974). In a ferroelectric, the orientation state of spontaneous electric polarization can be altered by the application of an electrical field. In a ferromagnet, the orientation state of magnetization in domains can be switched by the application of a magnetic field; in a *ferroelastic*, the direction of spontaneous strain in a domain can be switched by the application of mechanical stress. Such transitions are described as ferroic transitions. In all these cases, the boundaries of domains are moved by the application of a force in order to accomplish changes in orientation. $BaTiO_3$ is a typical ferroelectric, CrO_2 a ferromagnet, while $CaAl_2Si_2O_8$ is ferroelastic. The properties for which directionality changes in the above examples, namely electric polarization, magnetic polarization and elastic strain, are primary quantities in the sense that their magnitudes directly determine the free energy of the system. Ferroics governed by the switchability of these properties are therefore called *primary ferroics*.

In *secondary ferroics*, these properties occur as induced quantities. The orientation states differ in derivative quantities, which characterize the induced effects. Thus, the induced electric polarization is characterized by dielectric susceptibility, k_{ij}, induced magnetic polarization by magnetic susceptibility, χ_{ij}, and the induced strains by the elastic compliance, C_{ijkl}. The orientation states in secondary ferroics, therefore, differ in k_{ij}, χ_{ij} and C_{ijkl}; these are tensor quantities, and the rank of the tensor is equal to the number of the subscripts. The induced effects such as polarization or magnetization can also result from cross-coupled effects such as stress-induced polarization (*piezoelectric*), stress-induced magnetization (*piezomagnetic*) or as a combined effect of two types of fields, such as in elastoelectric or elastomagnetic effects. The directional change can then be visualized as occurring in the corresponding derivative quantities. In Table 6.8, we have listed different types of ferroic effects. The classification scheme of ferroics into primary, secondary and so on has a thermodynamic origin.

Table 6.8. *Primary and secondary ferroics*

	Ferroic property	Switching field	Example
Primary ferroics			
Ferroelectric	Spontaneous polarization	Electric field	$BaTiO_3$
Ferromagnetic	Spontaneous magnetization	Magnetic field	CrO_2
Ferroelastic	Spontaneous strain	Mechanical stress	$CaAl_2Si_2O_8$
Secondary ferroics			
Ferrobielectric	Dielectric susceptibility	Electric field	$SrTiO_3$
Ferrobimagnetic	Magnetic susceptibility	Magnetic field	NiO
Ferrobielastic	Elastic compliance	Mechanical stress	α-quartz
Ferroelastoelectric	Piezoelectric coefficients	Electric field and mechanical stress	NH_4Cl
Ferromagnetoelastic	Piezomagnetic coefficients	Magnetic field and mechanical stress	$FeCO_3$
Ferromagnetoelectric	Magnetoelectric coefficients	Magnetic field and electric field	Cr_2O_3

The Gibbs free energy of a crystalline solid can be written in the differential form as

$$dG = -SdT + P_i dE_i + M_i dH_i + \varepsilon_{ij} d\sigma_{ij} \qquad (6.59)$$

where the small variation in the free energy of a mole of material (dG) is the sum of the thermal (SdT) term and terms arising from various forces. Here, P_i and E_i stand for electric polarization and electric field respectively, M_i and H_i are magnetic polarization (magnetization) and magnetic field respectively, while ε_{ij} and σ_{ij} represent the elastic strain and the corresponding mechanical stress respectively; the subscripts i and j run from 1 to 3. When we examine ferroic phenomena of interest under isothermal conditions, the first term in equation (6.59) is zero. Remembering that both spontaneous and induced effects have to be taken into account, the difference in the Gibbs free energies of two orientation states is given by

$$\begin{aligned} \Delta G &= \Delta P_i^s E_i + \Delta M_i^s H_i + \Delta \varepsilon_{ij}^s \sigma_{ij} \\ &+ \tfrac{1}{2}[\Delta k_{ij} E_i E_j + \Delta \chi_{ij} H_i H_j + \Delta C_{ijkl} \sigma_{ij} \sigma_{kl}] \\ &+ 2[\Delta \alpha_{ij} E_i H_j + \Delta d_{ijk} E_i \sigma_{jk} + \Delta Q_{ijk} H_i \sigma_{jk}] \end{aligned} \qquad (6.60)$$

Here Δ represents the difference in the relevant quantity between orientation states II and I and 's' stands for spontaneous quantities. That is, $\Delta\varepsilon_{ij}^s = \varepsilon_{ij}^s(\text{II}) - \varepsilon_{ij}^s(\text{I})$ etc., are the differences in the ij components of ε^s etc., in the states II and I. When no external force is acting on the system, $\Delta G = 0$ and the orientation states are energetically degenerate.

Several possible ferroic phenomena become evident from equation (6.60), depending upon the dominance of particular terms. In a material which has a large value of spontaneous polarization, other terms become unimportant and the free energy in an electric field is governed by the expression

$$\Delta G = \Delta P_i^s E_i = \Delta P^s E \tag{6.61}$$

where E represents the electrical field in a chosen direction (say z). Similar expressions can be written for primary ferroics involving spontaneous magnetization and spontaneous strain when they are the dominant quantities and interact with the corresponding external fields. When $\Delta P_i^s = \Delta M_i^s = \Delta \varepsilon_{ij}^s = 0$, the quantities which determine the ΔG values arise from terms in the two sets of square brackets in equation (6.60). The first set of quantities is Δk_{ij}, $\Delta \chi_{ij}$ and ΔC_{ijkl}. When $\Delta k_{ij} \neq 0$ and $\Delta \alpha_{ij}$ and Δd_{ijk} make little or negligible contribution, as in a ferrobielectric, the free energy is determined by the expression

$$\Delta G \approx \tfrac{1}{2}\Delta k_{ij}E_i E_j \approx \Delta k E^2 \tag{6.62}$$

Dominance of the other terms gives rise to similar expressions for other secondary ferroics.

A ferroelectric transition from one orientation state to another (observed through hysteresis loops) is electrically a first-order transition. The order of the transition between domain states in ferroics is simply the sum of the exponents of the field terms in the free-energy expression. Ferrobielastic, ferrobimagnetic, ferroelastoelectric, ferromagnetoelectric and other such transitions are all second-order. We should note here that the various coefficients are in themselves interdependent. For example, spontaneous polarization is associated with a field that in turn produces a strain through strong coupling to the lattice, thus activating electrostrictive or piezoelectric coefficients. Wherever a spontaneous polarization exists, therefore, spontaneous strain would also be present with it and vice versa. Whenever spontaneous polarization and spontaneous strain produce orientation states fully independent of each other, both ferroelastic and ferroelectric properties can be fully realized and the materials are referred to as fully ferroelectric–fully ferroelastic materials. Other interrelations of a similar nature can be visualized from the interdependence of various quantities implicit in equation (6.60).

At high temperatures, ferroelectric materials transform to the *paraelectric* state (where dipoles are randomly oriented), ferromagnetic materials to the paramagnetic state, and ferroelastic materials to the twin-free normal state. The transitions are characterized through order parameters (Rao & Rao, 1978). These order parameters are characteristic properties parametrized in such a way that the resulting quantity is unity for the ferroic state at a temperature sufficiently below the transition temperature, and is zero in the nonferroic phase beyond the transition temperature. Polarization, magnetization and strain are the *proper order parameters* for the ferroelectric,

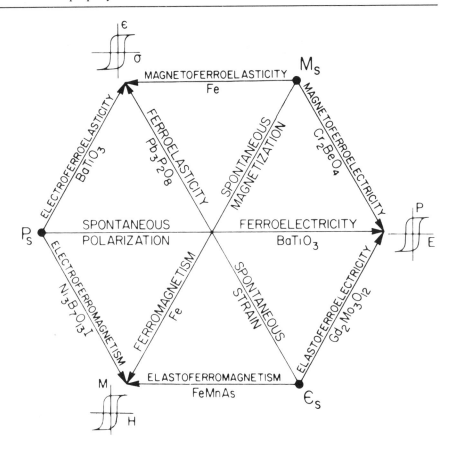

Figure 6.54   Diagram illustrating several types of order parameters involved in proper and improper ferroics. (After Newnham & Cross, 1981.)

ferromagnetic and ferroelastic transitions respectively. Whenever transitions are governed by the expected variations of these order parameters, they are called *proper ferroics*. As we noted earlier, the coupled nature of ferroic phenomena is such that the order parameter which determines the transition can often be different from the 'proper' order parameter. Ferroics where the order parameter does not represent a 'proper' property are called *improper ferroics*. In terbium molybdate, for example, the order parameter is a condensed optic mode (Dorner *et al.*, 1972). The optic mode causes a spontaneous strain, which in turn causes a spontaneous polarization through piezoelectric coupling; terbium molybdate is therefore both an improper ferroelectric and an improper ferroelastic.

A hexagonal representation of proper and improper primary ferroics as proposed by Newnham & Cross (1981) is given in Fig. 6.54. The order parameter for proper ferroics appears on the diagonals of the hexagon, while the sides of the hexagon represent improper ferroics. They indicate the cross-coupled origin of ferroic phenomena. An improper primary ferroic in this classification is distinguished from a true secondary ferroic by the appearance of the prefix 'ferro' only with the primary ferroic quantity and not for both of the coupled quantities. Thus, the term magnetoferroelectric (e.g. Cr_2BeO_4) implies that the material is an improper ferroic, whereas the term ferromagnetoelectric (e.g. Cr_2O_3) would mean that the material is a secondary ferroic.

Ferroelectrics.   Among the 32 crystal classes, 11 possess a centre of symmetry and are centrosymmetric and therefore do not possess polar properties. Of the 21 noncentrosymmetric classes, 20 of them exhibit electric polarity when subjected to a stress and are called piezoelectric; one of the noncentrosymmetric classes (cubic 432) has other symmetry elements which combine to exclude piezoelectric character. Piezoelectric crystals obey a linear relationship $P_i = g_{ij}F_j$ between polarization P and force F, where g_{ij} is the piezoelectric coefficient. An inverse piezoelectric effect leads to mechanical deformation or strain under the influence of an electric field. Ten of the 20 piezoelectric classes possess a unique polar axis. In nonconducting crystals, a change in polarization can be observed by a change in temperature, and they are referred to as *pyroelectric* crystals. If the polarity of a pyroelectric crystal can be reversed by the application on an electric field, we call such a crystal a ferroelectric. A knowledge of the crystal class is therefore sufficient to establish the piezoelectric or the pyroelectric nature of a solid, but reversible polarization is a necessary condition for ferroelectricity. While all ferroelectric materials are also piezoelectric, the converse is not true; for example, quartz is piezoelectric, but not ferroelectric.

One of the important characteristics of ferroelectrics is that the dielectric constant obeys the *Curie–Weiss law* (equation 6.48), similar to the equation relating magnetic susceptibility with temperature in ferromagnetic materials. In Fig. 6.55 the temperature variation of dielectric constant of a single crystal of $BaTiO_3$ is shown to illustrate the behaviour. Above 393 K, $BaTiO_3$ becomes paraelectric (dipoles are randomized). Polycrystalline samples show less-marked changes at the transition temperature.

Barium titanate, which crystallizes in the perovskite structure, has cubic symmetry above 393 K, with Ba^{2+} in the body centre and TiO_6 octahedra in the corners. It undergoes a transformation to a tetragonal structure at 393 K, to an orthorhombic structure at 278 K, and to a rhombohedral structure at 183 K (Fig. 6.55). Relative to the cubic phase, elongation occurs along one of the edges ([100] direction) in the tetragonal phase, along one of the face diagonals ([110] direction) in the orthorhombic phase, and along one of the body diagonals ([111] direction) in the rhombohedral phase. The Ti^{4+} ion moves in these three directions successively as the crystal is cooled from the cubic phase (which has Ti^{4+} in the centre of the octahedron). Besides the dielectric constant and polarization, heat capacity and other properties also show anomalous changes at the three phase transitions of $BaTiO_3$ (Fig. 6.55).

Polarization of a ferroelectric material varies nonlinearly with the applied electric field. The *P–E* behaviour is characterized by a *hysteresis loop* and observation of the hysteresis loop is the best evidence for the existence of ferroelectrcity in a material. The hysteresis loop has its origin in the rearrangement of domains under the influence of an applied elecric field. Generally, the domains are randomly distributed, giving a net zero polarization. Under an applied field or mechanical stress, favourably oriented *domains*

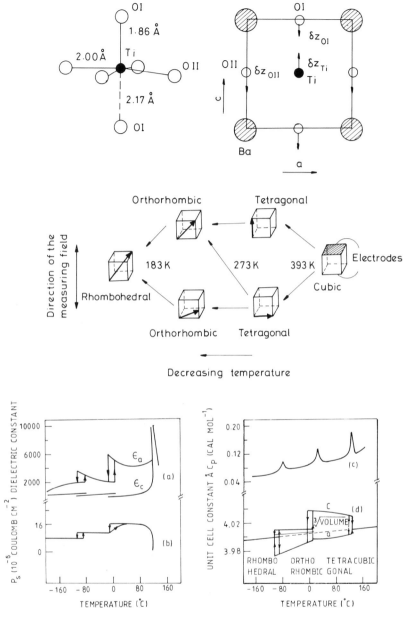

Figure 6.55   Distortion of TiO_6 octahedron in the tetragonal $BaTiO_3$ (top) and possible orientations of the polar axis when an electric field is applied along the pseudo-cubic (001) direction of $BaTiO_3$ (middle). Polar axes are shown by arrows inside each cube. Phase transitions in $BaTiO_3$ accompanied by changes in (a) dielectric constant; (b) spontaneous polarization; (c) heat capacity and (d) lattice dimensions (bottom).

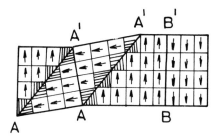

Figure 6.56   Domain walls in tetragonal $BaTiO_3$. AA' is a 90° wall and BB' is a 180° wall.

grow at the expense of the less favourably oriented domains until a single domain configuration is obtained. The domain structure itself is related to the crystallography of the ferroelectric phase with respect to the paraelectric phase. Thus, in the tetragonal phase of $BaTiO_3$, adjacent domains may have their polar axes making angles of 90° or 180° (Fig. 6.56).

Many of the ferroelectric materials exhibit softening of certain vibrational modes. *Soft-mode* behaviour of ferroelectric materials has been investigated in detail by employing Raman spectroscopy and neutron scattering, and the subject has been reviewed by Blinc & Zeks (1974).

A variety of materials exhibit ferroelectricity (Jona & Shirane, 1962; Subbarao, 1974). These include oxides of perovskite structure (e.g. $BaTiO_3$), hydrogen-bonded solids (e.g. Rochelle salt, KH_2PO_4), tungsten bronze type structures, pyrochlores, simple salts (e.g. $(NH_4)_2SO_4$, $NaNO_2$, KNO_3), alums, organic compounds (e.g. thiourea, glycine sulphate) and binary compounds as simple as HCl, FeS, GeTe and V_3Si. Hydrogen-bonded ferroelectrics like KH_2PO_4 display deuterium isotope effects (with T_c increasing on deuterium substitution), showing the important role of hydrogen bonding. It has been recently found that even amorphous samples of materials such as $LiNbO_3$ show ferroelectric-like dielectric anomaly (Varma *et al.*, 1985).

Antiferroelectric materials show superstructure in the antipolar phase as well as a dielectric constant anomaly at T_c. Changes in other physical properties are also observed at T_c when the antiferroelectric phase transforms to the paraelectric phase. Antiferroelectric materials, however, do not show the P–E hysteresis loop. Since the energy difference between antiferroelectric and ferroelectric states is rather small, application of a large electric field, mechanical stress or compositional variation can induce antiferroelectric materials to become ferroelectric. $PbZrO_3$ with the perovskite structure is a typical antiferroelectric. Other examples are $NaNbO_3$, $BiFeO_3$, $PbCo_{0.5}W_{0.5}O_3$, $NH_4H_2PO_4$ and $Cu(HCOO)_2·4H_2O$.

In recent years, a large number of materials exhibiting other interesting properties (possibly of technological value) besides ferroelectricity has been reported. Typical paired properties with examples are:

Ferroelectric–ferroelastic, $Gd_2(MoO_4)_3$, $KNbO_3$,
Ferroelectric–antiferromagnetic, $YMnO_3$, $HoMnO_3$,

Ferroelectric–ferromagnetic, $Fe_3B_7O_{13}Cl$, $Bi_9Ti_3Fe_5O_{27}$,
Antiferroelectric–antiferromagnetic, $BiFeO_3$, $Cu(HCOO)_2 \cdot 4H_2O$,
Ferroelectric–semiconducting, FeS, reduced $SrTiO_3$, $YMnO_3$,
Ferroelectric–superconducting, $SrTiO_3$, GeTe, V_3Si.

Primary ferroics.   As mentioned earlier, the orientation states in ferroelectrics correspond to polarization directions in domains. We can classify ferroelectric materials on the basis of the magnitude of the Curie constant C (eqn 6.48). Large values of C ($\sim 10^5$ K) are evident in oxide ferroelectrics, particularly those containing ions Ti^{4+}, Nb^{5+} and W^{6+} and those containing lone-pair ions Pb^{2+}, Bi^{3+} and Sb^{3+}. Owing to their electronic structure, these ions promote distortions, giving rise to spontaneously polarized structures. The second category, where C is around 10^3 K, comprises mainly order–disorder type ferroelectrics where rotational ordering of dipolar ions (as in $NaNO_2$) or positional ordering of protons as in KH_2PO_4 is seen. The third category, with very low values of C (~ 1–10 K), comprises improper ferroelectrics such as $Gd_2(MoO_4)_3$, where the dimensionality of atomic displacements increases in the same order as the decrease in the C value. Whereas in titanates the displacements are one-dimensional, giving rise to strong dipole moments, in hydrogen phosphates and $NaNO_2$ dipoles are ordered (two-dimensional displacements) in planes, decreasing the net effect of polarization. In improper ferroelectrics such as molybdates, displacements are three-dimensional. It may be noted in passing that in ferroelectrics the direction of spontaneous polarization is capable of being switched from one state to another by a coercive field that does not exceed the dielectric breakdown strength.

Ferroelectric–paraelectric transitions can be understood on the basis of the Landau–Devonshire theory using polarization as an order parameter (Rao & Rao, 1978). The ordered ferroelectric phase has a lower symmetry, belonging to one of the subgroups of the high-symmetry disordered paraelectric phase. The exact structure to which the paraelectric phase transforms is, however, determined by energy considerations.

Domain patterns in ferroics are affected by applied fields. In ferroelastics, domain walls (twin lamellae) can be moved by the application of stress. Spontaneous strains associated with ferroelastics are of the order of a few parts per thousand, and the coercive stresses not greater than 10^4–10^5 N cm^{-2}. Ferromagnetics, by the definition given above, encompass both ferromagnets (e.g. CrO_2, γ-Fe_2O_3) and ferrimagnets (e.g. magnetite) since they both possess spontaneous magnetization. Magnetic domains such as those in haematite can be observed by using the Faraday effect, and observations of domain-wall movements can be made under applied magnetic field.

Secondary ferroics. $SrTiO_3$ (which is an incipient ferroelectric) and $NaNbO_3$ (which is an antiferroelectric) may be considered to be ferrobielectric. The dielectric anisotropy in antiferroelectric domains can give rise to high values of induced polarization that are orientationally different in different domains. Thus, they give rise to domain rearrangement under applied fields. Quartz is a classic example of *ferrobielasticity*. The twin structure in α-quartz is governed by rules of twin morphology. However, when an external stress is applied the induced strains differ in different twins; as a consequence, the differential of the induced strains gives rise to twin-wall movement in the stress field. Stress-induced movement of Dauphine twins (180° twins) in quartz was observed by Thomas & Wooster (1951) and Aizu (1970). Nickel oxide (NiO), a well-known antiferromagnetic material, is also ferrobimagnetic and ferroelastic.

Ammonium chloride is considered to be a ferroelastoelectric. In the low-temperature (~ 247 K) ordered CsCl phase, NH_4Cl has domains in which values of d_{ijk} (elastoelectric or piezoelectric coefficients) are different, and when both an electric field and a mechanical stress are applied simultaneously, orientational switching occurs. *Ferromagnetoelasticity* is exhibited by $FeCO_3$ (siderite) and CoF_2. It appears that potential candidates for ferrobielectric, ferrobimagnetic and ferrobielastic behaviour may be generally expected from among those materials that are normally antiferroelectric, antiferromagnetic and antiferroelastic (those with internally compensated strains and possessing 180° twins) respectively. The antiferroic states suggest large susceptibilities, which can be expected to give rise to pronounced induced effects.

Improper ferroics. Improper ferroics occur as a result of cross-coupled phenomena. Typical examples of improper ferroics are electroferromagnetic nickel iodine boracite, $Ni_3B_7O_{13}I$, (Ascher *et al.*, 1966) and elastoferromagnetic lithium ammonium tartrate (Sawada *et al.*, 1977); Cr_2BeO_4 is a magnetoferroelectric (Newnham *et al.*, 1978). Upon cooling, the cubic boracite undergoes a transformation in which the Ni^{2+} ions produce an antiferromagnetic ordering. This phase, which is an antiferromagnetic–piezoelectric, transforms on further cooling to an orthorhombic phase (at 64 K), which is polar and develops a spontaneous polarization. As a consequence of the polarization effect, the neutral spin alignment is destroyed and a weakly ferromagnetic state results. Since nickel iodine boracite is ferromagnetic because of its ferroelectricity, it is referred to as an electroferromagnet – an improper ferroelectric whose order parameter is electric polarization. At its magnetic transition, tetragonal (paramagnetic) Cr_2BeO_4 transforms to an orthorhombic antiferromagnetic phase. The antiferromagnetic coupling causes slight shifts and rearrangements in the positions of the Cr^{3+} ions in such a way that a polar axis develops, giving rise to a spontaneous polarization. Cr_2BeO_4 is thus a ferroelectric where the driving force arises from magnetic interactions, and is therefore referred to as a magnetoferroelectric.

Relaxor ferroelectrics. These materials show diffuse transitions extending over a large temperature range that are associated with large changes in dielectric constant (Setter & Cross, 1981). The magnitudes of the peak dielectric constants decrease with increasing frequency and the peaks shift to higher temperatures, suggesting features of a relaxational mechanism. It is generally understood from optical and electrical measurements that microdomains arising from compositional fluctuations are responsible for the phenomenon. The importance of these relaxor ferroelectrics stems from the fact that in many of them large permittivities are associated with very low electrostrictive coefficients, a feature desirable in many engineering applications such as in pressure gauges.

Electrostriction. As distinct from inverse piezoelectric effect, electrostriction is a phenomenon in which the strain and the electrical field inducing the strain are related by $\varepsilon_{ij} = M_{ijk}E_k^2$, where M_{ijk} are electrostriction coefficients. Several relaxor ferroelectrics are known to possess low thermal expansivities, some of the good examples being $Pb(Mg_{1/3}Nb_{2/3})O_3$, 0.9 $Pb(Mg_{1/3}Nb_{2/3})O_3$, 0.1 $PbTiO_3$ and $Pb(Zn_{1/3}Nb_{2/3})O_3$ (Uchino *et al.*, 1981).

Electrooptic materials. The dependence of refractive index on the electric field or the lattice polarization is referred to as the electrooptic effect:

$$\Delta(1/n^2)_{ij} = \sum_k r_{ij,k}E_k + \sum_{k,l} k_{ij,kl}E_kE_l + \cdots \tag{6.63}$$

$$\Delta(1/n^2)_{ij} = \sum_k f_{ij,k}P_k + \sum_{k,l} g_{ij,kl}P_kP_l + \cdots \tag{6.64}$$

In equation (6.63), the first term on the right-hand side describes the linear or Pockels electrooptic effect, and the second the quadratic or Kerr electrooptic effect. In equation (6.64), $f_{ij,k}$ and $g_{ij,kl}$ are called polarization optic coefficients; large values of these coefficients are expected to be associated with large susceptibilities, since $r \propto (D-1)f$ and ferroelectrics are obvious candidates. Ferroelectric electrooptic crystals can be centric as in $LiNbO_3$, $LiTaO_3$ and $Ba_2NaNb_5O_{15}$ or acentric as in $KH_2PO_4(KDP)$, $NH_4H_2PO_4(ADP)$ and CsH_2AsO_4. Electrooptic composites have several advantages over single-crystal materials, such as ease of fabrication, low cost and better control of optic-axis orientation. Furthermore, it is possible to vary the effective birefringence by ferroelectric polarization and to tailor-make by appropriate ionic substitution. An electrooptic ceramic is a nearly transparent ferroelectric material, $Pb(Zr_{1-x}Ti_x)O_3$ (or PZT) being the most well-known example. The composition (Zr/Ti ratio) can be varied and Pb partly substituted by other ions (La, Bi, etc.). Hot-pressed PZT ceramics show electrically controlled light scattering (e.g. PBiZT) and birefringence (e.g. PLZT and PBiZT).

Composite ferroics. Ingenious experiments have been performed with composites made from a ferroic and another material (Newnham & Cross, 1981; Lynn *et al.*, 1981; Rittenmeyer *et al.*, 1982; Safari *et al.*, 1982). For example, in a piezoelectric like PZT, the piezoelectric voltage coefficient g can be defined for a given direction (say $Z = 33$); thus, $g_{33} = d_{33}/(\varepsilon_o k_{33})$ where d and k stand for piezoelectric coefficient and dielectric susceptibility respectively. There are situations (e.g. in hydrophones) where g_{33} is required to be high. Further, when a hydrostatic pressure is applied, the equation for g gets modified as $\bar{g} = \bar{d}/\varepsilon_o \bar{k}$ and $\bar{d} = d_{33} + 2d_{31}$ ($\bar{k}$ has a similar expression); $\bar{d}$ is effectively zero under hydrostatic pressure since $d_{33} = -2d_{31}$. The

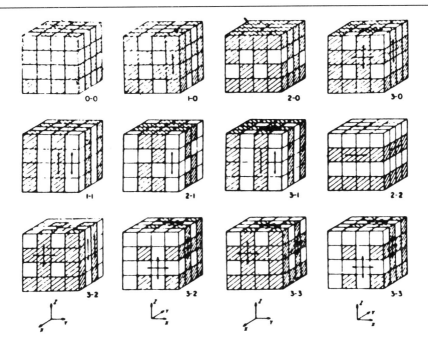

Figure 6.57   Ten connectivity patterns for a diphasic solid. Each phase has zero-, one-, two-, or three-dimensional connectivity to itself. In the 3–1 composite, for example, the shaded phase is three-dimensionally connected and the unshaded phase is one-dimensionally connected. Arrows are used to indicate the connected directions. Two views of the 3–3 and 3–2 patterns are given because the two interpenetrating networks are difficult to visualize on paper. The views are related by 90° counterclockwise rotation about Z. (After Newnham & Cross, 1981.)

composite ferroic strategy is to design a ceramic–polymer composite in such a way that d_{31} is eliminated and $\bar{k}$ is reduced. The reduction in $\bar{k}$ is obvious since introducing a polymer filler phase reduces the volume fraction of the high dielectric constant piezoelectric. The method by which reduction (or elimination) of d_{31} is achieved is as follows (Safari *et al.*, 1982): the ceramic is exposed to pressure only in the $\langle 33 \rangle$ direction, and being the less compressible phase (as compared to the polymer) it takes up the entire pressure in the $\langle 33 \rangle$ direction upon itself. No hydrostatic pressure is found in the $\langle 11 \rangle$ and $\langle 22 \rangle$ directions directly due to the ceramic since the polymer absorbs all the stress. On the other hand, $\bar{k} = fk_{33}$, where f is the volume fraction of the ceramic. The $\bar{g}$ value would be $d_{33}/(\varepsilon_o fk_{33})$, a value much higher than when a cubic piece of piezoelectric is used.

In the above example PZT is connected to itself in one direction while the polymer phase is interconnected in all three dimensions (1–3 connectivity). Other types of *connectivities* in ceramic–polymer diphasic composites (such as 2–3 or 3–3 composites) can be visualied, as in Fig. 6.57. In a 2–3 connected diphasic composite, we can, in principle, exploit two different properties of the ceramic in two different directions, or the same property in an additive manner. 3–3 composites have been made by the so-called replamine lost-wax technique. A natural template for 3–3 connectivities is provided by goniopara coral. Corals are first filled with casting wax and later the carbonate (coral) structure is dissolved away in acid. The negative blanks are now invested with PZT in place of the coral carbonate. It is sintered, during which process the wax is lost. The sintered product is filled with the polymer.

6.11   Amorphous solids

The structure and properties of amorphous solids have been investigated extensively for many years. Although there are many aspects of the amorphous state that we do not understand (e.g. the nature of the glassy state and the glass transition), these studies have thrown much light on this important class of solids. Ion transport in the glassy state is quite interesting (Angell, 1990). In highly ionic glasses, it seems necessary to invoke vacancy-mediated transport with a broad definition of a vacancy. The absence of an ion from a local minimum energy site is recognized as a vacancy and this acts as a source of polarization for the surrounding ions. Although the absence of a three-dimensional lattice makes it difficult to evaluate all the energy terms arising out of long-range interactions by direct summation, it is possible to devise approximate methods to overcome such deficiencies. A detailed calculation on thioborate glasses has shown the relevance of polarization energy terms ignored in previous calculations (Rao *et al.*, 1993). The importance of medium range ordering is being recognized in discussing ion transport. It is, however, not yet clear whether AgI-based glasses owe their high conductivity to the formation of AgI within the respective glass matrices. The a.c. conductivity behaviour of ionic glasses has been investigated in detail (Funke, 1993). The polarization relaxation is nonexponential in several glass systems. Efforts have been made to explain such dielectric responses through universal power laws. Models where the glass structure is taken into account (e.g. presence of non-bridging oxygens and energetically equivalent position for occupancy of cations) seem to provide insights to the origin of the exponential behaviour of dielectric response (Elliott, 1988). Dielectric studies of glasses over a wide range of frequencies (from 0.01 to 10^9 Hz) have been useful in examining the dynamics of the glassy state. The studies show that the stretching exponent and the frequency exponent are often related to each other, although their temperature dependence is different.

Pressure induces amorphization in SiO_2, SnI_4, $AlPO_4$, $LiKSO_4$ and ice, which are all built up of tetrahedral structural units (McNeil & Grimsditch, 1992). The transformation in SiO_2 under pressure is irreversible, leading to a compacted anisotropic amorphous solid. In the case of SnI_4 and $AlPO_4$, the transformation is reversible. The reversible transition is suggestive of a highly correlated movement of atoms or a rearrangement of groups under pressure. In the case of SiO_2, densification appears to occur by packing of chains in a particular direction causing anisotropy. Hypersonic velocity measurements indicate the presence of an arrested transformed phase. The transformation in $AlPO_4$ is more dramatic giving rise to a memory effect (Kruger & Jeanloz, 1990).

In investigations of structures of amorphous solids, MAS NMR spectroscopy has been used effectively (see Chapter 2). Structures of borates, silicates and phosphates have been advantageously studied by this technique. An important advance in this area is the development of a double rotor technique, or the dynamic angle spinning method, which enables the study of oxygen nuclei and their environment in glasses (Farnan *et al.*, 1992). Valuable results on Si–O–Si bond angle distributions and populations of bridging and non-bridging oxygens in glasses have been obtained.

Molecular dynamics has been used extensively in recent times to examine the structure and dynamics in the glassy state (Barrat *et al.*, 1991). Since the mixed nature of bonding is more a rule than the exception in the glassy state, in the absence of proper potentials to simulate a covalent bond, it is difficult to carry out proper molecular dynamics simulations of glasses. Molecular dynamics coupled with semi-empirical quantum chemical calculations is not altogether satisfactory. In spite of such limitations, equilibrium structures, local and long-range correlations and other aspects of glasses have been examined by molecular dynamics.

The observation of a *p–n* transition in lead-containing germanium chalcogenide glasses is significant since it enables preparation of *n–p* junctions made out of pure glasses with no further doping requirement (Tohge *et al.*, 1987). Pb, Bi, and Sn in their lower oxidation states (s^2 ions) are associated with interesting electronic properties in glasses containing them, giving rise to mixed bonding characteristics, electronic conduction and unusual structural features.

6.12   Liquid crystals

We briefly discussed the origin and structure of liquid crystals in Section 4.13. The last decade has witnessed a surge of interest in liquid crystals because of their applications in display devices (devices that convert an electrical signal into visual information). The design of liquid crystal (LC) devices relies on the relation between the molecular structure and the phase behaviour (relative smectic–nematic tendency, T_{NI} etc.) as well as the physical properties of the liquid crystals (Chandrasekhar, 1994).

It is important to understand the relationship between the molecular structure and the type of liquid-crystalline state exhibited, especially in order to elucidate the smectic and nematic tendencies of LC systems (Gray, 1983). The question is not simply whether molecules will pack with their long axes parallel, but whether they will do so in a manner such that their ends lie in a plane (smectic tendency). There is evidence to show that, in general, lengthening the alkyl chain of a terminal *n*-alkyl or *n*-alkoxy substituent in a homologous series of mesogens (molecules capable of forming liquid crystals) promotes smectic behaviour through a mutual affinity of neighbouring alkyl chains. For instance, in the series

there is no smectic phase up to $n = 4$. For $n \geqslant 5$, a smectic phase (S_c) appears whose stability region increases with increasing n.

The question of predicting the type of smectic polymorph that a given mesogen would adopt is much more difficult. This essentially amounts to asking for details of the crystal structure on the basis of the molecular structure. Some correlation between the molecular structure and the tendency to form S_c (tilted disordered lamellar structure) is available. Terminal alkoxy groups in aromatic mesogens favour the formation of S_c phases, while their alkyl analogues often give rise to S_A phases where the molecules are not tilted. Tendency of the molecules to tilt with respect to the lamellar planes, and hence favour the S_c phase, is attributed to the existence of terminal outboard dipoles associated with alkoxy substituents.

More-definitive structural correlations have been established in the nematic–cholesteric systems. Following Gray (1983), we summarize them below:

(i) Addition of rings or multiple-bonded units extending the rigid core of the molecule increases the nematic–isotropic liquid transition temperature, T_{NI}:

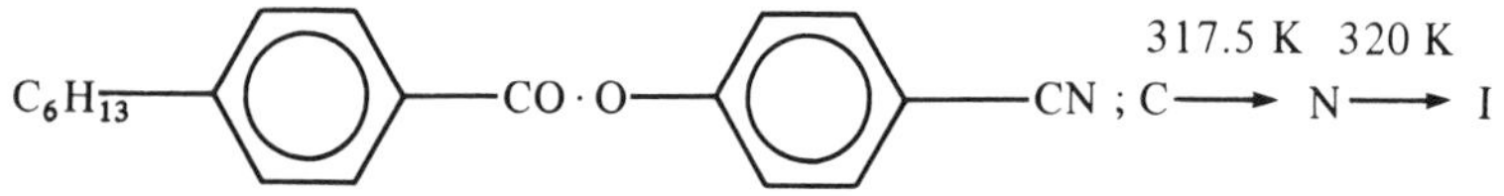

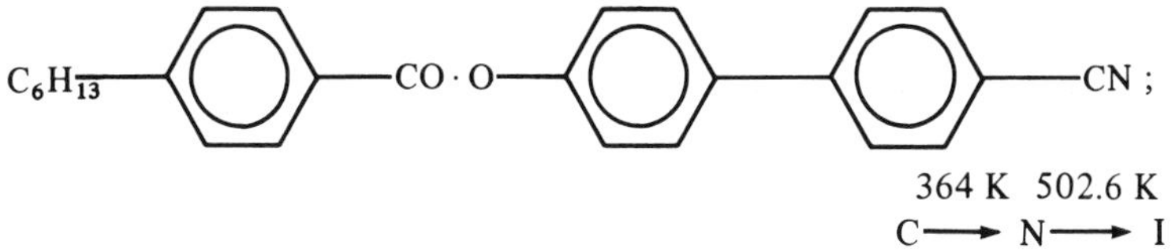

$$C_6H_{13}-\text{benzene}-CO\cdot O-\text{benzene}-CN \;; C \xrightarrow{317.5\,K} N \xrightarrow{320\,K} I$$

$$C_6H_{13}-\text{benzene}-CO\cdot O-\text{benzene}-\text{benzene}-CN \;;$$

$$C \xrightarrow{364\,K} N \xrightarrow{502.6\,K} I$$

(ii) In a homologous series consisting of n-alkyl or n-alkoxy terminal chains, T_{NI} decreases with increasing chain length; in addition, a regular alternation of T_{NI} between odd and even n-alkyl members noticed at the beginning of the series tends to disappear with increasing alkyl chain length.

(iii) Effect of terminal group on T_{NI} is given by,
$$Ph > NHCOCH_3 > CN > OCH_3 > NO_2 > Cl > Br >$$
$$N(CH_3)_2 > CH_3 > I > H.$$

(iv) Effect of linking groups on T_{NI} in a core structure can be illustrated with the general system

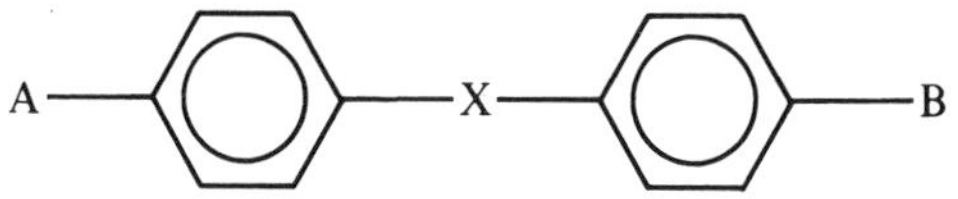

$$A-\text{benzene}-X-\text{benzene}-B$$

T_{NI} is maximum when X is a benzene ring (phenylene group) and minimum when X is a direct bond between the two benzene rings; T_{NI} for other linking groups falls in between.

(v) Substitution at lateral positions in the core structure generally decreases T_{NI}. For instance, in the series of carboxylic acids:

T_{NI} varies with X as follows: $H > F > CH_3 \approx Cl > Br > I \approx NO_2$. It is seen that variation of T_{NI} correlates with the size of X irrespective of the permanent C–X dipole.

(vi) Chain branching has a great diminishing effect on T_{NI} especially in the 1-position; the effect decreases as the substituent moves towards the free end of the chain, as illustrated by the following series:

R	T_{NI} (K)
$-CH_2CH_2CH_2CH_2CH_3$	409.5
$-\underset{\underset{CH_3}{\vert}}{C}HCH_2CH_2CH_2CH_3$	< 293
$-CH_2\underset{\underset{CH_3}{\vert}}{C}HCH_2CH_2CH_3$	385
$-CH_2CH_2\underset{\underset{CH_3}{\vert}}{C}HCH_2CH_3$	376
$-CH_2CH_2CH_2\underset{\underset{CH_3}{\vert}}{C}HCH_3$	392.5

Correlations between molecular structure and the LC state have been fruitfully employed to design 4-alkyl and 4-alkoxy-4'-cyanobiphenyls, which form the basis for electrooptic display devices. The reasoning behind the design of these nematogens is as follows. In a family of nematogens of the general formula

maximum stability requires that the two rings be linked directly (A = single bond), but this has the consequence of lowering T_{NI}. This can be countered by using a CN group for Y that is high in the nematic efficiency order. Using alkyl or alkoxy groups as X promotes nematic order and keeps the melting point low. Compounds so designed are good nematogens giving stable colourless nematic phases at room temperature, a typical example being

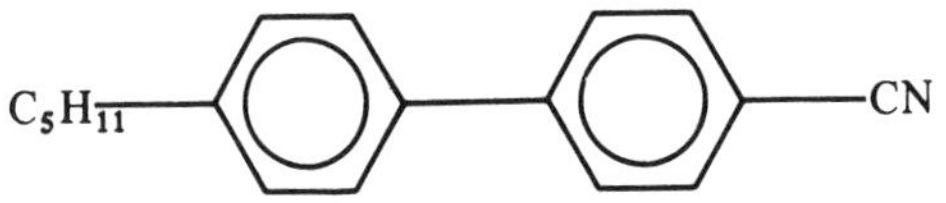

C_5H_{11} —⬡—⬡— CN

295 K   308 K
C ——→ N ——→ I

The benzene ring in nematogens can be substituted by alicyclic rings such as cyclohexane, bicyclo (2.2.2) octane and cubane. This has led to the discovery of cyclohexane and bicyclooctane analogues of cyanobiphenyls. Substitutions of a benzene ring by cyclohexane or bicyclooctane has the effect of increasing T_{NI}:

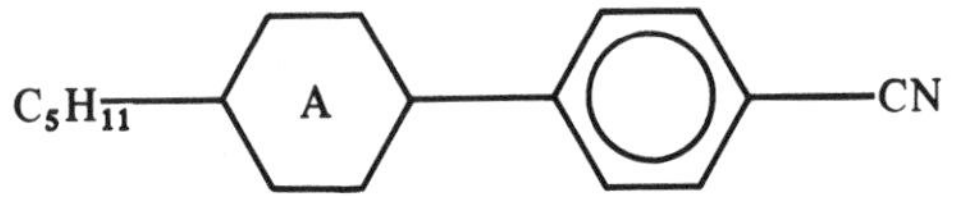

C_5H_{11} —⬡ A ⬡—⬡— CN

A	T_{NI} (K)
Benzene	308
Cyclohexane	328
Bicyclooctane	373

'On the contrary, replacement of a benzene, or a cyclohexane or a bicyclooctane ring by a cubane ring has the effect of depressing T_{NI} markedly. The effect has been explained in terms of ring flexibility and collinearity of the 1, 4 bonds. The bicyclooctane ring, being most flexible, assists in efficient packing of molecules, giving a stable nematic state. The high strain energy of the cubane ring (650.8 kJ mol^{-1}) makes it rigid and least flexible and hence favours the nematic state least.

For device purposes, molecules used should be stable to heat and light. The LC phases have to be stable over a wide temperature range around room-temperature. This can be achieved by using mixtures of mesogens. A mixture of cyanobiphenyls, one with alkyl substituent (C_5H_{11}) and the other with alkoxy substituent ($C_8H_{17}O$), has proved to be most useful.

Physical properties of liquid crystals are generally anisotropic (see, for example, du Jeu, 1980). The anisotropic physical properties that are relevant to display devices are refractive index, dielectric permittivity and orientational elasticity (Raynes, 1983). A nematic LC has two principal refractive indices, $n_{\parallel}$ and $n_{\perp}$, measured parallel and perpendicular to the nematic director respectively. The birefringence $\Delta n = n_{\parallel} - n_{\perp}$ is positive, typically around 0.25. The anisotropy in the dielectric permittivity which is given by $\Delta \varepsilon = \varepsilon_{\parallel} - \varepsilon_{\perp}$ is the driving force for most electrooptic effects in LCs. The electric contribution to the free energy contains a term that depends on the angle between the director $\vec{n}$ and the electric field $\vec{E}$ and is given by

$$F_E \simeq \tfrac{1}{2}\varepsilon_o \Delta\varepsilon (\vec{n}.\vec{E})^2 \tag{6.65}$$

In the absence of other constraints, free energy is minimized by a rotation of the director. There are two possibilities: when $\varepsilon_{||} > \varepsilon_{\perp}$ (positive anisotropy), the director orients parallel to the field and when $\varepsilon_{\perp} > \varepsilon_{||}$ (negative anisotropy), the orientation of the director is perpendicular to the field.

For a nematic LC, the preferred orientation is one in which the director is parallel everywhere. Other orientations have a free-energy distribution that depends on the elastic constants, K_{ii}. The orientational elastic constants K_{11}, K_{22} and K_{33} determine respectively splay, twist and bend deformations. Values of elastic constants in LCs are around 10^{-11} N so that free-energy difference between different orientations is of the order of 5×10^{-4} J m^{-2}, the same order of magnitude as surface energy. A thin layer of LC sandwiched between two aligned surfaces therefore adopts an orientation determined by the surfaces. This fact forms the basis of most electrooptical effects in LCs. Display devices based on LCs are discussed in Chapter 7.

Although only a few types of thermotropic liquid crystals had been described originally (uniaxial nematic with orientational order of rod-like molecules, cholesteric with chiral molecules and smectic with a periodic stacking of liquid layers), more than 20 types of thermotropic systems have been discovered (de Gennes & Prost, 1993). Meyer et al. (1975) noted that the symmetry of a chiral smectic C liquid crystal permits it to have transverse polarization in a liquid direction in the smectic layers. It was then demonstrated that a surface-stabilized book-shelf structure can be used for fast switching display devices with memory (Clark & Lagerwall, 1980). A large number of ferroelectric liquid crystals have since been characterized (see for example, Khened et al., 1991). There are no commercial ferroelectric displays as yet, but some interesting new phases such as the TGB (Goodby et al., 1989), antiferroelectric and ferrielectric phases (Chandani et al., 1989) have been discovered. TGBA and TGBC (tilted but untwisted layers) phases, the latter with both commensurate and incommensurate structures, have been found (Nguyen et al., 1992). Bond-orientationally-ordered (BOO) hexatic phases have been found in free standing films (Pindak et al., 1981). In lyotropic liquid crystals, smectic colloids (lamellar phase with a 1 μm layer spacing) as well as the spongy phase (Porte et al., 1988) have been realized. Polymeric liquid crystals have gained considerable importance and Kevlar fibres with high strength and high modulus are indeed drawn from a liquid crystalline phase.

References

Abrahams, E., Anderson, P. W., Licciardello, D. C. & Ramakrishnan, T. V. (1979) *Phys. Rev. Lett.* **42**, 673.

Adler, D. (1975) in *Treatise on Solid State Chemistry* (ed. Hannay, N. B.) Vol. 2, Plenum Press, New York.

Aizu, K. (1970) *Phys. Rev.* **B2**, 754.

Alcacer, L. (ed.) (1980) *The Physics and Chemistry of Low Dimensional Solids*, Reidel, Dordrecht, Holland.

Allen, P. B. & Mitrovic, B. (1982) *Solid State Phys.* **37**, 2.

Anderson, P. W. (1958) *Phys. Rev.* **109**, 1492.

Anderson, P. W. (1959) *Phys. Rev.* **115**, 2.

Ando, T., Fowler, H. B. & Stern, F. (1982) *Rev. Mod. Phys.* **54**, 437.

Angell, C. A. (1990) *Chem. Rev.* **90**, 523.

Ascher, H., Reider, H., Schmid, H. & Stossell, H. (1966) *J. Appl. Phys.* **37**, 1404.

Balkanski, N. (ed.) (1971) *Light Scattering in Solids*, Flammarion, Paris.

Banus, M. D. & Reed, T. B. (1970) in *The Chemistry of Extended Defects in Nonmetallic Solids* (eds Eyring, L. & O'Keeffe, M.) North-Holland, Amsterdam.

Barbara, T. M., Gammic, G., Lyding, J. W. & Jonas, J. (1988) *J. Solid State Chem.* **50**, 173.

Barrat, J. L. & Klein, M. (1991) *Ann. Rev. Phys. Chem.* **42**, 23.

Bednorz, J. G. & Müller, K. A. (1986) *Z. Phys.* **B64**, 189.

Bell, A. E. & Caplin, A. D. (1975) *Contemp. Phys.* **16**, 375.

Bem, D. S., Houmes, J. D. & Zur Loye, H.-C. (1994) *Materials Science Forum* **152–3**, 183.

Benfield, R. E., Edwards, P. P. & Stacy, A. M. (1982) *J. Chem. Soc. Chem. Comm.* 525.

Beyerlin, R. A., Horowitz, H. S. & Longo, J. M. (1988) *J. Solid State Chem.* **72**, 2.

Bhide, V. G., Rajoria, D. S., Rao, G. R. & Rao, C. N. R. (1972) *Phys. Rev.* **B6**, 1021.

Blinc, R. & Zeks, B. (1974) *Soft Modes in Ferroelectrics and Antiferroelectrics*, North-Holland, Amsterdam.

Bloor, D. (1985) *Phil. Trans. Roy. Soc. London* **A314**, 67.

Bongers, P. F. (1975) in *Crystal Structure and Chemical Bonding in Inorganic Chemistry* (eds Rooymans, C. J. M. & Rabenau, A.) North-Holland, Amsterdam.

Bouchard, R. J., Gillson, J. L. & Jarett, H. S. (1973) *Mater. Res. Bull.* **8**, 489.

Brandow, B. H. (1977) *Adv. Phys.* **26**, 651.

Bronger, W. (1981) *Angew. Chem. Int. Edn (Engl.)* **20**, 52.

Brown, D. B. (ed.) (1980) *Mixed-Valence Compounds*, Reidel, Dordrecht, Holland.

Burdett, J. K. (1980) *J. Amer. Chem. Soc.* **102**, 450.

Burdett, J. K. (1982) in *Adv. Chem. Phys.* Vol. XLIX (eds Prigogine, I. & Rice, S. A.) Wiley-Interscience, New York.

Burdett, J. K. (1985) *Inorg. Chem.* **24**, 2244.

Burdett, J. L. & Rosenthal, G. L. (1980) *J. Solid State Chem.* **33**, 173.

Burdett, J. K., McLarnan, T. J. & Haaland, P. (1981) *J. Chem. Phys.* **75**, 5774.

Burdett, J. K. & McLarnan, T. J. (1982) *Inorg. Chem.* **21**, 1119.

Burdett, J. K. & Kulkarni, G. V. (1988) *J. Amer. Chem. Soc.* **110**, 5361.

Burdett, J. K. & Mitchell, J. F. (1990) *J. Amer. Chem. Soc.* **112**, 6571.

Burdett, J. K. & Mitchell, J. F. (1993) *Inorg. Chem.* **32**, 5004.

Burdett, J. K., Marians, C. & Mitchell, J. F. (1994) *Inorg. Chem.* **33**, 1848.

Cahn, R. W. (1980) *Contemp. Phys.* **21**, 43.

Canadell, E. & Whangbo, M. H. (1990) *Inorg. Chem.* **29**, 2256.

Canadell, E. & Whangbo, M. H. (1991) *Chem. Rev.* **91**, 965.

Canadell, E., Jobic, S., Brec, R., Rouxel, J. & Whangbo, M. H. (1992) *J. Solid State Chem.* **99**, 189.

Chainani, A., Mathew, M. & Sarma, D. D. (1992) *Phys. Rev.* **B46**, 9976.

Chandari, A. D. L., Gorecka, E., Ouchi, Y., Takezoe, H. & Fukuda, A. (1989) *Jpn J. Appl. Phys.* **28**, 1265.

Chandrasekhar, S. (1994) *Liquid Crystals*, Cambridge University Press, 2nd Ed.

Chern, M. Y. & DiSalvo, F. J. (1990) *J. Solid State Chem.* **88**, 459.

Chern, M. Y. & Vennos, D. A. & DiSalvo, F. J. (1992) *J. Solid State Chem.* **96**, 415.

Chien, J. C. W. (1984) *Polyacetylene: Chemistry, Physics and Materials Science*, Academic Press, New York.

Clark, N. A. & Lagerwall, S. T. (1980) *Appl. Phys. Lett.* **36**, 899.

Clark, R. J. H. (1977) *Ann. N. Y. Acad. Sci.* **313**, 672.

Coey, J. M. D., Roux-Buisson, H. & Brusetti, R. (1976) *J. Physique, Colloq.* **C4**, 11.

Coleman, L. B., Cohen, M. J., Sandman, D. J., Yamagishi, F. G., Garito, A. F. & Heeger, A. J. (1973) *Solid State Comm.* **12**, 1125.

Collins, B. T., Ramanujachary, K. V., Greenblatt, M., McCarrol, W. H., McNally, P. & Waszczak, J. V. (1988) *J. Solid State Chem.* **76**, 319.

Corish, J., Jacobs, P. W. M. & Radhakrishna, S. (1977) in *Surface and Defect Properties of Solids* (Senior Reporters Roberts, M. W. & Thomas, J. M.) Vol. 6, The Chemical Society, London.

Cotton, F. A. (1971) *Chemical Applications of Group Theory*, 2nd Edn, Wiley-Eastern, New Delhi.

Creutz, C. & Taube, H. (1973) *J. Amer. Chem. Soc.* **95**, 1086.

Datars, W. R., Razavi, F. S., Gillespie, R. J. & Ummat, P. (1985a) *Phil. Trans. Roy. Soc. London* **A314**, 115.

Datars, W. R., Ummat, P. K. & Gillespie, R. J. (1985b) *Mater. Res. Bull.* **20**, 865.

Day, P. (1975) *J. Amer. Chem. Soc.* **97**, 1588.

Day, P. (1979) *Acc. Chem. Res.* **12**, 237.

Day, P. (1981) *Int. Rev. Phys. Chem.* **1**, 149.

Day, P. (1983) *Chem. Britain* **19**, 306.

Day, P. (1985) *Phil. Trans. Roy. Soc. London* **A314**, 145.

de Gennes, P. G. & Prost, J. (1993) *The Physics of Liquid Crystals* (2nd Ed), Oxford.

de Jongh, L. J. & Miedema, A. R. (1974) *Adv. Phys.* **23**, 1.

Demazeau, G., Langlais, F., Portier, J., Tressaud, A. & Hagenmuller, P. (1980) *High Pressure Science and Technology* **1**, 579.

De Vries, A. B. & Haas, C. (1973) *J. Phys. Chem. Solids* **34**, 651.

DiSalvo, F. J. (1990) *Science* **247**, 649.

Dorner, B., Axe, J. D. & Shirane, G. (1972) *Phys. Rev.* **B6**, 1950.

Doumerc, J. P., Pouchard, M. & Hagenmuller, P. (1985) in *The Metallic and the Nonmetallic States of Matter* (Edwards, P. P. & Rao, C. N. R. eds.), Taylor & Francis, London.

du Jeu, W. H. (1980) *Physical Properties of Liquid Crystalline Materials*, Gordon & Breach, New York.

Duwez, P. & Johnson, W. L. (1978) *J. Less-Common Metals* **62**, 607.

Edwards, P. P. & Sienko, M. J. (1983) *Int. Rev. Phys. Chem.* **3**, 83.

Edwards, P. P. & Rao, C. N. R. (eds) (1995) *Metal–Insulator Transitions Revisited*, Taylor & Francis, London.

Edwards, P. P., Ramakrishnan, T. V. & Rao, C. N. R. (1995) *J. Phys. Chem* **99**, 5228.

Edwards, S. F. & Anderson, P. W. (1975) *J. Phys.* **F5**, 965.

Elder, S. H., Doerrer, L. H., DiSalvo, F. J., Parise, J. B., Guyomard, D. & Tarascon, J. M. (1992) *Chem. Mater.* **4**, 928.

Elliott, S. R. (1984) *Physics of Amorphous Materials*, Longman, London.

Elliott, S. R. (1988) *Solid State Ionics* **27**, 131.

Emin, D. (1977) *Phil. Mag.* **35**, 1189.

Englman, R. (1972) *The Jahn-Teller Effect in Molecules and Crystals*, Wiley, London.

Etemad, S., Heeger, A. J. & MacDiarmid, A. G. (1982) *Ann. Rev. Phys. Chem.* **33**, 443.

Eyring, L. (1974) in *Solid State Chemistry* (ed. Rao, C. N. R.) Marcel Dekker, New York.

Eyring, L. (1979) in *Handbook on the Physics and Chemistry of Rare Earths* (eds Gschneidner Jr, K. A. & Eyring, L.) North-Holland, Amsterdam.

Falicov, L. M., Hanke, W. & Maple, M. P. (eds) (1981) *Valence Fluctuation in Solids*, North-Holland, Amsterdam.

Farnan, I., Grandinetti, P. J., Baltisberga, J. H., Stebbins, J. F., Wener, U., Eastman, M. A. & Pines, A. (1992) *Nature* **358**, 31.

Ford, P. J. (1982) *Contemp. Phys.* **23**, 141.

Franzen, H. F. (1978) *Prog. Solid State Chem.* **12**, 1.

Friebel, C. & Reinen, D. (1974) *Z. Anorg. Allgem. Chem.* **407**, 193.

Fujimori, A., Hase, J., Namatame, M., Fujishima, Y., Tokura, Y., Eisaki, H., Uchida, S., Takegahara, K. & de Groot, F. M. F. (1992) *Phys. Rev. Lett.*, **69**, 1796.

Funke, K. (1993) *Prog. Solid State Chem.* **22**, 111.

Fyne, P. J., Day, P., Hutchings, M. T., Depinna, S., Cavenett, B. C. & Pynn, R. (1984) *J. Phys.* **C17**, L245.

Ganguly, P. & Rao, C. N. R. (1984) *J. Solid State Chem.* **53**, 193.

Ganne, M., Dion, M. & Boumaza, A. C. R. (1986) *C.R. Acad. Sci. Ser.* **2**, 302, 635.

Gehring, G. A. & Gehring, K. A. (1975) *Rep. Prog. Phys.* **38**, 1.

Gillespie, R. J., Brown, I. D., Datars, W. R., Morgan, K. R., Tun, Z. & Ummat, P. K. (1985) *Phil. Trans. Roy. Soc. London* **A314**, 105.

Goodby, J. W., Waugh, M. A., Stein, S. M., Chin, E., Pindak, R. & Patel, J. S. (1989) *Nature* **337**, 449.

Goodenough, J. B. (1963) *Magnetism and the Chemical Bond*, John Wiley, New York.

Goodenough, J. B. (1967) *Colloques Internationaux de CNRS*, No. 157, Paris.

Goodenough, J. B. (1971) *Prog. Solid State Chem.* **5**, 149.

Goodenough, J. B. (1974) in *Solid State Chemistry* (ed. Rao, C. N. R.), Marcel Dekker, New York.

Goodenough, J. B. & Longo, J. M. (1970) *Landolt-Börnstein Tabellen*, New Series, III/4a, Springer-Verlag, Berlin.

Gopalakrishnan, J., Murugesan, T., Hegde, M. S. & Rao, C. N. R. (1979) *J. Phys.* **C12**, 5255.

Gopalakrishnan, J., Uma, S., Vasanthacharya, N. Y. & Subbanna, G. N. (1995) *J. Amer. Chem. Soc.* **117**, 2353.

Gray, G. W. (1983) *Phil. Trans. Roy. Soc. Lond.* **A309**, 77.

Greenblatt, M. (1988) *Chem. Rev.* **88**, 31.

Greenblatt, M. (1994) *Int. J. Mod. Phys.* **7**, No. 23 & 24, Special issue on Oxide Bronzes.

Greene, R. L., Mayerle, J. J., Schumaker, R., Castro, G., Chaikin, P. M., Etemad, S. & LaPlaca, S. J. (1976) *Solid State Comm.* **20**, 943.

Hagenmuller, P. (1983a) *Zeit. Chem.* **23**, 1.

Hagenmuller, P. (1983b) *Proc. Indian Acad. Sci. (Chem. Sci.)* **92**, 1.

Hagenmuller, P. (ed.) (1985) *Solid Inorganic Fluorides*, Academic Press, New York.

Hagenmuller, P., Reau, J. M., Lucat, C., Matar, S. & Villeneue, G. (1981) *Solid State Ionics* **3/4**, 341.

Hatfield, W. E. (ed.) (1978) *Molecular Metals*, Plenum Press, New York.

Hauser, J. J. & Waszczak, J. V. (1984) *Phys. Rev.* **B30**, 5167.

Haven, Y. (1978) in *Solid Electrolytes* (eds. Hagenmuller, P. & Van Gool, W.) Academic Press, New York.

Hess, H. F., De Conde, K., Rosenbaum, T. F. & Thomas, G. A. (1982) *Phys. Rev.* **B25**, 5578.

Hodgson, J. N. (1971) *Optical Absorption and Dispersion in Solids*, Chapman & Hall, London.

Hoffmann, B. M. & Ibers, J. A. (1983) *Acc. Chem. Res.* **16**, 15.

Hoffmann, R. (1988) *Solids and Surfaces: A chemist's view of bonding in extended structures*, VCH, New York.

Hoffmann, R. & Zheng, C. (1985) *J. Phys. Chem.* **89**, 4175.

Holtzberg, F. (1980) *Phil. Mag.* **B42**, 491.

Holtzberg, F., McGuire, T. R. & Methfessel, S. (1970) *Landolt-Bornstein Tabellen*, New Series, III/4a, Springer-Verlag, Berlin.

Honig, J. M. (1981) in *Preparation and Characterization of Materials* (eds. Honig, J. M. & Rao, C. N. R.) Academic Press, New York.

Honig, J. M. (1982) *J. Solid State Chem.* **45**, 1.

Honig, J. M. (1986) *Proc. Ind. Acad. Sci. (Chem. Sci.)* **96**, 391.

Honig, J. M. & Van Zandt, L. L. (1975) *Ann. Rev. Mater. Sci.* **5**, 225.

Honig, J. M. & Spalek, J. (1986) *Proc. Ind. Nat. Sci. Acad.* **52A**, 32.

Hubbard, J. (1963) *Proc. Roy. Soc. Lond.* **A276**, 238.

Hughbanks, T. & Hoffmann, R. (1983) *J. Amer. Chem. Soc.*, **105**, 1150.

Huisman, R., de Jong, R., Haas, C. & Jellinek, F. (1971) *J. Solid State Chem.* **3**, 56.

Hulliger, F. (1968) *Structure and Bonding*, **4**, 83.

Hulliger, F. (1977) *Structural Chemistry of Layer Type Phases*, D. Reidel, Dordrecht, Holland.

Hurd, C. M. (1982) *Contemp. Phys.* **23**, 469.

Hush, N. S. (1975) *Chem. Phys.* **10**, 361.

Ikeda, M., Gschneidner, Jr, K. A., Beaudry, B. J. & Altzmany, U. (1980) *Solid State Comm.* **36**, 657.

Ingram, M. D., Moynihan, C. T. & Lesikar, A. V. (1980) *J. Non-Cryst. Solids*, **38 & 39**, 371.

Jacobs, H. & von Pinkowski, E. (1989) *J. Less Comm. Metals.* **146**, 147.

Jadhao, V. G., Singru, R. M., Rao, G. R., Bahadur, D. & Rao, C. N. R. (1976) *J. Phys. Chem. Solids.* **37**, 113.

Janke, E., Hutchings, M. T., Day, P. & Walker, P. J. (1983) *J. Phys.* **C16**, 5959.

Jayaraman, A., Bucher, E., Dernier, P. D. & Longinnotti, L. D. (1973) *Phys. Rev. Lett.* **31**, 700.

Jayaraman, A., Dernier, P. D. & Longinnotti, L. D. (1975) *High Temperatures-High Pressures*, **71**.

Jellinek, F. (1968) *Inorganic Sulphur Chemistry* (ed. Nickless, G.) Elsevier, Amsterdam.

Jellinek, F. (1972) *MTP International Review of Science*, Inorg. Chem. Vol. 5, Butterworths, London.

Jerome, D. (1985) *Phil. Trans. Roy. Soc. London* **A314**, 69.

Jhans, H. & Honig, J. M. (1981) *J. Solid State Chem.* **38**, 112.

Jobic, S., Deniard, P., Brec, R., Rouxel, J., Jouanneaux, A. & Fitch, A. N. (1991) *Z. Anorg. Allgem. Chem.* **598/599**, 199.

Jobic, S., Brec, R. & Rouxel, J. (1992) *J. Solid State Chem.* **96**, 169.

Johnson, D. C., Benfield, R. E., Edwards, P. P., Nelson, W. R. & Vargas, M. D. (1985) *Nature* **314**, 231.

Johnston, D. C., Prakash, H., Zachariasen, W. H. & Viswanathan, R. (1973) *Mater. Res. Bull.* **8**, 777.

Jona, F. & Shirane, G. (1962) *Ferroelectric Crystals*, Pergamon Press, Oxford.

Ju, H. L., Kwon, C. L., Li, Q., Greene, R. L. & Venkatesan, T. (1994) *Appl. Phys. Lett.* **21**, 233.

Kanamori, J. (1959) *J. Phys. Chem. Solids*, **10**, 87.

Keller, H. J. (ed.) (1975) *Low Dimensional Cooperative Phenomena*, Plenum Press, New York.

Keller, H. J. (ed.) (1977) *Chemistry and Physics of One-dimensional Metals*, Plenum Press, New York.

Khened, S. M., Prasad, S. K., Shivkumar, B. & Sadashiva, B. K. (1991) *J. de Physique II*, **1**, 171.

Kleiner, W. H. (1967) *MIT Lincoln Laboratory Solid State Research*, Report No. 3.

Klitzing, K. V., Dorda, G. & Pepper, M. (1980) *Phys. Rev. Lett.*, **45**, 494.

Kittel, C. (1976) *Introduction to Solid State Physics*, 5th edn Wiley-Eastern, New Delhi.

Koelling, D. D. (1981) *Rep. Prog. Phys.* **53**, 517.

Koerber, G. G. (1962) *Properties of Solids*, Prentice-Hall, Englewood-Cliffs, New Jersey.

Kojima, N., Hasegawa, M., Kitagawa, H., Kikegawa, T. & Shimomura, O. (1994) *J. Am. Chem. Soc.*, **116**, 11368.

Krebs, H. (1975) *Prog. Solid State Chem.* **9**, 269.

Krogmann, K. (1969) *Angew. Chem. Int. Edn. (Engl.)* **8**, 35.

Kruger, M. B. & Jeanloz, R. (1990) *Science* **249**, 647.

Kuwamoto, H., Honig, J. M. & Appel, J. (1980) *Phys. Rev.* **B22**, 2626.

Lee, S. C. & Holm, R. H. (1990) *Angew. Chem. Int. Ed. Engl.* **29**, 840.

Le Flem, G., Delmas, C., Menil, F., Niel, M., Cros, C., Fouassier, C. & Pouchard, M. (1977) *J. Physique Colloq.* **C7**, 262.

Ledsham, R. D. & Day, P. (1981) *J. Chem. Soc. Chem. Comm.* 921.

Lewis, J. & Green, M. L. H. (1982) *Phil. Trans. Roy. Soc. London* **A308**, 1.

Little, W. A. (1964) *Phys. Rev.* **134**, A1416.

Long, J. R., Williamson, A. S. & Holm, R. H. (1995) *Angew. Chem. Int. Ed. Engl.* **34**, 226.

Lynn, S. Y., Newnham, R. E., Klicker, K. A., Rittenmyer, K., Safari, A. & Schulze, W. A. (1981) *Ferroelectrics* **38**, 955.

MacDiarmid, A. G. & Heeger, A. J. (1980) *Syn. Met.* **1**, 101.

MacDiarmid, A. G., Mammone, R. J., Kaner, R. B. & Porter, S. J. (1985) *Phil. Trans. Roy. Soc. London* **A314**, 3.

Maeno, Y., Hashimoto, H., Yoshida, K., Nishizaki, S., Fujita, T. Bednorz, J. D. & Lichtenberg, F. (1994) *Nature* **372**, 532.

Mahesh, R., Mahendiran, R., Raychaudhuri, A. K. & Rao, C. N. R. (1995) *J. Solid State Chem.* **114**, 297; **120**, 204 and the references cited therein.

Maple, M. B. & Fisher, F. (eds) (1982) *Topics in Current Physics* **34**, Springer-Verlag, Berlin.

Mar, A. & Ibers, J. A. (1991) *J. Solid State Chem.* **92**, 352.

Mar, A., Jobic, S. & Ibers, J. A. (1992) *J. Amer. Chem. Soc.* **114**, 8963.

Matthias, B. T. (1955) *Phys. Rev.* **97**, 74.

Mattis, D. C. (1981) *The Theory of Magnetism I*, Springer-Verlag, Berlin.

Mayoh, B. & Day, P. (1972) *J. Amer. Chem. Soc.* **94**, 2885.

McCarley, R. E. (1982) *Phil. Trans. Roy. Soc. Lond.* **A308**, 141.

McCormack, M., Jin, S., Tiefel, T. H., Fleming, R. M., Phillips, J. M. & Ramesh, R. (1995) *Appl. Phys Lett.* **64**, 3045.

McNeil, L. E. & Grimsditch, M. (1992) *Phys. Rev. Lett.* **68**, 83.

Meerschaut, A. (1982) *Ann. Chim. Fr.* **7**, 131.

Messmer, R. P. (1977) in *Semi-empirical Methods of Electronic Structure Calculation*: Part B (ed. Segal, G. A.) Plenum Press, New York.

Meyer, R. B., Liebert, L., Strzelecki, L. & Keller, B. (1975) *J. de Physique Lett.* **36**, 69.

Miller, J. S. (ed.) (1982) *Extended Linear Chain Compounds*, Plenum Press, New York.

Miller, J. S. & Epstein, A. J. (eds) (1978) *Ann. N.Y. Acad. Sci.* 313.

Mitzi, D. B., Field, C. A., Harrison, W. T. A. & Guloy, A. M. (1994) *Nature,* **369**, 467.

Mitzi, D. B., Field, C. A., Schlesinger, Z. & Laibowitz, R. B. (1995a) *J. Solid State Chem.* **114**, 159.

Mitzi, D. B., Wang, S., Field, C. A., Chess, C. A. & Guloy, A. M. (1995b) *Science* **267**, 1473.

Moore, M. W. & Day, P. (1985) *J. Solid State Chem.* **59**, 23.

Mott, N. F. (1949) *Proc. Phys. Soc.* **62A**, 416.

Mott, N. F. (1956) *Canadian J. Phys.* **34**, 1356.

Mott, N. F. (1967) *Adv. Phys.* **16**, 49.

Mott, N. F. (1972) *Phil. Mag.* **26**, 1015.

Mott, N. F. (1974) *Metal-Insulator Transitions*, Taylor & Francis, London.

Mott, N. F. (1983) *Int. Rev. Phys. Chem.* **4**, 1.

Mott, N. F. & Davis, E. A. (1979) *Electronic Processes in Non-crystalline Materials*, Clarendon Press, Oxford.

Muetterties, E. L., Rhodin, T. N., Band, E., Brucker, C. F. & Pretzer, W. R. (1979) *Chem. Rev.* **79**, 91.

Muraleedharan, K., Srivastava, J. K., Marathe, V. R., Vijayaraghavan, R., Kulkarni, J. A. & Darsane, V. S. (1985) *Solid State Commun.* **55**, 363.

Murugesan, T., Ramesh, S., Gopalakrishnan, J. & Rao, C. N. R. (1981) *J. Solid State Chem.* **38**, 165.

Mydosh, J. A. (1975) *AIP Conf. Proceedings* **24**, 131.

Mydosh, J. A. (1982) in *Magnetism in Solids. Some Current Topics* (eds. Cracknell, A. P. & Vaughan, R. A.) SUSSP Publications, Edinburgh University.

Newnham, R. E. (1974) *Amer. Mineralogist* **57**, 906.

Newnham, R. E. (1975) *Structure-Property Relations*, Springer-Verlag, New York.

Newnham, R. E., Kramer, J. J., Schulze, W. A. & Cross, L. E. (1978) *J. Appl. Phys.* **49**, 6088.

Newnham, R. E. & Cross, L. E. (1981) in *Preparation and Characterization of Materials* (eds. Honig, J. M. & Rao, C. N. R.), Academic Press, New York.

Nguyen, H. T., Bouchta, A., Navailles, L., Barios, P., Isaert, N., Twieg, R. J., Maaronbti, A. & Destrade, C. (1992) *J. de Physique II*, **2**, 1889.

Nomura, S. (1978) *Landolt-Börnstein Tebellen*, New Series, III/12a, Springer-Verlag, Berlin.

Nye, J. F. (1957) *Physical Properties of Crystals*, Oxford University Press.

Peierls, R. E. (1955) *Quantum Theory of Solids*, Oxford University Press.

Piepho, S. B., Krausz, E. R. & Schatz, P. N. (1978) *J. Amer. Chem. Soc.* **100**, 2996.

Pindak, R., Moncton, D. E., Davey, S. C. & Goodby, J. W. (1981) *Phys. Rev. Lett.* **46**, 1135.

Porte, G., Marignan, J., Basserean, P. & May, R. (1988) *J. de Physique.* **49**, 511.

Portier, J. (1976) *Angew. Chem. Int. Edn. (Engl.)* **15**, 475.

Portier, J., Reau, J. M., Matar, S., Soubeyroux, J. L. & Hagenmuller, P. (1983) *Solid State Ionics* **11**, 83.

Prassides, K. & Day, P. (1984) *J. Chem. Soc. Faraday II* **80**, 85.

Raccah, P. M. & Goodenough, J. B. (1967) *Phys. Rev.* **155**, 932.

Ramdani, A., Gleitzer, C., Gavoille, C., Cheetham, A. K. & Goodenough, J. B. (1985) *J. Solid State Chem.* **60**, 269.

Rao, C. N. R. & Subba Rao, G. V. (1974) *Transition Metal Oxides: Crystal Chemistry, Phase Transition and Related Aspects*, NSRDS-NBS Monograph 49, National Bureau of Standards, Washington D.C.

Rao, C. N. R., Bhide, V. G. & Mott, N. F. (1975) *Phil. Mag.* **32**, 1277.

Rao, C. N. R. & Pisharody, K. P. R. (1976) *Prog. Solid State Chem.* **10**, 207.

Rao, C. N. R., Parkash, P., Bahadur, D., Ganguly, P. & Nagabhushana, S. (1977) *J. Solid State Chem.* **22**, 353.

Rao, C. N. R. & Rao, K. J. (1978) *Phase Transitions in Solids*, McGraw-Hill, New York.

Rao, C. N. R., Sarma, D. D., Vasudevan, S. & Hegde, M. S. (1979) *Proc. Roy. Soc. London* **A367**, 239.

Rao, C. N. R., Ganguly, S. Swamy, H. R. & Oxton, I. A. (1981) *J. Chem. Soc. Faraday II*, **77**, 1825.

Rao, C. N. R., Buttrey, D., Harrison, G., Ganguly, P. & Honig, J. M. (1984) *J. Solid State Chem.* **51**, 266.

Rao, C. N. R. & Ganguly, P. (1985) in *Localization and Metal-Insulator Transitions* (eds. Alder, D. & Fritzsche, H.), Plenum Press, New York.

Rao, C. N. R. & Ganguly, P. (1986) *Solid State Commun.* **57**, 5.

Rao, C. N. R., Ganguly, P., Singh, K. K. & Mohanram, R. A. (1988) *J. Solid State Chem.* **72**, 14.

Rao, C. N. R., Rao, K. J., Ramasesha, S., Sarma, D. D. & Yashonath, S. (1992) *Annual Reports (Section C)*, The Royal Society of Chemistry, London.

Rao, C. N. R., Vijayakrishnan, V., Aiyer, H. N., Kulkarni, G. U. & Subbanna, G. N. (1993) *J. Phys. Chem.* **97**, 11157.

Rao, K. J., Estournes, C., Levasseur, A., Shastry, M. C. R. & Menetrier, M. (1993) *Phil. Mag.* **B67**, 389.

Raychaudhuri, A. K. (1991) *Phys. Rev.* **B44**, 8472.

Raynes, E. P. (1983) *Phil. Trans. Roy. Soc. London* **A309**, 167.

Reinen, D. (1979) *Structure and Bonding* **37**, 1.

Rittenmeyer, K., Strout, T., Schulze, W. A. & Newnham, R. E. (1982) *Ferroelectrics* **41**, 189.

Robbins, D. J. & Day, P. (1973) *Chem. Phys. Lett.* **19**, 529.

Robin, M. B. & Day, P. (1967) *Adv. Inorg. Chem. Radiochem.* **10**, 247.

Rogers, D. B., Shannon, R. D., Sleight, A. W. & Gillson, J. L. (1969) *Inorg. Chem.* **8**, 841.

Röhler, J. & Kaindl, G. (1980) *Solid State Comm.* **36**, 1055.

Rouxel, J. (1992) *Acc. Chem. Res.* **25**, 328.

Rouxel, J. (1993) *Comments Inorg. Chem.* **14**, 207.

Rouxel, J., Meerschaut, A., Guemas, L. & Gressier, P. (1982) *Ann. Chim. Fr.* **7**, 445.

Safari, A., Halliyal, A., Newnham, R. E. & Lachman, I. M. (1982) *Mater. Res. Bull.* **17**, 301.

Sarma, D. D. (1995) in *Metal-Insulator Transitions Revisited* (eds Edwards, P. P. & Rao, C. N. R.) Taylor-Francis, London.

Sarma, D. D., Krishnamurthy, H. R., Nimkar, S., Ramasesha, S., Mitra, P. P. & Ramakrishnan, T. V. (1992) *Pramana – J. Phys.* **38**, L531.

Sato, M., Grier, B. H., Shirane, G. & Akahane, T. (1982) *Phys. Rev.* **B25**, 6876.

Sawada, A., Udagawa, M. & Nakamura, T. (1977) *Phys. Rev. Lett.* **39**, 829.

Sayer, M., Chen, R., Fletcher, R. & Man Singh (1975) *J. Phys.* **C8**, 2059.

Sayetat, F., Ghedira, M., Chenavas, J. & Marezio, M. (1982) *J. Phys.* **C15**, 1627.

Schlenker, C., Dumas, J., Escribe-Filippini, Guyot, H., Marcus, J. & Fourcandot, G. (1985) *Phil. Mag.* **B52**, 643.

Schöllhorn, R. (1983) in *Inclusion Compounds* (eds Atwood, J. L., Davies J. E. D. & MacNicol, D. D.) Academic Press, New York.

Schultz, A., Williams, J. M., Miro, N. D., MacDiarmid, A. G. & Heeger, A. J. (1978) *Inorg. Chem.* **17**, 646.

Setter, N. & Cross, L. E. (1981) *Ferroelectrics* **37**, 551.

Shukla, A. K., Vasan, H. N. & Rao, C. N. R. (1981) *Proc. Roy. Soc. London* **A376**, 619.

Soos, Z. G. & Klein, D. J. (1976) in *Treatise on Solid State Chemistry* (ed. Hannay, N. B.) Vol. 3, Plenum Press, New York.

Soos, Z. G. & Ramasesha, S. (1983) *Phys. Rev. Lett.* **51**, 2374.

Stewart, G. R. (1984) *Rev. Mod. Phys.* **56**, 755.

Stryjewski, E. & Giordano, N. (1977) *Adv. Phys.* **26**, 487.

Subbarao, E. C. (1974) in *Solid State Chemistry* (ed. Rao, C. N. R.) Marcell Dekker, New York.

Subba Rao, G. V. & Shafer, M. W. (1979) *Intercalated Layered Materials* (ed. Levy, F.) D. Reidel, Dordrecht, Holland.

Subba Rao, G. V. & Balakrishnan, G. (1984) *Bull. Mater. Sci.* **6**, 283.

Subramanian, M. A., Aravamudan, G. & Subba Rao, G. V. (1983) *Prog. Solid State Chem.* **15**, 55.

Subramanian, M. A., Torardi, C. C., Johnston, D. C., Pannetier, J. & Sleight, A. W. (1988) *J. Solid State Chem.* **72**, 24.

Subramanyam, S. V. (1981) in *Preparation and Characterization of Materials* (eds Honig, J. M. & Rao, C. N. R.) Academic Press, New York.

Subramanyam, S. V. & Naik, H. (1985) in *The Metallic and Non-metallic States of Matter* (eds Edwards, P. P. & Rao, C. N. R.) Taylor & Francis, London.

Swinnea, J. S. & Steinfink, H. (1982) *J. Solid State Chem.* **41**, 114, 124.

Taher, S. M. & Gruber, J. B. (1981) *Mater. Res. Bull.* **16**, 1407.

Takano, M. & Takeda, Y. (1983) *Bull. Inst. Chem. Res.*, Kyoto University, Japan **61**, 406.

Thomas, L. A. & Wooster, W. A. (1951) *Proc. Roy. Soc. London* **A208**, 43.

Thornton, G., Tofield, B. C. & Hewat, A. W. (1986) *J. Solid State Chem.*, **61**, 301.

Tohge, N., Matsuo, H. & Minami, T. (1987) *J. Noncryst. Solids*, **95–96**, 809.

Torrance, J. B. (1979) *Acc. Chem. Res.* **12**, 79.

Torrance, J. B., Shafer, M. W. & McGuire, T. R. (1972) *Phys. Rev. Lett.* **29**, 1168.

Torrance, J. B. & Silverman, B. D. (1977) *Phys. Rev.* **B15**, 788.

Tressaud, A. & Dance, J. M. (1977) in *Adv. Inorg. Chem. Radiochem.* (eds. Emeleus, H. J. & Sharpe, A. G.) Vol. 20, Academic Press, New York.

Tressaud, A. & Dance, J. M. (1982) *Structure and Bonding* **52**, 87.

Tsai, P. P., Potenza, J. A. & Greenblatt, M. (1987) *J. Solid State Chem.* **69**, 329.

Uchino, K., Nomura, S., Cross, L. E. & Newnham, R. E. (1981) *Ferroelectrics* **38**, 825.

Underhill, A. E. (1985) *Phil. Trans. Roy. Soc. London* **A314**, 125.

Urushibara, A., Moritomo, Y., Arima, T., Asamitsu, A., Kido, G. & Tokura, Y. (1995) *Phys. Rev. B* **51**, 14103.

Van der Heide, H., Hemmel, R., Van Bruggen, C. F. & Haas, C. (1980) *J. Solid State Chem.* **33**, 17.

Van Zandt, L. L. & Honig, J. M. (1974) *Ann. Rev. Mater. Sci.* **4**, 191.

Varma, K. B. R., Harshvardhan, K. S., Rao, K. J. & Rao, C. N. R. (1985) *Mater. Res. Bull.* **20**, 315.

Vasantacharya, N. Y., Ganguly, P., Goodenough, J. B. & Rao, C. N. R. (1984) *J. Phys. C* **17**, 2745.

Wachter, P. (1980) *Phil. Mag.* **B42**, 497.

Wang, E. & Greenblatt, M. (1988) *J. Solid State Chem.* **76**, 340.

Whangbo, M. H. & Schneemeyer, L. F. (1986) *Inorg. Chem.* **25**, 2424.

Whangbo, M. H. & Canadell, E. (1988) *J. Amer. Chem. Soc.* **110**, 358.

Whangbo, M. H., Canadell, E., Foury, P. & Pouget, J. P. (1991) *Science* **252**, 96.

Wickelhaus, W., Simon, A., Stevens, K. W. H., Brown, P. J. & Ziebeck, K. R. A. (1982) *Phil. Mag.* **B46**, 115.

Wigner, E. P. (1938) *Trans. Faraday Soc.* **34**, 678.

Williams, J. M., Kini, A. M., Wang, H. H., Carlson, K. D., Geiser, U., Montgomery, L. K., Pyrka, G. J., Watkins, D. M., Kommers, J. M., Boryschuk, S. J., Croush, A. V. S., Kwok, W. K., Schirber, J. E., Overmyer, D. L., Jung, D. & Whangbo, M. H. (1990) *Inorg. Chem.* **29**, 3272.

Williams, P. M. (1976) *Optical and Electrical Properties of Layered Materials* (ed. Lee, P. A.) D. Reidel, Dordrecht, Holland.

Wilson, J. A. (1972) *Adv. Phys.* **21**, 143.

Wilson, J. A. (1985) in *The Metallic & Nonmetallic States of Matter* (eds Edwards, P. P. & Rao, C. N. R.) Taylor & Francis, London.

Wilson, J. A. & Yoffe, A. D. (1969) *Adv. Phys.* **18**, 193.

Yvon, K. (1979) in *Current Topics in Materials Science*, **3** (ed. Kaldis, E.) North-Holland, Amsterdam.

Zaanen, J., Sawatsky, G. A., Allen, J. W. (1985) *Phys. Rev. Lett.* **55**, 418.

7 Fashioning solids for specific purposes: aspects of materials design

7.1 Introduction

We discussed several interesting and useful properties of solids in the previous chapter. Tailor-making materials of desired properties is an integral part of modern solid state chemistry and an account of the subject would be incomplete without reference to it. To do justice to any one class of materials or to any one type of application, we would have to deal extensively with materials design, devices and other technological details. Since this would be beyond the scope of this monograph, we shall briefly present a few of the selected materials applications.

An area of great technological relevance today is *high-performance ceramics* (Bowen, 1980; Katz, 1980; Sanders, 1984). The term 'ceramic' itself refers to an inorganic, nonmetallic material processed or consolidated at high temperatures. Some ceramic materials (e.g. cubic BN, SiC, Si_3N_4) are hard, oxidation-resistant and light. They can often be made from abundantly available raw materials and are less expensive than metals. The shortcomings of ceramics are that they are generally brittle and difficult to make in large complex shapes. Uniform submicron powders required for making high-technology ceramics are made by the sol–gel process or from the vapour phase. Fibres made of Si_3N_4, SiC or carbon are toughened by depositing metals or silicon around them; lithium aluminium silicate glass ceramics are reinforced with SiC fibre. Ceramic materials are used in heat engines (e.g. Si_3N_4), as materials of fabrication, as thermistors, as substrates for electronics (e.g. barium silicate) and in magnetic and other applications. The materials used in thermistors can have negative ($NiFe_2O_4$ and $MgFe_2O_4$ for temperature control and measurement) or positive ($MgAl_2O_4$ and Zn_2TiO_4 for heating elements and switches) temperature coefficients of resistance. Hard ferrites ($BaFe_{12}O_{19}$) and soft ferrites ($MnFe_2O_4$) have a variety of applications in memory devices, recording tapes and transformer cores. Many ceramics are ionically conducting and find applications in battery systems. It is interesting that the light tiles of the space shuttle are made of composite ceramics (silica and aluminium borosilicate fibres). A predominant proportion of materials dealt with by solid state chemists are ceramics and it is important that greater attention is paid to the area of high-performance ceramics. Of all the ceramic materials, the high-temperature superconductors have created the biggest sensation. After the discovery of 30 K superconductivity in a cuprate of the La–Ba–Cu–O system, (Bednorz & Müller, 1986), a variety of other superconducting cuprate materials have been prepared and characterized, the highest superconducting T_c reached hitherto being around 150 K. Superconducting transitions up to 35 K have been reached in $Ba_{1-x}K_xBiO_3$ and also alkali fullerides. We shall discuss these superconducting cuprates and other oxides in some detail in Section 7.7 and alkali fullerides in Section 7.8.

Solid materials showing anomalous negative thermal expansion are technologically important in electronic, optical and structural applications (Sleight, 1995). Anisotropic solids such as cordierite ($Mg_2Al_4Si_5O_{18}$), β-eucryptite ($LiAlSiO_4$) and NZP [$NaZr_2(PO_4)_3$] show positive thermal expansion in one or two dimensions and negative thermal expansion in the other two or one dimensions. Since NZP has a tubular structure, the material expands on heating in the c-direction and contracts in the a-, b-directions. The net result is very low-volume thermal expansion (Roy et $al.$, 1989). ZrV_2O_7 and its hafnium analogue crystallizing in isotropic NaCl-like structure show an unusual thermal contraction from 423 K to its decomposition point, 1073 K. No other material with such a strong negative thermal expansion over such a wide temperature range is known to date. The $ZrP_{2-x}V_xO_7$ solid solutions show isotropic negative thermal expansion (Sleight, 1995; Korthius et $al.$, 1995).

Other important aspects that we shall review are optical materials, organic materials, polymers, photoelectrochemistry and hydrogen-storage materials. Organic materials are gaining importance because of the emerging area of molecular electronics. Amongst the organic materials, special mention must be made of organic ferromagnets (with or without metal ions in them), organic conductors and superconductors, non-linear materials and luminescent materials. We shall examine these materials with some examples.

7.2   Fast ion conductors

There is a large, growing family of ionic solids in which certain ions exhibit unusually rapid transport. These materials have come to be known as fast ion conductors (FICs). In some cases, the rapid ion transport is accompanied by appreciable electronic conduction as well. There is tremendous interest in the science and technology of fast ion conductors in view of their potential use as electrodes or electrolyte materials in electrochemical energy conversion devices (Hagenmuller & van Gool, 1978; Chandra, 1981; Goodenough, 1984).

In order for a solid to show fast ion conduction, it must satisfy the following criteria: it must have a high concentration of potential charge carriers, a high concentration of vacancies or interstitial sites, and a low activation energy for ion hopping. The presence of a set of energetically equivalent sites partly occupied by the mobile ions and satisfying the condition $c(1-c) \neq 0$ (where c is the fraction of sites occupied) is essential. Crystal chemical considerations can be applied to identify structures that meet the first two criteria, but prediction of activation energy is not so easy. The conductivity is essentially given by the expression

$$\sigma = (\gamma Nq^2/kT)c(1-c)l^2 v_0 \exp(-\Delta H_m/kT) \qquad (7.1)$$

where ΔH_m is the enthalpy of migration of the mobile ion, N the density of energetically equivalent sites, l the jump distance, c the concentration of occupied sites and v_0 the jump frequency; $\gamma = \frac{1}{6}z\exp(\Delta S_m/k)$, where z is the number of nearest-neighbour sites.

Fast ion conduction is not a discovery of recent times; as early as 1914, Tubandt & Lorenz had observed it in certain silver compounds. For example, these workers found that the conductivity of AgI just below the melting point was actually about 20% greater than the conductivity of molten AgI. Solids in which fast ion conduction has been established include several stoichiometric compounds (AgI, $RbAg_4I_5$, AgSI) as well as nonstoichiometric compounds such as β-alumina, $Na_{1+x}Al_{11}O_{17+x/2}$, calcia-stabilized zirconia, $Zr_{1-x}Ca_xO_{2-x}$, and Li_xTiS_2. The last shows both ionic and electronic conductivity. Fast ion conduction has been found in a number of glasses containing Li^+ or Ag^+ (Rao, 1981).

Fast ion conduction can occur in one, two or three dimensions, depending on the crystal structure. One-dimensional ionic conduction is found, for example, in $LiAlSiO_4$ (β-eucryptite), which has a hexagonal structure where the lithium ions are found in channels parallel to the c-axis. Two-dimensional conduction occurs in compounds crystallizing in layer structures, typical examples being β-aluminas and Li_xTiS_2. Three-dimensional conduction is found in Li_3N as well as in complex structures consisting of interlocking polyhedra, such as $Na_3Zr_2PSi_2O_{12}$ (NASICON, sodium superionic conductor). Analogous lithium ion conductors (LISICON) based on solid solutions of Li_2ZnGeO_4 and Li_4GeO_4 are also known (Bruce & West, 1982; Mazumdar et al., 1984). Fig. 7.1 shows the conductivity behaviour of a number of fast ion conductors in both crystalline and glassy states.

The conducting ion sublattice in FICs is generally considered 'molten'. The molten sublattice model for fast ion conduction was first proposed by Strock (1936) on the basis of structural and thermodynamic data for AgI. In most FICs, the entropy of the phase transition to the FIC state is larger than the entropy of melting. For example, in AgI the entropy of the transition at 420 K from the β-form to the α-form (FIC state) is 14.7 J deg^{-1} mol^{-1}, whereas the entropy of melting at 861 K is only 11 J deg^{-1} mol^{-1}.

Structural studies of fast ion conducting solids show that the conducting ions are distributed statistically over a large number of sites and exhibit large vibrational amplitudes. In α-AgI, the two silver ions per unit cell are largely distributed among the $12d$ (space group $Im3m$) tetrahedral sites (Fig. 7.2), and the activation energy for migration of Ag^+ ions is hardly 0.05 eV. Electron density maps reveal a small amount of electron density at the $24h$ sites also, which bridge the $12d$ positions. Electrical transport behaviour of fast ion conductors can be interpreted in terms of the point defect model, as for ordinary ionic conductors, but the ionic conduction of several FIC solids shows a negative deviation from the Arrhenius-type relation at high temperatures. O'Keeffe (1973) has suggested that the conductivity data of FICs can be fitted into an equation of the type $\sigma = bT - a$, applicable to liquids.

Framework (skeleton) structures of oxides have been identified for fast ion conduction of Na^+ and other ions (Goodenough et al., 1976). One-, two- or three-dimensional space is interconnected by large bottlenecks in these oxide hosts. While the tungsten bronze and β-alumina structures contain one- and two-dimensional interstitial space, the hexagonal framework of $NaZr_2(PO_4)_3$ has a three-dimensional

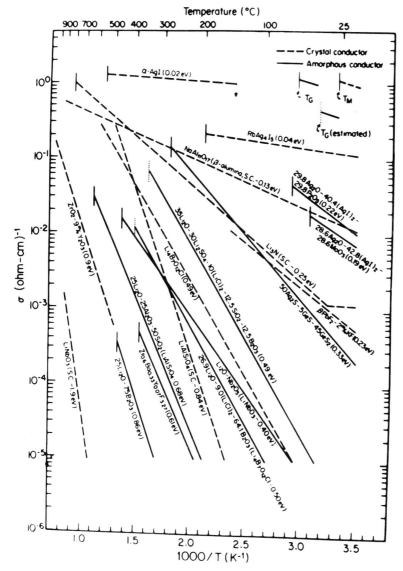

Figure 7.1   Ionic conductivity of crystalline and amorphous fast ion conductors. (After Tuller *et al.*, 1980.)

interstitial space (Fig. 7.3) where, along the c-axis, the vacant trigonal–prismatic sites, p, the octahedral Zr^{4+} sites, Z, and the octahedral sites available to Na^+ ions, M, are ordered as–Z–p–Z–M–Z. For each filled M site, there are three empty M_0 sites forming hcp layers perpendicular to the c-axis. In $NaZr_2(PO_4)_3$, the M sites are filled and the M_0 sites are empty. In $Na_4Zr_2(SiO_4)_3$, which has the same framework, both M and M_0 sites are filled. Both are poor Na^+ conductors but the solid solutions $Na_{1+x}Zr_2P_{3-x}Si_xO_{12}$ are Na^+ conductors, with an activation energy of $\sim 0.3\,eV$ when $x \approx 2.0$ (NASICON). Diffusion paths and single particle potentials of Na^+ ion in

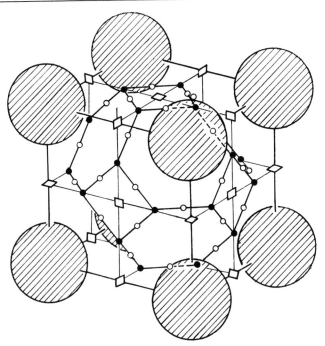

Figure 7.2   Structure of α-AgI, showing interstitial cation sites: large hatched circles, iodide ions; squares, octahedral 6b sites; solid circles, tetrahedral 12d sites; open circles, 24h sites.

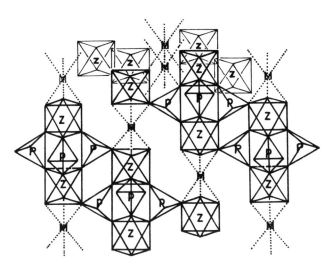

Figure 7.3   Structure of NaZr$_2$(PO$_4$)$_3$. (After Goodenough, 1984.)

NASICON have been studied (Kohler & Schulg, 1985). LISICON type FICs are found to show negative activation volumes at high pressures (Bose *et al.*, 1984).

Sodium β-alumina has received considerable attention, triggered by the discovery of Weber & Kummer (1967) that it can be used as a solid electrolyte in the

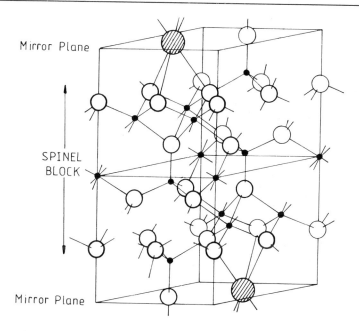

Figure 7.4 Structure of sodium β-alumina. Only half the unit cell is shown. Large hatched circles represent sodium, open circles oxygen, and small filled circles aluminium.

sodium–sulphur battery. The novelty of this material arises from the fact that, although its melting point is around 2200 K, several amp cm^{-2} of current can be passed across it at 570 K. The key to rapid Na$^+$ transport in this solid lies in its crystal structure (Fig. 7.4) and the low potential energy path this structure provides for Na$^+$ migration. In general, β-aluminas, $M_{1+x}Al_{11}O_{17+x/2}$ (M = Na$^+$, K$^+$, Ag$^+$ etc.) are layered non-stoichiometric compounds. For sodium β-alumina $x = 0.20$ to 0.30. The hexagonal structure of this solid (Fig. 7.4) consists of two close-packed spinel-like blocks of four oxygens thick along the c-axis; the two blocks are separated by a mirror plane in which the density of oxygen atoms is only $\frac{1}{4}$ of that in the spinel blocks. The mirror planes contain mobile sodium ions. Na$^+$ transport is essentially two-dimensional in this structure. Nonstoichiometry in this solid arises from the possibility of accommodating variable quantities of oxygen and sodium ions in the conducting planes. Whittingham & Huggins (1971) measured the ionic conductivity of sodium β-alumina and compared the value with the diffusion coefficient obtained by Yao & Kummer (1967). Typical values are: $D(cm^2s^{-1}) = 2.4 \times 10^{-4} \exp(-0.165/kT)$ and σT (K ohm^{-1}cm^{-1}) = $2.37 \times 10^3 \exp(-0.164/kT)$. The Haven ratio (Section 6.3.2) is 0.610, which compares well with the value of 0.600 calculated assuming an interstitialcy mechanism for the migration of sodium. β''-alumina, which is a phase stabilized by the presence of magnesium, $Na_{1+x}Mg_xAl_{11-x}O_{17}$, is a better ionic conductor than β-alumina (Farrington & Briant, 1979). The conductivity and activation energy of these aluminas vary with the stoichiometry and arrangement of defects. Defect structures of β-aluminas have been investigated by electron microscopy (Ganapathi *et al.*, 1985).

There are several other oxide ion conductors. Among them, brownmillerite-like $Ba_2In_2O_5$ (Goodenough *et al.*, 1990), anion-deficient Aurivillius phases such as $Bi_2VO_{5.5}$ (Varma *et al.*, 1990; Abraham *et al.*, 1990) and Sr- and Mg-doped $LaGaO_3$ (Ishihara *et al.*, 1994) are recent additions. The dynamics of oxide-ion motion in $Ba_2In_2O_5$ investigated by O^{17} NMR experiments (Adler *et al.*, 1994) reveals that oxygens become mobile first within the tetrahedral layers at the order–disorder transition (1198 K); above 1348 K, when the structure becomes cubic the entire population of oxide ions becomes mobile. Nominal $La_{0.9}Sr_{0.1}Ga_{0.8}Mg_{0.2}O_3$, exhibiting a high oxide ion conductivity that is independent of oxygen pressure down to 10^{-20} atm., is intriguing, since it is not obvious how the oxygen vacancies arise in this material.

Difluorides such as PbF_2 with the fluorite structure exhibit fast ion conduction due to facile F^- ion transport (Section 6.4.5). An interesting structure showing Li^+ conduction is that of Li_3N (Rabenau, 1978). Conduction is two-dimensional. Cooperative basal plane excitations involving the rotation of six Li^+ ions by 30° about a common N^{3-} ion to edge positions (positions midway between N^{3-} ions in the Li_2N layer) seem to be responsible for conduction in this nitride. In the fluorite structure, a rotation by 45° of a single cube of F^- ions seems to be involved. The Zintl alloy LiAl is also a lithium-ion conductor.

The discovery of a number of fast ion conducting glasses is significant (Tuller *et al.*, 1980). AgI is a component of many of the Ag^+-containing glassy FICs; the Li^+ FIC glasses are generally formed in network structures. Examples of glassy FICs are Li_2O–B_2O_3(42.5–57.5), Li_2O–SiO_2(33.3–66.7), AgI–Ag_2MoO_4(75–25), and ZrF_4–BaF_2–ThF_4(58.7–31.3–10); compositions are given in parentheses. Besides the ease of fabrication, glassy fast ion conductors offer advantages of property manipulation through composition control. Mechanism of conduction in FIC glasses is more complex than in crystalline state.

Hitherto we have dealt with model FICs that are mostly useful as solid electrolytes. The other class of compounds of importance as electrode materials in solid state batteries is mixed electronic–ionic conductors (with high ionic conductivity). The conduction arises from reversible electrochemical insertion of the conducting species. In order for such a material to be useful in high-energy batteries, the extent of insertion must be large and the material must sustain repeated insertion–extraction cycles. A number of transition-metal oxide and sulphide systems have been investigated as solid electrodes (Murphy & Christian, 1979).

FICs are useful as electrochemical sensors, electrolytes and electrodes in batteries and in solid state displays (Farrington & Briant, 1979; Ingram & Vincent, 1984). If a FIC material containing mobile M ions separates two compositions with different activities of M, a potential is set up across the FIC that can be related to the difference in the chemical activities of M. By fixing the activity on one side, the unknown activity on the other can be determined. This principle forms the basis of a number of ion-selective electrodes; LaF_3 doped with 5% SrF_2 is used for monitoring fluoride ion concentration in drinking water. Similarly, calcia-stabilized-zirconia is used in cells of the type

$$Pt(s), O_2(g)|ZrO_2-CaO|O_2(g), Pt(s)(ref.)$$

to monitor oxygen concentrations in vehicle exhaust emissions, carburizing furnace atmospheres or in molten steel.

The most promising application of FICs is in solid state batteries. Two kinds of batteries may be distinguished: (i) small *primary* cells, where long lifetime and no self-discharge are essential requirements, and (ii) rechargeable *secondary* batteries, when high energy density is the main criterion. A battery of the first type used as a cardiac pacemaker is based on lithium iodide solid electrolyte. Here, the requirement is low power and continuous operation for long periods (~ 15 years). A typical pacemaker cell consists of a lithium anode, lithium iodide electrolyte and poly-2-vinylpyridine–I_2 complex as cathode. High energy density batteries are important from the point of view of alternate energy sources, being useful for vehicular transport and load-levelling in power stations. An important landmark in the development of high energy density batteries is the Na/S battery, which employs Na–β alumina as the solid electrolyte and molten sodium and sulphur as electrodes (for a description of these advanced batteries, see Tofield *et al.*, 1978). The cell voltage (2.08 V) is derived from the chemical reaction between sodium and sulphur to produce sodium polysulphide. The theoretical energy density of the Na/S battery is quite high ($\sim 750 \, W \, h \, kg^{-1}$) as compared with the energy density of $170 \, W \, h \, kg^{-1}$ of an ordinary lead–acid battery. The main disadvantage is the use of molten sodium and sulphur, which are hazardous when exposed to air and moisture. Current interest is focussed on lithium–metal sulphide batteries which make use of organic electrolytes and solid solution electrodes (which are mixed electronic–ionic conductors). Whittingham (1976) developed a Li/TiS_2 battery which makes use of an organic electrolyte for transport of lithium. The battery has a high energy density ($480 \, W \, h \, kg^{-1}$) and the added advantage of room-temperature operation.

A wide range of mixed conductors has been investigated as electrodes in batteries, which include fluorinated graphite, alkali metal β-ferrites (isostructural with β-aluminas), nonstoichiometric silver sulphides, tungsten and vanadium bronzes and alloys such as Li_xAl and Li_xSi. Crystallographically aligned Ni/ZrO_2 composite made by reduction of a directionally solidified NiO/ZrO_2 eutectic (Revcolevschi & Dhalenne, 1985) may find use as an electrode material. An all-polymer solid state battery based on doped polyacetylene, $(CHNa_y)_x$ |sodium iodide–polyethylene oxide| $(CHI_z)_x$, has been developed.

One of the earliest applications of fast ion conductors was the use of stabilized zirconia as the oxide ion electrolyte in high-temperature *fuel cells*. Fuel cells are electrochemical devices that convert chemical energy of a fuel-burning reaction such as

$$H_2(g) + \tfrac{1}{2}O_2(g) \rightarrow H_2O(g)$$

into electrical energy. Fuel gas (e.g. H_2 or CO) circulates through the inner side of the electrolyte tube, and oxygen or air is led through the outer side. The cell requires operation temperatures around 1200 K. During power generation, oxygen picks up electrons at the air electrode, forming O^{2-} ions that pass through the electrolyte tube and react at the fuel electrode with H_2 or CO, producing H_2O or CO_2. The electrons released in the oxidation process flow through the external circuit back to the cathode.

A combination of a fuel cell, where a fuel is burnt, and a water electrolysis cell, where hydrogen is produced, is a new concept of a clean energy cycle. The main limitation of the fuel cell is the high working temperature (>1000 K) required for its operation. The key to the whole problem lies in finding a suitable electrolyte material that can function at moderately low temperatures (400–600 K). Since it is almost impossible to find a low-temperature oxide ion conductor, efforts are being directed towards developing fast H^+ ion conducors. $HNbO_3$ (Sen & Sen, 1983) and uranyl phosphate (Skou et al., 1983) are reported to be good proton conductors.

Methanol is a convenient liquid fuel that can be either blended with petrol or burnt directly in an engine. Utilization of methanol can be envisaged for automotive transport, should a cheap, reliable, long-lived methanol–air fuel cell be developed. Two principal materials problems must be overcome before such a cell can be realized in the market place: a proton electrolyte capable of cheap manufacture and stable to about 570 K and a catalytic anode for the conversion of methanol and water via the reaction

$$CH_3OH + H_2O \rightarrow CO_2 + 6e^- + 6H^+.$$

Imaginative fuel-cell designs must be coupled with this materials effort.

Fast ion conduction has been made use for developing electrochromic displays. WO_3 is a pale yellow solid that becomes deep blue when a small amount of Na is incorporated, owing to formation of Na_xWO_3. Since the change occurs reversibly and very fast, the material has been used in displays.

7.3   Photoelectrochemistry

As the era of inexpensive fossil fuels draws to a close, interest in harnessing alternative sources of energy is increasing markedly. Among the renewable sources of energy, solar energy is not only the most abundant but also clean. Its conversion to useful forms of energy is, however, beset by problems such as low flux density and intermittent insolation. Even the most efficient *solar energy conversion* devices (e.g. photovoltaic silicon solar cells, efficiency $\sim 15\%$) are quite uneconomical for large-scale use because of the manufacturing cost and associated encapsulation, installation and land costs. In this context, conversion of solar energy to chemical energy (e.g. H_2 from water) or electrical energy using semiconductor electrodes in a photoelectrochemical cell (PEC) has attracted much attention. The subject has been excellently reviewed in the literature (Nozik, 1978, 1980; Bard, 1980; Heller, 1981; Aruchamy et al., 1982; Kutal, 1983; Wrighton, 1983).

Photoelectrochemical effect involves production of a voltage and an electric current when light falls on a semiconductor electrode immersed in an electrolyte solution and connected to a counter electrode (Becquerel, 1839). Working with germanium electrodes, Brattain & Garrett (1955) showed that the *Becquerel effect* was due to the formation of a semiconductor–electrolyte junction. The idea of using an illuminated

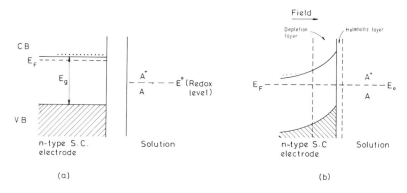

Figure 7.5   Schematic representation of an *n*-type semiconductor-solution electrolyte junction showing the formation of depletion layer, band bending and Helmholtz layer; (a) before immersion and (b) after immersion in solution.

semiconducting electrode in an electrolytic cell to produce chemical products or electrical power arose from their work, but the feasibility was not demonstrated until Fujishima & Honda (1972), working with *n*-type TiO_2 photoanodes, showed that oxidation of water to oxygen could be achieved at a significantly more negative potential than the standard redox potential of the H_2O/O_2 couple.

In order to understand photoelectrochemistry involving *semiconductor electrodes*, a knowledge of the energetics of semiconductor–electrolyte junctions is essential. When a semiconductor is dipped into an electrolyte, equilibration of Fermi levels (chemical potentials) occurs by transfer of majority charge carriers. This produces a *space-charge layer* (also called depletion layer) in the semiconductor. As a result, the conduction and valence bands are bent such that a potential barrier is established; the barrier prevents further electron transfer (Fig. 7.5). A charged layer (*Helmholtz layer*) also exists in the electrolyte, which consists of charged ions from the electrolyte adsorbed on the solid electrode surface.

When the semiconductor–electrolyte junction is illuminated with light, photons of energy greater than the band gap (E_g) are absorbed in the depletion layer, creating electron-hole pairs, which separate under the influence of the electric field present in the space-charge region. Light-induced production and subsequent separation of electron-hole pairs tends to bring the Fermi level back to its original position before the junction was established. Photogenerated minority carriers are injected into the electrolyte to drive a redox reaction. For *n*-type semiconductors, holes are injected into the electrolyte, which can result in anodic oxidation, while for *p*-type semiconductors, electrons are injected into the electrolyte to produce a cathodic reduction reaction. Photogenerated majority carriers leave the semiconductor *via* an ohmic contact in both the cases and are injected at the counter electrode to drive a redox reaction inverse to that occurring at the semiconductor photoelectrode.

Two types of PECs must be distinguished, depending upon the nature of the redox reaction. If the reaction at the anode is different from that at the cathode, a net chemical change in the electrolyte results and the cell is called a *photoelectrolysis cell*: here, the

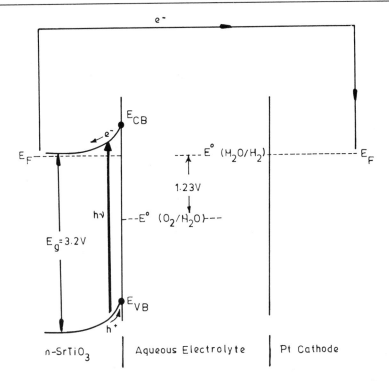

Figure 7.6   Photoelectrolysis of water using SrTiO$_3$ as anode. (After Wrighton, 1983.)

incident optical energy is converted to chemical free energy of the cell reaction. If, on the other hand, oxidation at the anode is simply reversed at the cathode, no net chemical change occurs in the electrolyte. In this case, the cell behaves as an electrochemical *photovoltaic cell*, where the incident optical energy is converted to electrical energy (Gerischer, 1975, 1979).

Photoelectrolysis cells.   A photoelectrolysis cell can be constructed in two ways. One is to make use of a semiconductor electrode and a metal counter electrode (*Schottky cell*); the semiconductor could be *n*- or *p*-type. In the other configuration, both the electrodes are semiconductors, one *n*-type and the other *p*-type. Photoelectrolysis of water using an *n*-type semiconductor anode (SrTiO$_3$) is illustrated in Fig. 7.6. The cell can operate to produce H$_2$ and O$_2$ from H$_2$O without any additional energy input, other than the light energy, to excite electrons in SrTiO$_3$. Unfortunately, since the band gap is large (~ 3.2 eV), only a small portion ($< 5\%$) of solar energy spectrum having energy greater than the band-gap energy can be converted to chemical energy by this cell. Efficient solar conversion requires band gaps of the semiconductor electrodes in the range 1.3 ± 0.3 eV.

Photoelectrolysis cells of the *p–n* variety can be constructed using the same semiconductor material, one doped *p*-type and the other *n*-type or with two different (*n*- and *p*-) semiconductors. A photoelectrolysis cell reported by Leygraf *et al.* (1982), making use of polycrystalline n- and p-doped Fe_2O_3, is an example of the former type. The cell consists of Mg^{2+}- and Si^{4+}-doped Fe_2O_3 pellets which are cemented back to back. Suspended in aqueous $NaOH$ or Na_2SO_4 and illuminated with visible light, the diode evolves H_2 and O_2. In a p–n cell, both the electrodes are illuminated and two photons are absorbed to produce a net electron-hole pair. The pair consists of a minority hole from the n-electrode and a minority electron from the p-electrode. An important advantage of the p–n cell is that for a given cell reaction, it is possible to use smaller band-gap materials; since the photocurrent increases with decreasing band gap, the efficiency increases. The enhanced electron-hole pair potential available from a p–n cell can also eliminate the bias potential usually required in Schottky type cells. This was, indeed, shown to be the case in a n-TiO_2/p-GaP cell in which photoelectrolysis of water could be achieved in zero bias with simulated sunlight (Nozik, 1976). Such cells could also be used for carrying out other reactions that produce chemically useful products. Thus, CO_2 can be reduced with p-GaP electrodes to methanol, and chloride can be photooxidized to Cl_2 using n-TiO_2.

Photoelectrolysis of water is the most attractive reaction for solar energy conversion, but it has not been possible to achieve this efficiently (Harris & Wilson, 1978). Recent work (Gantron *et al.*, 1980; Lemasson *et al.*, 1982) shows that it is possible to increase the efficiency of the n-TiO_2 anode either by making it oxygen-deficient or by alloying it with VO_2. An efficiency of about 21% has been recorded using $TiO_{2-x}(x \sim 7 \times 10^{-4})$ electrodes and 335 nm radiation. Use of $Ti_{1-x}V_xO_2$ solid solutions is more efficient because the band gap can be reduced to ~ 2.0 eV. Photocathodes such as p-InP or p-$GaAs(E_g \approx 1.4$ eV$)$ are intrinsically more stable than n-type photoanodes in electrochemical cells. Heller (1981) has shown that surface recombination of electrons and holes can be decreased by chemical modification of the electrode surface. Thus, adsorption of Ru^{3+} ions at the electrode surfaces can remarkably increase the efficiency.

Electrochemical photovoltaic cells.   In an electrochemical photovoltaic cell, the two electrode reactions are the inverse of each other: $R + h^+ \rightarrow O^+$ (anode) and $O^+ + e^- \rightarrow R$ (cathode), where R and O^+ stand for reduced state and oxidized state respectively. No net change occurs in the electrolyte. The incident optical energy increases the free energy of electrons in the semiconductor electrode, which could be tapped as electrical energy when the cell is in operation.

Electrochemical (liquid–junction) photovoltaic cells are essentially similar to the photoelectrolysis cells. The design of a *liquid-junction photovoltaic cell* making use of an n-type semiconductor electrode is illustrated in Fig. 7.7. When the semiconductor electrode is placed in an electrolyte and subjected to a positive bias, the region of the semiconductor near the surface becomes depleted in electrons. When light of sufficient energy is absorbed in this depletion layer, the electric field is able to separate the photogenerated electron (in the conduction band) from the hole (left behind in the valence band). The direction of the field is such as to drive the hole to the surface, where it is transferred to the reduced form R of the redox couple O/R in solution. The electron is transferred around the external circuit, performing work $e.V$, finally reducing O^+ at

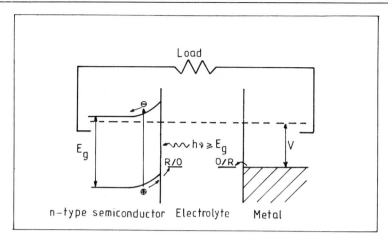

Figure 7.7   Schematic representation of a liquid-junction photovoltaic cell using an *n*-type semiconductor. R/O is the redox couple in the electrolyte.

the metal counter electrode. Thus, no net chemical change occurs in the cell but light energy $h\nu$ is converted to electrical work eV. The open circuit voltage of such a cell, V_{oc}, which is the maximum voltage obtained at zero current, corresponds to the difference between the flat-band potential (conduction-band edge) of the semiconductor and the redox potential of O/R in solution. Maximum current is obtained when the two electrodes are shorted. Between these two extremes, there is a current-potential region where useful work can be obtained.

Although the principles of this cell have long been known, practical devices for solar energy conversion have proved difficult because of the following problems: (i) photocorrosion of the semiconductor electrodes. This can be avoided by fast oxidation kinetics of R, which minimizes the concentration of holes at the semiconductor surface; (ii) trapping of holes at the surface states near the conduction band, which are ultimately destroyed by capturing an electron from the conduction band; (iii) strong chemisorption of R and O at the semiconductor affects the kinetics of the electrode reactions unfavourably and decreases the overall efficiency. Several efforts are being made to overcome these limitations.

Photocorrosion can be prevented by adding a redox couple to the electrolyte whose potential is more favourable than the decomposition potential such that the redox reaction occurs preferentially. When *n*-CdS is used as photoanode in aqueous electrolytes, the electrode is photocorroded since the reaction, $CdS + 2h^+ \rightarrow S + Cd^{2+}$, occurs readily. By adding NaOH and sodium polysuphide to the electrolyte (Ellis *et al.*, 1976), photocorrosion is prevented. The S^{2-}/S_n^{2-} redox couple preferentially scavenges the photoholes. At the anode, sulphide is oxidized to polysulphide (free sulphur) and free sulphur is reduced back at the dark cathode. Similarly *n*-Si anodes have been stabilized by using a nonaqueous electrolyte containing a ferricinium/ferrocene redox couple (Legg *et al.*, 1977; Chao *et al.*, 1983). Unfortunately, a similar stabilization technique cannot be applied to photoelectrolysis cells. Some examples of electrode

Table 7.1. *Typical examples of semiconductor-based electrochemical photovoltaic cells operating in an aqueous medium*

Semiconductor electrode	Predominant redox couple	Maximum experimental efficiency (%)
n–CdS	S^{2-}/S_n^{2-}	1–2
n–CdSe	Se^{2-}/Se_n^{2-}	7–8
n–CdTe	Te^{2-}/Te_n^{2-}	10
n–GaAs	Se^{2-}/Se_n^{2-}	9–10
n–Si	$Fe(cp)_2^a/Fe(cp)_2^+$	2

^{a}cp is $(\pi - C_5H_5)$

materials and redox couples used in liquid-junction photovoltaic cells are given in Table 7.1.

Compared to solid state photovoltaic cells, liquid-junction photovoltaic cells possess important advantages: (i) potential barriers (band bending) are easily established by simply dipping the electrode in the electrolyte; (ii) polycrystalline semiconductor films can be used without drastic decrease in efficiency and (iii) band bending can be altered by adjusting the redox potential of the electrolyte. The last factor is, however, limited by the electrode stability. If the potential of the electrolyte is too positive in the case of n-electrodes (or too negative for p–electrodes), electrode decomposition may become more favourable than the cell reaction. Recent reports indicate that liquid-junction photovoltaic cells may very soon become competitive to solid state solar cells. Using n-GaAs$_{1-x}$P$_x$ in acetonitrile and ferrocene–ferrocinium redox couple, Gronet & Lewis (1982) have reported an efficiency of 13.0% in natural sunlight, this being the highest efficiency reported for a liquid-junction photovoltaic cell.

Photochemical conversion of solar energy in homogeneous/microheterogeneous media. Chemical conversion of solar energy can also be achieved by photochemical reactions occurring in homogeneous or colloidal solutions. The strategy essentially involves simulating the photosynthesis of carbohydrates in nature, i.e. to drive a chemical reaction uphill so that the photoproducts have higher free energy than the reactants. The extra energy can be tapped by reversing the reaction. Photochemical cleavage of water into elements is attractive for this purpose:

$$H_2O(l) \xrightarrow{h\nu} H_2(g) + \tfrac{1}{2}O_2(g); \ \Delta G_{298}^{\circ} = 238 \, kJ \, mol^{-1}$$

The reaction requires a sensitizer because water itself does not absorb above 200 nm. Transition-metal complexes exhibiting photoredox properties are useful as sensitizers for this reaction. The main problem, however, is that most transition-metal complexes involve one electron transfer per photon absorbed, whereas the water-splitting reaction

$$H_2O + 2e^- = 2OH^- + H_2$$
$$2H_2O = 4H^+ + O_2 + 4e^-$$

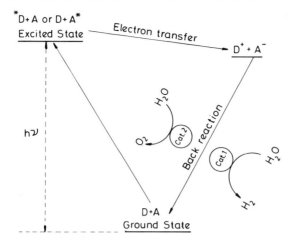

Figure 7.8   Photosensitized cleavage of water based on bimolecular photoredox reaction between an electron donor D and an acceptor A. Cat. 1 and cat. 2 are charge storage catalysts for reduction and oxidation of water, respectively.

is a multielectron process. Consequently, photosensitized water-splitting requires the presence of a charge-storage catalyst that mediates multielectron changes. In essence, the catalyst accumulates the proper number of electrons for delivery to the reactants.

An approach to water-splitting based on photoredox chemistry is shown in Fig. 7.8 (Kutal, 1983). Light absorption initiates an electron-transfer reaction between a donor D and an acceptor A, resulting in the formation of energy-rich species D^+ and A^-. The reducing power of A^- is coupled to the reduction of H_2O by a suitable catalyst, while the catalysed oxidation of H_2O is accompanied by D^+. Ru $(bipy)_3^{2+}$ is an example of a photosensitizer that performs the role of D^+ and A^- when excited by the charge-transfer absorption at 450 nm. The excited state of the bipyridyl complex, $*Ru(bipy)_3^{2+}$, which has an energy of 2.1 eV above the ground-state, can function both as a reducing agent and an oxidizing agent. Reductive quenching of $*Ru(bipy)_3^{2+}$ produces a powerful reductant, $Ru(bipy)_3^+$, and oxidative quenching generates a strong oxidant, $Ru(bipy)_3^{3+}$.

Grätzel and coworkers (1981) have reported a successful water-splitting system making use of solar energy (Fig. 7.9). A ruthenium (II) complex is used as photosensitizer, n-methyl viologen as electron relay and TiO_2 (particles) loaded with Pt and RuO_2 as a bifunctional catalyst. Upon photoexcitation, the ruthenium complex is excited and oxidatively quenched by methyl viologen. The latter injects an electron into the conduction band of platinum where it is used to evolve hydrogen from water. The sensitizer gives up the hole to RuO_2, which is used to evolve oxygen. Colloidal TiO_2 plays a vital role in the process; oxygen generated at its surface is reduced to O_2^- and is bound to the colloidal particle, thus providing a practical means of separating hydrogen and oxygen evolved. Hydrogen is evolved when the system is illuminated; oxygen stored on TiO_2 can be removed in the dark by adding sodium phosphate.

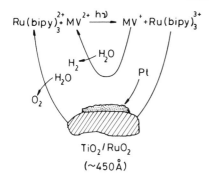

Figure 7.9   Schematic representation of visible light-induced decomposition of water using RuO_2 and Pt as redox catalysts codeposited on colloidal TiO_2. (Following Grätzel, 1981.)

Grätzel and coworkers have developed another interesting system for splitting H_2S. The system makes use of colloidal CdS loaded with ultrafine RuO_2 (0.1–0.5%). When the catalyst particles are dispersed in aqueous H_2S and illuminated with visible light, H_2 is generated with the deposition of elemental sulphur. For a discussion of photochemical splitting of water in colloidal and microheterogeneous systems, the article of Kiwi *et al.* (1982) is recommended.

Photochemical decomposition of water into its constituents has made significant progress in two directions: photoelectrochemical cells making use of one or two semiconductor electrodes and homogeneous photochemical reactions in solutions. The apparent gap between the two approaches is being bridged by the use of colloidal metals and semiconductor suspensions in solution and by the use of photochemical diodes that consist of particles of the two electrodes in direct contact as a single entity, not connected to an external supply. It appears that this 'wireless' photoelectrolysis could be the ultimate solution for large-scale solar energy conversion. Research in this area is, however, plagued by problems of irreproducibility (see, for example, Yesodharan & Grätzel, 1983).

Photochemical conversion of solar energy in artificial systems (as in natural photosynthesis) is essentially based on electron-transfer between an excited photosensitizer (A*) and an appropriate electron acceptor (B)

$$A^* + B \rightarrow A^+ + B^-$$

The light energy stored in the radical ion pair is utilized to split water or to produce other high-energy chemicals. While impressive progress has been made in increasing the efficiency of this charge separation reaction (Fox & Chanon, 1988; Norris & Meisel, 1989), inhibition of the exothermic back-reaction to neutral species,

$$A^+ + B^- \rightarrow A + B$$

is a major challenge. A strategy to prevent this back-reaction is to immobilize the donor and the acceptor in a porous sol–gel silica glass (Slama-Schwok *et al.*, 1992). By using immobilized pyrene as the photosensitized donor and methyl viologen as the electron acceptor, through the mediation of a mobile charge carrier (N, N′-tetramethylene 2,2′-bipyridinium bromide), exceedingly long-lived (up to a few hours) charge-separated pairs have been produced.

7.4   Magnetic materials

Uses of ferrite spinels in various applications (e.g. telephones, computers, microwave applications) are well known. Many magnetic alloys have been made in recent years (Chin, 1980; White, 1985). Efforts are being made to design molecular polymetallic systems exhibiting ferromagnetic order (Kahn, 1985). High-spin polycarbenes have been examined as models for organic ferromagnets (Iwamura, 1986).

Rare earth–cobalt intermetallic compounds of the formulae $LnCo_5$ and Ln_2Co_{17} (Ln = rare earth) constitute high-performance permanent magnetic materials (Nesbitt & Wernick, 1973; Wallace, 1973; Menth *et al.*, 1978). They are characterized by very high values of intrinsic coercivity (up to 3 million amperes per metre), extremely high magnetocrystalline anisotropy and maximum (BH) energy up to $240 \, kJ \, m^{-3}$. $SmCo_5$ is the most outstanding member of the $LnCo_5$ family, possessing the largest value of magnetocrystalline anisotropy ($17 \, MJ \, m^{-3}$) and large coercive force ($1600 \, kA \, m^{-1}$). Ln_2Co_{17} phases have higher saturation magnetization than $LnCo_5$ and hence higher (BH) products (Wallace & Narasimhan, 1980). $LnCo_5$ crystallizes in the hexagonal $CaCu_5$ structure while the Ln_2Co_{17} structure can be viewed as derived from the $LnCo_5$ structure by the ordered replacement of $\frac{1}{3}$ of the Ln atoms by pairs of cobalt atoms. Magnetism of Ln–Co compounds arises from interatomic exchange between the spins of the two sublattices and the spin–orbit coupling in the rare-earth atom. In the lighter rare earths, the spins of Ln and Co are aligned parallel and thus saturation magnetization is large. In the heavier rare-earth intermetallics, the spins are antiparallel, leading to low values of saturation magnetization. Magnetocrystalline anisotropy has its origin in the itinerant electrons of the cobalt sublattice and the crystal field of the rare-earth atoms.

Commercially, single-phase $SmCo_5$ and precipitation-hardened $Sm_2(Co, Cu)_{17}$ are important magnetic alloys. To achieve high coercivity, fine powders ($1–10 \, \mu m$) of $SmCo_5$ are aligned and compacted in a strong magnetic field and sintered. The mechanism of coercivity involves domain nucleation or wall pinning at grain boundaries. Thus high coercivity requires minimization of the existence of domain walls and this is achieved by having the alloy in the form of fine particles. Substitution of samarium by Ce or Nd lowers saturation magnetization and magnetocrystalline anisotropy as well as T_c. Gd-substitution decreases the temperature coefficient of remnance. Partial substitution of Co by Cu in $SmCo_5$ can increase coercivity on suitable heat treatment. Copper substituted magnets need not be ground into fine particles to develop magnetic hardness. Electron microscopy has revealed that low-temperature heat-treatment results in a homogeneous precipitation of the second phase, probably of Sm_2Co_{17} type, which is coherent with the $SmCo_5$ structure. Substitution of cobalt by iron, zirconium and hafnium has enabled optimisation of saturation magnetization, coercivity and (BH) product. Exceptionally large (BH) products and coercivity of the alloys render them useful in devices where small size and superior performance are required (Chin, 1980). Thus magnets for electronic wrist-watches, magnetic couplings and travelling wave tubes are made from Sm–Co alloys; small size d.c. and synchronous motors also make use of these alloys. In general, rare-earth–cobalt permanent magnets are useful in a number of low-volume specialized applications.

Certain alloys reported recently in the Fe–Ln–B (Ln = La, Tb or Nd) systems possess useful magnetic properties that render them competitive to $SmCo_5$ in permanent magnet applications. For example, amorphous ribbons of composition $(Fe_{0.82}B_{0.18})_{0.90}Tb_{0.05}La_{0.05}$ produced by melt-spinning develop large hysteresis and high coercivity when annealed around 920 K; annealing produces microcystals of several phases, which include Fe_3B, Ln_6Fe_{23} and $Ln_3Fe_{21}B$ (Das & Koon, 1983). The most powerful permanent magnet ever known was discovered in 1985. This is in the Fe–Nd–B system involving $Nd_2Fe_{14}B$ which has an unusual tetragonal structure with a high saturation magnetization and coercivity. The structure of this solid has been determined and the material is already being produced commercially; considerable research is now underway on various aspects of this extraordinary magnetic material (Herbst et al., 1985; Koon et al., 1985; Robinson, 1984; White, 1985). Details can be found in the *Proceedings of the Eighth International Workshop on Rare Earth Magnets and their Applications* held in Dayton in May 1985, the *Proceedings of the Workshop Meeting on Nd–Fe Permanent Magnets* published by the Commission of the European Communities and many other publications. The discovery of $Nd_2Fe_{14}B$ as a perma-nent magnet material gave impetus to research on Ln_2Fe_{17} (Ln = rare earth) and related materials. Some Ln_2Fe_{17} alloys had high saturation magnetization, but the Curie temperatures were lower due to antiferromagnetic Fe–Fe coupling. Interstitial hydrogen increases the T_c of Nd_2Fe_{17} and Sm_2Fe_{17} (Buschow, 1988). Alloying with carbon increases T_c and also improves the anisotropy (Zhong et al., 1990). Introduction of nitrogen is found to increase T_c markedly besides improving anisotropy (Coey & Sun, 1990).

The development of a solid state memory device referred to as '*magnetic bubble*' memories for use in computers and other data storage systems, based on the movement of magnetic domains through the crystal, has made a great impact (Nielsen, 1976, 1979; Giess, 1980; Blunt, 1983). The formation of magnetic bubbles is shown schematically in Fig. 7.10. In a thin film of a magnetic material with appropriate magnetization and an axis of uniaxial magnetic anisotropy perpendicular to the plane of the film, the magnetic domains formed in the film would consist of a serpentine array of stripe domains magnetized up and down as in Fig. 7.10(a). On the application of a magnetic field (bias field) perpendicular to the plane of the film, one set of domains will grow at the expense of the other (see Fig. 7.10(b)). With increase in the applied field, the domains shrink and become cylindrical (Fig. 7.10(c)). The cylindrical domains are called *magnetic bubbles* since their behaviour is described by equations similar to those applicable in the case of soap bubbles. In optically transparent materials such as garnets and ferrites, bubbles can be observed directly using the Faraday effect. Bubbles

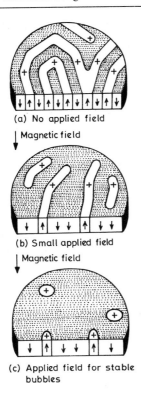

(a) No applied field

Magnetic field

(b) Small applied field

Magnetic field

(c) Applied field for stable
bubbles

Figure 7.10   Formation of magnetic bubbles (schematic). (After Blunt, 1983.)

move towards regions of lower bias field and this is made use of in memory devices. The stored information, in the form of bubble domains, moves through a stationary support medium; in a tape recorder, on the other hand, the stored information, in the form of stationary domains, is on a moving tape. Bubble memory devices of several megabits capacity are now being made.

Orthoferrites, hexaferrites, metallic alloys such as $GdCo_5$ can all support bubbles, but only rare-earth iron garnets, $Ln_3Fe_5O_{12}(Ln = $ rare earth) have provided the required flexibility. In these cubic materials, the rare-earth ion is in the dodecahedral site and the Fe^{3+} ions distributed between octahedral (2) and tetrahedral (3) sites are coupled antiparallely. In yttrium iron garnet (YIG), $Y_3Fe_5O_{12}$, the magnetization is due to the excess tetrahedral Fe^{3+} ions giving rise to a magnetization (1760 G) that is too large for sustaining bubbles of $1 \mu m$–$8 \mu m$ diameter. The magnetization is controlled by chemical manipulations and typical compositions employed for different bubble diameters are: $Sm_{0.4}Y_{2.6}Ga_{1.2}Fe_{3.8}O_{12}$ (8 μm), $Sm_{0.18}Lu_{0.19}Y_{1.73}Ca_{0.9}Ge_{0.9}$ $Fe_{4.1}O_{12}$ (6 μm) and $Sm_{0.51}Lu_{0.42}Y_{1.21}Ca_{0.86}Ge_{0.70}Si_{0.16}Fe_{4.14}O_{12}$ (2–3 μm). Liquid-phase epitaxy is employed to grow thin garnet films on the substrate (usually, nonmagnetic $Gd_3Ca_5O_{12}$); for this purpose, a dipping technique from a solution of the garnet oxides dissolved in molten PbO–B_2O_3 flux is employed.

Liquid magnets (*ferrofluids*) exhibit interesting properties of value to technology (Rosensweig, 1982; Popplewell, 1984). Since cooperative magnetism is a property of the solid state, a magnetic fluid cannot be produced by simply melting a magnetic solid. Ferrofluids are, however, obtained by dispersing fine magnetic particles in a liquid medium and preventing them from agglomerating. There are two forces which would tend to agglomerate magnetic particles in a suspension; they are the short-range van der Waals and the long-range magnetic forces. When the particles are small ($\sim 100\,\text{Å}$ diameter), magnetic forces turn out to be negligible. Attraction due to van der Waals forces is prevented by coating the magnetic particles with a monolayer of surfactant. An example of a ferrofluid is Fe_3O_4 dispersed in oil to which oleic acid is added as surfactant.

In the presence of a magnetic field, a ferrofluid experiences a body force – a force experienced throughout the given volume. Thus, a stream of ferrofluid can be deflected and the surface of a ferrofluid distorted by a magnetic field. Particles in ferrofluid are smaller than a single domain and hence there are no domain walls. But the magnetization is not uniform throughout the particle because of surface anisotropy. Ferrofluids exhibit novel properties, which include levitation of nonmagnetic and magnetic objects and characteristic instabilities such as the one at the interface between a ferrofluid and an ordinary (transparent) liquid in the presence of a magnetic field perpendicular to the interface. The instability at the interface takes the form of a labyrinthine pattern which is similar to the 'wiggly' domain patterns of magnetic bubble memory devices.

For ferrofluids, the usual form of the Bernoulli equation which relates the pressure of an incompressible liquid to its kinetic and potential energy must be modified to include the magnetic pressure induced by an applied magnetic field. The modified equation reads:

$$P + \tfrac{1}{2}\rho v^2 + \rho g h - \mu_o \int_o^H M dH = \text{constant} \tag{7.2}$$

where P is the pressure, v is the streamline velocity and $\rho g h$ is the gravitational potential energy. The last term describes the magnetic pressure on a ferrofluid due to an external magnetic field. Equation (7.2) describes the behaviour of magnetic fluids in external magnetic fields. For example, when the pressure and gravitational energy are constant, a ferrofluid experiences an increase in velocity as it flows through a magnetic field. The effect is to decrease the cross-section of a horizontal jet of ferrofluid. Similarly, levitation of magnetic and nonmagnetic objects in a ferrofluid can be understood in terms of the modified Bernoulli's equation.

Ferrofluids find several technical applications, which include rotary shaft seals, damping fluids, accelerometers, optical switches and shutters, and magnetic inks. The earliest application was in the construction of wear-free seals for rotary shafts. A similar seal is used in computer disc drives. Using ferrofluids, high-pressure differential seals which work in multiple stages have been constructed. When the seal is pressurized, each stage develops a momentary leak into the next chamber. The total pressure that can be contained is the sum of the pressure differences across all the stages. Pressure seals which work at ~ 5000 rpm under a differential pressure of ~ 470 kPa are common. In loudspeakers, a ferrofluid surrounding the voice coil damps unwanted sound resonances and dissipates heat, improving the quality of sound. A ferrofluid accelerometer uses a magnet levitated in a ferrofluid as detector. A jet of ferrofluid can be deflected in a magnetic field to form letter characters. New printing techniques based on ferrofluid inks are being developed; the technique is useful in invisible printing, where characters can be read with a magnetic reader. Ferrofluids also find medical applications. Cranial aneurysms which involve rupture of arteries in the brain may be treated without surgery using ferrofluid. A ferrofluid is held by a magnetic field so as to isolate the rupture from the blood-stream, allowing the rupture to heal naturally. One important property of ferrofluids that may find application in future is the temperature dependence of magnetization, especially near the Curie temperature, where the change is large.

7.5   Hydrogen storage materials

Hydrogen is one of the most attractive forms of nonconventional energy and can be produced by several means. One of the possible means of hydrogen production is by photoelectrolysis of water using solar energy (Section 7.3). Much of the interest in hydrogen as a fuel, the so-called hydrogen economy, stems from the idea that cheap and abundant electrical energy (from nuclear fusion or from solar photovoltaics) could be used to electrolyse water to produce hydrogen in large quantities. Use of hydrogen as a fuel presents storage problems. Hydrogen can be stored as a gas, liquid or solid. The technology of gaseous and liquid storage is well-known, but the low boiling point (20.4 K) and low density (0.071 kg litre^{-1}) make it difficult to store it as a liquid. Storage of hydrogen as a gas at high pressures cannot be improved beyond the 150-atmosphere storage tanks commonly used. Moreover, there is always the danger of hydrogen forming explosive mixtures with air (Hindenburg syndrome). Storage of hydrogen in the form of a solid compound (hydride) is an alternative possibility that is attractive for various reasons; it is safe, storage densities as high as those achieved in liquid form are possible, there is no loss on storage, and storage is reversible. Moreover, no energy expenditure is required for storage as hydride. In fact, energy released during the hydriding reaction can be exploited. Hydrogen storage in the solid form has been extensively reviewed in the literature (see, for example, Reilly & Sandrock, 1980; Cohen & Wernick, 1981; Buschow & van Mal, 1982; Angus, 1981; Ivey & Northwood, 1983).

The ideal requirements of a metallic hydrogen storage material are low atomic mass, rapid kinetics of hydrogen sorption and ability to store large quantities of hydrogen reversibly with an equilibrium partial pressure of about 1 atmosphere around room temperature. The hydrogen sorption behaviour is represented in the form of pressure–composition isotherms (Fig. 7.11). The initial absorption results in a sharp increase in hydrogen pressure. At a specific pressure, the hydriding reaction begins and the material absorbs large quantities of hydrogen at nearly constant pressure. In this region the metal hydride (β phase) and the metal saturated with hydrogen (α phase) coexist. When all the α phase is converted to β phase, further increase in pressure results in small change in the hydrogen/metal ratio. The pressure corresponding to the equilibrium

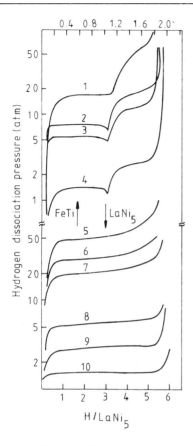

Figure 7.11   Pressure–composition isotherms of LaNi$_5$ hydride and FeTi hydride. 1, 343 K; 2, 313 K; 3, 303 K; 4, 273 K; 5, 413 K; 6, 393 K; 7, 373 K; 8, 333 K; 9, 313 K; 10, 293 K. (After Cohen & Wernick, 1981.)

$\alpha + H_2 \rightleftharpoons \beta$ is called the plateau pressure. When the pressure is lowered below the plateau pressure, the reaction is reversed and the material releases hydrogen. The sorption is not strictly reversible, being accompanied by hysteresis in several instances. The hydriding reaction is exothermic and the reverse reaction is endothermic.

Although many metals form hydrides, the thermodynamics and kinetics are unfavourable to meet the requirements of an ideal hydrogen storage material. Two types of behaviour are common with metals. Metals like Pd, V, Nb and Ta absorb hydrogen reversibly but the hydrogen-to-metal ratio is small; on the other hand, electropositive metals such as the alkaline earths, rare earths, Ti and Zr absorb large quantities of hydrogen in forming hydrides but the reaction is nearly irreversible, rendering these solids useless for hydrogen storage. Combination of these two classes of metals in the form of intermetallics has, however, proved very useful. Some of the important hydrogen storage materials are given in Table 7.2 together with the characteristics of the hydrogen storage reaction. Intermetallics useful in hydrogen storage are of the stoichiometry AB, A$_2$B, AB$_2$ and AB$_5$ where A is Ti, Ca, La, Ce, Sm,

Table 7.2. *Intermetallic compounds useful in hydrogen storage*

Compound	x^a	Plateau pressure (atm)	Heat of reaction (kJ mol^{-1} H$_2$)
LaNi$_5$	5–6	2.9(313 K)	− 31.0
SmCo$_5$	2.5	3.3(293 K)	− 32.7
CeNi$_3$	3	0.09(323 K)	–
FeTi	1–2	5(303 K)	− 28.0
ZrMn$_2$	3.6	0.01(293 K)	− 53.2
Zr$_2$Cu	1.2	0.20(1073 K)	− 120

$^a x$ denotes number of hydrogen atoms per formula unit of the intermetallic.

Mg or Zr, while B is Fe, Ni, Mn, Cu or Al. Among these, LaNi$_5$ and FeTi are the two candidates seriously being considered for hydrogen storage at present.

The design of suitable intermetallics is not straightforward. First not all combinations of two metals do form intermetallics in all cases; for instance no intermetallics are known between rare earths and Nb, Mo, Ta or W. Miedema (1973) has proposed a model to predict whether or not combination of two metals will form an intermetallic compound. The model, however, does not predict the composition or the crystal structure of the intermetallics. Knowledge of the phase diagram is essential in this regard. Intermetallic compounds are usually prepared by melting the parent materials in an oxide crucible (Al$_2$O$_3$, MgO or ThO$_2$) in an induction furnace. Levitation-melting or arc-melting is also used to avoid the possibility of contamination by reaction with the crucible material. If the vapour pressure of one of the constituents is high at elevated temperatures (Mg, Zn or Cd), sealed-tube techniques using tantalum or molybdenum tubes are resorted to.

LaNi$_5$ and FeTi, besides being able to store relatively large amounts of hydrogen, absorb and desorb hydrogen rapidly at convenient pressures and temperatures (Fig. 7.11). The heat of hydride formation is about 30 kJ mol^{-1} of hydrogen. This heat must be removed while storing and supplied while releasing hydrogen from the alloy. The thermodynamic stability of the intermetallic relative to its hydride determines the usefulness of the material for hydrogen storage. Miedema *et al.* (1977) have found that the enthalpy of hydride formation is opposite to the enthalpy of intermetallic compound formation in a series of isostructural AB$_n$ intermetallics. This leads to the rule of reversed stability, viz. the less stable the parent intermetallic, the more stable the hydride is and vice versa. When the hydride is stable, its equilibrium hydrogen (plateau) pressure is low. Thus of LaNi$_5$($\Delta H_f^\circ = -60$ kJ mol^{-1}) and LaCu$_5$($\Delta H_f^\circ = -101$ kJ mol^{-1}), the nickel compound is a better hydrogen storage material. A linear relationship exists between the logarithm of plateau pressure and the unit cell volume of LaNi$_5$-type phases (Fig. 7.12), which is of great help in selecting hydrides for application.

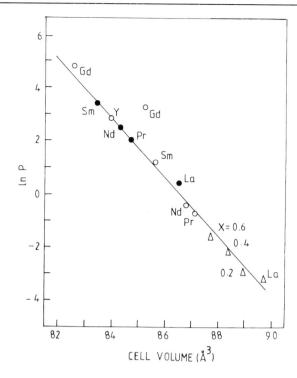

Figure 7.12   Dependence of the hydrogen equilibrium pressure on the unit-cell volume of various LnNi$_5$-type compounds (Ln = rare earth). Open circles, LnCo$_5$; closed circles, LnNi$_5$; open triangles, LaCo$_{5-5x}$Ni$_{5x}$. (Following Buschow & van Mal, 1982.)

Alloys of LaNi$_5$ type crystallize in the hexagonal CaCu$_5$ structure, which provides a number of interstitial sites (nine) for hydrogen absorption. The composition of the normal hydride of LaNi$_5$ is LaNi$_5$H$_6$, but under high pressures a hydride of composition LaNi$_5$H$_{8.35}$ has been obtained where all the interstitial sites are probably occupied. Hydrogen absorption–desorption in LaNi$_5$ is a topotactic reaction, occurring without drastic change of structure (the space group changes from $P6/mm$ to $P31m$ and the lattice expands by about 25% when the hydride LaNi$_5$H$_6$ is formed). FeTi (CsCl structure) forms two hydrides, FeTiH$_1$ and FeTiH$_2$; the structure of the hydrides is different from the parent intermetallic structure. An important feature of LaNi$_5$ and its analogues is that they can be substituted with other 3d metals or group IIIA and IVA metals at the nickel site, or with rare earths or calcium at the lanthanum site. Substitution drastically changes the equilibrium pressure of the hydride. LaNi$_4$Al, for example, has a plateau pressure of 0.002 atm, while GdNi$_5$ has a pressure of 150 atm at room temperature; LaNi$_5$ has a pressure of 2.9 atm. Thus, by suitable substitution, it is possible to tailor-make intermetallics with specific hydrogen storage characteristics (van Mal et al., 1974; Pourarian & Wallace, 1982).

Sorption of hydrogen by intermetallics requires prior 'activation'. LaNi$_5$ can be activated by simply exposing it to hydrogen at a pressure of a few atmospheres. During the activation, the sample cracks and splinters vigorously. After hydrogen is absorbed and desorbed through a few cycles, the material becomes a fine powder (10–100 μm). It is more difficult to activate FeTi and considerable over-pressures as well as high temperatures (400–600 K) are required for the initial absorption. The microscopic processes that occur during activation and deactivation were not known until recently. Mössbauer, magnetic and XPS measurements (Shenoy et al., 1980; Stucki & Schlapbach, 1980) have revealed that activation produces microcrystals of $3d$ metals at the surface which probably help dissociate H_2 molecules, enabling rapid absorption.

Hydrogen storage in the form of intermetallic hydrides has several technical applications (see Journal of Less Common Metals, Vols. 73 and 74) which include production of ultrapure hydrogen, separation of deuterium from hydrogen, use of hydrogen fuel from the hydride in internal combustion engines, and use of intermetallics as electrodes in protonic batteries or in combination with fuel cells (Cohen & Wernick, 1981). Among the various applications, use of FeTi hydride as a source of fuel for powering motor vehicles, as developed by the Daimler–Benz Company, the HYCSOS (hydrogen conversion and storage system) chemical heat pump developed by the Argonne National Laboratory and the use of metallic hydrides in load-levelling in power stations are noteworthy. The use of intermetallic hydrides in cars, however, has to compete with sodium–sulphur batteries for similar use. The HYCSOS chemical heat pump makes use of two different storage materials such as $LaNi_5$ and $CaNi_5$ that require different temperatures to produce the same plateau pressure (Sheft et al., 1980). Low grade thermal energy is used to decompose the metal hydride (A) with a higher free energy of dissociation. The hydrogen released is absorbed at an intermediate temperature and stored as the second hydride (B) which has a lower free energy of dissociation. In the heat-pump mode of operation, outdoor heat is used to decompose the second hydride and to reabsorb the hydrogen at the same intermediate temperature as the first hydride. The heat of absorption of the first hydride can now be used for space heating. By rejecting the intermediate temperature heat of absorption to the outdoors, the heat-pump cycle can be used for space cooling. The approach can be used to build a hydrogen absorption refrigerator whose sole energy source is heat at about 370 K, so that it is ideally suited for solar-powered air conditioning. The use of hydrides for load-levelling in power stations is based on the idea that hydrogen produced electrolytically during off-peak hours can be stored as hydride (FeTi hydride) and the hydrogen can be subsequently converted to electricity using fuel cells during peak hours (Reilly, 1978).

There are several problems in the use of intermetallics for hydrogen storage. First of all, even small quantities of gases like CO and H_2O can reduce the hydrogen storage capacity of $LaNi_5$ and FeTi. Since the intermetallic hydrides are metastable with respect to other stable phases (for example, $LaNi_5H_x$ is metastable with respect to LaH_2 and Ni) there is an intrinsic degradation of the material on repeated cycling. In addition, heat liberation and absorption, volume changes associated with hydrogen absorption, and the fine particle nature of the storage materials all pose problems in the efficient use of intermetallics.

Intermetallics, which store hydrogen reversibly, can function as the negative electrodes in a nickel–metal hydride battery, the positive electrode being $Ni(OH)_2$. The half-cell reactions on charge and discharge are:

$$M + H_2O + e^- \underset{\text{discharge}}{\overset{\text{charge}}{\rightleftharpoons}} MH + OH^-$$

$$Ni(OH)_2 + OH^- \underset{\text{discharge}}{\overset{\text{charge}}{\rightleftharpoons}} NiOOH + H_2O + e^-$$

Batteries based on metal hydrides show the promise of becoming power sources for electric vehicles in the future (Ovshinsky *et al.*, 1993). Such a battery constructed of nontoxic recyclable materials can operate at room-temperature.

7.6   Amorphous materials

A variety of materials of vital technological importance are amorphous or quasicrystalline. Apart from window panes and vessels, glasses and other amorphous substances find a number of applications. A promising photovoltaic material for conversion of solar energy to electricity is amorphous silicon. Properties of magnetic, dielectric and other materials rendered amorphous promise several applications as well (Zallen, 1983; Elliott, 1984). The ease of fabrication of glassy materials makes them attractive for use in a variety of situations. We do indeed have glasses which are transparent in any desired region of the electromagnetic spectrum ($300 \, nm$–$300 \, cm^{-1}$), glasses which are semiconducting, glasses which behave like ferroelectrics and glasses with controlled magnetic behaviour. Ferroelectric glasses and glasses with controlled anisotropy of properties are areas worthy of exploration. Some superconducting glasses (e.g. $Pb_{0.9}Cu_{0.1}$ and $La_{0.8}Au_{0.2}$) have been discovered. Many of the metallic glasses are ferromagnetic. The observation of ferromagnetism and superconductivity in glassy systems shows that long-range order is not a necessary condition for these important properties. One of the exciting applications of glasses is in replacing the copper telephone cables by optical fibre light-guides. In Table 7.3, we list some of the important applications of amorphous substances in advanced technology. In addition, fast ion-conducting glasses may become valuable for use in batteries and fuel cells.

The optical properties of amorphous solids are interesting. These solids are optically isotropic. Furthermore, the sharp features present in crystal spectra are absent in the spectra of amorphous solids even at low temperatures. The overall features in the electronic spectra of amorphous solids (broad band maxima) are, however, not unlike those of crystals, reflecting the importance of short-range order in determining these characteristics. The optical absorption edges of amorphous materials are not sharp and there is an exponential tail in the absorption coefficient (Fig. 7.13) associated with the intrinsic disorder.

The difference between amorphous and crystalline solids shows up more clearly in phonon excitations than in electronic excitations. Contributions from the entire phonon density of states appear in the first-order Raman and infrared spectra of amorphous solids. All modes in elemental amorphous semiconductors are active in the infrared.

Table 7.3. *Technological applications of amorphous materials*

Nature of amorphous solid	Example	Application
Semiconductors	(a) Silicon (from silane) and $Si_{0.9}H_{0.1}$	Photovoltaic solar cells
	(b) $Te_{0.8}Ge_{0.2}$	Computer memory elements based on amorphous–crystalline transition (induced by electric field)
Metallic glasses	(a) High Fe or Co containing glasses (e.g. $Fe_{0.7}P_{0.2}Co_{0.1}$)	Low-loss ferromagnetic ribbons for use in transformer cores
	(b) TiCu	Hydrogen storage
Chalcogenide glasses	As_2Se_3	Xerography based on photoconducting property
Silica-based glasses	SiO_2 with 10% GeO_2	Optical fibres for communication
Fast ion-conducting glasses	NASIGLAS, $Na_{1+x}Zr_{2-x/3}Si_xP_{3-x}O_{12-2x/3}$	Solid electrolytes in batteries and fuel cells.

In Fig. 7.14, we show the imaginary part of the dielectric constant of crystalline and amorphous Si, obtained by Kramers–Kronig analysis of the ultraviolet reflectivity spectra. Both show the absorption band due to bonding orbital → antibonding orbital transition, but the absorption spectrum of amorphous silicon is smooth while the crystal shows structure. This difference in the absorption bands has implications for their use as *solar cell* materials. Amorphous Si is a better candidate for large-area solar cells for power generation (crystalline Si can be used only when cost is of no consideration, e.g. as space probes), since its absorption coefficient is much larger in the 1–3 eV region. The optical band gaps of crystalline and amorphous silicon are about the same ($\sim$ 1 eV), but there are no restrictive selection rules to inhibit light absorption by the amorphous variety starting just above 1 eV.

Another important optical application of amorphous solids is in the use of SiO_2-based glasses in *optical fibre* communication (MacChesney, 1981; Goodman, 1983). This depends on the high transparency in the infrared window below the band gap. The propagation of light down the fibre is by a process of total internal reflection. For long-distance applications, the fibre must have a very low light loss. Two possible sources of loss are scattering (from density fluctuations following Rayleigh's law) and absorption (due to electronic and vibration excitations). The wavelength dependence of overall loss looks like a V-shaped curve (Fig. 7.15). Rayleigh scattering is minimized by using long-wavelength radiation and the low-loss window is then determined by the

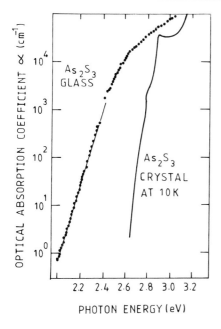

Figure 7.13   Comparison of the optical absorption edges of amorphous and crystalline As_2S_3. (After Zallen *et al.*, 1971.)

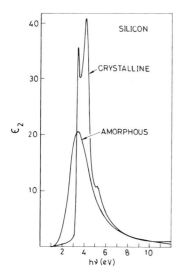

Figure 7.14   Imaginary part of the dielectric constant ε_2 as a function of photon energy, hv, for amorphous and crystalline silicon. (After Pierce & Spicer, 1972.) Since ε_2 is proportional to the optical absorption coefficient α, ε_2 is a dimensionless measure of optical absorption.

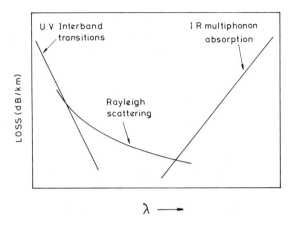

Figure 7.15   Schematic loss spectrum of an optical fibre.

intersection of the Rayleigh curve with the multiphonon absorption edge (1.6 μm in SiO$_2$ with a theoretical loss of 0.2 dB km^{-1}). Actual loss in SiO$_2$ fibres is much greater ($\sim$ 10 dB km^{-1}) owing to absorption by OH groups.

Losses can be reduced by making fibres of materials for which the multiphonon edge lies at longer wavelength than SiO$_2$. This can be done by using materials with heavier atoms or weaker force constant. Halide glasses (based on ZrF$_4$ along with BaF$_2$, LaF$_3$ etc.) and Ge- or Sb-substituted SiO$_2$ have been tried; GeO$_2$–Sb$_2$O$_3$ glasses have losses of 5 dB km^{-1}. The difficulty lies in making very pure fibres which do not crystallize.

Chalcogenide glasses such as WSe$_2$ show unusual structural changes when irradiated by light (Chopra et al., 1981; Elliott, 1983; Owen et al., 1985). The nature and cause of the photostructural changes are not entirely clear, but the photoinduced changes are accompanied by changes in optical and other properties. The photostructural changes can be removed by annealing at T_g. These effects have several technological implications, including in optical storage and lithography.

For using lithium batteries (which generally have high energy densities) under extreme conditions, more durable and better conducting electrolytes are necessary. Salt-in-polymer electrolytes discovered by Angell et al. (1993) seem to provide the answer. Polypropylene oxide or polyethylene oxide is dissolved in low melting point mixtures of lithium salts to obtain rubbery materials which are excellent lithium-ion conductors at ambient temperatures.

Incorporation of nitrogen in oxide glasses is known to improve properties such as hardness and refractive index. Grande et al. (1994) have prepared pure nitride glasses in the Li$_3$N–Ca$_3$N$_2$–P$_3$N$_5$ system by employing high pressures and found them to have very high refractive indices (1.97–2.0), hardness and glass transition temperatures (> 970 K).

7.7   High-temperature oxide superconductors

The highest superconducting transition temperature known until 1986 was 23 K and it seemed as though this barrier would not be broken in spite of much effort lasting several decades since the discovery of superconductivity. The materials investigated during this period include metal oxides as well. Thus, the $BaPb_{1-x}Bi_xO_3$ (Sleight et al., 1975) and $Li_{1+x}Ti_{2-x}O_4$ (Johnston et al., 1973) systems showed superconductintg transitions around 13 K. The discovery of 30 K superconductivity in an oxide of the La–Ba–Cu–O system by Bednorz & Müller (1986), however, changed the picture. This oxide had the quasi two-dimensional K_2NiF_4 structure. A variety of superconducting oxides, especially cuprates, have since been synthesized and characterized (Sleight, 1988; Cava, 1990; Armstrong & Edwards, 1991; Rao, 1991 a, b; Rao & Ganguli, 1995; Raveau et al., 1991), the highest transition temperature as of today being 155 K in $HgBa_2Ca_2Cu_3O_{8+\delta}$ (Nunez-Regueiro et al., 1993), which is not far from the lowest temperature recorded on earth (183 K). Studies of the various families of cuprates have shown some interesting commonalities and unifying features. Superconducting properties of the cuprates have been related to certain structural and electronic parameters. Although there is no simple relation between the superconducting transition temperature and any specific structural feature of the cuprates, the various correlations help to understand these materials better and to design newer ones. We shall examine some of the structural features as well as the significant structure–property relations in the superconducting oxides in this section.

7.7.1   Structural features of cuprate superconductors

Many cuprate families have been discovered in the last eight years. The major families of cuprates are: (a) $La_{2-x}A_xCuO_4$ (A = alkaline earth) possessing the tetragonal $K_2NiF_4(T)$ structure (Fig. 7.16); (b) $LnBa_2Cu_3O_{7-\delta}$ (Ln = Y or rare earth other than Ce, Pr and Tb) referred to as the 123 type (Fig. 7.17) and the related $LnBa_2Cu_4O_8$ (124) and $Ln_2Ba_4Cu_7O_{15}$ (247) cuprates containing perovskite layers with CuO_2 sheets as well as Cu-O chains; (c) $Bi_2(Ca, Sr)_{n+1}Cu_nO_{2n+4}$ containing two BiO layers and perovskite layers with CuO_2 sheets (Fig. 7.17); (d) $Tl_2A_{n+1}Cu_nO_{2n+4}$ and $TlA_{n+1}Cu_nO_{2n+3}$ (A = Ca, Ba, Sr, etc) containing Tl-O layers and perovskite layers with CuO_2 sheets (Fig. 7.17); (e) Lead-based superconducting cuprates such as $Pb_2Sr_2LnCu_3O_8$ containing CuO_2 sheets and Cu(I)-O sticks; (f) Tl, Bi and Pb cuprates containing fluorite layers and CuO_2 sheets; (g) Mercury cuprates of the type $HgCa_{n-1}Ba_2Cu_nO_{2n+2+\delta}$ and (h) Infinite-layer cuprates such as $ACuO_2$ (A = Ca, Sr, Ba) and $Ln_{1-x}A_xCuO_2$ (Ln = Nd, Pr; A = Sr, Ba) depicted in Fig. 7.18. Recently several oxyanion cuprate derivatives have been prepared and characterized and some of them are superconducting; typical of the anions are CO_3^{2-}, PO_4^{3-}, BO_4^{5-}, BO_3^{3-} and NO_3^{1-} (Greaves, 1994; Rao et al., 1993). All these cuprates have holes as charge carriers (excess positive charge on Cu as in Cu^{3+} or on oxygen as in O^{1-}). The only well established superconducting cuprates where electrons are charge carriers are $Nd_{2-x}M_xCuO_4$ (M = Ce, Th) and related compounds with the tetragonal T' structure possessing square-planar CuO_4 units instead of the octahedra in the T structure (Fig. 7.16). Matching between the Ln-O bond length and $\sqrt{2}$ of the Cu-O bond length determines the relative stabilities of T and T' structures and also of the orthorhombic structure of these oxides related to K_2NiF_4 structure. Holes can be doped when the

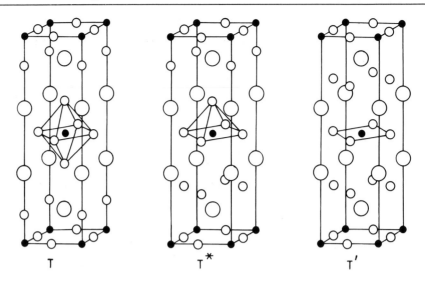

Figure 7.16   Structures of T, T^* and T' types of tetragonal structures of cuprates of the type Ln_2CuO_4 (Ln = rare earth). The T^* structure with CuO_5 square-pyramids is obtained by combining a half of the T unit cell with a half of the T' cell. The three types of Cu-O polyhedra in the three structures typify those found in the superconducting cuprates.

Cu-O bonds are under compression as in the T structure, while electrons can be doped when the Cu-O bonds are under tension as in the T' structure.

The cuprates can be described on the basis of certain structural features common to many of them. For example, the structures of $La_{2-x}A_xCuO_4$, $Bi_2(Ca, Sr)_{n+1}Cu_nO_{2n+4}$ and the thallium cuprates can be considered to be intergrowths of oxygen-deficient perovskite layers, $ACuO_{3-x}$, with AO type rock-salt layers (Fig. 7.19). The cuprates contain different types of Cu-O polyhedra with the hole supercon-ductors necessarily having CuO_5 or CuO_6 units and the electron-superconducting $Nd_{2-x}M_xCuO_4$ containing only CuO_4 square-planar units (Figures 7.16 and 7.17). Thus, the essential feature of the cuprates can be considered to be the presence of CuO_2 sheets with or without apical oxygens. The mobile charge carriers in the cuprates are in the CuO_2 sheets. All the cuprates have charge reservoirs as exempli-fied by the Cu-O chains in the 123 and 124 cuprates and the TlO, BiO and HgO layers in the other cuprates.

That the CuO_2 sheets are the seat of high-temperature superconductivity is demonstrated by the fact that intercalation of iodine between BiO layers in the bismuth cuprates does not affect the superconducting transition temperature while introduction of fluorite layers between the CuO_2 sheets adversely affects superconductivity. In the different series of cuprates with varying number of CuO_2 sheets studied hitherto, the T_c reaches a maximum when $n = 3$ except in single thallium layer cuprates where the maximum is at $n = 4$. The infinite layered cuprates, where the CuO_2 sheets are separated by alkaline-earth and other cations, show T_cs in the 40–110 K range (Azuma

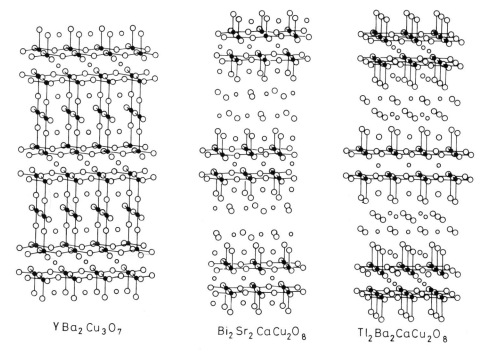

$YBa_2Cu_3O_7$ $Bi_2Sr_2CaCu_2O_8$ $Tl_2Ba_2CaCu_2O_8$

Figure 7.17   Structures of $YBa_2Cu_3O_7$, $Bi_2CaSr_2Cu_2O_8$ and $Tl_2CaBa_2Cu_2O_8$. Notice the presence of the CuO_2 sheets (containing CuO_5 square-pyramids) and of the Cu-O chains in $YBa_2Cu_3O_7$. In $YBa_2Cu_4O_8$ there are two Cu-O chains in the unit cell compared to one in $YBa_2Cu_3O_7$.

et al., 1992). Superconductivity in these materials appears to be due to the presence of Sr-O defect layers corresponding to the insertion of $Sr_3O_{2\pm x}$ blocks (Zhang *et al.*, 1994).

Based on the interplanar Cu-Cu distances, one can classify cuprates into two categories (Nobumasa *et al.*, 1990). In one category, r(Cu-Cu) lies between 3.0 and 3.6 Å with T_cs varying between 50 and 133 K and in another it is between ~ 6 and 12.5 Å, encompassing superconductors with lower T_c (< 50 K) (Fig. 7.20), except $Tl_2Ba_2CuO_6$ and $HgBa_2CuO_{4.065}$ with T_cs of ~ 90 K. In the first category with r(Cu-Cu) < 3.6 Å, the T_c increases as the Cu-Cu distance decreases. In the 2222-type fluorite-based superconductors, there are three copper oxygen sheets, $[CuO_2\text{-}CuO_\delta\text{-}CuO_2\text{-Fluorite-}CuO_2\text{-}CuO_\delta\text{-}CuO_2]$, each block of three sheets separated by a Ln_2O_2 fluorite layer. The r(Cu-Cu) relevant to this compound would be the distance between the CuO_2 sheets across the fluorite layer (~ 6.2 Å) and not the distance between two neighbouring sheets. Accordingly, these cuprates exhibit low T_cs (45–50 K). Thus, the distance between the CuO_2 sheets carrying charge carriers is a factor in determining the value of T_c, indicating the presence of some interaction among the closely spaced CuO_2 sheets although the cuprates have quasi two-dimensional character.

The nature of the charge reservoirs in the cuprates determines the carrier concentration and the ease of charge-transfer to the CuO_2 sheets. Covalent charge reservoirs can

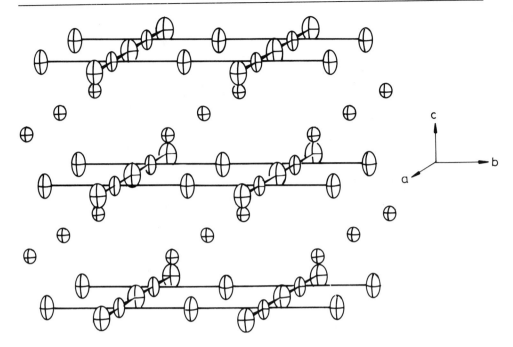

Figure 7.18   Structure of $Ca_{0.86}Sr_{0.14}CuO_2$ having planar CuO_2 sheets separated by Ca, Sr (From Siegrist *et al.*, 1988).

redistribute charge effectively through the apical oxygen of the CuO_5 square pyramids giving rise to high T_cs. Ionic charge reservoirs, on the other hand, would be less flexible with regard to the charge states and do not favour high T_cs. Structural mismatch as well as disorder in the reservoirs also adversely affect the superconducting properties. The covalency of the Hg-O bond could be related to the high T_c of the Hg cuprates.

Oxygen stoichiometry and ordering play a crucial role in determining the structure and properties of cuprates. The dependence of the structure and properties of $YBa_2Cu_3O_{7-\delta}$ on oxygen content has been studied in detail. Thus, $YBa_2Cu_3O_{7-\delta}$ which is orthorhombic ($c \simeq 3b$) with a T_c of 90 K when $0.0 \lesssim \delta \lesssim 0.25$, assumes another orthorhombic structure ($c \neq 3b$) when $0.3 \lesssim \delta \lesssim 0.4$ with a T_c of 50–60 K (Fig. 7.21 and 7.22). When $\delta = 1.0$, all the oxygens in the CuO chains are depleted and the structure becomes tetragonal and the material is nonsuperconducting. When $\delta = 0.5$ ($T_c \sim 45$ K), there is an ordered arrangement of oxygen vacancies with the presence of fully oxidized (O_7) and fully reduced (O_6) chains alternately. The variation of T_c with δ (Fig. 7.23) shows a plateau region around 60 K ($\delta \approx 0.3$–0.4); the valence of Cu(2) in the CuO_2 sheets also varies similarly. The compositions showing 60 K superconductivity are metastable and transform to a 124-type phase on heating at low temperatures (Nagarajan & Rao, 1993).

Oxygen-excess La_2CuO_4 is biphasic consisting of the stoichiometric antiferromagnetic phase and an oxygen-excess superconducting phase (Jorgensen, 1991). In the bismuth cuprates, excess oxygen in the BiO layers (one oxygen for every four or five unit

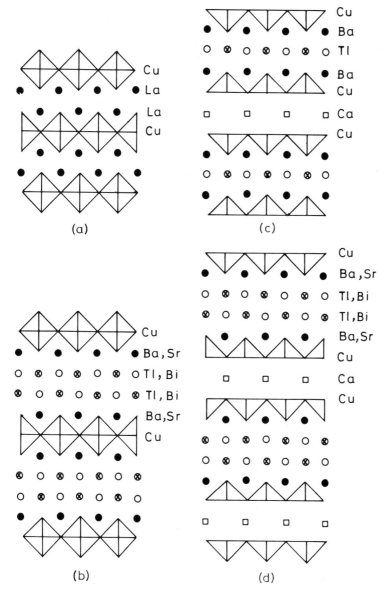

Figure 7.19   Schematic representation of the structures of (a) La_2CuO_4, (b) $Bi_2Sr_2CuO_6$ and $Tl_2Ba_2CuO_6$, (c) $TlCaBa_2Cu_2O_7$ and (d) $Bi_2CaSr_2Cu_2O_8$ and $Tl_2CaBa_2Cu_2O_8$, showing intergrowth of rock-salt and perovskite layers. Oxygens are shown as open circles and Bi and Tl by crosses.

cells) gives rise to incommensurate modulation (LePage *et al.*, 1989). Modulation-free superconducting bismuth cuprates have been made (Manivannan *et al.*, 1991) by replacing one Bi^{3+} by Pb^{2+}. In $HgBa_2CuO_{4+\delta}$, oxygen-excess in the Hg plane is necessary to render it superconducting.

Many of the cuprate families have an antiferromagnetic insulator member at one end

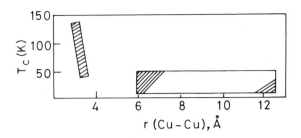

Figure 7.20   Variation of T_c with the interplanar Cu-Cu distance. Experimental points are in the shaded regions.

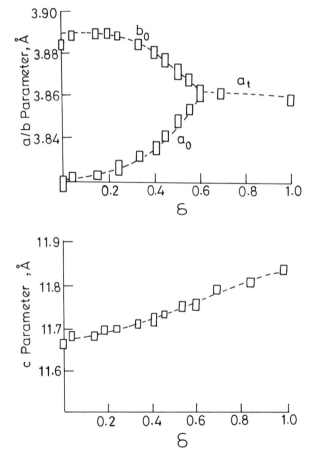

Figure 7.21   Variation of unit cell parameters of $YBa_2Cu_3O_{7-\delta}$ with δ showing the change from orthorhombic to tetragonal structure.

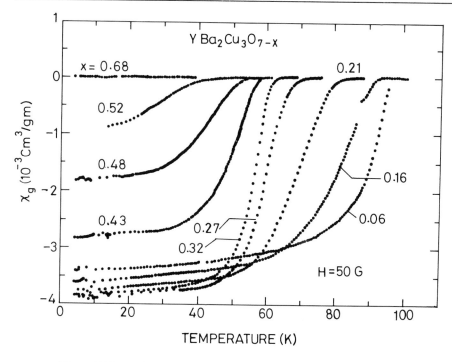

Figure 7.22   Meissner effect in $YBa_2Cu_3O_{7-\delta}$ (From Johnston *et al.*, 1987).

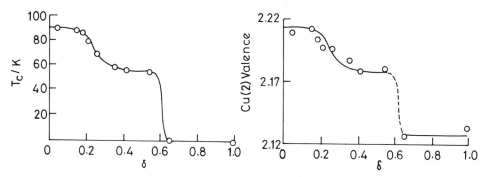

Figure 7.23   Variation of the T_c and the in-plane Cu(2) valence in $YBa_2Cu_3O_{7-\delta}$ with δ (From Cava *et al.*, 1990).

of the composition (e.g. La_2CuO_4 in $La_{2-x}A_xCuO_4$, $YBa_2Cu_3O_6$ in $YBa_2Cu_3O_{7-\delta}$ and $TlYSr_2Cu_2O_7$ in $TlY_{1-x}Ca_xSr_2Cu_2O_{7-\delta}$). What is more interesting is that all the cuprates are at a metal–insulator boundary. The resistivities data of cuprates do indeed fall between those of insulators and metals (Fig. 7.24). Some of them undergo a metal–nonmetal transition as a function of composition (e.g. $Bi_2Ca_{1-x}Y_xSr_2Cu_2O_8$ and $Tl\,Y_{1-x}Ca_xSr_2Cu_2O_7$ with change in x).

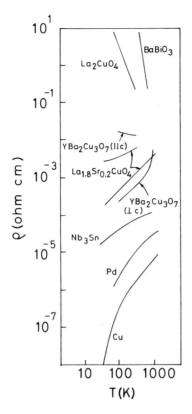

Figure 7.24   Resistivity data of cuprates in comparison with those of traditional metals and insulators.

7.7.2   *Relation between T_c and the hole concentration in cuprates*

As mentioned earlier, a majority of the cuprates have holes as charge carriers. These holes are created by the extra positive charge on copper (e.g. Cu^{3+}) or on oxygen (e.g. O^{1-}). The excess positive charge can be represented in terms of the formal valence of copper, which in the absence of holes will be $+2$ in the CuO_2 sheets. In hole superconducting cuprates, it is generally around $+2.2$. In electron superconductors, it would be less than $+2$ as expected. The actual concentration of holes, n_h, in the CuO_2 sheets in $La_{2-x}A_xCuO_4$, $YBa_2Cu_3O_7$ and Bi cuprates is readily determined by redox titrations. In the 123 cuprates, the concentration of mobile holes in the CuO_2 sheets can be delineated from that in the Cu-O chains. Determination of n_h in the thallium cuprates poses some problems, but in single Tl-O layer cuprates, chemical methods have been developed to obtain reasonable estimates (Hermann & Yakhmi, 1994). Generally, T_c in a given family of cuprates reaches a maximum value at an optimal value of n_h as shown in Fig. 7.25, with a maximum around $n_h \sim 0.2$ in most cuprates (Rao *et al.*, 1991, Rao & Ganguli, 1995). Single-layer thallium cuprates also show this behaviour. In $Tl_{1-y}Pb_yY_{1-x}Ca_xSr_2Cu_2O_7$ where the substitution of Tl^{3+} by Pb^{4+} has an effect opposite to that due to the substitution of Y^{3+} by Ca^{2+}, the T_c reaches a maximum at an

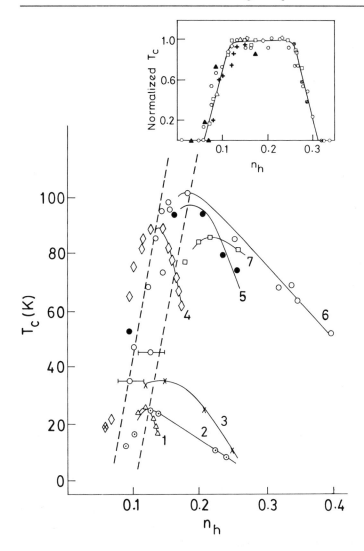

Figure 7.25   Variation of T_c with hole concentration n_h in superconducting cuprate families (from Rao & Ganguli, 1995). 1, 2, 6 and 7, Bi cuprates; 5, 123 cuprates; 3, $La_{2-x}Sr_xCuO_4$, 4, Tl cuprates. The variation of normalized T_c with n_h is shown in the inset (From Zhang & Sato, 1993).

optimal value of $(x-y)$, which is a measure of the hole concentration. By suitably manipulating x and y, the T_c of this system can be increased up to 110 K. Another way of representing the variation of T_c with n_h (Zhang & Sato, 1993) is to plot the reduced T_c (T_c observed/maximum value of T_c) against n_h as shown in the inset of Fig. 7.25. The ends of the plateau region in this curve correspond to insulating (possibly antiferromagnetic) and metallic regimes of the materials. Notice that the points in the underdoped region (low n_h values) in Fig. 7.25 fall close to a straight line. Deviations occur in the overdoped region possibly because of the smaller effective mass of the charge carriers.

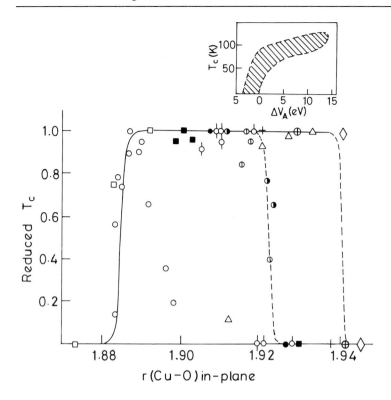

Figure 7.26   Plot of reduced T_c with in-plane $r(Cu\text{-}O)$ distance (From Rao & Ganguli, 1995). Inset shows the variation of T_c with ΔV_A (after Ohta *et al.*, 1990).

An unusual feature of the cuprate superconductors is the anomalous suppression of superconductivity in $La_{2-x}Ba_xCuO_4$ and related phases when the hole concentration x is near 1/8. A possible explanation is a dynamical modulation of spin and charge giving antiferromagnetic 'stripes' of copper spins periodically separated from the domains of holes. Neutron-diffraction evidence has been presented in the case of $La_{1.6-x}Nd_{0.4}Sr_xCuO_4$ ($x = 0.12$) which is a static analogue of the dynamical stripe model (Tranquada *et al.*, 1995). It appears that spatial modulations of spin and charge density are related to the superconductivity in these oxides.

The Cu-O bonds in the CuO_2 sheets involve an antibonding π interaction, and doping with holes reduces the bond distance. The in-plane Cu-O bond distance, $r(Cu\text{-}O)$, therefore reflects the hole concentration and a variation of T_c with $r(Cu\text{-}O)$ represents an alternative way of examining the T_c–n_h relation. In cuprates where n_h cannot be readily determined, the T_c vs. $r(Cu\text{-}O)$ plots show maxima at an optimal distance. The value of $r(Cu\text{-}O)$ is around half that of the a-parameter in most cuprates. Whangbo *et al.* (1989) find three distinct T_c–$r(Cu\text{-}O)$ relations depending on the cation located above and below the CuO_2 sheets, with each exhibiting a T_c maximum at an optimal value of the distance. If we plot the reduced T_c against $r(Cu\text{-}O)$, we get the curves shown in Fig. 7.26 where the highest T_c values occur in the 1.89–1.94 Å range

(Rao & Ganguli, 1995). When r(Cu-O) < 1.88 Å, the material would be metallic; those with r(Cu-O) > 1.94 Å are certainly insulating, but there are different insulating boundaries for the different cation families somewhat similar to those found by Whangbo *et al.* However, if we consider only the maximum T_c value in each cuprate family and the corresponding r(Cu-O), we get a curve which peaks around $r \approx 1.92$ Å (Rao & Ganguli, 1995).

7.7.3   Importance of the apical Cu-O distance, Madelung potentials and bond valence sums

Cuprates which are hole superconductors possess apical oxygens which act as the link between the charge reservoirs and the CuO_2 sheets. (Note that electron superconductors such as $Nd_{2-x}Ce_xCuO_4$ with the T' structure contain only CuO_4 units without apical oxygens). In $YBa_2Cu_3O_{7-\delta}$, the T_c–r(Cu2-O1) relation mirrors the T_c–δ relation (Cava *et al.*, 1990). In $YBa_2Cu_4O_8$, T_c increases with pressure from 80 K to 90 K, as the apical Cu-O distance decreases (Nelmses *et al.*, 1992). In a series of cuprates with a varying number of CuO_2 sheets, (e.g., $TlCa_{n-1}Ba_2Cu_nO_{2n+3}$), the apical (Cu-O) distance decreases with the increase in n, while T_c increases linearly with the decrease in the apical distance. The slope of the T_c vs. the apical Cu-O distance plot is nearly the same for $TlCa_{n-1}Ba_2Cu_nO_{2n+3}$, $Tl_{0.5}Pb_{0.5}Ca_{n-1}Sr_2Cu_nO_{2n+3}$ and $HgCa_{n-1}Ba_2Cu_nO_{2n+3}$, all of them having a single rock-salt layer (TlO, $Tl_{0.5}Pb_{0.5}$O or HgO_δ). The $Tl_2Ca_{n-1}Ba_2Cu_nO_{2n+4}$ family also shows increasing T_c with the decreasing apical Cu-O distance, but with a different slope. The mercury-based superconductors have larger apical distances compared to the other cuprates and also show a large pressure dependence of T_c.

The importance of the Madelung site potential in relation to the hole conductivity of cuprate superconductors was first pointed out by Torrance & Metzger (1989). Two classes of cuprates have been delineated depending on the difference in the Madelung site potential for a hole on a Cu site and that on an oxygen site, ΔV_M. Those with high ΔV_M (≥ 47 ev) are metallic and superconducting, while those with lower ΔV_M are semiconducting with localized holes. It has been found instructive to correlate T_c simultaneously with ΔV_M and the in-plane Cu-O bond length in the CuO_2 planes (Muroi, 1994).

The difference in the Madelung site potential for a hole in the apex and the in-plane oxygen positions, ΔV_A, provides a measure of the position of the energy level of the p_z-orbital on the apical oxygens (Ohta *et al.*, 1990). When the maximum values of T_c of hole-doped superconductors are plotted against ΔV_A, one finds that nearly all the cuprates are located on a curve (with some width), the cuprates with large ΔV_A exhibiting high T_cs (see the inset of Fig. 7.26). It appears that the energy level of the apical oxygen plays a significant role in the electronic states of the doped holes because of the stability of the local singlet state made up of two holes in the Cu $3d_{x^2-y^2}$ and O $2p$ orbitals in the CuO_2 sheet. The local singlet is well defined and stable when the energy level of the apical oxygen atom is sufficiently high.

The bond–valence sum is a measure of the total charge on an atom in a structure. Its value changes with oxygen doping, cation substitution or applied pressure indicating

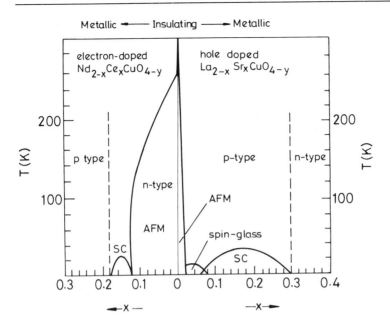

Figure 7.27   Symmetry in the phase diagrams of electron and hole superconductors $Nd_{2-x}Ce_xCuO_{4-y}$ and $La_{2-x}Sr_xCuO_{4-y}$ (after Maple, 1990).

the occurrence of charge-transfer within the structure. One can define $V_- = 2 + V_{Cu2} - V_{02} - V_{03}$ where V_- is the total excess charge in the planes (O2 and O3 are the oxygens in the plane) and $V_+ = 6 - V_{Cu2} - V_{02} - V_{03}$. The T_c vs. V_- plot for $YBa_2Cu_3O_{7-\delta}$ is similar to the T_c vs. δ plot; the plot of T_c against V_+ is linear (Tallon & Williams, 1990).

7.7.4   Other high T_c oxides

We have hitherto discussed cuprate superconductors with holes as charge carriers. Cuprates of T' structure (Fig. 7.16) such as $Nd_{2-x}Ce_xCuO_4$ and $Nd_2CuO_{4-x}F_x$ are electron superconductors. There is however some symmetry in the phase diagrams of the electron and hole superconductors (Maple, 1990) as can be seen from Fig. 7.27.

$Ca_2CuO_2Cl_2$, an analogue of La_2CuO_4 where the apical oxygens of CuO_6 octahedra are replaced by chlorine atoms, has been hole-doped by substitution of Na for Ca. Superconductivity with T_cs upto 26 K is found in $(Ca, Na)_2CuO_2Cl_2$ samples (Hiroi et al., 1994; Hiroi, 1995). $Sr_2CuO_2F_{2+x}$, another La_2CuO_4 analogue, also shows superconductivity with a maximum T_c of 46 K (Al-Mamouri et al., 1994). These results suggest that hole superconductivity in cuprates can be realized when the apical oxygen on copper is replaced by halogens.

Spin-$\frac{1}{2}$ ladder compounds such as $(VO)_2P_2O_7$ and $Sr_4Cu_2O_6$ containing two legs of metal–oxygen–metal chains, when appropriately doped, are supposed to form singlet pairs leading to spin-gapped metallic phases with superconductivity or CDW behaviour at low temperatures (Dagotto et al., 1992; White et al., 1994). $La_2Cu_2O_5$ is a double-legged spin-ladder compound which has been synthesized at high pressures

(Hiroi & Takano, 1995). Hole-doped derivatives of this cuprate of the type $La_{2-x}Sr_xCu_2O_5$ do not, however, show superconductivity, but undergo an insulator–metal transition.

Most of the infinitely layered cuprate superconductors are prepared under high pressures. An interesting series of cuprates of the general formula $(Cu,C)_m(m+1)(n-1)(n)$ represented by $(Cu_{0.5}C_{0.5})_mBa_{m+1}Ca_{n-1}Cu_nO_{2(m+n)+1}$ have been prepared at high temperatures and pressures. The $m=1$ and $m=2$ members show T_cs above 100 K (Takayama-Muromachi, 1995). The carbonate derivatives of cuprates are specially interesting amongst the various oxyanion derivatives. Thus, $Sr_2CuO_2(CO_3)$ is antiferromagnetic (T_N, 280 K), but some of the cuprates of a similar nature are superconducting (Nakata & Akimitsu, 1995), typical examples being $Ba_2Ca_{n-1}Cu_n(CO)_3O_y$ with $n=3$ ($T_c = 120$ K), $(Ba_{1-x}Sr_x)_2Cu_{1+y}O_{2+2y+z}(CO_3)_{1-y}$ ($T_c \sim 40$ K) and $Sr_2CuO_2(CO_3)_{1-x}(BO_3)_x$ ($T_c \sim 35$ K).

Other than cuprates, many other oxides show superconductivity. Generally, the T_c is well below 20 K, except in $Ba_{1-x}K_xBiO_3$ which has a T_c of 30 K (Cava et al., 1988). $La_{2-x}Sr_xNiO_4$ was once suspected to be superconducting, but subsequent studies have proved it otherwise (Sreedhar & Rao, 1990).

7.7.5   Mechanism of superconductivity in cuprates

In Fig. 7.28, we show the single-site energy levels of the $[CuO_2]$ unit cell of cuprates. The one d hole of Cu^{2+} (or the d^9 level) lies lowest at energy ε_d^h with respect to the no-hole or the d^{10} configuration. To add another d hole (Cu^{3+} or d^8) costs extra energy ($\varepsilon_d^h + U$), $U(\simeq 8\text{ eV})$ being the single site Mott–Hubbard repulsion. The large value of U would invalidate non-interacting electron band theory, leading to the insulating

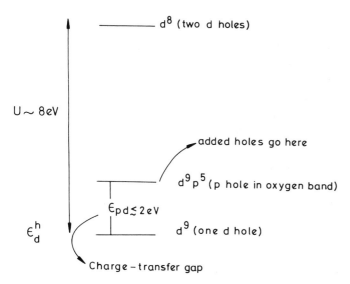

Figure 7.28   Relative energy levels of Cu and oxide ions in the CuO_2 sheets of cuprates.

behaviour of the stoichiometric one d hole per site material (e.g., La_2CuO_4). A feature that is distinctive in cuprates (as well as in many other transition-metal oxides) compared to one-band Mott insulators is the presence of oxygen p-hole levels, ε_p^h, between d^8 and d^9 energies. The minimum energy required to transfer a d hole to an unoccupied site is $(\varepsilon_p^h - \varepsilon_p^d) \simeq 2\,eV$, the charge-transfer being to the same linear combination of p_x, p_y oxygen hole orbitals with local $d_{x^2-y^2}$ symmetry. When holes are added to La_2CuO_4 by the substitution of trivalent La with divalent Sr, they go into the oxygen p-hole band, and hop from site to site, both directly (with hopping amplitude $t_{pp} \simeq 0.7\,eV$) and through an admixture with the d states ($t_{pd} \simeq 1\,eV$). All the cuprates (hole-superconductors) have a fraction x (with x ranging from 0.05 to 0.30) of mobile holes in the plane per $[CuO_2]$ unit cell and may be considered to be doped charge-transfer insulators in the strong local correlation or the Mott limit.

From the phase diagram in Fig. 7.27 we see that the antiferromagnetic order of La_2CuO_4 is destroyed rapidly by doping, with T_N falling from over 210 K to 0 K as x changes from 0 to about 2%. Beyond this, there is an insulating or a barely metallic spin-glass phase at zero temperature up to $x \simeq 0.07$. The ground state is either insulating or superconducting. If the holes move, the insulating antiferromagnetic state is destroyed and a superconducting state obtains. The normal metal region above T_c is rather unusual in the range of $0.05 < x \simeq 0.30$. Overdoped $La_{2-x}A_xCuO_4$ ($x > 0.3$) appears to be a conventional metal.

One of the unique features of the cuprates is the relatively small Cu-O charge-transfer energy. It is therefore of significance to relate this property with superconductivity. X-ray photoemission spectra of cuprates in the Cu $2p_{3/2}$ region show a characteristic two-peak feature. The peak around 933 eV (main peak) is due to a final state primarily with $3d^{10}$ character, while the weaker intensity peak at about 941 eV binding energy is mainly due to a $3d^9$ final state (satellite). Model calculations of the Cu $2p$ core-level photoemission within a CuO_4 square-planar cluster show that the I_s/I_m ratio (relative intensity of the satellite to the main peak) is related to Δ/t_{pd} where Δ is the charge-transfer excitation energy and t_{pd} is the hybridization strength between the Cu $3d$ and the O $2p$ orbitals. Thus, any relation between the experimentally observed I_s/I_m ratio with n_h would suggest a link between n_h (T_c) and the Δ/t_{pd} ratio. It is indeed found that I_s/I_m ratio is related to the hole concentration n_h in many of the superconducting cuprates (Santra et al., 1991). Fig. 7.29 shows how in three series of cuprates, the I_s/I_m ratio decreases monotically with the increase in magnitude of hole doping. The variation is smooth across the insulator–superconductor–metal boundaries. Calculations show that the I_s/I_m ratio increases with the Δ/t_{pd} ratio in the cuprates. Accordingly, at a given n_h value where the T_c is maximum, increasing I_s/I_m is accompanied by an increase in T_c. The observation of a maximum T_c around an optimal value of n_h or Cu-O distance can therefore be traced to the optimal Δ/t_{pd} value or of the Cu-O charge-transfer energy.

The ab-plane dc electrical resistivity of cuprate superconductors above T_c is approximately linear in temperature over a wide range of several hundred degrees (Fig. 7.30). The slopes are nearly the same ($\simeq 1\,\mu\Omega\,cm/K$) to within a factor of two for a wide range of single-crystal specimens. The extrapolated zero temperature intercept can be

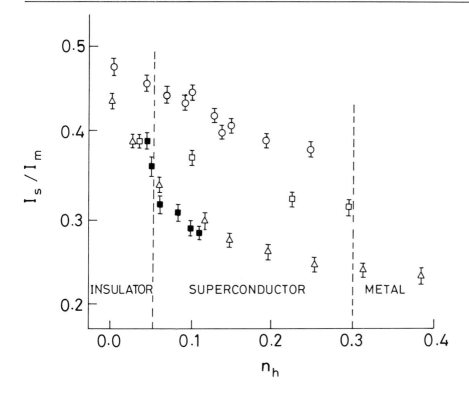

Figure 7.29   Variation of the I_S/I_M ratio with hole concentration (n_h): $La_{2-x}Sr_xCuO_4$ (triangles), $BiPbSr_2Y_{1-x}Ca_xCu_2O_\delta$ (squares) and $Bi_2Ca_{1-x}Ln_xSr_2Cu_2O_8$ (circles) (from Rao, 1991a)

small. A linear temperature dependence of resistivity in the range 10–400 K is unusual indeed. Thus, the resistivity of clean metals due to electron–phonon interaction rises initially as T^n with n ranging from 3 to 5. A number of properties are consistent with the linear temperature dependence of resistivity (Verma *et al.*, 1989), some of them being the low-frequency background in Raman scattering, the ac conductivity and the weakly temperature-dependent nuclear spin relaxation rate. This behaviour suggests that low-energy electronic excitations are not well defined unlike in conventional metals. The Hall effect in the normal state of the superconducting cuprates is also unusual, being strongly temperature-dependent. For example, the ratio (σ/σ_H) is proportional to T^2, σ being the electrical conductivity and σ_H, the Hall conductivity. Unusual disorder effects have been seen at low temperatures in $Bi_2Sr_2CuO_6$ which has a linear resistivity, but no superconducting transition. At temperatures well below 5 K, inelastic processes are a small perturbation on the elastic random scattering of carriers, and weak localization effects become noticeable (e.g., in the upturn of the resistivity). There is a negative orbital magnetoresistance which can be fitted to a weak localization form, from which one can extract an inelastic decay time for carriers. All these results indicate that the lowest energy ($\varepsilon < 0.01$ eV) electronic excitations in cuprates are

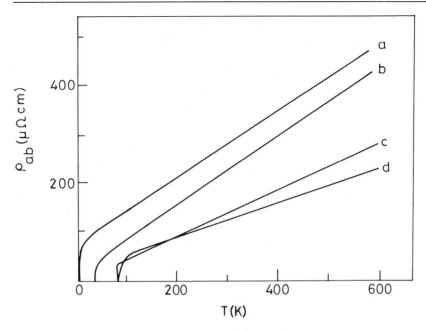

Figure 7.30   The *ab* plane resistivity in cuprate superconductors as a function of temperature: (a) Bismuth cuprate (2201); (b) $La_{1.85}Sr_{0.15}CuO_4$; (c) Bismuth cuprate (2212); (d) $YBa_2Cu_3O_7$.

unlike those of any known metal. It is believed that the key to high-temperature superconductivity is in understanding the unusual normal state properties. Many of these unusual properties as well as the small isotope effect on T_c cannot be understood in a straightforward manner on the basis of the BCS theory (see Section 6.3).

Superconductivity in cuprates is due to the coherence between electron pairs as confirmed by experiments which show that the magnetic flux is quantized in units of $(hc/2e)$. There is disagreement as to what causes pairing. In conventional superconductors, pairing is due to retarded phonon-induced attraction. There is no evidence for large electron–phonon coupling in cuprates giving rise to high T_c. Theories for cuprate superconductors fall into two categories. In the conventional approaches, one looks for Bosonic excitations which have the right energy scale and coupling to electrons and the metallic state is conventional. Bosonic excitations include excitons, plasmons, magnons and spin fluctuations. Such a system generally has a normal metallic state above T_c unless radical assumptions are made about the boson spectrum and the electron–boson coupling. In other theories, it is assumed that because of strong local correlations as well as quantum effects, novel ground states arise.

Anderson (1987, 1991) has proposed several new ideas to account for the superconductivity in cuprates. According to Anderson, high T_c superconductivity in cuprates is a strong correlation effect connected with the formation of a coherent assembly of singlet spin pairs. The quantum-mechanical overlap of a large number of such singlet configurations with different pair lengths and pair members renders such a resonating valence bond (RVB) state highly stable. Doping the RVB system with holes leads to neutral spin-like excitations (spinons) and charged hole-like excitations (holons). The question to be answered relates to the degree to which these mutually interacting excitations are approximated by relatively weakly coupled spinons and holons. Zhang & Rice (1988) have pointed out that a realistic two-band (oxygen p – copper d band) model reduces to a one-band Hubbard model, in the limit of low doping. Several types of spin-liquid ground states, depending on the spin-pairing amplitudes and relative phases, have been proposed by several workers (Ramakrishnan, 1992). One strategy has been to explore the possible spin-liquid solutions in the undoped Mott insulator, hoping that they describe the actual spin-liquid ground states in the presence of holes (Balachandran *et al.*, 1990). Some authors have explored very simple models to understand superconductivity in cuprates. For example, Burdett (1993) has examined superconducting cuprates in terms of a local thermodynamic minimum with mixed electronic character. In spite of the tremendous effort of a large number of theoreticians, we are not yet in a position to fully understand the superconducting and normal state properties of cuprates.

7.8   Organic materials

Organic solids exhibit a variety of interesting electronic and optical properties, which, in some instances, have found technological applications. Low-dimensional organic solids of high electrical conductivity as well as organic liquid crystals were discussed in Chapter 6. The area of *molecular electronics* based on organic solids has indeed become viable (Munn, 1984). Organic solids show photoconductivity and photovoltaic effects as well as charge storage and release. Electronic processes in organic crystals have been excellently reviewed in an authoritative text by Pope & Swenberg (1982). We shall briefly examine some important aspects of organic magnetic and conducting materials, as well as superconductors in addition to some of the *polymeric materials.*

Polymer science and technology have produced several 'wonder' materials of modern times. Among the new generation of polymer materials, *Kevlar* (*p*-phenylene diamine–terephthalic acid copolymer) is truly remarkable for its high strength, high modulus and low density, rendering it useful in various applications that include automobile tyres, cables for off-shore drilling platforms, and bulletproof vests. Aromatic polyamides of the Kevlar type are highly crystalline rigid-chain polymers that form extended chain crystals. Completely amorphous polyamides are also known; an example is the polyamide formed from terephthalic acid and mixed tetramethyl hexamethylene diamines, which is marketed commercially as Trogamid T. This material is transparent and glassy and has better retention of mechanical properties at elevated temperatures than many crystalline polyamides. Insulating polymers find use as engineering plastics, frequently replacing metals in mechanical and structural functions. They replace metals because they can be readily fabricated in complex shapes, possess better resistance to corrosion and are lighter in weight. One of the earliest engineering plastics that is still in use is nylon 66, prepared from hexamethylene diamine and adipic acid. Polymers containing aromatic rings such as polyphenylene ethers and polyphenylene sulphides possess greater structural rigidity, thermal and oxidative stability, and high softening temperatures (Anderson *et al.*, 1980). Aromatic polymer composites, consisting of polyether ether-ketone (PEEK) and carbon fibre, developed by ICI, England, possess a remarkable combination of properties such as light weight, high-temperature performance and fire resistance, which renders them useful structural materials.

Introducing metal atoms into the polymeric structure (*organometallic polymers*) increases the rigidity and stability, and if the metal atoms are arranged in a chain, high electrical conductivity is possible. Conducting organometallic polymers have recently been made: polymeric metal phthalocyanines can be spun into fibres from strongly acidic media. The wet-spun fibres can be doped to conductivities of $1–2 \, ohm^{-1} cm^{-1}$. Metal atoms in polymers can result in interesting optical properties. Complexes of Eu and Tb with 4-methyl-4'-vinyl bipyridine, when copolymerized with methyl methacrylate, show efficient fluorescence in the 500–600 nm region on excitation with 335 nm radiation. A novel application of organometallic polymers is the possibility of their use as pre-heat shields for targets in laser fusion experiments (Sheats *et al.*, 1984).

Different polymers can be combined to yield a variety of composite multipolymer systems. Combinations can be made in layered sheets and films, homogeneous and heterogeneous blends, interpenetrating polymer networks, bicomponent fibres and so on. Properties of *composite polymers* are roughly additive, but synergistic interactions can yield properties and performances superior to those of individual components. Multilayer films made of alternating layers of two transparent polymers with different refractive indices can exhibit interesting optical effects similar to the iridescent colours of butterflies and fish. *Polymer 'alloys'* possessing a better combination of properties than the constituents have been made. Polystyrene and polypropylene oxide are miscible in all proportions, and the blends have higher tensile strengths than either component. Multicomponent polymer systems have been discussed by Alfrey, Jr. & Schrenk (1980).

Several organic solids with high conductivity ($10^2–10^3 \, ohm^{-1} cm^{-1}$ at room temperature) in the metallic regime were described earlier. Doped polyacetylenes (e.g. with halogens, AsF_5, perchloric acid, etc., as dopants, giving *n*-type materials) could become useful in photovoltaic devices. Reversible electrochemical doping of polyacetylene enables fabrication of a rechargeable battery (MacInnes *et al.*, 1981). A stable conducting polymer that can be easily processed would have a major technological impact; although this goal is yet to be realized, progress is being made towards it.

Electrical conductivity of many organic solids is markedly affected by the absorption of gases. This property can be exploited to develop *gas sensors*. For example, phthalocyanine films can be used as NO_2 sensors with a sensitivity of 1 ppm (Wright *et al.*, 1983). Polydiacetylenes possess strongly anisotropic structure and optical properties; the material shows metal-like reflectivity when light is polarized parallel to the chain direction, and blue reflectance when light is polarized perpendicular to the chain. Some of the polydiacetylenes show efficient second harmonic generation (Gartio & Singer, 1982).

Photoconducting polymers such as poly (*N*–vinylcarbazole) are used in xerography or electrophotography, since the photogenerated charge carriers can travel through the polymer film with relative facility before getting immobilized or trapped (Roberts,

1980). Some of the insulating polymers such as poly (1, 1-difluoroethylene) meet the requirements of *electret* materials (the electric analogue of a permanent magnet); they respond to applied electric fields by limited movement of charge, and the charge does not leak back when the field is removed.

Polymer light emitting diodes (LEDS) and thin-film transistors have been receiving considerable attention recently due to interest in developing ultrathin computer monitors and television sets (flat-panel displays). The most common polymer LED consists of two electrodes sandwiching a layer of the light emitting polymer PPV (poly-*p*-phenylene vinylene) first reported by Burroughes *et al.* (1990). The efficiency of this system has been improved significantly by various means (Gustaffson & Cao, 1992; Greenham *et al.*, 1993). Most of the PPV materials emit green–yellow light. A blue polymer LED has been made (Yang *et al.*, 1993). Halls *et al.* (1995) have shown that an interpenetrating network formed from a phase-segregated mixture of two semiconducting PPV-related polymers provides both the spatially distributed interfaces necessary for efficient charge photogeneration and the means for separately collecting the electrons and holes. Thin films of these polymer mixtures promise to be good as large-area photodetectors.

7.8.1 *Molecular magnets*

Classical magnets are extended solids where cooperative interaction between d or f spins on metallic sites gives rise to spontaneous magnetization (cf. Section 6.3.1). Magnetic interaction between molecular species where at least one type of spin site is located on s/p orbitals can also give rise to spontaneous magnetization below a critical temperature $T_c > 0K$. Although molecular ferromagnetism was predicted in the 1960s, experimentally it was realized only in 1986 in the charge-transfer salt, $[Fe^{III}(C_5Me_5)_2]$ (TCNE) (Me = methyl; TCNE = tetracyanoethylene) (Miller *et al.*, 1986) and in the chain compound, CuMn (pbaOH)·$3H_2O$ (pbaOH = 2-hydroxy –1,3 – propylenebisoxamato) (Pei *et al.*, 1986). The T_cs are in the 4.6 – 4.8 K range in these two systems. Since then, considerable effort has been directed towards designing *molecular magnets* exhibiting higher T_cs (Kahn, 1992; Miller & Epstein, 1994). The approaches essentially make use of one or more of the following mechanisms for spin coupling: (i) intramolecular ferromagnetic spin coupling through orthogonal orbitals within a molecule or ion, (ii) intermolecular spin coupling through pairwise configuration interaction and (iii) dipole–dipole through-space interaction.

Design of molecular magnets based on mechanism (i) has been pursued by Kahn *et al.* The strategy is to employ bridging groups such as the various oxalate derivatives to form extended chains wherein metal centres of different spins (e.g. Cu(II) and Mn(II) alternate. $[Cu^{II}Mn^{II} (pbaOH)]·3H_2O$ is a typical example of this kind. Interestingly, dehydration of a water molecule from this compound raises the T_c to 30 K (Nakatani *et al.*, 1991). A complex compound containing three different spin carriers, Mn(II), Cu(II) and the organic radical cation, 2-(4-N-methylpyridinium) – 4,4,5,5-tetramethyl-imidazoline-1-oxyl-3-oxide, that crystallizes in an interlocking necklace-like structure orders magnetically below 22.5 K (Stumpf *et al.*, 1993). A one-dimensional chain compound consisting of a bisnitroxide radical and Mn^{II} hexafluoroacetylacetonate is a recent example of a molecular metamagnet that orders at 5.5 K (Inoue & Iwamura, 1994).

Cyanide ion is a better communicator of spins between adjacent metallic centres than organic bridging ligands. Accordingly higher T_cs have been achieved in mixed-valent Prussian blue type compounds such as $[Cr_5(CN)_{12}] \cdot 10H_2O$ ($T_c = 240$ K) and $Cs_{0.75}[Cr_{2.175}(CN)_6] \cdot 5H_2O$ ($T_c = 190$ K) (Mallah et al., 1993). The idea of having a large number of unpaired electrons residing in orthogonal orbitals which are ferromagnetically coupled through spinless bridging halide ions has been employed by Day (1979, 1993) to design the transparent ferromagnets, Rb_2CrCl_4 and its derivatives exhibiting T_cs greater than 50 K. However, both Prussian blue and Rb_2CrCl_4 are extended solids and are not really 'molecular' magnets. The ferromagnetic Zintl phase $Ca_{14}MnBi_{11}$ containing isolated ions also belongs to this category (Kuromoto et al., 1992).

Charge-transfer salts containing linear chains of alternating metallocenium donors (D) and cyanocarbon acceptors (A) are among the well-characterized molecular magnets to date exhibiting T_cs up to ~ 350 K (Miller & Epstein, 1994), ($Fe^{III}(C_5Me_5)$) (TCNE) with a T_c of 4.8 K being the first member discovered. Replacement of Fe(III) by Mn(III) in this material leads to an increase of T_c (8.8 K), whereas replacement by Cr(III) decreases the T_c to 3.65 K. Surprisingly a vanadium derivative of empirical formula, $V(TCNE)_x \cdot yCH_2Cl_2$, prepared by reaction of bis(benzene) vanadium (0) with TCNE, shows a room-temperature T_c (Manriquez et al., 1991).

Mathoniere et al. (1994) and Graham et al. (1995) have synthesized mixed-valent solids of the general formula $AM^{II}M^{III}(C_2O_4)_3$ where A is an organic cation such as $N(n\text{-}C_4H_9)_4$ or $P(C_6H_5)_4$ and M is Fe or Cr. The structure consists of honeycomb-like hexagonal layers of M atoms bridged by C_2O_4 groups, the layers being interleaved with organic cations. While the M = Cr compounds exhibit short-range antiferromagnetic correlations below 100 K and no phase transition to a long-range magnetically ordered state down to 2 K, the M = Fe compounds behave as ferrimagnets below 35 K. The tetrabutylammonium iron compound shows an unusual negative magnetization at low temperatures. The oxalato iron (III) complex, β''-$(BEDT\text{-}TTF)_4[(H_2O) Fe(C_2O_4)_3] \cdot C_6H_5CN$ is superconducting with a $T_c = 8.5$ K (Graham et al., 1995). This solid appears to be the first example of a superconductor containing localized magnetic moments arising from $3d$ electrons. The strategy of using cyanide bridges for the realization of a novel ferromagnetic material ($T_c = 18.6$ K), $[Ni(en)_2]_3[Fe(CN)_6]_2 \cdot 2H_2O$, possessing a rare rope-ladder structure has been reported by Ohba et al. (1994). This compound is synthesized by the simple reaction of $K_3[Fe(CN)_6]$ and trans-$Ni(en)_2Cl_2$ in aqueous solution. The structure is remarkable in that three cyano groups of each $[Fe(CN)_6]^{3-}$ bridge the trans-$[Ni(en)_2]^{2+}$ in the meridonal mode forming the rope-ladder structure.

In the above examples, a transition-metal atom is invariably present. As such, one might believe that the unpaired d electrons of the transition metal are the main cause of ferro/ferrimagnetic ordering, the organic/molecular species only playing a secondary role. There are a few examples of molecular magnets without the transition-metal atoms, albeit showing low T_cs. These are p-nitrophenyl nitronyl nitroxide ($T_c = 0.6$ K)

(Tamura *et al.*, 1991) and 1,3,5,7-tetramethyl-2,6-diazaadamantane-2,6-dioxyl ($T_c = 1.48\,K$) (Chiarelli *et al.*, 1993), both containing nitroxyl groups on which unpaired electrons are located. Dipole–dipole through space spin–spin interaction is the likely mechanism for ferromagnetic ordering in these materials.

Based on the knowledge that the ground states of the oxygen molecule and carbenes (: CH_2 and its derivatives) are triplets, Mataga (1968) suggested that large planar alternate hydrocarbons consisting of *meta*-substituted diphenyl carbene units would have a ferromagnetic ground state in accordance with Hund's rule and spin polarization. Iwamura and coworkers (1993) have prepared several high-spin polycarbene radicals with $S = 4$, 6 and 9. While magnetization confirms high-spin ground states, the temperature dependence of the susceptibility shows only antiferromagnetic coupling. A stable diradical (I) with a robust triplet ground state (having a large energy gap between the triplet ground and singlet excited states) has been prepared (Inoue & Iwamura, 1995).

I

I shows a magnetic moment of 2.71 μ_B between 100 and 300 K. Since there is no specific short contact between the molecules, there is no ferromagnetic interaction in the solid state. Among the nonmetallic molecular magnets known, $C_{60}\cdot$TDAE where TDAE = $(Me_2N)_2\,C{=}C(NMe)_2$, which orders into an itinerant ferromagnetic state with a T_c of 16.1 K is noteworthy (Allemand *et al.*, 1991). $C_{60}\cdot$TDAE is a soft ferromagnet and is a low-dimensional (possibly Heisenberg) system (Rao & Seshadri, 1994). It does not show remnance in the magnetization curve.

7.8.2   *Organic superconductors*

The prediction of high-temperature superconductivity in a hypothetical one-dimensional organic conductor containing polarizable side chains by Little (1964, 1965) turned out to be somewhat unrealistic because it ignored the fact that a one-dimensional metal cannot exist at 0K due to Peierls distortion. Little's hypothesis, however, triggered a great deal of research on organic conductors and superconductors. We have discussed earlier conducting organic solids such as TTF–TCNQ and polyacetylene in Sections 6.7 and 6.8. It was also mentioned that superconductivity is found in $(TMTSF)_2X$ type charge-transfer salts (TMTSF = tetramethyl tetraselenafulvalene). $(TMTSF)_2PF_6$ with a T_c of 0.9 K was the first organic superconductor to be discovered (Jerome *et al.*, 1980). Several $(TMTSF)_2X$ compounds with X$=$AsF$_6$, SbF$_6$, TaF$_6$, ReO$_4$, FSO$_3$ and ClO$_4$ have since been found to be superconducting with T_cs in the range of 2.1 to 0.4 K at applied pressures of 6.5–12 kbar (Bechgaard, 1982). Pressure is required to suppress the metal–insulator transition arising from an antiferromagnetic (spin-density wave) ordering in these materials. The structure of these materials consists of stacking of planar organic molecules on the top of each other like pancakes, with the inorganic anions filling the space in between the stacks. Such a structure minimizes the Coulomb repulsion and favours orbital overlap, giving rise to quasi-one-dimensional conductivity along the stack axis.

Superconductivity is now known in about 40 charge-transfer (CT) salts, the majority of which are derived from bis-(ethylene dithio)-tetrathiafulvalene (ET). The current record of highest T_c of 12.8 K is held by $(ET)_2Cu[N(CN)_2]Cl$ (Williams $et\ al.$, 1991). In most of the superconducting ET-salts, the transition to the superconducting state occurs at atmospheric pressure.

ET charge-transfer salts are generally synthesized by electrochemical crystallization where the organic donor is oxidized in the presence of appropriate anions in a H-type electrochemical cell using Pt electrodes. Slow electrochemical synthesis using small currents ($\sim 1\,\mu A\,cm^{-2}$) yields good quality single crystals suitable for structural studies and physical property measurements. Besides the conventional anions, halides, ClO_4^-, ReO_4^-, PF_6^-, AsF_6^- etc., polymeric anions such as $[Ag_4(CN)_5^-]_n$, $[Hg_3Cl_8^{2-}]_n$, $[Cu(NCS)_2^-]_n$ and $\{Cu[N(CN)_2Br^-]\}_n$ are employed as counteranions in the synthesis. With a given anion, ET donor forms several crystalline phases; the ET-I_3 system, for instance, forms as many as 14 different phases. Among them, the β- and the κ-phases of the $(ET)_2X$ compounds are noteworthy. The β-phases adopt a layered structure where the molecules are oriented parallel to each other, making short S ... S intermolecular contacts in a honeycomb-like pattern that provides pathways for conduction electrons. The κ-phases also crystallize in a layered structure where the molecules pack to give face-to-face dimers, oriented approximately at right angles to their neighbours giving rise to a two-dimensional conducting network.

In both the normal and superconducting states of organic superconductors, the electrical conductivity increases with the decreasing temperature above T_c and the superconducting state is characterized by zero resistance and flux expulsion. The mechanism of superconductivity in these materials is yet an open question. Heat capacity measurements indicate that if the mechanism is BCS-like, a rather strong coupling would be required in the high T_c compounds of this family. The physics and chemistry of organic superconductors have been discussed by Jerome (1991) and Williams $et\ al.$ (1991).

An important recent discovery in organic superconductivity is the observation of high T_cs in doped fullerides. In fact, the highest T_cs in organic superconductors are found in this family. We shall briefly review these materials here (Rao & Seshadri, 1994). The ability of the fullerenes to act as electron acceptors results in the formation of different types of salts. Alkali and alkaline-earth fullerides show interesting electronic properties (Murphy $et\ al.$, 1992). Photoemission studies, probing the occupied electronic states, as well as inverse photoemission measurements, probing the unoccupied electronic states allow the direct monitoring of the nature of electron doping into the C_{60} levels (Weaver, 1992). The valence-band of solid C_{60} is derived from a five-fold degenerate h_u orbital. This orbital is well-stabilized and removing electrons from C_{60} is therefore not easy. However, the three-fold degenerate t_{1u} and t_{1g} levels are quite low-lying, making C_{60} a good acceptor. Exposing C_{60} to alkali metal vapour results in filling of the t_{1u} level. The Fermi energy is pinned to the top of the filled level, so that the spectral manifold is shifted to lower energy with increasing filling. Since the level is three-fold degenerate, half-filling corresponds to three electrons and accordingly K_3C_{60} is a metal which becomes superconducting below 19 K. Further filling gives A_6C_{60} (e.g. Cs_6C_{60}) which, since it has a fully-filled t_{1u} LUMO, is insulating. The structures of C_{60}, K_3C_{60} and Cs_6C_{60} as determined by high-resolution powder X-ray diffraction and allied techniques are cubic and the alkali metal atoms sit in the tetrahedral and octahedral voids of the fcc-C_{60} structure. With alkaline-earth metals, the t_{1g} orbital derived band is also partially filled. Thus, Ca_5C_{60}, Sr_6C_{60} and Ba_6C_{60} are superconducting (Kortan $et~al.$, 1992).

It is believed that the mechanism for high-temperature superconductivity in the doped fullerides is rather simple. Firstly, the fulleride superconductors are isotropic with cubic lattices. High-symmetry systems resist low-temperature distortions that give rise to magnetic and/or insulating ground-states. Secondly, the C_{60} cage resists distortions that would lift the degeneracy of the t_{1u} and t_{2g} levels, giving filled bands. Finally, the intramolecular vibrations or phonons are very strong (of the order of 1000 K). If these phonons couple with the conduction electrons to give rise to Cooper pairs within the framework of the traditional BCS theory, the high transition temperatures are not surprising. An important piece of evidence supporting the role of intramolecular phonons is that in a variety of A_3C_{60} superconductors, the transition temperature is a function of only the separation between C_{60} molecules and does not depend on the identity of the species in the A-site. Separating the molecules results in a sharpening of the density of states and a concomitant increase in T_c. This is shown in Figure 7.31 in terms of a plot of the superconducting transition temperature against the a-parameter of fcc-A_3C_{60}. Figure 7.31 combines ambient pressure data from alloying experiments in the A-site as well as high-pressure data on K_3C_{60} and Rb_3C_{60}. The highest T_c to date is ~ 35 K in Rb_2CsC_{60}. Despite being a more recent phenomenon, superconductivity in the doped fullerides seems to be a more tractable problem than in the low-dimensional copper oxides from a mechanistic viewpoint.

7.8.3   Photochromic materials

These materials can store information on a molecular scale. The main requirement here is a molecular transformation associated with a colour change that is truly reversible. A typical reaction is the dimerization of the diene (Guru Row $et~al.$, 1983)

$(R = COOCH_3)$

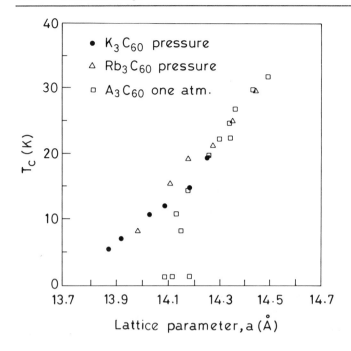

Figure 7.31   Variation of T_c with the cubic lattice parameter in doped fullerides.

The reaction, unfortunately, does not appear to be completely reversible. The conrotatory photochemical ring closure and opening in the reaction below seems to be attractive (Darcy *et al.*, 1981):

$$X = O, S, \text{ or } NCH_3$$
$$R = H, CH_3 \text{ or } C_6H_5$$

Organic photochromic materials find a host of applications in control and measurement of light intensity, production of visual contrast effects as well as information imaging and storage (Dürr, 1989; Dürr & Bouas-Laurent, 1990), the first two being currently the most sought after application. A change in the colour of a compound by the action of sunlight finds application where optical filters are used (e.g., sun glasses, optical lenses, glass windows and windshields for automobiles). Of the several types of compounds, spiropyrans (e.g.I) have been recognized to be suitable for photochromic spectacle lenses. Systems with virtually no fatigue have been found in the

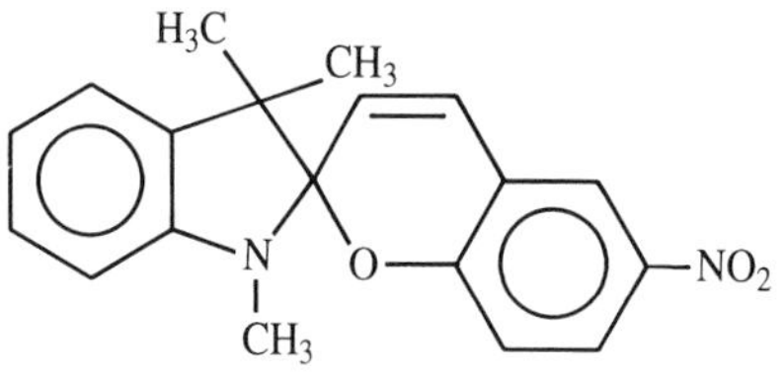

H₃C   CH₃ ... —NO₂   I

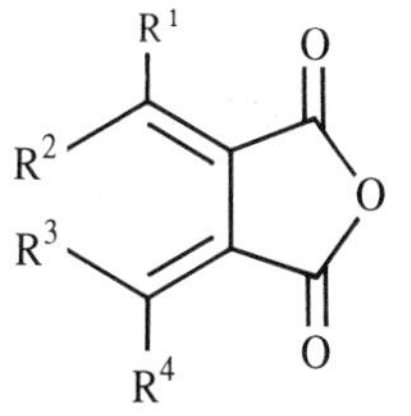

R¹ R² R³ R⁴ O O O   III

H₃C   CH₃ =N ... —OCH₃ N CH₃   II

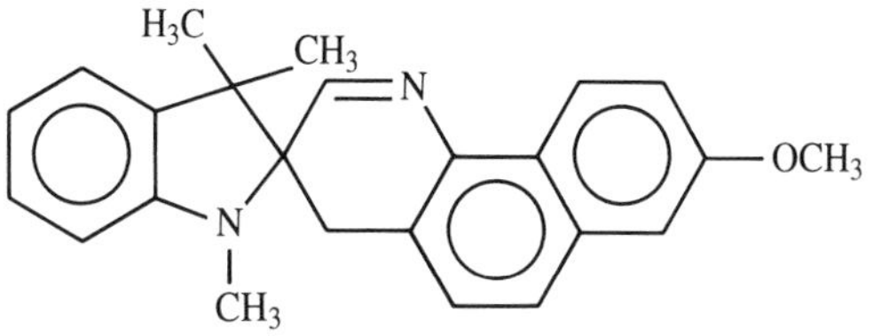

spirooxazine family (II) and these are already being used commercially for making sunglasses. Fulgides which are derivatives of dimethylene succinic anhydride (III) find application as photorefractive materials for optical lenses. Photochromic organic compounds have been reviewed by Heller (1991).

Various types of actinometers employed for the measurement of radiation intensities in the uv and visible regions are based on photochromic substances. Typical examples are Aberochrome 540 and Aberochrome 999 which belong to the fulgide series. Azobenzenes, fulgides or spiropyrans coupled with liquid crystalline polymers are used for information recording and in holography. Spirodihydroindoles are useful for frequency doubling of laser radiation.

An interesting futuristic application is in the field of molecular electronics where a one-dimensional molecular wire such as polyacetylene in combination with a suitable molecular switch, e.g. salicylideneanilines, would yield a molecular microchip whose information storage capacity would be about 10^6 times that of a conventional microchip. A new generation of high performance computer with memory elements of nanometre dimensions is visualized on the basis of such molecular microchips.

While with photochromic solids, light is used to cause colour change and detect it, with electrochromic solids, an electric field is used to change colour. Certain hemiquinones whose colour is externally switchable by an electric field are potentially useful as memory elements.

7.8.4 Hole burning

Molecules in crystals or dispersed in host lattices are often present in a range of environments, and this results in a broadening of the electronic absorption spectrum. Such an inhomogeneously broadened absorption band (envelope of transitions) may be considered as a superposition of several distinguishable sites. A narrow line laser can saturate one of the transitions under the envelope and the corresponding molecules will no longer take part in the absorption process. This phenomenon is referred to as *hole*

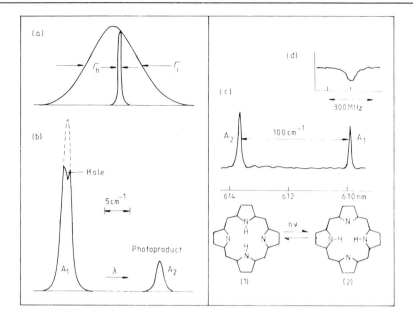

Figure 7.32   (a) Broadening of an electronic absorption band of a molecule due to an inhomogeneous environment; (b) illustration of a laser-induced photochemical hole burned in the 0–0 A_1 line of free-base porphyrin in n-octane at 2 K; (c) excitation spectrum of the 0–0 lines of the S_1–S_0 transition of the free base in n-hexane, showing a frequency difference ($\sim 100\,\mathrm{cm}^{-1}$) between the two tautomeric forms (1) and (2) of the free base in a single type of site. Irradiation into the line A_1 transforms it into A_2 and vice versa; (d) hole burned in line A_1 at 4.2 K. (After Williams, 1983.)

burning since it looks as though a hole is burnt in the absorption spectrum (Fig. 7.32). If the molecules excited by the laser undergo a photochemical transformation, the hole burnt in the absorption band will be accompanied by the absorption of the photoproduct at another frequency. Hole burning may be useful in optical storage of information (Castro, 1983). Some of the possible systems for hole burning are free-base porphyrin, phthalocyanine, s-tetrazine and chlorin, besides colour centres in alkali halides (and azides) and binary semiconductors; in the latter, bound exciton (zero phonon) transitions are involved in hole burning. The main materials requirements for hole burning are high quantum yield, absence of photobleaching, insensitivity to cycling and photochemical reversibility.

7.9   Langmuir–Blodgett films

Thin films of materials play an important role in modern electronics. Langmuir–Blodgett films (Mort, 1980) are prepared from surface-active molecules that adsorb at the surface of an aqueous solution. The surface is gently compressed to produce a close-packed monolayer, which can be coated on appropriate substrates by dipping at constant surface pressure. The layers so produced are of a precise thickness, and repeated dipping gives multilayers. For example, diacetylenes and other polymerizable

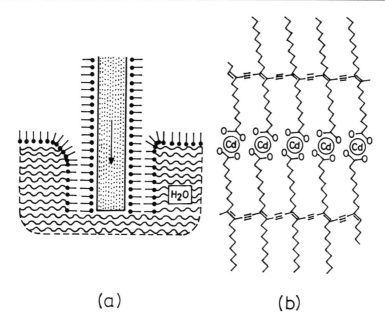

(a)   **(b)**

Figure 7.33   Langmuir-Blodgett film formed by polydiacetylene: (a) schematic diagram of Langmuir trough and (b) structure of a typical polydiacetylene film. (After Williams, 1983.)

units have been incorporated into Langmuir–Blodgett films, allowing the films to be selectively polymerized through a mask and the rest dissolved (Fig. 7.33). Such a film yields a *resist* for microelectronic circuitry. Specifically, the polydiacetylene films are used as guided wave structures, and in lithography with ultraviolet, X-rays or electron beams (with $1\,\mu$m resolution) on oxidised Si wafers. Magnetic films can be similarly prepared; Mn(II) stearate films have been investigated to explore two-dimensional magnetism (Pomerantz *et al.*, 1978). Langmuir–Blodgett films also find uses in integrated optics and as gas sensors. LB films and self-assembled monolayers have been widely reviewed in the recent literature where applications have also been discussed (Swallen, 1991; Mirkin & Ratner, 1992; also see special issue of *Langmuir*, 1994, vol. 10).

7.10   Liquid crystals

We discussed earlier (Section 6.12) structure–property relations in liquid crystals. Liquid-crystal displays (LCDs) are now commonplace in watches, calculators, cameras, hi-fi equipment, electronic games etc., and are predicted to find use in pocket-computers, microprocessor-based portable systems, portable televisions, and teletext systems remotely accessing data from a large computer. LCDs are based on the electro-optic properties of oriented nematics and their ability to switch orientation. The reorientation of the director (achieved by applying a voltage) results in a change of optical properties. There are three ways in which the change in orientation of a LC can be made visible. The change in refractive index can be observed by using two polarizers,

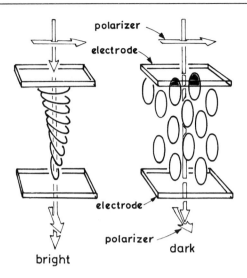

Figure 7.34   The twisted nematic cell in the absence of field (left) and in the field (right). (After Raynes, 1983.)

one to polarize and the other to analyse the light passing through the LC. The change in anisotropic light absorption associated with director reorientation can be observed with a polarizer or directly. The director reorientation can also be visualized by induced absorption produced by the addition of pleochroic dye molecules (Raynes, 1983).

Among the various LCDs, the three most widely used are (i) dynamic scattering display, (ii) twisted nematic display and (iii) guest–host display (Shanks, 1982).

Dynamic scattering displays, which are the first generation LCDs, make use of a nematic LC with a negative dielectric anisotropy ($\Delta\varepsilon = \varepsilon_{\parallel} - \varepsilon_{\perp} < 0$) doped with an ionic solute to increase conductivity. The display is transparent in the 'off' state and changes to a milky-white appearance when a voltage (> 8 V) is applied. The change is due to turbulent, vortex-like flow in the LC as a result of a conductivity anisotropy. This causes a spatial variation in the effective refractive index and hence scattering of incident light.

Most of the commercial LC displays are based on the *twisted nematic electrooptic effect*. They employ a nematic LC having a positive dielectric anisotropy and the display involves a pure dielectric reorientation; they are thus true field-effect displays requiring no charge flow in the LC. The basic structure and operation of a twisted nematic display cell is shown in Fig. 7.34. A thin layer ($6–15\,\mu$m) of LC material is confined between two glass panels that have on the interior surfaces a transparent electrically conducting coating (made of indium–tin oxide) through which an electric field can be applied across the LC. A surface alignment layer consisting of micro-grooves (e.g. rubbed polyimide) is overlaid on to the conducting layer. The groove directions in the inner surface layers are orthogonal to each other. This results in a twist of the nematic director by 90° on going from one boundary surface to the other (Fig. 7.34). Incident polarized light is wave-guided by the twisted nematic structure on passing through the cell and is transmitted. Mauguin, who discovered this remarkable property several years ago, showed that wave-guiding is effective only if $d\Delta n > 2\lambda$, where d is the thickness of the LC layer, Δn the anisotropy of refractive index and λ the wavelength of light. When an alternating voltage, somewhat larger than the threshold value, is applied, the director is reoriented to lie parallel to the applied field; the wave-guiding property is lost and the polarized light is not transmitted. As the director realigns along the electric field, effective Δn is lowered below the level required for guiding. Twisted nematic displays have become popular in consumer and professional electronics because they possess several advantages over conventional displays: (i) good display is achieved at low voltages ($\sim 2\,\mathrm{V}$) with small power consumption ($\sim 1\,\mu\mathrm{W\,cm^{-2}}$); (ii) devices can be made in compact flat panel geometry and (iii) devices exhibit good legibility in a wide range of illumination. The cell can be used in the transmission or reflection mode. An interesting variation of the twisted nematic effect is the production of a large-area electrooptic colour switch by making one of the polarizers a coloured dichroic type.

The third important display device making use of nematic LCs is called the *guest–host device* (Taylor & White, 1974; Scheffer, 1983). The LC material consists of a nematic with positive dielectric anisotropy doped with a chiral additive (cholesteric LC) and a pleochroic dye. The mixture behaves like a cholesteric LC, having a helical pitch that depends on the concentration of the chiral additive. The pleochroic dye molecules are aligned in the LC so that their long axes are parallel to the director (guest–host interaction). The dye absorbs light whose **E** vector lies along its long axis and the display shows the colour of the dye. When a voltage greater than the threshold value is applied, the helical structure is destroyed and the director realigns parallel to the applied field. The dye molecules are also realigned to lie parallel to the director. Light now passes through the cell with no absorption. Anthraquinone and tetrazine dyes, which are strongly dichroic and photochemically stable, are used in the guest–host display devices.

It can be safely predicted that applications of liquid crystals will expand in the future to more and more sophisticated areas of electronics. Potential applications of ferroelectric liquid crystals (e.g. fast shutters, complex multiplexed displays) are particularly exciting. The only LC that can show ferroelectric property is the chiral smectic C. Viable ferroelectric displays have however not yet materialized. Antiferroelectric phases may also have good potential in display applications. Supertwisted nematic displays of twist angles of around 240° and materials with low viscosity which respond relatively fast, have found considerable application. Another development is the polymer dispersed liquid crystal display in which small nematic droplets ($\sim 2\,\mu\mathrm{m}$ in diameter) are formed in a polymer matrix. Liquid crystalline elastomers with novel physical properties would have many applications.

7.11   Non-linear optical materials

Non-linear second-order optical properties such as second harmonic generation (SHG) and the linear *electrooptic effect* are due to the non-linear susceptibility in the relation between the polarization and the applied electric field. SHG involves the

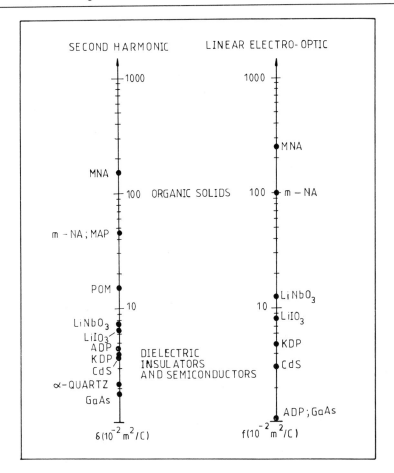

Figure 7.35   Comparison of non-linear optical figures of merit for organic and inorganic solids. δ is measured at $1.06\,\mu$m; f is measured at $0.633\,\mu$m except for GaAs where it is measured at $0.9\,\mu$m. (After Garito & Singer, 1982.© *Laser Focus/Electro-Optics* magazine, all rights reserved. (Reprinted with permission of PennWell Publishing, Advanced Technology Group, Littleton, Mass.)

doubling of frequency of light in a material; the linear electrooptic effect involves a change in the refractive index of the material under the applied field (see Sections 6.3 and 6.10). Electrooptic devices include optical switches and modulators. For SHG to occur, the crystal should be noncentrosymmetric; for maximum SHG, crystals should possess directions of propagation where the crystal birefringence cancels the natural dispersion, giving rise to equal indices of refraction at the fundamental and second harmonic frequencies (phase matching). Measures of material non-linearities are Mitter's δ and the polarization optic coefficient f, both of which are related to the non-linear susceptibility. In Fig. 7.35 we show the δ and f ranges in inorganic and organic materials (Garito & Singer, 1982) to illustrate that the organic and polymeric materials (MNA, m–NA, MAP) possess δ and f values that are orders of magnitude larger.

2-Methyl-4-nitroaniline (MNA) exhibits δ and f values that are 50 times larger than those of KDP. These large values of δ and f have their origin in the nature of the excited states of the π electron systems. The polarization of the π cloud in organic materials (as opposed to the displacement or rearrangement of nuclear coordinates in inorganic materials) gives rise to large non-linear effects, which make them good candidates for high-frequency applications. In organic solids, the crystal susceptibility can be considered to be an additive property of the microscopic second-order susceptibility, β, of the molecules. The large β of p-nitroaniline arises from the highly asymmetric, charge-correlated excited states of the π electron structure.

Optically non-linear single-crystal polymers can be made by the polymerization of the disubstituted diacetylene monomers in the solid state. Polymerization is initiated thermally or by irradiation with X-rays, electrons or UV light. The substituents R and R' in the diacetylene system, $R—C \equiv C\text{-}C \equiv C—R'$, are chosen to be optically non-linear, acentric organic substituents. Furthermore, the anisotropy afforded by the ordered polymer chains leads to large optical birefringence, which is of importance to phase matching. One of the examples of such polymer chain structures is 1,6-bis (2, 4-dinitrophenoxy)-2, 4-hexadiene which shows increasing SHG with increased polymerization; the second harmonic intensity of diacetylene polymers is very much greater than of $LiIO_3$, and several new polymeric structures continue to be examined (Williams, 1983). Besides single crystal polymers, polymer thin films are obtained by the Langmuir–Blodgett technique. The thin films offer the possibility of a variety of novel applications, including guided wave structure, lithography etc.

An important class of organics showing large second-order polarizabilities is the donor–acceptor substituted stilbenes. Among them, 4-[2-(4-dimethyl aminophenyl) ethenyl]-1-methyl pyridinium cation ($DAMS^+$) exhibits one of the largest known frequency doubling power (Marder et al., 1989). On intercalation of this organic cation into layered MPS_3 (M = Cd or Mn), composite materials exhibiting high SHG efficiency towards 1.34 μm laser radiation have been obtained (Lacroix et al., 1994). It is believed that on intercalation, the organic species spontaneously takes on a polar alignment in the layered host giving rise to the observed optical non-linearity.

The increasing interest in organic systems is because of their large NLO coefficients and the ultrafast response to the π-electrons. For example, the third harmonic generation (THG) coefficient of polyacetylene (PA) is the largest known for any material in the gap region away from resonances (Etemad & Baker, 1987). There are intramolecular donor–acceptor systems possessing large SHG coefficients (Chemla & Zyss, 1987). Among the organic solids, π-conjugated polymers offer promise as systems with large NLO coefficients, since the coefficients increase superlinearly with the length of conjugation over length scales of the order of the inverse of the degree of dimerization (Ramasesha & Das, 1990). The NLO response of PA systems with about 5% bond alternation would increase superlinearly up to a length of about 20 bonds, and thereafter get saturated showing a linear dependence on length. Since PA has a centre of inversion, it would not, however, exhibit a SHG response. The centre of inversion can be broken by introducing push-and-pull groups in the chain and such push–pull systems are good candidates for SHG applications. The superlinear dependence increases with the increse in push–pull strength. Typical long chain conjugated systems which have push-and-pull groups without centres of inversion are poly(2,3-diazabutadiene)s and polymethineimines. In 1,6-bis(2,4-dinitrophenoxy)-2,4-hexadiene, SHG increases with increased polymerization. Several other polymeric structures have been examined for NLO properties (Williams, 1983; Chemla & Zyss, 1987). Langmuir–Blodgett films of polymeric NLO materials offer novel applications. The third-order non-linear properties that have attracted attention are the THG response, electric-field-induced second harmonic generation (EFISH) and the Kerr, or the quadratic electro-optic effect. In PA, these properties depend upon the conjugation length.

For commercial NLO applications, inorganic materials showing SHG seem to be more suitable than organic ones because inorganic materials are more rugged and can be obtained as high-purity crystals. The possibility of ion-exchange is an added advantage because it enables fabrication of wave-guide structures. Accordingly, inorganics such as β-BaB_2O_4, $Ba_2NaNb_5O_{15}$, $LiNbO_3$ and $KTiOPO_4$ (KTP) represent the current workhorses of NLO technology. Among these, KTP has been used for SHG of 1064 nm infrared light and in wave-guides to produce phase-matched SHG in the blue. KTP is also attractive for other applications such as optical frequency mixing (sum and difference mixing) and in electrooptics for modulation of Q-switching (Eaton, 1991; Bierlein & Vanherzeele, 1989). KTP is also remarkable for its ability to accommodate a wide range of substitutions (Na, Rb or Tl for K; Sn, Zr or Nb for Ti and As for P) which enables important properties such as refractive indices to be varied systematically.

Third-order NLO effects, which include the Kerr effect, photorefractivity and third harmonic generation (THG) are potentially useful in futuristic applications in optical switching, image processing and optical computing, among others. Today, no third-order material applications are practical because the observed non-linearities fall short by 2 to 4 orders of magnitude of what is required for commercial devices. Size-quantized semiconductor materials such as CdS and GaAs clusters are among the best-known third-order NLO materials at present (Wang, 1991). New material breakthroughs are needed before promises of applications can become a reality in this area. Since no symmetry requirements govern third-order susceptibility, all kinds of materials, liquids, thin films, glasses and crystals are potential candidates. Eventually, practical considerations such as fabricability may become more important than the magnitude of the non-linearity.

7.12   Luminescent materials

Luminescent inorganic materials (phosphors) have been in common use for several decades, those related to ZnS having been discovered in the mid-nineteenth century. Phosphors continue to be an important commodity, and there are efforts to improve the efficiency and colour of existing phosphors as well as to make new ones (Hill, 1983; Hagenmuller, 1983). Small quantities of impurities markedly affect the luminescence characteristics of phosphors; for example one or two parts in 10^8 of copper dominates the luminescence of ZnS.

Alkaline-earth halophosphates have been used widely in the fluorescent-lamp industry. The role of the phosphor is to absorb the 254 nm radiation of a mercury lamp and emit light at the right wavelength in the visible region. Impurity atoms act as absorbing and emitting centres. $3Ca_3(PO_4)_2 \cdot CaF_2$ doped with Sb and Mn has such good luminescence characteristics; the energy absorbed by Sb is shared by both Sb and Mn emitting centres. Sr, Ba or Cd have been substituted for Ca and Cl for F to improve the performance. In recent years, phosphors emitting in narrow spectral bands have been produced by using rare-earth dopants. Eu^{2+} in barium magnesium aluminate gives a blue emission while Eu^{3+} in Y_2O_3 gives red emission. Ce^{3+} and Tb^{3+} ions have been used in host lattices to get green emission. Eu^{2+} doped in $BaAlF_5$, $BaSiF_6$ and BaY_2F_8 appears to give $^6P_{7/2}-^8S_{7/2}$ monochromatic emission. $Na_3Ce_{0.65}Tb_{0.35}(PO_4)_2$ has been found to be a very efficient phosphor since the emission of Ce^{3+} overlaps with the excitation ($^5D_4 - ^7F_5$) of Tb^{3+} very well.

Cathode ray tube phosphors with fast luminescence as well as persistent luminescence are needed in different situations. A variety of phosphors of the ZnS–CdS family are used for the purpose with Ag, Cu etc. as dopants. Other host lattices used are CaS, Zn_2SiO_4, Y_2O_2S and $Zn_3(PO_4)_2$; typical dopants are Mn, Ce, Eu and Tb.

A common application of inorganic phosphor materials is in fluorescent lighting. Here, the classic halophosphate has been replaced by mixtures of rare-earth activated phosphors resulting in higher light output and better colour rendering. Besides lighting, luminescent inorganic materials find application in a variety of high-technology areas which include X-ray medical radiography, tunable lasers, cathode ray phosphors for projection television and as scintillator materials in general (Blasse, 1994). For a general discussion of inorganic luminescent materials, see the review by Blasse (1988) and the references given therein.

Photoluminescent organic polymers are of considerable importance as they find application as materials for flexible light emitting diodes. For a system to be a good luminescent material, the lowest excited state must have a non-zero transition dipole to the ground-state (Kasha rule). There are very few exceptions to this rule and accordingly long chain polyenes are poor luminescent materials, since they have a two-photon state below the optical gap. The existence of a two-photon state below the optical gap is due to electron correlations lowering the energies of the covalent states, and raising the energies of the ionic states, thereby reversing the energy-level ordering predicted by non-interacting theories and leading to a weak dipole-allowed excitation as the lowest energy excitation in the polyene systems. Experimentally, poly-paraphenylenevinylene (PPV) exhibits strong fluorescence (Colaneri et al., 1990) although the difference between PPV and polyacetylene is essentially topological. The main difference between the two is that PPV has a larger effective bond alternation. The lowest excited state for small dimerization in the singlet manifold is a two-photon state while at large dimerization, it is a one-photon state. The other fluorescing polymers are polythiophene and some of the polydiacetylenes. In the latter, bond alternation is different from polyacetylene, the lowest excited state being a one-photon state while in the former there is no alternancy symmetry. In polyparaphenylene, the phenyl rings are not coplanar and the fluorescence is weak compared to PPV. Heterocyclic polymers

Table 7.4. *Typical solid state lasers*

Laser material	Emission (μm)
$Al_2O_3(Cr^{3+})$	0.6943
$CaF_2(Nd^{3+})$	1.046
$CaF_2(Sm^{3+})$	0.7085
$CaF_2(Ho^{3+})$	2.09
$CaWO_4(Nd^{3+})$	1.06
GaAs	0.84
GaP	0.72
$GaAs_xP_{1-x}$	0.64–0.86
InP	0.91
InAs	3.10
PbS	4.27
PbSe	8.53
$Sn_xPb_{1-x}Te$	9.5–28

Laser ions are mentioned in parentheses.

generally do not fluoresce strongly because of the high excitation energy (in the undoped state).

7.13   Laser materials

Typical laser materials where the lasing centres are the impurity ions in a host solid are listed in Table 7.4, along with semiconductor lasers; the emission wavelength is also shown in the table. Device-quality GaAs, InP and related semiconductors can be conveniently synthesized by chemical vapour deposition using metallo–organic compounds (Moss, 1983). Other laser materials of interest are laser windows and interference coatings (Newnam, 1982). Polycrystalline windows of LiF, NaCl, KCl and CaF_2 with enhanced mechanical properties have been made; rapid quenching increases the laser-damage threshold by a factor of 4.6. New infrared window materials of hafnium fluoride and fluorozirconate glasses have been developed. $LiYF_4$ is found to be a good ultraviolet laser host or window material.

Many optical elements comprise interference coatings that are prone to laser damage. Single-layer films of various materials have been produced by RF sputtering and other techniques to raise the damage thresholds. Typical materials used are TiO_2, $TiO_2 + SiO_2$ and $SiO_2 + MgF_2$. Owing to the unique properties of diamond, diamond-like carbon films produced by ion-beam deposition are being investigated. Pre-irradiation of alkali halide windows coated with As_2S_3, NaF etc., also brings about an increase in the damage threshold.

Nd-doped $YAl_3(BO_3)_4$ is a self frequency doubling laser material that produces $0.53\,\mu$m laser output from $1.06\,\mu$m. (Lu *et al.*, 1989). Accordingly, this material combines the functions of a conventional Nd-YAG laser and a SHG material like KTP. Another significant development in this field is the discovery of upconversion laser material, $BaY_{2-x}Yb_xF_8$ doped with Tm or Er, which has a fluorite-related structure. This material generates red, green, blue and even ultraviolet light from an infrared input. Infrared radiation is absorbed by Yb^{3+} ions and the energy transferred to Tm^{3+} by a cross-relaxation mechanism to populate a wide range of Tm^{3+} levels, the higher energy levels being accessed by successive excitations. Laser radiations of 456, 482, 512, 649 and 799 nm have been obtained at room-temperature (Thrash & Johnson, 1994; Owen et al., 1994). The fluoride lattice is important because it minimizes depopulation of Tm levels by lattice vibrations, enabling population inversion and laser action at a number of frequencies.

References

Abraham, F., Boivin, J. C., Mairesse, G. & Nowogrocki, G. (1990) *Solid State Ionics* **40–41**, 934.

Adler, S. B., Reimer, J. A., Baltisberger, J. & Werner, U. (1994) *J. Amer. Chem. Soc.* **116**, 675.

Alfrey, Jr. T. & Schrenk, W. J. (1980) *Science* **208**, 813.

Al-Mamouri, M., Edwards, P. P., Greaves, C. & Slaski, M. (1994) *Nature* **369**, 382.

Allemand, P., Khemani, K. C., Koch, A., Wudl, F., Holczer, K., Donovan, S., Grüner, G. & Thompson, J. D. (1991) *Science* **253**, 301.

Anderson, B. C., Barton, L. R. & Collette, J. W. (1980) *Science* **208**, 807.

Anderson, P. W. (1987) *Science* **235**, 1196.

Anderson, P. W. (1991) *Physica C* **185–189**, 11.

Angell, C. A., Lin, C. & Sanchez, E. (1993) *Nature* **362**, 137.

Angus, H. C. (1981) *Phys. Technol.* **12**, 245.

Armstrong, A. R. & Edwards, P. P. (1991) *Ann. Rep. Prog. Chem. Sect. C.*, **88**, 259.

Aruchamy, A., Aravamudan, G. & Subba Rao, G. V. (1982) *Bull. Mater. Sci.* **4**, 483.

Azuma, M., Hiroi, Z., Takano, M., Bando, Y. & Takeda, Y. (1992) *Nature* **356**, 225.

Balachandran, A. P., Ercolessi, E., Morandi, G. & Srivastava, A. M. (1990) *Hubbard Model and Anyon Superconductivity*, World Scientific, Singapore.

Bard, A. J. (1980) *Science* **207**, 139.

Bechgaard, K. (1982) *Mol. Cryst. Liq. Cryst.* **79**, 1.

Becquerel, E. (1839) *C.R. Acad. Sci.* **9**, 561.

Bednorz, J. G. & Müller, K. A. (1986) *Z. Phys.* **B64**, 189.

Bierlein, J. D. & Vanherzeele, H. (1989) *J. Opt. Soc. Am. B*, **6**, 622.

Blasse, G. (1988) *Prog. Solid State Chem.* **18**, 79.

Blasse, G. (1994) *Chem. Mater.* **6**, 1465.

Blunt, R. (1983) *Chem. Britain* **19**, 736.

Bose, D. N., Parthasarathy, G., Mazumdar, D. & Gopal, E. S. R. (1984) *Phys. Rev. Lett.* **53**, 1368.

Bowen, K. (1980) *Mater. Sci. Engg.* **44**, 1.

Brattain, W. H. & Garrett, C. G. (1955) *Bell. Syst. Tech. J.* **34**, 129.

Bruce, P. G. & West, A. R. (1982) *J. Solid State Chem.* **44**, 354.

Burdett, J. K. (1993) *Inorg. Chem.* **32**, 3915.

Burroughes, J. H., Bradley, D. D. C., Brown, A. R., Marks, R. N., Friend, R. H., Burn, R. L. & Holmes, A. B. (1990) *Nature* **347**, 539.

Buschow, K. H. J. (1988) *Rep. Prog. Phys.* **54**, 1123.

Buschow, K. H. J. & van Mal, H. H. (1982) in *Intercalation Chemistry* (eds Whittingham, M. S. & Jacobson, A. J.) Academic Press, New York.

Castro, G. (1983) in *Photochemistry and Photobiology* (ed. Zewail, A. H.) Harwood.

Cava, R. J. (1990) *Science*, **247** 656.

Cava, R. J., Batlogg, B., Krajewski, J. J., Farrow, R., Rupp, L. W. Jr., White, A. E., Short, K., Peck, W. F. Jr. & Kometani, T. (1988) *Nature* **332**, 814.

Cava, R. J., Hewat, A. W., Hewat, E. A., Batlogg, B., Marezio, M., Rabe, K. M., Krajewski, J. J., Peck, W. F. Jr. & Rupp, L. W. Jr. (1990) *Physica C* **165**, 419.

Chandra, S. (1981) *Superionic Solids: Principles and Applications*, North-Holland, Amsterdam.

Chao, S., Robbins, J. L. & Wrighton, M. S. (1983) *J. Amer. Chem. Soc.* **105**, 181.

Chemla, D. S. & Zyss, J. (eds) (1987) *Nonlinear Properties of Organic Molecules and Crystals*, Academic Press, New York.

Chiarelli, R., Novak, M. A., Rassat, A. & Tholence, J. L. (1993) *Nature* **363**, 147.

Chin, G. Y. (1980) *Science* **208**, 888.

Chopra, K. L., Harshvardhan, K. S., Rajagopalan, S. & Malhotra, L. K. (1981) *Solid State Comm.* **40**, 387.

Coey, J. M. D. & Sun, H. (1990) *J. Mag. Matls.* **87**, L251.

Cohen, R. L. & Wernick, J. H. (1981) *Science* **214**, 1081.

Colaneri, N. L., Bradley, D. D. C., Friend, R. H., Burns, P. L., Holmes, A. B. & Spangler, C. W. (1990) *Phys. Rev. B*, **42**, 11670.

Dagotto, E., Riera, J. & Scalapino, D. (1992) *Phys. Rev. B.* **45**, 5744.

Darcy, P. J., Heller, H. G., Strydom, P. J. & Whittal, J. (1981) *J. Chem. Soc. Perkin* **1**, 202.

Das, B. N. & Koon, N. C. (1983) *Met. Trans.* **14A**, 953.

Day, P. (1979) *Acc. Chem. Res.* **12**, 237.

Day, P. (1993) *Chem. Soc. Rev.* 51.

Dürr, H. (1989) *Angew. Chem. Int. Ed. Engl.* **28**, 413.

Dürr, H, & Bouas-Laurent, H. (eds) (1990) *Photochromism: Molecules and Systems*, Elsevier, Amsterdam.

Eaton, D. F. (1991) *Science* **253**, 281.

Elliott, S. R. (1983) *J. Non-Cryst. Solids* **59–60**, 899.

Elliott, S. R. (1984) *Physics of Amorphous Materials*, Longman, London.

Ellis, A. B., Kaiser, S. W. & Wrighton, M. S. (1976) *J. Amer. Chem. Soc.* **98**, 1635, 6855.

Etemad, S. & Baker, G. L. (1987) *Proceedings of the International Conference on Synthetic Metals*, Santa Fe, NM.

Farrington, G. C. & Briant, J. L. (1979) *Science* **204**, 1371.

Fox, M. A. & Chanon, M. (eds) (1988) *Photoinduced Electron Transfer*, Elsevier, Amsterdam.

Fujishima, A. & Honda, K. (1972) *Nature* **238**, 37.

Ganapathi, L., Subbanna, G. N., Gopalakrishnan, J. & Rao, C. N. R. (1985) *J. Mater. Sci.* **20**, 1105.

Gantron, J., Lemasson, P. & Marucco, J. F. (1980) *Faraday Discussions of the Chemical Society*, No. 70, 81.

Garito, A. F. & Singer, K. D. (1982) *Laser Focus* **18**, 59.

Gerischer, H. (1975) *J. Electroanal. Chem.* **58**, 263.

Gerischer, H. (1979) in *Topics in Applied Physics* Vol. **31**, Solar Energy Conversion (ed. Seraphin, B. O.) Springer-Verlag, Berlin.

Giess, E. A. (1980) *Science* **208**, 938.

Goodenough, J. B. (1984) *Proc. Roy. Soc. London* **A393**, 215.

Goodenough, J. B., Hong, H. Y-P & Kafalas, J. A. (1976) *Mater. Res. Bull.* **11**, 203.

Goodenough, J. B., Ruiz-Diaz, J. E. & Zhen, Y. S. (1990) *Solid State Ionics* **44**, 21.

Goodman, C. H. L. (1983) *Chem. Britain* **19**, 745 and the references therein.

Graham, A. W., Kurmoo, M. & Day, P. (1995) *J. Chem. Soc. Chem. Commun.* 2061

Grande, T., Holloway, J. R., McMillan, P. F. & Angell, C. A. (1994) *Nature* **369**, 43.

Grätzel, M. (1981) *Acc. Chem. Res.* **14**, 376.

Greaves, C. (1994) *Chem. Britain* **30**, 743.

Greenham, N. C., Moratti, S. C., Bradley, D. D. C., Friend, R. H. & Holmes, A. B. (1993) *Nature* **365**, 628.

Gronet, C. M. & Lewis, N. S. (1982) *Nature* **300**, 733.

Gustafsson, G. & Cao, Y. (1992) *Nature* **357**, 477.

Guru Row, T. N., Swamy, H. R., Acharya, K. R., Ramamurthy, V., Venkatesan, K. & Rao, C. N. R. (1983) *Tetrahedron Lett.* **24**, 3263.

Hagenmuller, P. (1983) *Proc. Indian Acad. Sci. (Chem. Sci.)* **92**, 1.

Hagenmuller, P. & van Gool, W. (eds) (1978) *Solid Electrolytes*, Academic Press, New York.

Halls, J. J. M., Walsh, C. A., Greenham, N. C., Marseglia, E. A., Friend, R. H., Moratti, S. C. & Holmes (1995) *Nature* **376**, 498.

Harris, L. A. & Wilson, R. H. (1978) *Ann. Rev. Mater. Sci.* **8**, 99.

Heller, A. (1981) *Acc. Chem. Res.* **14**, 154.

Heller, H. G. (1991) in *Electronic Materials from Silicon to Organics* (eds Miller, L. S. & Mullin, J. B.) Plenum, New York.

Herbst, J. F., Croat, J. J., Pinkerton, F. E. & Yelon, W. B. (1985) *J. Appl. Phys.* **57**, 4086.

Hermann, A. M. & Yakhmi, J. V. (eds) (1994) *Thallium-based High-Temperature Superconductors*, Marcel Dekker, New York.

Hill, C. G. A. (1983) *Chem. Britain* **19**, 723 and the references therein.

Hiroi, Z. (1995) *ISTEC Journal* **8**, 63.

Hiroi, Z., Kobayashi, N. & Takano, M. (1994) *Nature* **371**, 139.

Hiroi, Z. & Takano, M. (1995) *Nature* **377**, 41.

Ingram, M. D. & Vincent, C. A. (1984) *Chem. Britain* **20**, 235.

Inoue, K. & Iwamura, H. (1994) *J. Chem. Soc. Chem. Commun.* 2273.

Inoue, K. & Iwamura, H. (1995) *Angew. Chem. Int. Ed. Engl.* **34**, 927.

Ishihara, T., Matsuda, H. & Takita, Y. (1994) *J. Amer. Chem. Soc.* **116**, 3801.

Ivey, D. G. & Northwood, D. K. (1983) *J. Mater. Sci.* **18**, 321.

Iwamura, H. (1986) *Pure and Appl. Chem.* **58**, 187.

Iwamura, H. (1993) *Pure and Appl. Chem.* **65**, 57.

Jerome, D. (1991) *Science* **252**, 1509.

Jerome, D., Mazuad, A., Ribault, M. & Bechgaard, K. (1980) *J. Phys. (Paris) Lett.* **41**, L95.

Johnson, D. C., Prakash, H., Zachariasen, W. H. & Viswanathan, R. (1973) *Mater. Res. Bull.* **8**, 777.

Johnston, D. C., Jacobson, A. J., Newsam, J. M., Lewandowski, J. T., Goshorn, D. P., Xie, D. & Yelon, W. B. (1987) in *ACS Symposium Series* 351, pp. 136–51.

Jorgensen, J. D. (1991) *Physics Today*, **44**, 34.

Kahn, O. (1985) *Angew. Chem. Int. Edn (Engl.)* **24**, 834.

Kahn, O. (1992) *Molecular Magnetism*, VCH Publishers, New York.

Katz, R. M. (1980) *Science* **208**, 841.

Kiwi, J., Kalyanasundaram, K. & Grätzel, M. (1982) *Structure and Bonding* **49**, 39.

Kohler, H. & Schulg, H. (1985) *Mater. Res. Bull.* **20**, 1461.

Koon, N. C., Das, B. N., Rubinstein, B. & Tyson, J. (1985) *J. Appl. Phys.* **57**, 4091.

Kortan, A. R., Kopylov, N., Glarum, S., Gyorgy, F. M., Ramirez, A. P., Fleming, R. M., Thiel, P. A. & Haddon, R. C. (1992) *Nature* **355**, 529; **360**, 566.

Korthius, V., Khosrovani, N., Sleight, A. W., Roberts, N., Dupree, R. & Warren, W. W. (1995) *Chem. Mater.* **7**, 412.

Kuromoto, T. Y., Kauzlarich, S. M. & Webb, D. J. (1992) *Chem. Mater.* **4**, 435.

Kutal, C. (1983) *J. Chem. Educ.* **60**, 882.

Lacroix, P. G., Clement, R., Nakatani, K., Zyss, J. & Ledoux, I. (1994) *Science* **263**, 658.

Legg, K. D., Ellis, A. B., Bolts, J. M. & Wrighton, M. S. (1977) *Proc. Natl. Acad. Sci. USA* **74**, 4116.

Lemasson, P., Etman, M., Gantron, J. & Marucco, J. F. (1982) *Proc. 4th Int. Conf. on Photochemical Conversion and Storage of Solar Energy*, Jerusalem.

LePage, Y., McKinnon, W. R., Tarascon, J. M. & Barboux, P. (1989) *Phys. Rev.* **B40**, 6810.

Leygraf, C., Hendewerk, M. & Somorjai, G. A. (1982) *Proc. Natl. Acad. Sci. USA* **79**, 5739.

Little, W. A. (1964) *Phys. Rev.* **A134**, 1416.

Little, W. A. (1965) *Sci. Amer.* **212(2)**, 21.

Lu, B. S., Wang, J., Pan, H., Jiang, M., Liu, E. & Hou, X. (1989) *J. Appl. Phys.* **66**, 6052.

MacChesney, J. B. (1981) in *Preparation and Characterization of Materials* (eds Honig, J. M. & Rao, C. N. R.) Academic Press, New York.

MacInnes, Jr, D., Druy, M. A., Nigrey, P. J., Nairns, D. P., MacDiarmid, A. G. & Heeger, A. J. (1981) *J. Chem. Soc. Chem. Comm.* 317.

Mallah, T., Thiebaut, S., Verdaguer, M. & Veillet, P. (1993) *Science* **262**, 1554.

Mathoniere, C., Carling, S. G., Yusheng, D. & Day, P. (1994) *JCS Chem. Commun.* 1551.

Manivannan, V., Gopalakrishnan, J. & Rao, C. N. R. (1991) *Phys. Rev.* **B43**, 8686.

Manriquez, J. M., Yee, G. T., McLean, R. S., Epstein, A. J. & Miller, J. S. (1991) *Science* **252**, 1415.

Maple, M. B. (1990) *MRS Bull.* **15**, 60.

Marder, S. R., Perry, J. W. & Schaefer, W. P. (1989) *Science* **245**, 626.

Mataga, N. (1968) *Theor. Chim. Acta*, **10**, 372.

Mazumdar, D., Bose, D. N. & Mukherjee, M. L. (1984) *Solid State Ionics* **14**, 143.

Menth, A., Nagel, H. & Perkins, R. S. (1978) *Ann. Rev. Mater. Sci.* **8**, 21.

Miedema, A. R. (1973) *J. Less-Common Metals* **32**, 117.

Miedema, A. R., Buschow, K. H. J. & van Mal, H. H. (1977) *Electrochem. Soc. Conf. Proc.* **77-6**, 456.

Miller, J. S., Calabrese, J. C., Epstein, A. J., Bigelow, R. W., Zhang, J. H. & Reif, W. M. (1986) *J. Chem. Soc. Chem. Commun.* 1026.

Miller, J. S. & Epstein, A. J. (1994) *Angew. Chem. Int. Ed. Engl.* **33**, 385.

Mirkin, C. A. & Ratner, M. A. (1992) *Annu. Rev. Phys. Chem.* **43**, 719.

Mort, J. (1980) *Science* **208**, 819.

Moss, R. H. (1983) *Chem. Britain* **19**, 733.

Munn, R. W. (1984) *Chem. Britain* **20**, 518 and the references therein.

Muroi, M. (1994) *Physica C* **219**, 129.

Murphy, D. W. & Christian, P. A. (1979) *Science* **205**, 651.

Murphy, D. W., Rosseinsky, M. J., Fleming, R. M., Tycko, R., Ramirez, A. P., Haddon, R. C., Siegrist, T., Dabbagh, G., Tully, J. C. & Walstedt, R. E. (1992) *J. Phys. Chem. Solids* **53**, 1321.

Nagarajan, R. & Rao, C. N. R. (1993) *J. Solid State Chem.* **103**, 533.

Nakata, H. & Akimitsu, J. (1995) *ISTEC Journal* **8**, 59.

Nakatani, K., Bergerat, P., Codjovi, E., Mathoniere, C., Pei, Y. & Kahn, O. (1991) *Inorg. Chem.* **30**, 3977.

Nelmses, R. J., Loveday, E., Kaldis, E. & Karpinski, J. (1992) *Physica C* **172**, 311.

Nesbitt, E. A. & Wernick, J. H. (1973) *Rare Earth Permanent Magnets*, Academic Press, New York.

Newnam, B. E. (1982) *Laser Focus*, Vol. 18, No. 2, p. 53.

Nielsen, J. W. (1976) *IEEE Transactions on Magnetics*, MAG-12, 327.

Nielsen, J. W. (1979) *Ann. Rev. Mater. Sci.* **9**, 87.

Nobumasa, H., Shimizu, K. & Kawai, T. (1990) *Physica C*, **167**, 515.

Norris, J. R. & Meisel, D. (eds) (1989) *Photochemical Energy Conversion*, Elsevier, New York.

Nozik, A. J. (1976) *Appl. Phys. Lett.* **29**, 150.

Nozik, A. J. (1978) *Ann. Rev. Phys. Chem.* **29**, 189.

Nozik, A. J. (1980) *Faraday Discussions of the Chemical Society* No. 70.

Nunez-Regueiro, M., Tholence, J. L., Antipov, E. V., Capponi, J. J. & Marezio, M. (1993) *Science* **262**, 976.

Nuttall, C. J., Bellitto, C. & Day, P. (1995) *JCS Chem. Commun.* 1513.

Ohba, M., Maruono, N., Okawa, H., Enoki, T. & Latour, J. M. (1994) *J. Amer. Chem. Soc.* **116**, 11566.

Ohta, Y. Tohyama, T. & Mackawa, S. (1990) *Physica C*, **167**, 515.

O'Keeffe, M. (1973) in *Fast Ion Transport in Solids* (ed. van Gool, W.) North-Holland, Amsterdam.

Ovshinsky, S. R., Fetcenko, M. A. & Ross, J. (1993) *Science* **260**, 176.

Owen, A. E., Firth, A. P. & Even, P. J. S. (1985) *Phil. Mag.* **B52**, 347.

Owen, J. J., Cheetham, A. K., Nighman, N. A., Jarman, R. H. & Thrush, R. J. (1994) *J. Opt. Soc. Am. B*, **11**, 919.

Pei, Yu., Verdaguer, M., Kahn, O., Sletten, J. & Renard, J-P. (1986) *J. Amer. Chem. Soc.* **108**, 7428.

Pierce, D. T. & Spicer, W. E. (1972) *Phys. Rev.* **B5**, 3017.

Pomerantz, M., Dracol, F. H. & Segmuller, A. (1978) *Phys. Rev. Lett.* **40**, 246.

Pope, M. & Swenberg, C. E. (1982) *Electronic Processes in Organic Crystals*, Clarendon Press, Oxford.

Popplewell, J. (1984) *Phys. Technl.* **15**, 150.

Pourarian, F. & Wallace, W. E. (1982) *J. Less-Common Metals* **87**, 275.

Rabenau, A. (1978) *Adv. Solid State Phys.* **18**, 77.

Ramakrishnan, T. V. (1992) *J. Indian Inst. Sci.* **72**, 279.

Ramasesha, S. & Das, P. K. (1990) *Chem. Phys.* **145**, 343.

Rao, C. N. R. (1991a) *Phil. Trans. Royal Soc. London* **A336**, 595.

Rao, C. N. R. (ed) (1991b) *Chemistry of High Temperature Superconductors*, World Scientific, Singapore.

Rao, C. N. R., Gopalakrishnan, J., Santra, A. K. & Manivannan, V. (1991) *Physica C*, **174**, 11.

Rao, C. N. R., Nagarajan, R., Ayyappan, S., Mahesh, R. & Subbanna, G. N. (1993) *Solid State Commun.* **87**, 551.

Rao, C. N. R. & Seshadri, R. (1994) *MRS Bulletin* **19**(11), 28.

Rao, C. N. R. & Ganguli, A. K. (1995) *Chem. Soc. Rev.* **24**, 1.

Rao, K. J. (1981) in *Preparation and Characterization of Materials* (eds Honig, J. M. & Rao, C. N. R.) Academic Press, New York.

Raveau, B., Michel, C., Hervieu, M. & Groult, D. (1991) *Crystal Chemistry of High T_c Superconducting Copper Oxides*, Springer-Verlag.

Raynes, E. P. (1983) *Phil. Trans. Roy. Soc. London* **A309**, 167.

Reilly, J. J. (1978) in *Hydrides for Energy Storage* (eds Andersen, A. F. & Meeland, A. J.) Pergamon Press, Oxford.

Reilly, J. J. & Sandrock, G. D. (1980) *Scientific American* **242**, 98.

Revcolevschi, A. & Dhalenne, G. (1985) *Nature* **316**, 335.

Roberts, G. G. (1980) *Thin Solid Films* **68**, 135.

Robinson, A. L. (1984) *Science* **223**, 920.

Rosensweig, R. E. (1982) *Scientific American* **247**, 124.

Roy, R., Agrawal, D. K. & McKinstry, H. A. (1989) *Annu. Rev. Mater. Sci.* **19**, 59.

Sanders, H. J. (1984) *Chem. Eng. News* **62**(28), 26.

Santra, A. K., Sarma, D. D. & Rao, C. N. R. (1991) *Phys. Rev.* **B43**, 5612.

Scheffer, T. J. (1983) *Phil. Trans. Roy. Soc. London* **A309**, 189.

Sen, P. K. & Sen, S. (1983) *Solid State Ionics* **9** & **10**, 1015.

Shanks, I. A. (1982) *Contemp. Phys.* **23**, 65.

Sheats, J. E., Pittman, C. U. & Carraher, Jr, C. E. (1984) *Chem. Britain* **20**, 709.

Sheft, I., Gruen, D. M. & Lamich, G. J. (1980) *J. Less-Common Metals* **74**, 401.

Shenoy, G. K., Niarchos, D., Viccaro, P. J., Dunlap, B. D., Aldred, A. T. & Sandrock, G. D. (1980) *J. Less-Common Metals* **73**, 171.

Siegrist, T., Zahurak, S. M., Murphy, D. W. & Roth, R. S. (1988) *Nature* **334**, 231.

Skou, E., Anderson, I. G. K., Simonson, K. E. & Anderson, E. K. (1983) *Solid State Ionics* **9 & 10**, 1041.

Slama-Schwok, A., Ottolenghi, M. & Avnir, D. (1992) *Nature* **355**, 240.

Sleight, A. W. (1988) *Science* **242**, 1517.

Sleight, A. W. (1995) *Endeavour* **19**, 64.

Sleight, A. W., Gillson, J. L. & Bierstedt, P. E. (1975) *Solid State Commun.* **17**, 27.

Sreedhar, K. & Rao, C. N. R. (1990) *Mat. Res. Bull.* **25**, 1235.

Strock, L. W. (1936) *Z. Phys. Chem.* **B31**, 132.

Stucki, F. & Schlapbach, L. (1980) *J. Less Common Metals* **74**, 143.

Stumpf, H. O., Ouahab, L., Pei, Y., Grandjean, D. & Kahn, O. (1993) *Science* **261**, 447.

Swallen, J. W. (1991) *Annu. Rev. Mater. Sci.* **121**, 373.

Takayama-Muromachi, E. (1995) *ISTEC Journal* **8**, 57.

Tallon, J. L. & Williams, G. U. M. (1990) *J. Less-Common Metals* **164–5**, 70.

Tamura, M., Nakazawa, Y., Shiomi, D., Nozawa, K., Hosokoshi, Y., Ishikawa, M., Takahashi, M. & Kinoshita, M. (1991) *Chem. Phys. Lett.* **186**, 401.

Taylor, G. N. & White, D. L. (1974) *J. Appl. Phys.* **45**, 4718.

Thrash, R. J. & Johnson, L. F. (1994) *J. Opt. Soc. Am. B* **11**, 881.

Tofield, B. C., Dell, R. M. & Jensen, J. (1978) *Nature* **276**, 217.

Torrance, J. B. & Metzger, R. B. (1989) *Phys. Rev. Lett.* **63**, 1515.

Tranquada, J. M., Sternlieb, B. J., Axe, J. D., Nakamura, Y. & Uchida (1995) *Nature* **375**, 561.

Tubandt, C. & Lorenz, E. (1914) *Z. Phys. Chem.* **87**, 513.

Tuller, H. L., Button, D. P. & Uhlmann, D. R. (1980) *J. Non-Cryst. Solids* **40**, 93.

van Mal, H. H., Buschow, K. H. J. & Miedema, A. R. (1974) *J. Less-Common Metals* **35**, 65.

Varma, K. B. R., Subbanna, G. N., Guru Row, T. N. & Rao, C. N. R. (1990) *J. Mater. Res.* **5**, 2718.

Verma, C. M., Littlewood, P. B., Schmitt-Rink, S., Abrahams, E. & Ruckenstein, A. E. (1989) *Phys. Rev. Lett.* **63**, 1996.

Wallace, W. E. (1973) *Rare Earth Intermetallics*, Academic Press, New York.

Wallace, W. E. & Narasimhan, K. S. V. L. (1980) in *Science and Technology of Rare Earth Materials* (eds Subbarao, E. C. & Wallace, W. E.) Academic Press, New York.

Wang, Y. (1991) *Acc. Chem. Res.* **24**, 133.

Weaver, J. W. (1992) *J. Phys. Chem. Solids*, **53**, 1433.

Weber, N. & Kummer, J. T. (1967) *Adv. Energy Conv. Eng.* ASME Conf. Florida 913.

Whangbo, M-H., Kang, D. B., Torardi, C. C. (1989) *Physica C*, **158**, 371.

White, R. M. (1985) *Science* **229**, 11.

White, S. R., Noack, R. M. & Scalapino, D. (1994) *Phys. Rev. Lett.* **73**, 886.

Whittingham, M. S. (1976) *Science* **192**, 1126.

Whittingham, M. S. & Huggins, R. A. (1971) *J. Electrochem. Soc.* **118**, 1.

Williams, D. J. (ed.) (1983) *Nonlinear Optical Properties of Organic and Polymeric Materials*, ACS Symposium Series 233, American Chemical Society, Washington.

Williams, J. M., Schultz, A. J., Geiser, U., Carlson, K., Kini, A. M., Wang, H. H., Kwok, W-K, Whangbo, M-H. & Schirber, J. E. (1991) *Science* **252**, 1501.

Williams, J. O. (1983) *Chem. Britain* **19**, 713.

Wright, J. D., Chadwick, A. V., Meadows, B. & Miasik, J. J. (1983) *Mol. Cryst. Liq. Cryst.* **93**, 315.

Wrighton, M. S. (1983) *J. Chem. Educ.* **60**, 877.

Yang, Z., Sokolik, Y. & Karasz, F. E. (1993) *Macromolecules* **23**, 1188.

Yao, Y. Y. & Kummer, J. T. (1967) *J. Inorg. Nucl. Chem.* **20**, 2453.

Yesodharan, E. & Grätzel, M. (1983) *Helv. Chim. Acta* **66**, 2145.

Zallen, R. (1983) *The Physics of Amorphous Solids*, John Wiley, New York.

Zallen, R., Drews, R. E., Emerald, R. L. & Slade, M. L. (1971) *Phys. Rev. Lett.* **26**, 1564.

Zhang, F. C. & Rice, T. M. (1988) *Phys. Rev.* **B37**, 3759.

Zhang, H. & Sato, H. (1993) *Phys. Rev. Lett.* **70**, 1697.

Zhang, H., Wang, Y. Y., Zhang, H., Dravid, V. P., Marks, L. D., Han, P. D., Paynes, D. A., Radaelli, P. G. & Jorgensen, J. D. (1994) *Nature* **370**, 352.

Zhong, X. P., Radwinski, R. J., de Boer, F. R., Jacobs, T. H. & Buschow, K. H. J. (1990) *J. Mag. Mag. Matls.* **86**, 333.

8   Reactivity of solids

8.1   Introduction

The study of reactions involving solids is an important aspect of solid state chemistry from the point of view of understanding the influence of structure and imperfections on the chemical reactivity of solids. It is important to identify the factors governing solid state reactivity in order to be able to synthesize new solid materials with desired structure and properties. Solid state reactions differ from those in homogeneous fluid media in a fundamental respect; whereas reactions in the liquid or the gaseous state depend mainly on the intrinsic reactivity and concentration of the chemical species involved, solid state reactions depend to a large extent on the arrangement of the chemical constituents in crystals. The fact that the constituents are fixed in specific positions in crystals introduces a new dimension in the reactivity of solids, not present in other states of matter. In other words, chemical reactivity is determined more often by the crystal structure and defect structure of solids rather than by the intrinsic chemical reactivity of the constituents. This feature of solid state reactions is clearly brought out in topochemical and topotactic reactions. Most of the photochemical transformations of organic solids are controlled by crystal chemistry. We shall discuss organic solid state reactions in some detail and devote our attention to intercalation reactions of solids that have gained considerable importance recently. We shall briefly deal with some aspects of catalysis to point out how solid state chemistry plays a role in this crucial branch of chemical technology.

8.2   Nature of solid state reactions

It is convenient to classify solid state reactions (Gomes & Dekeyser, 1976) under the following categories: (i) solid $\rightarrow$ products, as in decomposition and polymerization reactions; (ii) solid + gas $\rightarrow$ products, as in oxidation; (iii) solid + solid $\rightarrow$ products, as in the formation of complex oxides from simpler components; (iv) solid + liquid $\rightarrow$ products, as in intercalation and (v) reactions of solid surfaces. Solid state reactions may involve one or more of the following elementary (rate-limiting) steps: (i) adsorption (and desorption) of gaseous species on the solid surface; (ii) chemical reaction on an atomic scale; (iii) nucleation of a new phase and (iv) transport phenomena through solids. In addition there are external factors such as temperature, ambient atmosphere, irradiation, etc. that affect the reactivity. Various types of inorganic solid state reactions and the factors affecting reactivity of solids have been reviewed in the literature (Hannay, 1976). Sintering and phase transitions also exhibit features similar to those of solid state reactions, but they cannot be considered as reactions in the chemical sense. In fact, in considering a transition to be polymorphic, one assumes that there is no chemical change.

Several factors affect solid state reactions, especially solid–solid reactions (Gomes & Dekeyser, 1976; Boldyrev, 1979, 1996). Of these, particle size, gas atmosphere and foreign additives are of importance. Increase of reactivity with decreasing particle size is well known in heterogeneous catalysis; small particle size also favours solid–solid reactions. Gas atmosphere can have significant effects on reactivity if the gas is also a component exchangeable between solid phases; the formation of a spinel such as $ZnFe_2O_4$ is favoured in the presence of air (rather than nitrogen). Doping with foreign additives has been found to affect reaction rates. The reaction $ZnO + CuSO_4 \rightarrow ZnSO_4 + CuO$ is favoured by doping ZnO with Li_2O. Since the reaction $Ge + 2MoO_3 \rightarrow GeO_2 + 2MoO_2$ involves electron transfer from Ge to MoO_3, n-type Ge is more reactive. Certain reactions are favoured if they are carried out at the phase transition temperature of one of the reactants (*Hedvall effect*); formation of $CoAl_2O_4$ is easier at the γ–α transition of Al_2O_3. Boldyrev (1979, 1996) has examined the factors affecting the total process rate, the reaction at a specific location (and in a required direction), and the reaction mechanism in solid state reactions. The method of preparation (prehistory) affects topochemical reactivity; the effect of mechanical treatment of solids is to introduce different types of defects, which in turn affect the total process rate. The ultimate aim is to control solid state reactivity not only in time but also in space. Factors affecting organic solid state reactions have been reviewed by Singh *et al.* (1994).

8.3   Reactions involving a single solid phase

Decomposition of inorganic solids and dimerization as well as polymerization of organic molecules in the solid state brought about by thermal or photochemical means belong to this category of reactions. Amongst such reactions, those of organic solids generally involve different structural principles and are discussed later in Section 8.8.

Decompositions of salts such as metal carbonates, oxalates, azides and ammonium perchlorate have been studied for quite some time and the underlying mechanisms are fairly well established. The crucial step in decomposition reactions, as established from kinetic studies, is *nucleation*. Kinetic data plotted as fraction of the solid decomposed vs. time show typical sigmoid curves (Fig. 8.1) consisting of at least three different periods: induction period (A), acceleration period (B) and decay period (C). These kinetic curves are interpreted in terms of a mechanism involving the initiation of the reaction at specific sites of the solid to give product nuclei (nucleation in region A), followed by their growth. Growth occurs by interfacial reaction at the product–reactant interface. As the reaction proceeds, the interface expands continuously, making the reaction accelerate up to the inflexion point B. Beyond this point, growing nuclei seem to overlap, leaving behind isolated chunks of unreacted material. The decreasing reaction rate in period C is likely to be due to the decrease in the interfacial area as the reaction goes towards completion. Various kinetic equations have been derived on the basis of this general model to account quantitatively for the kinetics of

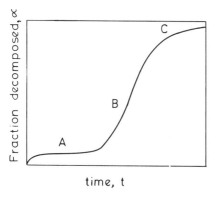

Figure 8.1   Typical solid decomposition curve showing fraction decomposed, α, as function of time, t.

solid decompositions. The equations differ from one another in terms of the exact mechanism assumed for nucleation and growth. In the *Avrami–Erofeev equation*

$$- \ln(1 - \alpha) = (kt)^n \tag{8.1}$$

where α is the fraction of reaction completed and k the rate constant, it is assumed that nucleation occurs randomly and isolated nuclei grow three-dimensionally. The *Prout–Tompkins equation*

$$\ln[\alpha/(1 - \alpha)] = kt \tag{8.2}$$

assumes that growth occurs by a chain-branching mechanism. No single kinetic equation generally fits the data over the entire course of a decomposition reaction. In the decay region C, the reaction frequently follows either the unimolecular law (8.3) or the contracting-cube law (8.4):

$$\ln(1 - \alpha)^{-1} = kt \tag{8.3}$$
$$[1 - (1 - \alpha)^{1/3}] = kt \tag{8.4}$$

Formation and growth of nuclei have been observed microscopically in single crystals in several decomposition and dehydration reactions.

Thermal decompositions are sensitive to the nature of pretreatment given to solids. Perhaps the earliest observation in this respect is that of Michael Faraday who noticed that the efflorescence of sodium carbonate was promoted by scratching the crystal with a pin. Modern interpretation of this observation is that dislocations produced by scratching provide sites for the reaction. Enhanced reactivity at dislocations has been demonstrated in several decomposition reactions (Thomas & Williams, 1971; Parasnis, 1970). Preirradiation of solids with neutrons, protons, or with UV-, X- or γ-radiation decreases the induction period and increases the rate of the decomposition without affecting the activation energy significantly. This observation has been taken to indicate that preirradiation increases the number of nuclei-forming sites, while retaining essentially the same mechanism. Although it is recognized that dislocations enhance the reactivity in several solid state processes, there is as yet no clear understanding of the exact role of dislocations. In general, enhanced reactivity at dislocations can arise from two factors. The core energy and the elastic strain energy associated with dislocations can aid in the nucleation process, bringing down the activation energy significantly. The abnormal stereochemistry of atoms or molecules at dislocations may also facilitate the attainment of a transition state that would otherwise be difficult to achieve in a perfect solid. In addition, there are factors such as trapping of electrons, holes and excitons in the neighbourhood of dislocations that may enhance the reactivity of thermally and photochemically stimulated decomposition of solids.

When the reactivity of a solid is controlled by the crystal structure, rather than by the chemical constituents of the crystal, the reaction is said to be *topochemically controlled*. The nature of products obtained in a decomposition reaction is frequently decided by topochemical factors, particularly when the reaction occurs within the solid without separation of a new phase (Thomas, 1974; Manohar, 1974). A *topotactic reaction* is a solid state reaction where the atomic arrangement in the reactant crystal remains largely unaffected during the course of the reaction, except for changes in dimension in one or more directions. Dehydration of $MoO_3 \cdot 2H_2O$ is a typical example of a topotactic reaction:

$$MoO_3 \cdot 2H_2O \rightarrow MoO_3 \cdot H_2O + H_2O \rightarrow MoO_3 + 2H_2O \qquad (8.5)$$

The two water molecules in $MoO_3 \cdot 2H_2O$ are structurally different (Fig. 8.2); one, in addition to oxygens, is coordinated to molybdenum, forming layers, and the other is held by hydrogen-bonding in between the layers. Günter (1972) found that the following topotactic relations exist between the dihydrate (D) and the monohydrate (M): $(010)_D \parallel (010)_M$; $[\bar{1}01]_D \parallel [001]_M$ and $[101]_D \parallel [100]_M$; between the monohydrate (M) and the anhydrous MoO_3(A), the relationships are: $(010)_M \parallel (010)_A$; $[100]_M \parallel [001]_A$ and $[001]_M \parallel [100]_A$. These relations indicate that the (010) plane remains essentially unchanged through the reaction. The layer structure is preserved in all three phases, with dehydration merely involving a collapse of the lattice along one of the crystallographic directions. A recent observation of importance is the fact that yellow $MoO_3 \cdot H_2O$ does undergo dehydration to the ReO_3-like structure under mild conditions; the metastable ReO_3-like MoO_3 transforms to the layered structure at high temperatures (Ganapathi et al., 1986). It has been shown that all compositions of $Mo_{1-x}W_xO_3 \cdot H_2O$ (including $x = 0$) undergo topochemical dehydration giving oxides of WO_3 structure (Ganapathi et al., 1986). Dehydration of layered $VOPO_4 \cdot 2H_2O$ and $VO(HPO_4) \cdot 0 \cdot 5H_2O$ to three-dimensional $VOPO_4$ and $(VO)_2P_2O_7$ respectively are also topotactic reactions of significance to catalytic oxidation of C_4 hydrocarbons to maleic anhydride (Tachez et al., 1982; Johnson et al., 1984).

Decomposition of WO_3, MoO_3, TiO_2, etc., yielding mixed-valence phases is another reaction of interest:

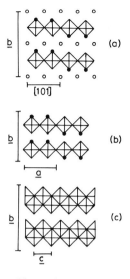

Figure 8.2 Topotactic dehydration of $MoO_3 \cdot 2H_2O$: (a) ($\bar{1}01$) projection of $MoO_3 \cdot 2H_2O$; (b) (001) projection of $MoO_3 \cdot H_2O$ and (c) (100) projection of MoO_3. Open circles denote interlayer water and filled circles, coordinated water. (After Günter, 1972.)

$$MO_n(s) \rightarrow MO_{n-x}(s) + \tfrac{1}{2}xO_2 \qquad (8.6)$$

where M = W, Mo, Ti, etc. In these reactions the oxygen loss does not occur randomly from the parent lattice to create isolated oxygen vacancies; instead oxygen loss is accommodated by collapse of the structure in specific crystallographic directions, creating crystallographic shear (cs) planes (Chapter 5). Decomposition of V_2O_5 to form V_6O_{13} involves the ($1\bar{3}0$) cs plane, and the two oxides intergrow (Colpaert *et al.*, 1973), just as the Magnéli phases W_nO_{3n-1} and W_nO_{3n-2}, based on (120) and (130) cs planes, intergrow in reduced WO_3 (Iguchi & Tilley, 1978).

8.4 Solid–gas reactions

Perhaps the most widely studied solid–gas reaction is the tarnishing of metals by reactive gases like oxygen and halogens (Hauffe, 1976; Smeltzer & Young, 1975). *Tarnishing* is essentially an oxidation reaction giving rise to solid products such as metal oxides and halides that form layers on the metal surface. The tendency of a metal to become tarnished in a particular atmosphere is governed by both thermodynamic and kinetic factors. The thermodynamic criterion is that the free-energy change in the reaction

$$M(s) + \frac{n}{2}X_2(g) \rightarrow MX_n(s) \qquad (8.7)$$

should be negative. This criterion is fulfilled in almost all the metal oxides, chalcogen-ides and halides, except those of the noble metals. There are certain metals, like aluminium, which resist tarnishing by oxygen and sulphur although the free energies of formation of the oxide and sulphide are negative. In order to understand such effects, we have to examine the kinetics and mechanism of solid–gas reactions.

The reaction of a metal with a gaseous reactant like oxygen consists of, first, a phase-boundary reaction forming a product layer on the metal surface. Further reaction depends on the nature of the product layer. If the layer is porous, it does not prevent access of the reacting gas to the metal and hence oxidation proceeds unimpeded; the reaction follows a linear law (rate is constant in time), irrespective of the thickness of the product layer. If, on the other hand, a compact nonporous product is formed, the reacting gas has no direct access to the metal; further reaction can proceed only by transport of at least one of the reactants through the product. The mechanism of metal oxidation crucially depends, therefore, on whether the product layer is nonporous or porous. A useful criterion in this regard is that of Pilling & Bedworth (1923), who suggested that a porous, nonprotective product layer is formed when the ratio of the molar volume of the product to that of the reacting solid is less than unity. Oxidation of alkali and alkaline-earth metals belongs to this category.

High-temperature oxidations forming fairly thick (> 1000 Å) product layers exhibit parabolic growth kinetics

$$\frac{\mathrm{d}x}{\mathrm{d}t} = \frac{k}{x}; \; x^2 = k_p t + C \tag{8.8}$$

where x is the thickness of the product layer at time t and k_p is the parabolic-rate constant. The *parabolic-rate* law is generally indicative of a mechanism involving rate-limiting diffusion of reactants through the product layer. When the oxidation occurs at fairly low temperatures, there is an initial rapid reaction followed by virtual cessation of the reaction, forming thin-film products (< 1000 Å). In thin-film products, the electroneutrality condition is not fulfilled, owing to the appearance of space charges (Cabrera & Mott, 1948), and so diffusion across the product layer under the influence of a chemical concentration gradient is not the rate-limiting step. Thin-film product growth follows a variety of kinetic laws such as logarithmic, linear, cubic and fourth power.

The oxidation of transition metals at high temperatures, giving thick product layers, follows parabolic growth kinetics. The reaction takes place by diffusion of either oxygen or metal. Oxides of metals in groups IIIB–VIB (with the exception of monoxides) grow by preferential transport of oxygen, while oxides of other transition metals grow by predominant transfer of metal ions.

The parabolic-rate law for the growth of thick product layers on metals was first reported by Tammann (1920), and a theoretical interpretation in terms of ambipolar diffusion of reactants through the product layer was advanced later by Wagner (1936, 1975). Wagner's model can be described qualitatively as follows: when a metal is

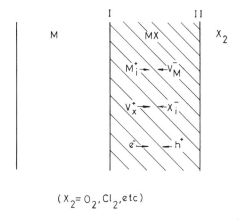

$(X_2 = O_2, Cl_2, etc)$

Figure 8.3   Schematic representation of oxidation of a metal M by a gas X_2.

exposed to oxygen at high temperatures, a compact thick product layer is formed which separates metal and oxygen (Fig. 8.3). Diffusion of reactants under the influence of concentration gradients in the product layer becomes rate-limiting. In principle, diffusion of the metal from interface I to II and of oxygen in the opposite direction is possible, but, in practice, diffusion of one of the reactants (either metal or oxygen) predominates, depending on the defect structure of the metal oxide in question. Wagner's model assumes that the transported species are ions and electrons/holes instead of neutral atoms.

There is a large body of data on the oxidation of metals that supports broad features of Wagner's model. The earliest example is the study of the oxidation of copper carried out by Wagner & Grünewald (1938). Although the metal forms two oxides, Cu_2O and CuO, the oxidation product is Cu_2O near atmospheric pressure and at high temperatures. Cu_2O is a p-type semiconductor containing copper vacancies, whose defect equilibrium is

$$O_2(g) \rightleftharpoons 2O + 4V_{Cu}^- + 4h^+ \qquad (8.9)$$

The oxidation is thought to proceed by reaction (8.9) at interface II, followed by subsequent diffusion of copper vacancies and holes from II to I; movement of copper vacancies determines the rate. Applying the mass-action law we obtain

$$[V_{Cu}^-] = [h^+] = \text{constant.} \ p_{O_2}^{1/8} \qquad (8.10)$$

The electrical conductivity of Cu_2O is expected to vary as $p_{O_2}^{1/8}$; experimentally it is found that the conductivity varies as $p_{O_2}^{1/7}$, suggesting that equation (8.10) roughly represents the true state of affairs in Cu_2O. Since the chemical potential of oxygen, μ_2, is given by $\frac{1}{2}RT \ln p_{O_2}$, the expression for the so-called rational-rate constant, k_r, is written as

$$k_r = \frac{1}{4}\left(\frac{RT}{F^2}\right)t_{Cu}\sigma^I\left[(p_{O_2}^{II})^{1/8} - (p_{O_2}^{I})^{1/8}\right] \tag{8.11}$$

where σ^I is the conductivity of Cu_2O in equilibrium with copper and t_{Cu} is the transference number of copper in Cu_2O. The experimental and calculated parabolic oxidation rate constants show amazingly good agreement. Copper oxidation becomes independent of oxygen pressure above 0.1 atm. This arises because of the formation of a layer of CuO on top of Cu_2O. Although the rate-determining step is still the diffusion of copper vacancies through Cu_2O, the concentration of copper vacancies at the Cu_2O/CuO interface is a fixed quantity dependent on the dissociation pressure of CuO and not on the external oxygen pressure.

Oxidation of zinc to zinc oxide is another example whose kinetics have been interpreted in terms of the Wagner model (Wagner & Grünewald, 1938). At 670 K, the reaction has been found to be independent of oxygen pressure between 0.02 and 1 atm. ZnO is a n-type semiconductor, having a stoichiometric excess of zinc accommodated as interstitials; the defect equilibrium could be represented as

$$ZnO = Zn_i^+ + e + \tfrac{1}{2}O_2(g) \tag{8.12}$$

The Wagner expression simplifies to

$$k_r = \frac{3}{2}\left(\frac{RT}{F^2}\right)t_{Zn}\sigma^{II}(p_{O_2}^{I} = \text{constant})^{-1/6} \tag{8.13}$$

Both ZnO and CuO are essentially electronic conductors. There are tarnishing reactions where the products are good ionic conductors (e.g. AgBr). In such instances, metal tarnishing is controlled by the mobility of electrons/holes through the product. Bromination of silver, forming AgBr, between 570 and 670 K is an example of this type; the reaction shows the following characteristics: (i) the kinetics follow a parabolic law with the rate constant $k \propto p^{1/2}$, (ii) the rate decreases by alloying silver with small quantities of divalent impurities like Cd, Zn or Pb and (iii) the rate increases by two orders of magnitude when silver and silver bromide are short-circuited through an external platinum wire. All these observations can be interpreted on the basis of the following mechanism: AgBr shows essentially ionic conduction arising from the Frenkel-type disorder of the silver sublattice:

$$V_i + Ag_{Ag} = Ag_i^+ + V_{Ag}^- \tag{8.14}$$

The parabolic growth of AgBr on Ag may be envisaged as occurring by the attack of bromine at interface II, forming Br^- ions and Ag vacancies; the vacancies are ionized partly, creating holes

$$\tfrac{1}{2}Br_2(g) = Br + V_{Ag}^- + h^+ \tag{8.15}$$

The hole concentration varies as the square root of bromine pressure. Formation of silver vacancies at interface II results in a concentration gradient in Ag_i across the product layer. The reaction proceeds by migration of Ag_i^+ from interface I to interface II and reverse migration of holes. The rate constant for silver bromination is approximately described by

$$k = \text{constant } [(p_{Br_2}^{II})^{1/2} - (p_{Br_2}^{I})^{1/2}] \tag{8.16}$$

where $p_{Br_2}^{II}$ and $p_{Br_2}^{I}$ are the bromine pressures at the $AgBr/Br_2$ and $Ag/AgBr$ interfaces respectively. Since $p_{Br_2}^{I} \ll p_{Br_2}^{II}$, the observed proportionality of rate constant k to the square root of the bromine pressure is consistent with the above expression for k. The influence of divalent metal additives is through V_{Ag} in equation (8.15). The increase in Ag/AgBr reaction rate when Ag and AgBr are short-circuited by an external Pt wire occurs because the rate-controlling hole mobility through AgBr is replaced by an easier electron flow in the opposite direction through an external circuit.

In contrast to the parabolic growth of products in metal–gas reactions at high temperatures, the low-temperature tarnishing of metals follows different kinetics. In several cases there is an initial period of extremely rapid oxidation to give thin films ($< 1000\,\text{Å}$) followed by virtual cessation of the reaction. When the films are thin, transport of reactants occurs in an electrochemical potential gradient arising from space-charge layers on the surface. The presence of an electrostatic field in a film during oxidation is established by surface potential measurements; its influence on the oxidation rate is confirmed by impressing an external potential difference on a growing film, which modifies the reaction rate. Furthermore, phenomena other than transport, such as nucleation and phase boundary reactions, may become rate-controlling; formation of products in the form of whiskers, blades and platelets also introduces complications in the early stages of oxidation. Accordingly, various kinetic laws, other than parabolic, such as linear, cubic, and logarithmic are found to be valid for thin-film oxidation. A linear rate law has been explained by Fehlner & Mott (1970) as arising from a rate-determining surface reaction between metal atoms at the surface and adsorbed gas; the cubic-rate law has been explained on the basis of field-aided transport through the product layer. The initial stages of oxidation, when gas-chemisorption and nucleation phenomena dominate, have been extensively investigated by Bénard & coworkers (1959) and by Gwathmey & Lawless (1960). An important result emerging from these studies is that different faces of a crystal exhibit different reactivities.

8.5   Solid–solid reactions

Two solids can react entirely within the solid phase, yielding solid products. Reactions of this type occurring between several inorganic solids such as metal oxides, halides, carbonates, sulphates and so on were studied some years ago by Hedvall (see Garner, 1955). They fall into two categories: addition reactions (e.g. $ZnO + Fe_2O_3 \rightarrow ZnFe_2O_4$) and exchange reactions (e.g. $ZnS + CdO \rightarrow CdS + ZnO$). The general mechanism of solid–solid reactions consists of the initial formation of one or more solid products that spatially separate the reactants. Subsequent reaction requires mass transport through product layers. When at least one of the reactants is in the form of a single crystal and the other in a compressed pellet form, the kinetics of the reaction follow a parabolic law analogous to the tarnishing of metals.

Solid–solid reactions are as a rule exothermic, and the driving force of the reaction is the difference between the free energies of formation of the products and the reactants. A quantitative understanding of the mechanism of solid–solid reactions is possible only if reactions are studied under well-defined conditions, keeping the number of variables to a minimum. This requires single-crystal reactants and careful control of the chemical potential of the components. In addition, a knowledge of point-defect equilibria in the product phase would be useful.

8.5.1   *Addition reactions*

The simplest cases of solid–solid addition reactions are those between ionic oxides AO and B_2O_3, forming AB_2O_4 spinels. Since the product is a ternary oxide, experimental conditions are uniquely defined if, in addition to temperature and pressure, the activity of oxygen is fixed. There are several careful studies of this reaction with single crystals reported in the literature (Schmalzried, 1971). The classical mechanism suggested by Wagner (see Schmalzried, 1976) for this reaction involves counter diffusion of cations through the product layer. Since the reaction occurs between ionic solids, it is reasonable to expect that the fluxes are provided by ions; moreover the fluxes due to differently charged ions must be coupled during the reaction to maintain electrical neutrality. The Wagner mechanism of counter diffusion of cations for spinel formation is shown in Fig. 8.4; other possible mechanisms are also indicated in this figure.

The various mechanisms may be grouped into two categories: (i) those involving ionic transport through the product layer only and (ii) those in which oxygen transport occurs through the gas phase in addition to cation transport through the product layer. The latter mechanism is possible only in cases where the reaction product exhibits electronic conduction and incorporates a transition-metal ion that can undergo oxidation-reduction.

In the formation of spinel oxides, the phase-boundary reaction at the AO/AB_2O_4 interface consists essentially of a homogeneous rearrangement of cations in the octahedral and tetrahedral sites of the cubic close-packed oxygen lattice, since the anion sublattice remains essentially the same in both the rocksalt (AO) and spinel structures. Accordingly, it is found that in the reaction between single-crystalline MgO and polycrystalline Cr_2O_3, the product growing at the MgO/Cr_2O_3 phase boundary is also mono-crystalline. What is surprising is that, in a similar reaction between polycrystalline NiO and single-crystalline Al_2O_3, the product formed at the $NiAl_2O_4/Al_2O_3$ phase boundary is also mono-crystalline. The reaction would involve two steps at this phase boundary: a change of anion-layer sequence from the hcp of α-Al_2O_3 to the ccp of spinel, followed by redistribution of cations. It has been suggested that a changeover of anion packing sequence occurs by a correlated motion involving dislocations rather than by diffusional motion.

The kinetics of spinel growth generally follow a parabolic-rate law when the

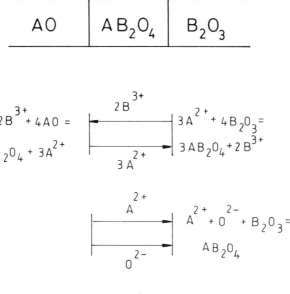

Figure 8.4   Reaction mechanisms for the formation of AB_2O_4 spinel by solid–solid reaction between AO and B_2O_3.

thickness of the product layer Δx exceeds $1\,\mu m$. The parabolic-rate law can be understood by assuming that a gradient in component activities (point-defects) exists across the product layer. The parabolic-rate constant has also been calculated with the knowledge of the self-diffusion coefficients of cations through the spinel and the Gibbs free energy of formation of the product. The calculation assumes certain reaction conditions to hold, but the agreement with experiment is quite good.

Inert markers have been used to obtain additional information regarding the mechanism of spinel formation. A thin platinum wire is placed at the boundary between the two reactants before the reaction starts. The location of the marker after the reaction has proceeded to a considerable extent is supposed to throw light on the mechanism of diffusion. While the interpretation of marker experiments is straightforward in metallic systems, giving the desired information, in ionic systems the interpretation is more complicated because the diffusion is restricted mainly to the cation sublattice and it is not clear to which sublattice the markers are attached. The use of natural markers such as pores in the reactants has supported the counter-diffusion of cations in oxide spinel formation reactions. A treatment of the kinetics of solid–solid reactions becomes more complicated when the product is partly soluble in the reactants and also when there is more than one product.

The foregoing treatment assumes that at least one of the reactants is a single crystal and the reactant/product geometry is well-defined. Many technologically important ceramic reactions, on the other hand, are usually carried out between polycrystalline powders. The reaction kinetics in these cases depend on several physical factors such as particle size, packing density, porosity, and so on. Jander (1927) and Carter (1961) have proposed models for powder reactions making several simplifying assumptions.

Efforts have been made to eliminate diffusion-control of solid–solid reactions by using superlattices of nanometric dimensions as reactants. Formation of $Cu_xMo_6Se_8$ from the superlattices of Cu, Mo and Se is one such example (Fister *et al.*, 1994). The results reveal that superlattice reactant geometry could be used to kinetically trap the ternary phases which are thermodynamically unstable with respect to the binary phases.

8.5.2   Exchange reactions

Exchange or double reactions in which two solids react to give two different products are more complicated and less completely understood. Two limiting mechanisms have been proposed, one by Jost (1937) and the other by Wagner (1938). Both the mechanisms assume that cations are far more mobile than anions. The Jost mechanism assumes that the reactants AX and BY are separated by the products BX and AY forming adjacent coherent layers in between. Progress of the reaction requires transport of A and B ions through the product layers, as is indeed the case in the reaction $MS + M'O \rightarrow M'S + MO$, where M = Pb or Zn and M' = Cd. Wagner found that many exchange reactions between solids are quite fast and cannot be explained in terms of the Jost mechanism, which requires mobility of the cations through both the products. Wagner suggested a closed ionic-circuit mechanism in which ions are required to move through their own products. The mechanism also does not require the products to be formed as adjacent layers; instead they form a mosaic structure that facilitates simultaneous diffusion of cations through their products. Metal exchange reactions such as $Cu + AgCl \rightarrow Ag + CuCl$ and $2Cu + Ag_2S \rightarrow 2Ag + Cu_2S$ are believed to follow the Wagner mechanism.

8.6   Solid–liquid reactions

Solid–liquid reactions are much more complex than solid–gas reactions and include a variety of technically important processes such as corrosion and electrodeposition. When a solid reacts with a liquid, the products may form a layer on the solid surface or dissolve into the liquid phase. Where the product forms a layer covering the surface completely, the reaction is analogous to solid–gas reactions; if the reaction products are partly or wholly soluble in the liquid phase, the liquid has access to the reacting solid, and chemical reaction at the interface therefore becomes important in determining the kinetics.

The simplest solid–liquid reaction is the *dissolution* of a solid in a liquid. The rate at which a solid dissolves in a liquid depends on the particular crystallographic plane (face) exposed. The effect of crystallographic planes on dissolution is clear from the observation that spherical single crystals acquire polyhedral shapes while dissolving. In the dissolution of zinc oxide in acids, the plane containing oxygens $(000\bar{1})$ is attacked more rapidly than the one containing zinc atoms (0001).

Like thermal decompositions, dissolution of solids is influenced significantly by dislocations. For example, etch pits are formed at sites where dislocations emerge on the crystal surface. It is for this reason that etching is a useful technique to render dislocations visible and even to determine their density. Measurements of the rate of growth of etch pits in $NiSO_4 \cdot 6H_2O$ have been made use of by Thomas $et\ al.$ (1971) to determine the lowering of activation energy for nucleation at dislocation sites. The correspondence between etch pits on one half of a freshly cleaved anthracene crystal and photodimerization reaction centres on the other half of the same crystal was shown by Thomas (1967).

Ion-exchange and acid-leaching are solid–liquid reactions. Ion-exchange reactions of several metal oxides have been discussed by Clearfield (1988). The exchange occurs mostly in aqueous solution or in molten salt media at relatively low temperatures and is accompanied by minimal reorganization of the structure of the parent solid. A typical example is the exchange of Li in $LiAlO_2$ with protons (Poeppelmeier & Kipp, 1988). This oxide crystallizes in three different structures, α, β and γ, of which only the α-modification, which adopts a rocksalt superstructure, undergoes facile exchange of Li^+ with H^+ in molten benzoic acid. The β- and γ-forms of $LiAlO_2$, where Li occurs in tetrahedral coordination, do not exhibit a similar ion-exchange. Unlike ion-exchange, acid-leaching of oxide materials may involve considerable reconstruction of the parent structure as shown in the case of chain silicates (Casey $et\ al.$, 1993) and tungstates like $LiAlW_2O_8$ (Bhuvanesh $et\ al.$, 1995).

Reactions between certain solids possessing layered structures (e.g. graphite, silicates, metal dichalcogenides such as TaS_2) and Lewis bases such as ammonia and pyridine, forming intercalation compounds, are discussed in the next section.

8.7   Intercalation chemistry

Intercalation refers to a solid state reaction involving reversible insertion of guest species G into a host structure [Hs]. The host provides an interconnected system of accessible unoccupied sites, $\square$. The reaction can be schematically represented as

$$x\mathrm{G} + \square_x[\mathrm{Hs}] \rightleftharpoons \mathrm{G}_x[\mathrm{Hs}] \tag{8.17}$$

The reaction is considered to be topotactic because the host matrix retains its integrity with respect to structure and composition in the course of intercalation and disintercalation. The first intercalation reaction (viz. intercalation of graphite by sulphate) was reported by Schauffautl as early as 1841, but it is only since the 1960s that intercalation chemistry has received wide attention, in view of its significance in a number of technically important problems such as reversible electrodes for high-energy density batteries, superconductivity and catalysis. Although intercalation traditionally referred to layered hosts, the subject nowadays also includes interactions in which the host is not layered (Thomas, 1983; Schöllhorn, 1984).

In addition to graphite, a wide variety of inorganic solids such as layered transition-metal oxides, chalcogenides, halides, oxyhalides, layered silicates, zeolites and even metal alloys have been used as host materials. Guest species range from simple atomic species such as hydrogen, alkali metals and halogens, through neutral molecules (H_2O, NH_3, long chain n-alkyl amines, amine oxides, heterocyclics) to complex organometallics such as cobaltocene, chromocene and dibenzene chromium. The subject has been extensively reviewed in the monographs edited by Levy (1979), and by Whittingham & Jacobson (1982).

8.7.1   *Graphite*

Graphite, with a van der Waals gap of 3.35 Å, is a unique host material in that it can be intercalated with both electron donors and electron acceptors. Alkali metals are among the most studied electron donors. The heavier alkali metals, K, Rb and Cs, having ionization energies lower than the electron affinity of graphite (4.6 eV), are easily intercalated into graphite. Alkali metal insertion results in expansion of the host structure along the c direction, the separation between the adjacent carbon planes being therefore dependent on the size of the alkali metal atom; the increase in c parameter is 0.38 Å for Li, 1.15 Å for Na, 2.0 Å for K, 2.3 Å for Rb and 2.60 Å for Cs. In addition to the geometrical change, an electron transfer is involved for the alkali metal to the graphite host as evidenced by the change from the diamagnetism of graphite to Pauli paramagnetism and a significant increase in the conductivity along the c direction. Several other physical properties also indicate that graphite is reduced when alkali metals are intercalated. Alkali intercalates of graphite are strongly coloured and take up hydrogen, presumably owing to formation of a hydride; for example, golden KC_8 is converted into blue $KH_{0.67}C_8$ in the presence of hydrogen. This ready reaction with hydrogen may be catalytically important, and is probably the reason for the formation of NH_3 when H_2 and N_2 are passed over the potassium intercalate of graphite (Sudo *et al.*, 1969).

A variety of electron acceptors such as HSO_4, NO_3, CrO_2Cl_2, CrO_3, MoO_3, Br_2 and metal halides like MoF_6, $FeCl_3$, $FeCl_2$ have been intercalated into graphite. In these compounds, graphite is the electron donor. The nature of electron transfer is revealed by Hall effect measurement, which shows p-type carriers. These graphitic salts are good ab-plane conductors, with conductivities comparable to that of metallic aluminium.

Structural studies reveal that the guest species intercalated into graphite are well-ordered, giving rise to *staging*. With the highest concentration, a monolayer of guest species is inserted between every pair of graphite layers (first–stage compound); with decreasing concentration, higher-stage intercalates are formed. In a second-stage compound, every alternate layer is occupied by the guest. In a third-stage compound, guest species are inserted between every third layer, and so on. Fig. 8.5 shows a schematic reprsentation of the first-, second- and third-stage intercalation compounds of graphite. One of the puzzling problems has been the mechanism of easy interconvertibility from one stage to the next. A domain model based on elastic forces, proposed by Daumas & Herold (1969), to account for staging has received direct support from electron microscopy (Thomas *et al.*, 1980). A study of $FeCl_3$ and $FeCl_2$ intercalates has revealed coexisting stages at the microstructural level (Fig. 8.6).

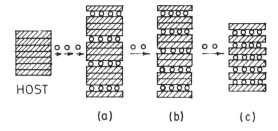

HOST

(a)   (b)   (c)

Figure 8.5 Schematic representation of 'staging' in graphite intercalation compounds: (a) third stage; (b) second stage and (c) first stage.

In the presence of liquid HF, graphite is found to be reversibly intercalated by fluorine. First-stage materials have the composition $C_{12}(HF)F_n (1 < n \leqslant 5)$, and the ultimate salt is the second-stage bifluoride $C_{12}HF_2$. Extensive studies on these intercalates by Mallouk *et al.*, (1985) have shown conductivities ranging from the metallic to the insulator regime.

8.7.2 *Transition-metal dichalcogenides*

Transition-metal dichalcogenides MX_2 of Groups IV, V and VI (M = Ti, Zr, Hf, V, Nb, Ta, Mo and W) possess hexagonal or rhombohedral layered structures. The building block of the structure of these compounds is the MX_2 sandwich, in which the transition metal M occupies either trigonal prismatic (TP) or octahedral (O) sites between two layers of close-packed chalcogen (X) atoms. These X–M–X sandwiches are stacked along the hexagonal *c*-axis. Bonding within the MX_2 units is strong (ionic covalent), whereas bonding between different MX_2 units is weak (van der Waals). It is this weak interlayer bonding and the associated van der Waals gap that make these chalcogenides two-dimensional materials exhibiting intercalation and other anisotropic physical properties. Various polymorphic modifications are known for the MX_2 chalcogenides, depending on whether the M atoms occur at the octahedral (or the trigonal prismatic) sites and the mode of stacking of the MX_2 layers along the *c*-direction.

Considerable attention has been devoted to the study of intercalation compounds of the dichalcogenides (Whittingham, 1978; Subba Rao & Shafer, 1979). Intercalation compounds of dichalcogenides can be divided into three categories: (a) compounds with Lewis base type molecules such as ammonia, *n*-alkylamines, pyridines etc.; (b) compounds with metal cations or molecular cations, Li, Na, K, etc., or $[(C_5H_5)_2Co]^+$ and (c) compounds containing both cations and neutral polar (solvated) molecules in the van der Waals gap.

A variety of organic intercalates of dichalcogenides (Jacobson, 1982) are known. The disulphides and diselenides form stable intercalates whereas the ditellurides do not. Intercalation results in an expansion of the lattice along the *c*-direction, the increase being characteristic of the guest. The stoichiometry and stability of the intercalates depend, however, on both the host and the guest molecule, highest stabilities being observed with $2HTaS_2$, $2HNbS_2$ and $1TTiS_2$.

Intercalation with organic compounds has been carried out by direct reaction of the

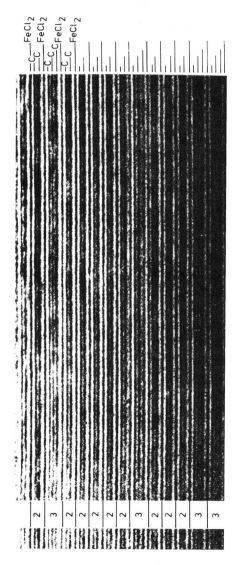

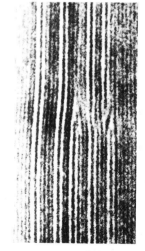

Figure 8.6 Electron-microscope images of graphite intercalates: (Top) image showing coexistence of second- and third-stage graphite–FeCl₂ intercalate; (Bottom) image of graphite–FeCl₃ intercalate showing interpenetration of differently staged regions. (Courtesy of G.R. Millward & J.M. Thomas, University of Cambridge.)

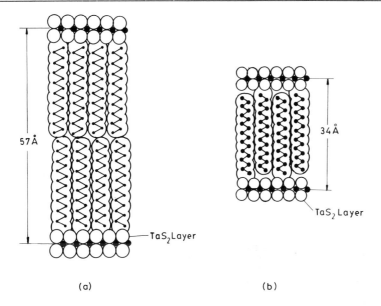

Figure 8.7   Structures of TaS_2-n-alkylamine intercalates: (a) bilayer and (b) monolayer configuration of the intercalated amine.

chalcogenide with the neat organic liquid, or in a solvent such as benzene or toluene. Progress of the reaction is followed by recording the lattice expansion using X-ray diffraction. A striking example of intercalation of this category is the reaction of n-alkylamies $(C_nH_{2n+1}NH_2)$, first reported for the superconducting host 2H TaS_2. The whole series of intercalates from $n = 1$ to $n = 18$ has been obtained by direct reaction. For values of n up to 4, the c-axis expansion is small and constant at $\sim 3.0\,\text{Å}$, probably indicating that the hydrocarbon chains lie parallel to the chalcogenide layers. For larger values of n, a definite stoichiometric phase, $A_{2/3}TaS_2$, is obtained whose c-parameter expansion increases linearly with the number of carbon atoms in the amine. In these, the hydrocarbon chain is perpendicular, or nearly so, to the chalcogenide layers. The stoichiometry and the c-axis expansion are indicative of a bilayer configuration of the amines (Fig. 8.7). For $n = 4$–9 amines, a bilayer structure is retained but the molecules are tilted at an angle of $56°$ to the c-direction. Staging is not a common phenomenon, although in stearamide intercalation in TaS_2 first- and second-stage compounds have been reported at 2:3 and 1:3 compositions respectively.

The dichalcogenides of group VI do not readily form intercalates with organic molecules. But n-alkylammonium compounds can be prepared by ion-exchange reactions of the hydrated sodium intercalation compounds such as $Na_{0.1}(H_2O)_{0.6}$ MoS_2. Alkylammonium compounds prepared by this route show lattice expansions that depend on the alkyl chain length, very similar to the MX_2-n-alkylamine systems described above. This raises the fundamental question whether protonated amines are the actual intercalated species in both the cases, involving reduction of the host. In fact, this constitutes the basis of the new model for the intercalation reaction in MX_2 proposed by Schöllhorn (1980).

Intercalation of pyridine (Py) and substituted pyridines in 2H NbS_2 and TaS_2 has been studied in some detail. The limiting composition is close to $(Py)_{0.5}MS_2$ for Ti, Nb and Ta chalcogenides, although smaller values are also known. Direct reaction of dry pyridine with 2H TaS_2 at 450–475 K gives two phases with slightly different c-axis, 23.7 Å and 24.1 Å. Electrochemical intercalation using TaS_2 as the cathode and pyridinium hydrochloride in methanol gives the 23.7 Å phase with hydrogen evolution. These pyridine intercalation compounds are of first-stage, every van der Waals gap being occupied by pyridine. Second-stage compounds can be prepared by controlling the pyridine concentration. Ammonia forms both first-stage and second-stage compounds with TaS_2. Lattice images clearly show repeat distances of 6, 12 and 18 Å between fringes of dark contrast, corresponding to TaS_2 layers, and of light contrast from empty or organic layers. Stacking faults are particularly prevalent in the second-stage material, in which random alternation of TaS_2 first-stage and second-stage regions was noticed (Moran *et al.*, 1971).

Some of the structural aspects of pyridine intercalates deserve mention. On intercalation, there is a change in the stacking sequence of 2H TaS_2 from AcA/BcB to AcA/AbA. Among the three possible orientations, one where the long CN axis is parallel to the dichalcogenide layer is favoured. Pyridine in both $(Py)_{1/2}NbS_2$ and $(Py)_{1/2}TaS_2$ seems to be ordered in two dimensions to give a rectangular superlattice with $2a\sqrt{3} \times 13a(a = a$ of $TaS_2)$. The volume of the supercell indicates that there is one pyridine molecule for every $2\frac{1}{6}TaS_2$ units, rather than for $2TaS_2$. This leads to the ideal composition $(Py)_{0.46}TaS_2$.

The nature of guest–host interaction in organic intercalates of dichalcogenides is still a matter of controversy. They are generally believed to be charge-transfer or donor–acceptor compounds, in which charge is transferred from the organic molecule to the empty (Ti, Zr, Hf) or half-filled (Nb, Ta) d_z^2 band of the chalcogenide. The group VI dichalcogenides with a filled low-lying d band do not directly intercalate. Evidence from a number of physical measurements supports charge transfer. The electric field gradient at Nb observed in Nb NMR of $(Py)_{1/2}NbS_2$ indicates a transfer of ~ 0.2 electrons to the niobium conduction band. Shifts of the nitrogen 1s level in the XPS also qualitatively indicate charge transfer to the host. The exact nature of interaction between the σ-lone pairs of the Lewis base (NH_3 or Py) and the orbitals of the host is difficult to visualize because the orientation of the guest molecules does not permit a direct overlap. The vibrational spectrum measured by inelastic neutron scattering is also not consistent with direct lone-pair σ-donation. To resolve this controversy, an alternative description of the bonding in pyridine and ammonia intercalation compounds has been proposed by Schöllhorn (1980). In this model the bonding is described as an electrostatic interaction between negatively charged host layers and intercalated cations, analogous to alkali and organometallic intercalates. For the case of pyridine–TaS_2 intercalates, the model may be formulated as

$$2 \text{ pyridine} \rightarrow \text{bipyridine} + 2H^+ + e\text{-}$$
$$x \text{ pyridine} + xH^+ \rightarrow x[PyH^+]$$
$$x[PyH^+] + (0.5 - x)Py + xe^- + TaS_2$$
$$\rightarrow [PyH^+]_x(Py)_{0.5-x}TaS_2.$$

The ionic model explains several experimental observations such as total volume decrease, shifts in nitrogen (1s) binding energy, charge transfer of about $0.25e^-$ per TaS_2 and sharp discontinuity in the voltage–charge curve at $0.25e^-/TaS_2$ during electrochemical intercalation. The ionic model also provides a unified description of bonding in all the intercalates of dichalcogenides.

A fascinating group of intercalates in metal dichalcogenides is that of organometallics such as cobaltocene and chromocene in TaS_2 (Dines, 1975). All the organometallics that have been intercalated so far are good reducing agents with low ionization energies (less than 6.2 eV). Accordingly, intercalation involves an electron transfer from the guest to the host. Direct evidence for electron transfer comes from magnetic studies; while the neutral cobaltocene is paramagnetic with one unpaired electron, it loses its magnetic moment when intercalated into TaS_2, indicating electron transfer.

The discovery of superconductivity in the molecular intercalates of layered dichalcogenides has led to considerable activity in this area, mainly in order to find out whether high superconducting-transition temperatures could be attained by intercalation. The failure to observe an enhancement of T_c leads to the belief that superconductivity is essentially confined to the two-dimensional metallic sheets in MX_2. Two experimental observations have confirmed this and suggest that interlaying coupling *via* Josephson coupling (tunnelling of superconducting pairs) between layers is not involved. The first indication came from the fact that T_c was unaffected by interlayer spacing as large as 60 Å in TaS_2 intercalation compounds. The second indication came from the isostructural cobaltocene and chromocene intercalates of TaS_2. The lattice expansions are identical in both the cases. However, intercalation results in diamagnetism in the case of cobaltocene and paramagnetism in the case of chromocene. Despite the presence of a localized magnetic moment in (chromocene)$_{1/4}TaS_2$, both the intercalates possess almost identical superconducting properties, showing clearly that interlayer Josephson coupling is not an essential ingredient of the superconductivity mechanism.

Among the metal atoms that are intercalated into MX_2 dichalcogenides, alkali and first-row transition metals have received considerable attention. In contrast to graphite, intercalates of electron acceptors are not known for the layered dichalcogenides. Monovalent cation intercalation compounds AMX_2 (A = monovalent cation, M = IVB or VB transition metal and X = S or Se) are important as possible electrode materials in solid state batteries. These are generally mixed conductors, the electronic part being associated with slabs of host material and the ionic part with the layers of the intercalate ions. Just as the M atom in MX_2 occurs in octahedral (O)(IT TiS_2) or trigonal prismatic (TP)(2H NbS_2, MoS_2) sites, the intercalated A atoms can also occur in O or TP sites. TP coordination is favoured for large alkali-metal ions and for large ionicity of the MX bond. Cu^+ and Ag^+ are found in tetrahedral sites. Staging is exhibited by AMX_2 when A is a large alkali-metal ion in TP sites. The intercalated A ions are especially ordered at low enough temperatures. Depending on the coordination geometry of the A ions, there are three possible two-dimensional sublattices for AMX_2: a triangular lattice for octahedrally coordinated A ions, a honeycomb lattice for trigonal prismatic A ions and a puckered honeycomb lattice for tetrahedral site A ions. In Na_xTiS_2, three different superlattices resulting from the ordering of TP coordinated sodium ions are found: two of them correspond to hexagonal (2×2) and $(\sqrt{3} \times \sqrt{3})$ lattices, while the third corresponds to $(2 \times \sqrt{3})$. These superlattices are seen at definite fractional compositions (x values).

Li_xTiS_2 is the single most important system in this family. Although in the earlier work it was reported as a single-phase system over the entire stoichiometry range $(0 \leqslant x \leqslant 1)$ (chemical potential varies rather smoothly with x), recent accurate measurement of chemical potentials with respect to composition has revealed pronounced structure at well-defined fractional compositions, indicating long-range order of Li^+. In addition, weak but sharp superstructure reflections were seen in diffraction patterns. Both $2a_0$ and $3a_0$ superstructures have been identified by diffuse X-ray analysis. Direct measurement of the entropy of intercalation provides additional evidence for ordered structures at $x = \frac{1}{12}, \frac{1}{7}, \frac{1}{4}$ and $\frac{1}{3}$ (Thompson, 1981).

Intercalation of both cationic and neutral (solvated) molecular species in MX_2 is particularly important when the reaction occurs in the presence of polar solvents, especially at low temperatures (Fig. 8.8). A simplified scheme of reactions which explains this type of intercalation has been proposed by Schöllhorn (1980):

$$[Hs] \overset{G,S}{\underset{-S}{\rightleftharpoons}} G_x^+(S)_y[Hs]^{x-} \rightleftharpoons G_x^+[Hs]^x$$

$$\updownarrow H_3O^+, H_2O \qquad\qquad (8.18)$$

$$H_x[Hs] \underset{H_2O}{\overset{+H_2O}{\rightleftharpoons}} (H_3O^+)_x(H_2O)_y[Hs]^{x-}$$

where Hs is the host, G the guest and S is the solvent. The above scheme should illustrate the complexity of topochemical intercalation reactions in solution when redox, solvation and exchange processes are involved. An important aspect of the solvated cations and the associated polyelectrolyte nature of these intercalation compounds is that intercalates of Lewis bases (L) can be considered as $(LH^+)_xL_y[Hs]^{x-}$, where the cation and solvated molecules differ only by an additional proton attached to the base.

8.7.3   *Other layered inorganic hosts*

Apart from the layered transition-metal dichalcogenides, there are a number of layered solids such as metal phosphorus trichalcogenides (e.g. MPS_3), metal chalcogenohalides, metal chalcogenocarbides and metal halides where intercalation has been studied. Transition-metal phosphorus trisulphides (Johnson, 1982) crystallize in a layered structure related to $CdCl_2$. By rewriting the formula as $M_{2/3}(P_2)_{1/3}S_2$, relationship with the $CdCl_2$ structure is seen; the sulphur atoms form a cubic close-packed array. Every alternate layer of octahedral sites between the close-packed layers is filled by transition-metal atoms and P_2 pairs. Perpendicular to the layer direction, the structure is held together by van der Waals forces, and it is in these van der Waals gaps that intercalation occurs. Another way of visualizing these solids is to consider them as metal salts of hexathiohypodiphosphate anion, $P_2S_6^{4-}$. MPS_3 phases are known for a number of

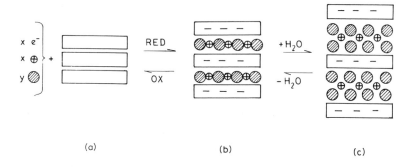

Figure 8.8   Formation of hydrated phases by topotactic redox intercalation reaction in aqueous electrolytes: (a) host lattice MX_2; (b) monolayer hydrate A_x^+ $(H_2O)_y[MX_2]^{x-}$ and (c) bilayer hydrate. (After Schöllhorn, 1980).

divalent metals such as Mg, Mn, Ni, Zn and Cd. A CdI_2 analogue of this structure is found among the selenides, $FePSe_3$, $MgPSe_3$, $MnPSe_3$ and so on. Just as with transition-metal dichalcogenides, three classes of guest species are known to be intercalated into MPS_3 and $MPSe_3$ hosts: organic molecules, alkali metals and organometallic compounds. Among the organic molecules, n-alkylamine intercalation into MPS_3 results in lattice expansion, indicating a bilayer configuration with the bulk of the alkyl chain oriented perpendicular to the layers. Structural details and the driving force for this type of intercalation are not clearly understood at present.

Alkali metals can be intercalated into MPS_3 either by chemical or electrochemical techniques. Chemical intercalation takes place when the host is reacted with an organo–alkali metal compound such as n-butyllithium or sodium napthalide. The reaction with n-butyllithium can be represented as:

$$xC_4H_9Li + MPS_3 \rightarrow Li_xMPS_3 + \frac{x}{2}C_8H_{18}$$

$NiPS_3$ gives a composition with lithium content as high as $Li_{4.5}NiPS_3$. Such large lithium contents found in chemical intercalation products probably result from an irreversible destruction of the host lattice to give a mixture of Li_2S and amorphous products. The possibility of electrochemical intercalation of lithium into MPS_3 makes them useful as cathode materials in secondary batteries. In $NiPS_3$, electrointercalation gives two single-phase regions of Li_xNiPS_3, $0 < x < 0.5$ and $0.5 < x < 1.5$. As long as x is less than 1.5, intercalation is electrochemically reversible. When the discharge is greater than $x = 1.5$, irreversible reaction takes place that results in the destruction of the host lattice.

The optical properties of MPS_3 and Li_xMPS_3 have been studied. The host phases are broad-band semiconductors with band gaps of 1.3–3.5 eV. The three compounds that are electrochemically active are $NiPS_3$, $FePS_3$ and $FePSe_3$, which possess the smallest band gaps. Upon intercalation with lithium, the transmittance decreases until the entire spectral region from 3300 to 500 nm is opaque. The d.c. resistivity behaviour parallels the optical properties. In $NiPS_3$ lithium intercalation decreases the resistivity by approximately eight orders of magnitude. The results suggest that an electron from lithium goes into a delocalized conduction band, resulting in increased electronic conductivity as well as free-carrier absorption and increased reflectivity in the optical spectra.

Magnetic susceptibility and Li^7 and P^{31} magnetic resonance results of $FePS_3$ and their lithium intercalates reveal that iron remains in the high-spin divalent state. In Li_xNiPS_3, electrons added by intercalation of lithium seem to go to the transition-metal $4s$ or higher empty sulphur orbitals. The disappearance of magnetism in Li_xNiPS_3 for $x > 0.5$ is not understood.

Among the large number of metal chalcogenohalides of the general formula MXY, only the oxyhalides of M = Ti, V, Cr, Fe and Al, and the thio- and seleno-halides of aluminium are known to undergo topochemical intercalation. Layered metal oxyhalides crystallize in four general structural types: (i) AlOCl structure in which Al^{3+} is four coordinated, (ii) FeOCl structure in which Fe^{3+} is six-coordinated, (iii) SmSI structure in which the metal is seven-coordinated and (iv) PbFCl structure in which lead is either eight- or nine-coordinated. Of these, only AlOCl and FeOCl structures function as host materials in intercalation (Halbert, 1982; Schafer-Stahl & Abele, 1980).

The guest species intercalated in dichalcogenide hosts are unsolvated alkali and transition-metal ions, solvated metal cations, organic and organometallic cationic species. For metal oxyhalide hosts, parallels are known for each class of these guests, although the range of hosts is limited to the transition-metal oxyhalides that have reducible M^{3+} cations. As in other hosts, the intercalation compounds can be prepared by chemical or electrochemical methods. Unsolvated Li^+ intercalation is reported for FeOCl, VOCl, and CrOCl. Solvated alkali-metal intercalation compounds, $M_x^+(H_2O)_y(FeOCl)^{x-}$ have been obtained by direct reduction of FeOCl using aqueous solutions of $M_4Fe(CN)_6$ (M = Li, Na, K, Cs). In these compounds, large expansions perpendicular to the layers have been found to occur. The solvated water molecules can be replaced by DMF, DMSO, and so on. Reduction of FeOCl by aqueous $FeSO_4$ in the presence of n-alkylammonium chlorides results in the formation of a series of compounds of the formula, $(RNH_3^+)_x(FeOCl)^{x-}$ in which the layer expansion increases with increasing length of the alkyl chain. Intercalation of metallocenes into FeOCl, VOCl and TiOCl has also been reported (Fig. 8.9). The more-easily reducible host, FeOCl, intercalates metallocenes with a higher ionization energy (e.g. ferrocene, 6.68 eV), illustrating an appropriate balance of redox properties of host and guest in the formation of these compounds.

8.7.4   *Insulating sheet structures*

Several insulating inorganic solids possessing sheet structures, for example, silicates belonging to the pyrophyllite family (Thomas, 1982), and acid phosphates (Alberti & Constantino, 1982; Clearfield, 1981) of some tetravalent metals form intercalation compounds with a variety of donor molecules; in these cases, intercalation does not involve a redox process, unlike in the cases of transition metal chalcogenides and

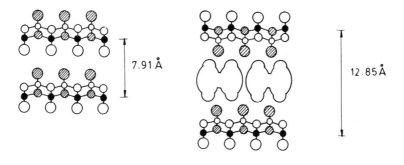

O	Cl at z = 0
⊘	Cl at z = 1/2
o	O at z = 0
⦰	O at z = 1/2
∘	Fe at z = 0
●	Fe at z = 1/2

Figure 8.9 A schematic representation of intercalation of cobaltocene into FeOCl. (Following Halbert, 1982.)

oxyhalides described in the previous section.

Pyrophyllites are typical examples of sheet silicates possessing the ideal formula $Al_2(Si_4O_{10})(OH)_2$ (see Section 1.5). They can be regarded as condensation–polymerization products of $Al(OH)_6$ sheets (O) and tetrahedral $Si_2O_3(OH)_2$ (T) sheets, the sequence of layers being TOT. Various substitutions of Si^{4+} and Al^{3+} are possible in these structures. When one-quarter of Si^{4+} in pyrophyllite is replaced by Al^{3+}, together with K^+ ions in the interlamellar space for charge compensation, the result is muscovite mica; $K[Al_2(AlSi_3O_{10})(OH)_2]$. In smectites (montmorillonite, hectorite, vermiculite, etc.), Li^+, Mg^{2+} and Fe^{2+} (or Fe^{3+}) replace Al^{3+} in octahedral layers, and Al^{3+} in turn replaces Si^{4+} in tetrahedral layers. The general formula of montmorillonites is $M_x[Al_{2-x}Mg_x](Si_4O_{10})(OH)_2$ and that of vermiculites, $M_x(Mg)_3[(Al_xSi_{4-x})O_{10}](OH)_2$, where M is a monovalent exchangeable cation. Important characteristics of these layered silicates are: (i) M ions occupying interlamellar space between the tetrahedral layers are exchangeable and (ii) substantial quantities of water present between the layers can be replaced (intercalated) by organic molecules. Smectites and vermiculites possess considerably large cation-exchange capacities, of the order of 100 milliequivalents per 100 g of the silicate. More importantly, the organic molecules intercalated into the interlamellar space in these silicates undergo unusual chemical reactions; it is this property that has been exploited to bring about novel chemical conversions of organic molecules using sheet silicates. Thermal decomposition, dimerization, oxidation–reduction, hydrogen-transfer, lactonization, ether formation, ester formation and conversion of primary amines to secondary amines are some of the reaction types that occur freely in sheet silicate intercalates. We shall cite a few specific examples: (i) Conversion of diprotonated 4, 4'-diamino-*trans*-stilbene montmorillonite to aniline by gentle heating of the intercalate:

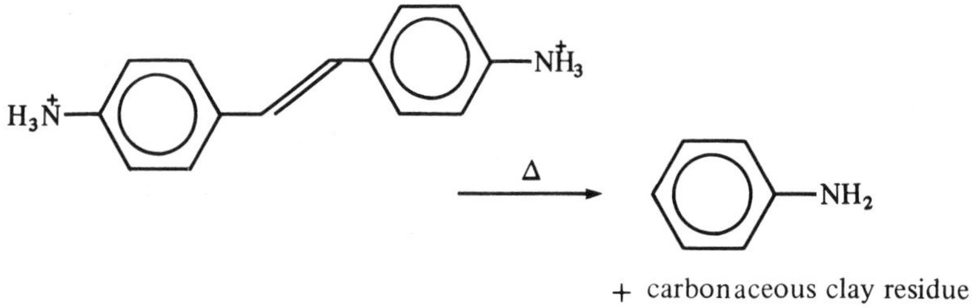

+ carbonaceous clay residue

(ii) Conversion of esters to alkenes and carboxylic acids occurs homogeneously at 670 K. But when the reactant is intercalated into smectites, the reaction is clean and occurs at much lower temperatures. For example, methylcyclohexane carboxylate intercalated into Al^{3+}-exchanged montmorillonite gave 98% conversion to cyclohexene and acetic acid at mild temperatures.

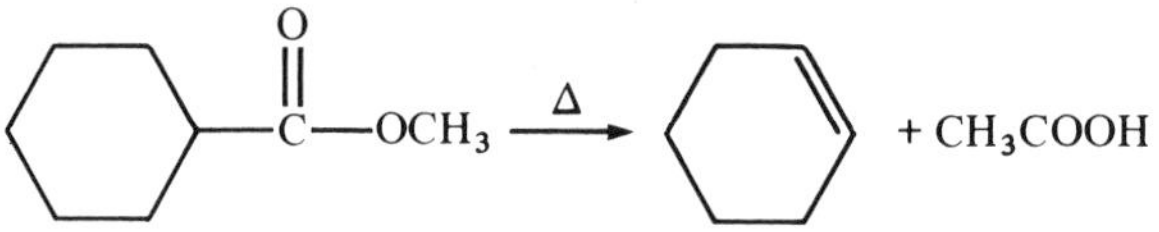

(iii) When 1, 1-diphenylethylene is refluxed in contact with Cu^{2+}-exchanged montmorillonite, 1-methyl-1,3,3,-triphenylindan is obtained in good yield:

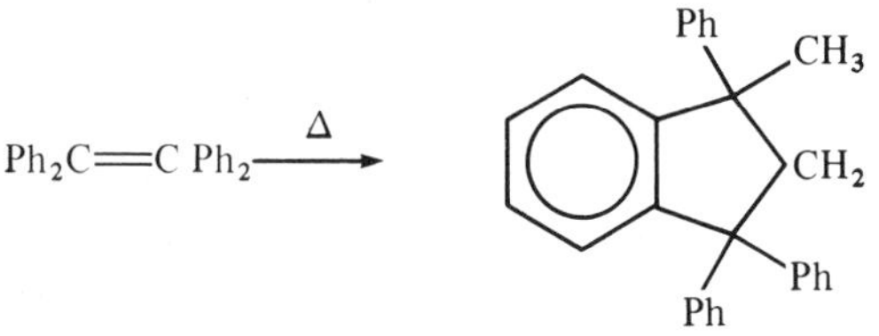

But the same 1, 1-diphenylethylene when reacted in the presence of fluorohectorite gives oxidized products, 1, 1-diphenylethane and benzophenone.

It is important to distinguish these novel reactions, which take place between the individual sheet silicates and involve the intercalation of one or more of the reactants, from those catalysed by clay mineral surfaces. While details of the mechanism of reactions involving intercalation in sheet silicates are not fully understood, it is generally known that acid sites present in sheet silicates (Brönsted, as well as Lewis) are involved in the reactions. Protonated intermediates such as carbonium or oxonium ions play a crucial role in such reactions.

Acid salts of tetravalent metals of the general formula, $M^{IV}(HXO_4)_2 \cdot nH_2O$ (M^{IV} = Ti, Zr, Sn; X = P, As) are another important class of insulating sheet materials that form intercalation compounds. The salts can be obtained in various crystalline forms, possessing fibrous, layered or three-dimensional structures. The layered acid salts in turn exist in two different modifications, α and β. The structure of the α-form of a typical acid salt, $Zr(HPO_4)_2 \cdot H_2O$, consists of a layer of zirconium atoms sandwiched between two sheets of phosphate groups. Three oxygens of each PO_4 are linked to three zirconium atoms so that each zirconium is octahedrally coordinated with six oxygens of six different phosphate groups. The fourth oxygen of each phosphate bears a replaceable proton; this can be replaced partly or fully by other cations. The layers so constituted are packed one over the other to produce a sheet structure. Because of the presence of strongly polar $\geqslant$P—OH groups, these materials can intercalate a number of proton acceptors such as ammonia, amines and alkanols. Owing to their novel properties, such as selectivity for certain cations, high stability in strongly acidic and or oxidizing media, high thermal stability and resistance to irradiation, acid salts of zirconium phosphate type have found use as inorganic ion exchangers, especially where organic resins cannot be employed because of their degradability.

There have been many interesting contributions in the last decade to the area of intercalation chemistry. Some of them are, intercalation-polymerization of organic molecules (e.g. aniline and pyrrole) in layered hosts such as V_2O_5 gel, FeOCl and 2:1 clays, exfoliation and restacking of layered solids such as MoS_2 in appropriate polar solvents to encapsulate organic/ organometallic guest species, oriented incorporation of hyperpolarizable organic molecules in layered $MnPS_3$ and $CdPS_3$ to induce large second-order optical nonlinearity and pillaring of robust molecular props in layered hosts to impart zeolite-like properties (See Section 3.2.3 for details).

8.7.5 *Framework hosts*

Rare-earth transition-metal alloys like $LaNi_5$, tunnel structures based on the WO_3 framework and zeolites are some of the three-dimensional hosts where intercalation has been intensively investigated. Reversible intercalation of hydrogen atoms by intermetallic compounds to give ternary hydrides such as $LaNi_5H_6$, TiFeH, $ThFe_3H_5$ has attracted attention in view of potential use of these solids for hydrogen storage (Section 7.5). Insertion of electron donors (Li, H) into three-dimensional oxide hosts such as WO_3, V_2O_5, etc., is also technically important as electrodes for solid state batteries and in electrochromic devices.

Among three-dimensional framework hosts forming intercalation compounds, zeolites (see Section 1.5) have attracted considerable attention because of their technical importance (Derouane, 1982; Dyer, 1984). Intracrystalline voids in anhydrous zeolites provide a strongly polar environment that can be filled with polar or ionic species to increase the crystal energy. Treatment of zeolites (e.g. Na–Y zeolite) with vapours of sodium results in the formation of a red product consisting of Na_4^{3+} intercalated into the zeolite cavities (Thomas, 1983). The locations of these clusters inside the zeolites are not yet known, but they seem to be buried within the rather inaccessible cavities inside zeolites, with the result that ordinary solvents do not reach them. Such intercalation of ionic species in zeolites may have implications in nuclear-waste treatment and storage.

Host–guest chemistry involving three-dimensional framework hosts has been exploited in recent years to impart novel properties. For instance, nanometre-sized semiconductors confined in the cavities of zeolite hosts exhibit quantum size effects in their electronic properties (Stucky & MacDougall, 1990). CdS clusters have been synthesized within the cavities of zeolite Y (Herron *et al.*, 1989). At low concentrations, the clusters are located as tetramer units in the sodalite cages of the host. The isolated clusters are distinguished by their unique optical properties that are considerably different from those of bulk CdS. The absorption threshold of the cluster is around 350 nm compared to 540 nm of bulk CdS. Moreover, the intense emission seen in bulk CdS is completely wiped out even at 4 K in the clusters encapsulated into the zeolite. The optical properties drastically change at higher CdS loadings. The absorption edge shifts to 420 nm, an excitonic feature evolves around 350 nm and a strong emission appears. These changes are ascribed to the formation of superclusters by a percolative process within the host. The unique optical properties which are different from those of conventional colloidal CdS imply a strong host–guest interaction. Other semiconductors such as GaP, elemental Se and Te, and even metal oxides such as those of tungsten and molybdenum have been quantum-confined in zeolite hosts (Stucky, 1992; Ozin *et al.*, 1992). Current research in this direction is motivated by the possibility of using quantum confinement to create optical transistors and optical data storage devices, among others.

Zeolite hosts have been used to encapsulate conducting polymers. For instance, aniline has been polymerized within the channels of mordenite and zeolite Y (Enzel & Bein, 1989). The same strategy has been employed to produce conducting polyaniline filaments within the pores of MCM-41 host (Wu & Bein, 1994a). Oriented filaments of polyaniline encapsulated into 30 Å mesopores of the host exhibit significant electronic conduction. Conducting graphite-like carbon wires encapsulated into the channels of the same host, produced by the pyrolysis of intercalated polyacrylonitrile, exhibit a microwave conductivity that is 10 times that of bulk carbonized polyacrylonitrile. Fabrication of such ordered conducting filaments (molecular wires) in nanometre-sized channels of inorganic hosts is a notable step towards the development of nanometre electronics (Wu & Bein, 1994b).

Another interesting aspect of host–guest chemistry is the possibility to perform 'ship-in-the-bottle' synthesis of organic molecules within the restricted regions of zeolite hosts. An example of this kind is the regioselective photodimerization of enones within the supercages of X- and Y-faujasites (Lem *et al.*, 1993). Photochemical reactions of organic molecules within the constrained regions of zeolite hosts have been discussed by Ramamurthy *et al.* (1992). Host–guest chemistry has been used to direct alignment of ensembles of hyperpolarizable molecules to induce macroscopic optical nonlinearity. Thus, when *p*-nitroaniline is absorbed in ALPO-5 (Cox *et al.*, 1988), the noncentric host forces polar alignment of the guest molecules imparting the property of second harmonic generation to the host–guest complex.

The brief overview of intercalation chemistry presented above shows that the subject encompasses a fascinating variety of solid state reactions between two- or three-dimensional host materials and guest atoms and molecules. Not all host material that exhibit intercalation are covered in this presentation. For instance, we have not included intercalation in β-aluminas and Chevrel phases (see Section 1.10), and reactions of linear-chain systems such as $NbSe_3$ and $AFeS_2$ incorporating lithium and similar metals have not been discussed. The subject of intercalation has received wide attention because of its relevance in a variety of technically important problems. As discussed in Chapter 7, fast ion transport in sodium β-aluminas has led to their use as solid electrolytes in the Na—S battery; reversible intercalation of lithium in TiS_2 has found application in cathodes in high-energy density batteries. Hydrogen bronzes of some transition-metal oxides are useful in nonemissive electrochromic display systems. Graphite intercalates are important in catalytic processes; similarly MoS_2 and WS_2 intercalated with Ni/Co are the major hydrodesulphurization catalysts. Intercalates of TaS_2 and Chevrel phases are high-temperature superconductors.

8.8   Reactions of organic solids

Chemical reactivity of organic solids differs from that of inorganic and metallic solids in some respects (Singh *et al.*, 1994). In contrast to metals and most of the compounds, organic compounds crystallize as molecular solids (Chapter 1). The role of imperfections on reactivity is also different; point defects, which play an important role in the reactions of other classes of solids, do not play as significant a role in the reactivity of organic solids, although they may contribute to diffusion. Energies required for the formation of vacancies and interstitials in organic solids are of the same order of magnitude as their enthalpy of sublimation, and the concentration of point defects would therefore be negligible even near the melting point. Line defects and planar defects play a more important role in the reactivity of organic solids.

In the reactivity of organic solids, crystallographic and environmental factors are far more important than the intrinsic reactivity of the molecule. It has long been recognized that the reactivity of a crystal of an organic compound could be different from the reactivity of the free molecules constituting the crystal. Libermann found as early as 1889 that *trans*-cinnamic acid dimerizes in the solid state while it merely isomerizes in solution. The following facts have emerged from the investigations of organic solid state reactions over the last two decades (Schmidt, 1971; Cohen & Green, 1973; Cohen, 1975; Morawetz, 1977; Thomas, 1974, 1979): (i) the products of an organic solid state reaction are fewer in number and frequently different from those obtained from the same reaction in fluid medium; (ii) certain reactions proceed far more rapidly in the solid state than in the liquid state and (iii) organic solids invariably crystallize in many polymorphic forms and their reactivities are quite different, for while certain polymorphs react to give distinct products, others do not react at all. Striking illustrations of the differences in reactivity between solid state and homogeneous media are shown in Fig. 8.10.

In the absence of defects, the reactivity of organic solids is mainly determined by molecular packing. Reactions in which the crystal structure holds sway over intrinsic molecular reactivity are said to be 'topochemically controlled' (Thomas, 1974). A classic example of a *topochemically controlled organic reaction* in the solid state is the photodimerization of *trans*-cinnamic acids studied by Schmidt *et al.* (see Ginsburg,

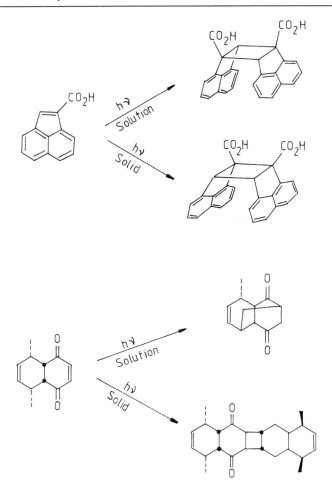

Figure 8.10   Illustration of the differences in reactivity of organic molecules in the solid state and in a homogeneous medium. (Following Thomas, 1979.)

1976). The monomer acids (e.g. *o*-ethoxy-*trans*-cinnamic acid) crystallize in three polymorphic forms: α, β and γ. In all three forms the monomers are packed in one-dimensional stacks, adjacent stacks being connected by hydrogen bonding. The three structures differ from one another in the mode of stacking and the repeat distance between molecules in the stack. The α-polymorph, where the molecules are centrosymmetrically related with a separation of about 4 Å between the double bonds, yields a centrosymmetric dimer (α-truxillic acid). The β-polymorph, in which the adjacent molecules are translationally equivalent (mirror-symmetry relationship), having the double bonds within 4 Å separation, gives β-truxinic acid as the product. In both cases the dimers are already 'preformed' in the monomer crystals, the photoreaction merely converting favourably oriented double bonds into cyclobutane rings (Fig. 8.11). The γ-form in which the molecules are not favourably disposed both in terms of molecular orientation and separation between olefinic double bonds is light-stable. As already

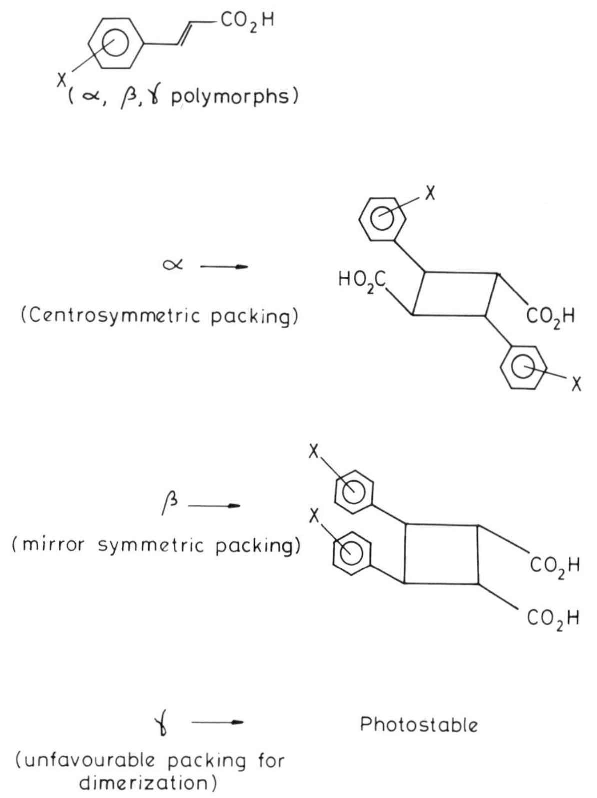

Figure 8.11   Photoreactivity of *trans*-cinnamic acid polymorphs.

pointed out, when the compound is irradiated in solution or melt, mere *trans–cis* isomerization occurs without dimerization.

Dimerization of *trans*-cinnamic acids is but one example of topochemically controlled (2 + 2) photocycloaddition reactions of disubstituted ethylenes in the solid state. Photoreactivity requires that the double bonds be parallel and separated by $\leqslant 4\,\text{Å}$ in the reactant crystals (see later); the stereochemistry of the photoproducts (cyclobutane derivatives) directly reflects the molecular arrangement in the parent crystals. Another important example of (2 + 2) photocycloaddition is the photopolymerization of 2,5-distyryl-pyrazine (DSP) (Hasegawa *et al.*, 1968). Molecular packing in one of the polymorphs of DSP (Fig. 8.12) is favourable for photochemical polymerization through (2 + 2) cycloaddition. That the reaction is topochemically controlled is evident from the fact that vapour-grown DSP, which crystallizes in a different structure, does not photopolymerize. A crystallographic study (Nakanishi *et al.*, 1972) of the photopolymerization of DSP has shown that both the monomer and the polymer crystallize in the same structure (space group *Pbca*). The polymer forms a separate phase within the monomer crystal exhibiting three-dimensional orientational relationship with the parent. Nakanishi *et al.* (1973, 1975, 1980) have exploited the reaction to produce an extensive range of cyclobutane polymers. The reaction in all the cases is

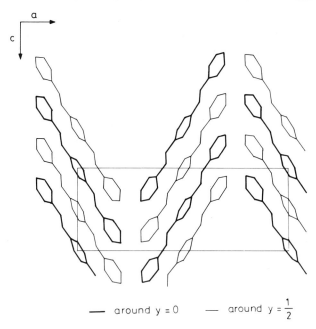

— around y = 0   — around $y = \frac{1}{2}$

Figure 8.12   Packing of distyryl-pyrazine molecules in the solid state (viewed down the *b*-axis) that favours facile photopolymerization. (Following Jones & Thomas, 1979.)

topochemical in the sense that the product is preformed in the monomer crystals (having the reactive double bonds collinear within 4 Å separation), and photopolymerization occurs by means of a sequence of dimerizations forming cyclobutane rings as repeating units. Thermal depolymerization is also a topochemical reaction. Electron microscopy (Jones & Thomas, 1979) has revealed no evidence for heterogeneity in the DSP photopolymerization; there was also no indication of dislocations serving as preferential sites for initiation of the reaction. Lahav and coworkers have exploited reactions of this type for the synthesis of chiral polymers (Addadi *et al.*, 1980) (see later). (2 + 2) photopolymerization is potentially useful in holographic interferometry (Mizuno *et al.*, 1977).

An important aspect of organic solid state reactions is concerned with the deliberate design of organic molecules so that they pack in a particular manner exhibiting predictable reactivity in the solid state (*crystal engineering*). The strategy involves choosing appropriate systems and making specified substitutions in order that they crystallize in desired structures. The key to crystal engineering lies in understanding the nature of delicate intermolecular forces that control packing and structure of organic solids. For example, intermolecular nonbonded interactions such as $C = O \ldots Cl$, $C = O \ldots Ph$ or $Cl \ldots Cl$ are known to steer molecules to pack in a specific manner. The fact that many chlorosubstituted aromatic compounds crystallize with a short ~ 4 Å unit cell axis has been used to engineer crystal structures with a ~ 4 Å separation between molecules (β-packing) (see, for example, Desiraju, 1984*a*). Substituents such as hydroxy, methyl, acetoxy and methoxy have also been investigated for the purpose of engineering organic solids. Methyl and chloro substituents appear to be interchangeable giving isomorphous crystals (Jones *et al.*, 1981). Acetyl group seems to favour β-packing as seen in coumarins (Gnanaguru *et al.*, 1985).

Among the noncovalent interactions involved in engineering organic solids (Desiraju, 1989), hydrogen bonding deserves special mention. Rules to predict hydrogen-bonding arrangements in single- and two-component crystals have been proposed (Leiserowitz & Hagler, 1983; Etter, 1990). The formation of intramolecular six-membered ring hydrogen bonds in preference to intermolecular hydrogen bonds is one of the rules. The other is that the best proton donors and acceptors, remaining after intramolecular hydrogen-bond formation, engage in intermolecular hydrogen bonds to one another. Cocrystals of 2-aminopyrimidine-carboxylic acid, where the best donor (acid OH) is paired with the best acceptor (ring N) illustrate the latter rule. Similarly, the formation of a microporous organic solid consisting of close-packed two-dimensional hydrogen-bonded networks of I (Venkataraman *et al.*, 1994) and of molecular tapes consisting of melamine and barbituric acid derivatives (Lehn *et al.*, 1990, 1992; Zerkowski *et al.*, 1994) illustrate the role of hydrogen bonding in designing novel organic solids.

I

An interesting example of crystal engineering and topochemical reactivity is provided by 2-benzyl-5-benzylidene-cyclopentanone (BBCP) derivatives which undergo single crystal–single crystal photodimerization (Jones *et al.*, 1980). In the monomer

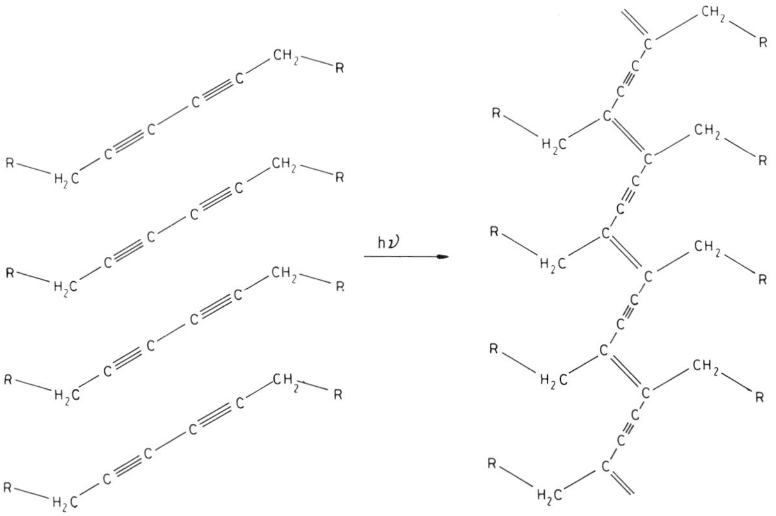

Figure 8.13   Polymerization of diacetylene in the solid state.

crystals, the double bonds are separated by less than 4 Å. The essentially diffusionless nature of the reaction was shown by a crystal structure analysis of partially reacted crystals of the p-MeBBCP-p-ClBBCP solid solution (Theocharis *et al.*, 1984). Computations with the aid of atom–atom potentials have helped to rationalize the photoreactivity of BBCP derivatives (Thomas *et al.*, 1985). The influence of substituents on the mode of molecular packing can be understood by this approach. The crystal structure of α-benzylidene-γ-butyrolactone has been determined with the aid of such calculations and solid state photoreactivity (Kearsley & Desiraju, 1985). The (2 + 2) photodimerization of cinnamylidine-cyanoacetate is another example of topochemical reaction with the possibility of reversible photochromism (Guru Row *et al.*, 1983). Although it has been generally believed that properly oriented double bonds located within 4.2 Å is essential for photoreactivity, studies (Gnanaguru *et al.*, 1985) on substituted coumarins have revealed that photodimerization occurs even where the reactive double bonds are not parallely oriented and are separated by more than 4.2 Å.

Polymerization of diacetylene (Fig. 8.13) is one of the most elegant examples of the topochemical principle. Wegner (1971, 1979) showed that diacetylene monomers, $R-C \equiv C-C \equiv C-R$, polymerize in the solid state by a 1, 4-addition reaction at the diacetylene group to produce a polymer that can be represented by the mesomeric structures:

$$[=(R)C-C \equiv C-C(R)=]_n \leftrightarrow [-(R)C=C=C=C(R)-]_n$$

Colourless diacetylene monomer crystals can be polymerized under heat, ultraviolet, X-ray or γ-ray irradiation to form single-crystal, highly coloured polyacetylenes. The solid state reaction transforms the entire monomer crystal to polymer crystal without phase separation; the polymer forms a solid solution with the monomer over the entire monomer-to-polymer conversion range. When R = p-toluene sulphonate, for instance, complete solid solution occurs between the monomer and the polymer and the reaction rate exhibits a dramatic autocatalytic effect (the rate increases by two orders of magnitude as the reaction proceeds). The autocatalysis has been attributed to the influence of strain on chain initiation and propagation (Baughman, 1978). Recent results indicate the importance of local strain in controlling individual propagation steps in both diacetylene and distyrylpyrazine polymerization (Niederwald et al., 1983). Interestingly, if the CH_3 group of p-toluene sulphonate is replaced by a Cl atom, the diacetylene monomer does not polymerize. The loss of reactivity is attributed to a strong attractive interaction between Cl and adjacent phenyl rings, separating the potentially reactive carbon atoms.

Until recently, the mechanism of polymerization of diacetylenes was not fully established. Work of Gross & Sixl (1982) has shown that the reaction follows a radical mechanism involving paramagnetic triplet (diradical) and quintet (dicarbene) intermediates. One of the important conclusions of this study is that optical polymerization at low temperatures gives a much lower chain length ($\sim$ 10 units) than thermal polymerization ($\sim$ 10000 units). This may be because potential energy barriers involved in chain termination can be surmounted by photons but not thermally. Polydiacetylenes have several potential applications. They exhibit strongly anisotropic optical properties; thus reflection is metal-like when light is polarized parallel to the chain direction, and blue when light is polarized perpendicularly. The extent of polymerization determines the colour. Certain polydiacetylenes show strong second harmonic generation. Polydiacetylenes can be incorporated in Langmuir-Blodgett films (Section 7.9). Single crystals of polydiacetylene prepared by polymerization of 1,6-di(N-carbazolyl)2,4-hexadiyne exhibit a high degree of internal perfection and possess remarkable mechanical stiffness and strength. The material has great potential for use as reinforcing fibre in composites (Galiotis et al., 1984). Polymerization of diacetylenes has been theoretically studied by the fragment formalism (Burdett, 1980). The results show that the reaction is photochemically allowed, but experimentally the reaction occurs thermally as well.

Systematic studies of topochemical reactions of organic solids have led to the possibility of asymmetric synthesis via reactions in chiral crystals. (A chiral crystal is one whose symmetry elements do not interrelate enantiomers.) (Green et al., 1979; Addadi et al., 1980). This essentially involves two steps: (i) synthesis of achiral molecules that crystallize in chiral structures with suitable packing and orientation of reactive groups and (ii) performing a topochemical reaction such that chirality of crystals is transferred to products. The first step is essentially a part of the more general problem of crystal engineering. An example of such a system where almost quantitative asymmetric induction is achieved is the family of unsymmetrically substituted dienes:

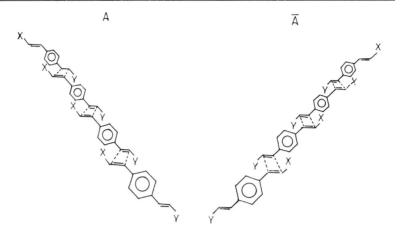

Figure 8.14   Packing of unsymmetrically substituted diene molecules in a crystal, favouring photopolymerization leading to asymmetric synthesis. (After Addadi *et al.*, 1980.)

The system was chosen after careful consideration of the relationship between molecular structure and crystal structure: symmetrically disubstituted benzene derivatives are known to crystallize in structures that undergo photopolymerization. Chiral R_1 groups such as *sec*-butyl induce crystallization in chiral space groups; *sec*-butyl group also eliminates the packing that would lead to reaction of equivalent double bonds. Nonequivalence of the double bond is ensured by the nitrile function. The ethyl ester (R_2 = ethyl; R_1 = *sec*-butyl) of the diene crystallizes in the space group *P1*, where the unsymmetrically substituted diene molecules are packed in such a way (Fig. 8.14) that the two nonequivalent double bonds are parallel to each other at a distance $\leqslant 4$ Å along an infinite stack. When this solid is irradiated at 278 K with $\lambda > 310$ nm, dimers, trimers and oligomers (up to molecular weight 10 000) are formed with nearly 100% quantitative enantiomeric yield.

Conformationally controlled gas–solid reactions involving organic solids are known to result in *asymmetric synthesis*. An example is the addition of bromine to nonchiral alkenes crystallizing in chiral space groups (Addadi *et al.*, 1980). A topochemically controlled gas–solid reaction with a single crystal of the alkene is expected to yield the enantiomeric dibromides in unequal quantities. Reaction of bromine with p, p'-dimethylchalcone crystallizing in $P2_12_12_1$ space group did indeed yield one of the enantiomers in excess (about 20%) (Fig. 8.15). The enantiomeric conformation of p, p'-dimethylchalcone, which gives an excess of chiral dibromide, is shown in Fig. 8.15(bottom) (Rabinovich & Hope, 1975).

Addadi and coworkers (1985) have shown that chirality can be induced by the use of selective additives which modify crystal morphology. Thus, when (R, S)-threonine is crystallized in the presence of impurities, enantiomeric excess greater than 95% is readily obtained. The strategy also affords a means of determining the absolute configuration of the additive molecule.

One of the new strategies in organic solid state chemistry is to employ host–guest

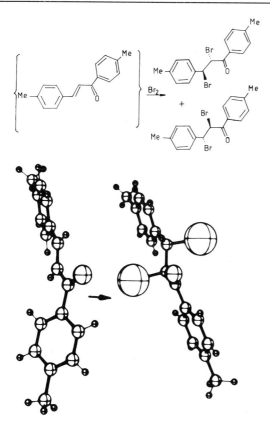

Figure 8.15   Reaction of p, p'-dimethylchalcone in crystalline state to yield unequal amounts of enantiomeric dibromides (top); conformation of p, p'-dimethylchalcone, which yields an excess of the enantiomeric dibromide (bottom). (After Addadi et al., 1980.)

molecular inclusion complexes of certain steroids to perform stereospecific and regiospecific reactions (Addadi et al., 1980). Deoxycholic acid (DCA) and apocholic acid (APA) incorporate a wide variety of guest molecules such as hydrocarbons, alcohols, esters and acids. In these inclusion compounds, only certain parts of the steroid molecule are exposed to the guest, in such a way that reactions are selective. A common structure adopted by DCA inclusion complexes is orthorhombic ($P2_12_12_1$) and encloses channels. The size and shape of the channels are adjusted according to the nature of the guest species. A typical example of a stereospecific reaction involving DCA di-t-butyl peroxymonocarbonate is shown in Fig. 8.16. Photolysis of the complex yields two products; one of the perester oxygens is transferred to position 5 of the steroid in the reaction. The topochemical nature of the reaction is supported by the observation that the reaction occurs neither in solution, nor in the molecular complex of APA, where the channel dimensions are different. Inclusion complexes of hydroquinone with hydrazine react in the solid state with aromatic esters to yield hydrazides (Toda et al., 1995).

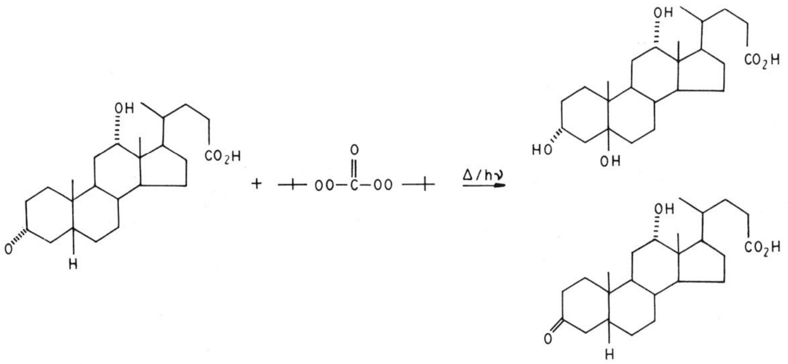

Figure 8.16   Reaction of deoxycholic acid (DCA)-di-t-butyl peroxymonocarbonate inclusion complex, giving stereospecific products. (After Addadi *et al.*, 1980.)

The presence of a polar axis confers anisotropic activity to organic crystals (Curtin & Paul, 1981; Desiraju, 1984b). Common polar (noncentrosymmetric) space groups adopted by organic crystals are $P2_12_12_1$ and $P2_1$. While chiral crystals must have polar directions, polar crystals need not be chiral. Anisotropic reactivity is seen for instance in the reaction of ammonia with p-bromobenzoic anhydride, which crystallizes with a polar axis; the polar axis directs the reaction. p-bromobenzoic anhydride is chiral as well as polar. Chirality of the crystalline anhydride has been exploited to resolve a racemic gaseous amine; the chiral crystal preferentially reacts with one of the enantiomers of the amine. Thus when p-bromobenzoic anhydride crystals are exposed to vapours of racemic phenylethylamine, the resulting amide contains one of the enantiomers in excess.

In view of the potential technological importance of noncentrosymmetric organic crystals, several approaches have been evolved to artificially achieve noncentrosymmetry, which include electric field poling of polymers, self-assembly of molecular layers, Langmuir–Blodgett assembly of films and host–guest interaction in noncentrosymmetric hosts (Marder *et al.*, 1994). Prediction and/or control of the three-dimensional structure of crystals, given only the information of molecular properties, however, remains difficult at present.

An assumption implicit in the topochemical principle is that the crystal lattice provides a cradle to the molecules, inhibiting substantial changes in their conformation and relative positions, but otherwise lets the molecule behave normally in its reaction. Investigations of McBridge (1983) have shown that the environment in the crystal acts not merely as a passive cradle, but provides considerable anisotropic stress (equivalent to tens of thousands of atmospheres). The stress manifests itself in unusual reactivity. We have already referred to the influence of stress resulting in autocatalytic effect in the thermal polymerization of diacetylenes. Direct evidence for the presence of stress on molecules during reaction in the solid state is seen in the infrared spectra of undecanoyl peroxide. Successive spectra of the sample annealed at high temperatures after photolysis show a steady shift of the asymmetric stretch to lower frequencies. The shift

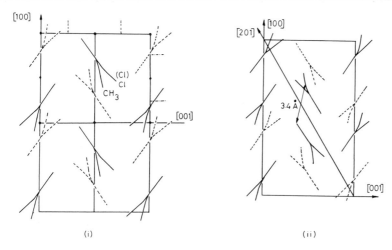

Figure 8.17   (i) Projection of the crystal structure of 1, 8-dichloro-9-methylanthracene in the *ac*-plane; (ii) molecular arrangement in the same plane after the translation $(201)\frac{1}{2}[010]$ which brings neighbouring molecules to favourable geometry for the formation of *trans*-dimer. (After Thomas, 1974.)

is attributed to relaxation of the anisotropic local stress, which can be as large as ~ 40 kbar.

The foregoing discussion of organic solid state reactions may convey the impression that crystal structure is the only factor that controls chemical reactivity and imperfections are unimportant. This is not entirely true because there are several reactions where one comes across 'wrong' products or 'wrong' reactivity, neither of which can be explained on the basis of structural considerations (Thomas, 1979). Thus, anthracene and 1,8-dichloro-9-methyl anthracene are not expected to photodimerize on the basis of structural consideration but they do. 9-cyano anthracene and 1,8-dichloro-9-methyl anthracene are expected to yield *cis*-dimers but yield *trans*-dimers. It is believed that imperfections contribute to such unexpected photoreactivity of solids. In considering the role of imperfections on reactivity, a distinction must be made between cases where excitation-energy transfer is efficient and where it is not. In *trans*-cinnamic acid transfer is inefficient, so photoreaction takes place at the site of photon absorption. Consequently structural imperfections are unimportant. In anthracene and its derivatives, where photon energy transfer is efficient, the excitation energy is trapped at imperfection sites and causes reaction there.

In the case of anthracene, the stable monoclinic phase transforms under stress to a triclinic phase in which molecules are favourably oriented for dimerization to occur. Although the triclinic phase has not been isolated as a pure phase, its structure has been established using low-temperature electron microscopy and atom–atom potential calculations (Jones & Thomas, 1979). In 1,8-dichloro-9-methyl anthracene, isolated dislocations with $(201)\frac{1}{2}[010]$ translation bring the molecules to the required geometry (Fig. 8.17) to facilitate photodimerization. 1, 5-dichloroanthracene is an interesting case. Instead of the expected 100% head-to-head dimers, photoreaction yields 80%

head-to-head and 20% head-to-tail dimers. An orientational defect involving a molecular rotation about one of the crystallographic axis is postulated to explain the formation of the unexpected head-to-tail dimer.

Organic solid state chemistry has come a long way since the pioneering work of Schmidt and Cohen in the early 1960s on the photodimerization of *trans*-cinnamic acids. Today, the impact of organic chemistry on solid state science is substantial not only in the design of new materials exhibiting diverse properties such as photochromism, metallic conductivity, superconductivity, ferromagnetism and nonlinear optical response, but also in pharmaceutical chemistry where the understanding of the organic solid state seems to enable better design of drug molecules (Byrn *et al.*, 1994). The reader is referred to the August 1994 issue of *Chemistry of Materials* devoted to organic solid state chemistry and to other reviews and books (Desiraju, 1989; Singh *et al.*, 1994).

8.9   Heterogeneous catalysis

Heterogeneous catalysis is a field of great technological importance, forming the backbone of the petroleum and chemical industries (Stinson, 1983). In terms of tonnage, catalysts contribute a very high proportion of solid materials manufactured today. Although impressive advances have been made in this field in terms of new processes and new catalysts, little is known about surface reactions at the molecular level. Techniques to investigate such reactions have become available in recent years; besides surface techniques, electron microscopy and NMR spectroscopy are especially useful in the characterization of catalysts (see Rao, 1984; Rao, 1985; Thomas & Lambert, 1980; Thomas, 1994; see also Chapter 2). The information obtained by these techniques is of great value for heterogeneous catalysis in understanding the relationship between reactivity of a surface and its structure and composition. In spite of these advances in characterization techniques, we are still not in a position to provide cook-book recipes for synthesizing catalysts for specific reactions; we are also unable to characterize surface species involved in reactions completely. Part of the problem arises because catalysts are mostly complex, polycrystalline (or amorphous) solids, while model studies are generally carried out on idealized surfaces.

Heterogeneous catalysis is a vast subject and we do not intend to discuss it in detail here. Inasmuch as solid state chemistry has contributed significantly in understanding the catalytic properties of solids and in designing new catalysts (Grasselli & Brazdil, 1985; Thomas & Zamaraev, 1992), we shall present a brief and selective discussion of a few heterogeneous catalysts of current interest, together with some basic ideas of heterogeneous catalysis.

A variety of solids is used as catalysts: metals, alloys, clays, metal oxides, sulphides, nitrides, carbides and so on. Catalysts may be single-phase substances or multiphasic mixtures; they may be crystalline, microcrystalline or even amorphous. Catalysts can be electrically insulating, semiconducting or metallic. Some examples of heterogeneously catalysed reactions are given in Table 8.1. Two aspects of catalysts are important: *activity* and *selectivity*. *Activity* refers to the ability of catalysts to accelerate chemical reactions so that equilibrium is achieved rapidly. The degree of acceleration

Table 8.1. *Some examples of heterogeneously catalysed reactions and catalysts*

Reaction	Catalyst
Cracking of hydrocarbons in presence of hydrogen	Zeolites
Hydrodesulphurization of petroleum	Cobalt and molybdenum on alumina support
Methanation of CO and H_2	Nickel supported on alumina
Ammonia synthesis	Ferric, aluminium and calcium oxides with potassium oxide promoter
Fischer–Tropsch synthesis (hydrogenation of CO)	Iron and cobalt promoted with K_2O and ThO_2
Oxidation of CO in automobile exhaust	Platinum or palladium on alumina
Polymerization of ethylene and propylene	Trialkyl aluminiums and titanium(IV) halides (Ziegler–Natta catalyst)
Ammoxidation of propylene	Bismuth molybdates, uranyl antimonate

can be as much as 10^{10} times in certain cases. A dramatic illustration of the catalytic activity is provided by the reaction of hydrogen and oxygen in the presence of platinum to form water with explosive violence; although the reaction is thermodynamically allowed, the mixture of gases can be stored indefinitely without reaction in the absence of the catalyst. *Selectivity* refers to the ability of catalysts to direct reactions to yield particular reaction product(s) to the exclusion of others. An example of selectivity is the dehydrocyclization of *n*-heptane to toluene. Although *n*-heptane can undergo a variety of reactions such as hydrogenolysis, dehydrogenation and isomerization, platinum selectively catalyses the dehydrocyclization reaction. Another example is the reaction of propylene and oxygen over bismuth molybdate to yield acrolein; the same reaction over bismuth oxide produces benzene and hexadiene. Selectivity in heterogeneous catalysis was the subject of a Faraday Discussion of the *Royal Society of Chemistry, London* (1981) and of a special issue of *Chemical Reviews* (1995). A catalyst is conventionally defined as a substance that accelerates a chemical reaction without itself being consumed; nevertheless, catalysts used in commercial processes do undergo alteration and require regeneration. For example, the zeolites used as catalysts for petroleum cracking have effective lifetimes of only a few seconds but are regenerated continually.

A heterogeneously catalysed reaction involves several steps (Mady *et al.*, 1976): (i) mass transport of fluid reactants to the surface, (ii) chemisorption of reactants on the surface, (iii) diffusion and chemical reaction at the surface and (iv) desorption and diffusion of products from the surface. Step (iii), involving the formation of surface intermediates, is the key step. Formation of surface intermediates, which ultimately give rise to products, was first proposed by Sabatier (see Burwell, 1973) and strikingly demonstrated by Sachtler & Fahrenfort (1960) in the decomposition of formic acid

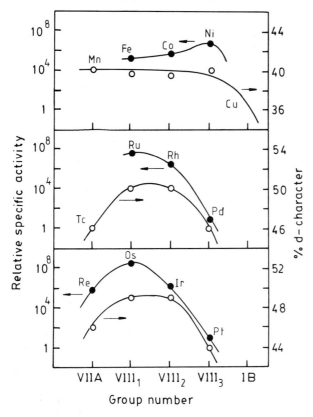

Figure 8.18   Correlation between catalytic activity of transition metals towards ethane hydrogenolysis and percentage *d*-character. The closed points represent activities and open points percentage *d*-character. (Following Sinfelt, 1977.)

catalysed by various transition metals. A concept often used in heterogeneous catalysis is that of the so-called *active sites*, proposed by Taylor (1952). The basic idea is that catalytic activity of a solid is restricted to certain specific sites on the surface. Active sites have come to be associated with surface irregularities or defects. The presence of unique atomic sites of low coordination or different valency on the surfaces that are active in chemical reactions has been demonstrated in several cases. Thus dissociation of hydrogen requires no activation energy at steps on platinum surfaces, while it requires a definite activation energy at terraces on the same surface (Salmeron et al., 1979). On oxide surfaces, point defects may be the active centres. In Ziegler–Natta catalysts employed in the stereospecific polymerization of propylene, active centres are Ti^{3+} sites having chlorine vacancy in α-$TiCl_3$ (Mady et al., 1976). Similarly W^{3+} ions seem to be active sites in WS_2, used for the hydrogenation of benzene.

One of the aims of research in catalysis is to identify the factors that contribute to the catalytic activity of solids. Among the various factors, a distinction is usually made between the *geometric factor* and the *electronic factor*. According to Balandin (1969), who emphasized the role of the geometric factor, activity depends on the presence of a correctly spaced array (multiplet) of atoms on the solid surface with which the reactant molecule interacts. Somorjai (1981) has found evidence for the geometric factor, though in a different sense, in his studies on the reactivity of the platinum surface. Surface irregularities such as steps and kinks on high-index planes of platinum single crystals (identified through LEED) contribute to reactivity. For several reactions involving hydrocarbons, kinked and stepped surfaces are found to be more active than the flat (111) planes. Hydrocarbon reactions occurring on metal surfaces are of two types: those that are sensitive to changes in catalyst particle morphology and those that are not. Methanation $(3H_2 + CO \rightarrow CH_4 + H_2O)$ is an example of a surface-insensitive reaction, while hydrogenolysis (e.g. $C_2H_6 + H_2 \rightarrow 2CH_4$) is sensitive. Recent work of Goodman and coworkers (Goodman, 1984) has shown that not only is the methanation reaction rate independent of the Miller index, (100) or (111), of nickel single crystals, but also that the rates are similar for both single crystals and high surface-area catalysts (Ni/Al_2O_3), showing that the reaction is insensitive to the surface structure of the catalyst. Boudart (1969) subdivides supported catalysts as structure-sensitive or -insensitive. A catalytic reaction is considered to be structure-sensitive if the rate changes markedly with the particle size (or the crystal face in the case of single crystals).

The electronic factor in catalysis refers to corelations between bulk electronic properties of solids and their catalytic activity. For transition-metal catalysts, catalytic activity has been correlated with percentage d-character of the metallic bond as defined by Pauling (1949). Sinfelt (1975) has found a fairly good correlation between percentage d-character and hydrogenolysis activity of second- and third-row transition metals (Fig. 8.18); however, comparison of activities of elements in different transition-metal rows (e.g. Pt and Ni) shows poor correlation. His conclusion is that while there is a degree of correlation between hydrogenolysis activity and percentage d-character, this parameter alone is not adequate.

There have been many attempts to relate bulk electronic properties of semiconductor oxides with their catalytic activity. The electronic theory of catalysis of metal oxides developed by Hauffe (1966), Wolkenstein (1960) and others (Krylov, 1970) is based on the idea that chemisorption of gases like CO and N_2O on semiconductor oxides is associated with electron-transfer, which results in a change in the electron transport properties of the solid oxide. For example, during CO oxidation on ZnO a correlation between change in charge-carrier concentration and reaction rate has been found (Cohn & Prater, 1966).

Correlations between catalytic activity and a variety of bulk properties of semiconductors have been reported: (i) the average band gap of III–V and II–VI semiconductors and activity towards hydrogenation of isopropanol; (ii) enthalpy of oxides and their activity towards oxidation of propylene and (iii) number of d-electrons (and crystal field stabilization energy) or $3d$-metal oxides and their activity towards N_2O decomposition. The last correlation, due to Dowden (1972), is important since it provides a connection between heterogeneous catalysis and coordination chemistry of transition-metal compounds. A correlation between the catalytic activity of transition-metal sulphides towards hydrodesulphurization of aromatic compounds and the position of the transition metal in the periodic table has been made by Whittingham &

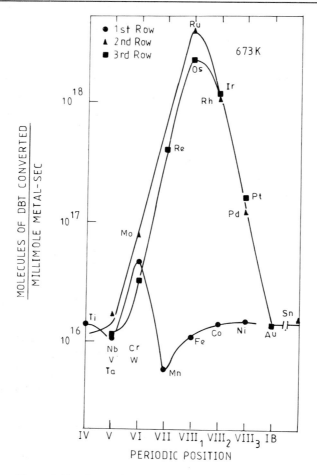

Figure 8.19   Catalytic activity of transition-metal sulphides for hydrodesulphurization of dibenzothiophene. (After Whittingham & Chianelli, 1980.)

Chianelli (1980) (Fig. 8.19). The most active metal sulphides are those containing the maximum number of d-electrons. Extensive investigations of several transition-metal-MoS_2 catalysts supported on Al_2O_3 have shown the presence of an active 'transition-metal-Mo-S' state where the transition metal (Co or Ni) forms short bonds with sulphur (Bouwens *et al.*, 1990; Kulkarni & Rao, 1991).

Despite various attempts, no single universal correlation between bulk properties and catalytic activity of solids has been found. It is now recognized that the geometric factor and the electronic factor cannot be separated from one another and that catalytic activity should be considered along with catalyst selectivity to arrive at an understanding of heterogeneous catalysis (Sachtler, 1981).

8.9.1   Typical solid catalysts

Transition metals are used as catalysts for a variety of reactions: hydrogenation, hydrogenolysis and isomerization of hydrocarbons (group VIII metals), oxidation of alkenes (Ag), ammonia synthesis (Fe, Ru, Os), Fischer–Tropsch synthesis (Co, Fe) and so on (Sinfelt, 1975). Platinum is one of the most versatile catalysts, used in both oxidizing and reducing conditions. Thus, it catalyses dehydrocyclization, isomerization and hydrogenation of hydrocarbons, oxidation of NH_3 and CO as well as finds use as an electrocatalyst in electrochemical cells. Metal catalysts are usually dispersed as fine particles on supports such as SiO_2 and Al_2O_3. The dispersion is to ensure that a large catalyst surface is available for reaction. A variation of a metal catalyst is two or more metals dispersed on the same support (bimetallic catalysts) (Sinfelt, 1975, 1977). Typical examples of bimetallic catalysts are Pt–Re and Pt–Ir dispersed on alumina, employed as catalysts for reforming of hydrocarbons (a reaction used in refineries to improve the octane rating of gasoline). Bimetallic catalysts are to be distinguished from the more common alloy catalysts because the former consist of clusters of only a few atoms of both metals, which are finely dispersed on the support; moreover such bimetallic catalysts include combinations of metals that do not normally form alloys in the bulk. In the Cu–Ni system, the presence of Cu promotes the ready reduction of Ni to the metallic state. Investigations of other bimetallic catalysts have shown the formation of interesting alloy phases on catalyst surfaces. For example, in the $NiFe/Al_2O_3$ system, the reducibility of Fe as well as Ni is greater than in the corresponding monometallic catalyst. FCC and BCC Ni–Fe alloy phases which are ferromagnetic or superparamagnetic are formed depending on the composition and heat treatment given to the precursor (Rao et al., 1992).

Metal alloys have long been investigated as catalysts with a view to understanding the electronic factor in catalysis. Earlier it was thought that catalytic properties could be systematically varied by changing the alloy composition. However, this was not to be the case because surface composition of alloys was found to be different from the bulk composition in many instances. An important aspect of alloy catalysis is the specificity towards certain reactions. Thus, the activity of nickel towards ethane hydrogenolysis is markedly decreased by alloying with copper, while the activity towards cyclohexane dehydrogenation remains constant over a wide range of bulk composition in Ni–Cu alloys (Fig. 8.20) (Sinfelt, 1977). The difference is rationalized in terms of the surface composition of the alloys. Ethane hydrogenolysis is faster by many orders of magnitude on nickel than on copper. Since the surface of Ni–Cu alloys is enriched with copper relative to the bulk, the decrease in catalytic activity with Cu concentration is understandable.

Amorphous (rapidly quenched) metals and alloys have been investigated as catalysts (Schlögl, 1985). It has been found that adsorption characteristics of carbon monoxide on metallic glasses are different from those on crystalline materials. For example CO is found to dissociate readily on the surface of $Ni_{76}B_{12}Si_{12}$ metglass, but it is always molecularly adsorbed on metallic nickel (Prabhakaran & Rao, 1985).

Catalysis of metal-cluster compounds is a field of considerable interest (Lewis & Green, 1982; Haggin, 1982). Metal clusters (Section 6.6) containing aggregates of metal atoms surrounded by ligands provide model systems that simulate surface intermediates formed by chemisorption of molecules on metal surfaces (Muetterties et al., 1979). An example of a cluster catalyst is $Os_3(CO)_{12}$, which was found to be active in the

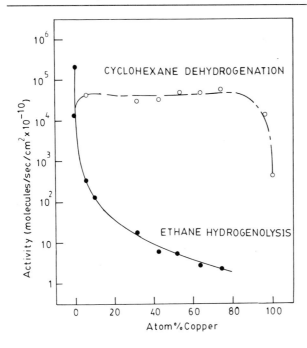

Figure 8.20   Activity of Ni–Cu alloys towards cyclohexane dehydrogenation and ethane hydrogenolysis. (Following Sinfelt, 1977.)

hydrogenation of CO to yield alkenes and alkanes. Osmium-based cluster catalysts showed marked selectivity for C_2 formation, and the product distribution was nearly independent of the H_2/CO ratio in the feed. Metal clusters may be regarded as intermediate between mononuclear metal complexes and small metal particles dispersed on inert supports. In Cu/ZnO methanol synthesis catalysts, copper is present in the form of small clusters and also as Cu^{1+}, the relative proportion of the two depending on the conditions of preparation of the catalyst (Arunarkavalli et al., 1993).

    Although it is generally believed that there is no interaction between the metal and the support in supported-metal catalysts, there is growing evidence to show that the support can influence catalytic reactions in many instances (Sleight, 1980). One such instance is the *spillover* of adsorbed species from the metal to the support. For example, hydrogen adsorbed on metals (e.g. Pt) undergoes dissociation and spills over to the support (e.g. Al_2O_3 or SiO_2), where it reacts with organic molecules adsorbed there (Sermon & Bond, 1974). More importantly, electronic and catalytic properties of metals can be remarkably altered by interaction with certain supports. For example, chromium exists in the low-spin Cr^{2+} state on SiO_2 and in the high-spin state on Al_2O_3. *Strong metal-support interaction (SMSI)* was first noticed with group VIII metals supported on TiO_2 (Tauster et al., 1978, 1981). The main features of SMSI are the suppression of H_2 and CO adsorption and radical changes in the activity and selectivity of reactions such as hydrogenation of CO. Thus Pt/TiO_2 reduced above 750 K is 10 times as active as Pt/Al_2O_3 and 100 times as active as Pt/SiO_2 towards methanation. SMSI catalysts exhibit superior activity and selectivity towards Fischer–Tropsch (CO–H_2) synthesis; such a behaviour of SMSI catalysts is probably due to unusual CO and H_2 sorption characteristics.

There has been considerable effort to understand the mechanism of SMSI. Experimental characterization of SMSI catalysts showed considerable morphological changes accompanied by reduction of the support oxide; in Pt/TiO_2, formation of Ti_4O_7 is revealed by electron diffraction. A charge-transfer model based on $X\alpha$ molecular orbital calculations was proposed by Horsley (1979) to explain SMSI. The interaction seems to arise from a charge-transfer from the reduced cation (Ti^{3+}) to the adjacent metal atom. Evidence for such a charge-transfer is provided by XPS studies of Pt/TiO_2 and $Pt/SrTiO_3$ (Bahl et al., 1980). In both the cases, chemical shift of the $Pt(4f)$ core level indicated substantial transfer of negative charge to the platinum atoms. Work by Dumesic and coworkers (Santos et al., 1983; Jiang et al., 1983) has shown that SMSI is due to diffusion of titania to the metal surface. Studies of SMSI in Ni/TiO_2 (Takatani & Chung, 1984; Rangarao & Rao, 1990) have revealed that hydrogen reduction at 700 K of nickel film deposited on TiO_2 results in the segregation of $TiO_x (x \sim 1.0)$ onto the nickel surface; reduced titania appears to diffuse rapidly through nickel. It seems that both physical coverage of nickel surface by TiO_x and chemical interaction between nickel and reduced titania on the surface contribute to SMSI behaviour. Similar migration of titania onto the metal surface has been reported with other transition metals as well (Ko & Gorte, 1984). SMSI is not special to TiO_2 supports alone. It has been reported with other transition-metal oxide supports such as Nb_2O_5 and Ta_2O_5; there appears to be a correlation between SMSI and the reducibility of support oxides. Although a complete understanding of SMSI is probably yet to emerge, there is no doubt that SMSI catalysts constitute one of the important hallmarks of heterogeneous catalysis, having far-reaching consequences in our understanding of catalytic behaviour of transition metals.

A variety of metal oxides (e.g. V_2O_5) have been employed for oxidation reactions, besides noble metals (e.g. Pt and Ag). Auto-exhaust catalysts employ metals such as Rh, Pd and Pt besides CeO_2 and other oxides. The use of metal oxide catalysts for oxidation reactions has been discussed quite widely in the literature (Grasselli & Brazdil, 1985). Perovskite oxides of the type $CaMnO_{3-\delta}$ and $La_{1-x}A_xMO_3$ (A = Ca, Sr; M = Co, Mn) are excellent candidates as oxidation catalysts. The 14-electron oxidation of butane to maleic anhydride is effectively carried out over phosphorus vanadium oxide catalysts of the type $VOPO_4$ (Centi et al., 1988).

Among the oxide catalysts, bismuth molybdates that catalyse selective oxidation and ammoxidation of propylene to yield acrolein and acrylonitrile have received considerable attention (Grasselli & Burrington, 1981):

$$CH_3CH = CH_2 + O_2 \rightarrow CH_2 = CHCHO + H_2O$$
$$CH_3CH = CH_2 + NH_3 + \tfrac{3}{2}O_2 \rightarrow CH_2 = CHCN + 3H_2O$$

Both the reactions involve formation of an allylic ($CH_2 \ldots CH \ldots CH_2$) intermediate. Catalytic activity towards this reaction requires sequential reduction and oxidation of the catalyst. With bismuth molybdates as catalyst, chemisorbed hydrocarbon is oxidized by the lattice oxygen; the active sites are replenished by rapid diffusion of lattice oxygens from the bulk to the catalyst surface. The catalysis cycle is completed by reducing gaseous oxygen and incorporating it in the form of anions in the lattice. Experiments with O^{18} have shown that the oxygen in the product originates primarily from the catalyst; since lattice oxygen is readily available, the rate of propylene oxidation is independent of the partial pressure of oxygen. Sleight (1977) has suggested that the reduction occurs at molybdenum sites and reoxidation at bismuth sites. The degeneracy (mixing) of Mo ($4d$) and Bi ($6p$) levels in the catalyst allows electrons added to Mo($4d$) during reduction to be available at bismuth sites for reoxidation. Another requirement for selectivity is that the structure should be able to accommodate easy removal and addition of oxygen. Coordinately unsaturated molybdenum at the surface, which can undergo easy interconversion between octahedral and tetrahedral coordination, seems to confer selectivity on bismuth molybdate catalysts (Grasselli *et al.*, 1981). Among the three bismuth molybdates, $Bi_2Mo_3O_{12}$ possesses a defect scheelite structure (tetrahedrally coordinated molybdenum) and γ-Bi_2MoO_6 crystallizes in a layered structure with octahedrally coordinated molybdenum. The structure of $Bi_2Mo_2O_9$ is related to scheelite. The selectivity and activity of the catalysts towards alkene oxidation are essentially constant with composition from $Bi_2Mo_3O_{12}$ to Bi_2MoO_6. It is probable that the coordination of molybdenum at the catalyst surface is different from that of the bulk solid.

Catalytic properties of transition-metal oxides also make them good candidates as *gas sensors*. Thus, the oxidizing power of oxide surfaces can be used to make gas sensors. For an oxide to be a good sensor, the reaction should occur fast and produce a change in the electrical resistivity or capacitance of the oxide. Percentage conversion and yield are not important (as in catalysis), but sensor selectivity is a relevant criterion. We list below the gas-sensing properties of a few oxides: ZnO (ethanol), Fe_2O_3 (CO or hydrocarbons depending on whether it is α or γ form), bismuth molybdates (NH_3), V_2O_5 (hydrocarbons). Many of the conducting transition-metal oxides are good *electrocatalysts*. Typical of these are $La_{1-x}Sr_xCoO_3$, $Pb_2(Ru_{2-x}Pb_x)O_{7-y}$ and RuO_2.

8.9.2   Zeolites

Perhaps the most important oxide catalysts are zeolites, which are extensively used in petroleum and chemical industries for catalytic cracking and hydrocracking, alkene and alkane isomerization and aromatic alkylation (Rabo, 1976; Dwyer, 1984). The most remarkable catalytic property of zeolites is their shape-selectivity; for instance starting from a variety of reactants such as methanol, ethanol, hexadecane, glycerides, unsaturated long-chain alcohols and acids, certain zeolites can produce a relatively narrow spectrum of intermediate-sized hydrocarbons (aliphatics peaking at C_3 and C_4 and aromatics in the C_6 to C_{10} range) whose composition corresponds closely to that of gasoline.

Zeolites are aluminosilicates enclosing channels and cages of molecular dimensions (see Section 1.5). We have discussed the structure of zeolites derived from sodalite units (Fig. 1.22), which result in large cavities (cages) connected by apertures (pores) of variable size. In the so-called pentasil family of high-silica zeolites (ZSM-5 and ZSM-11) (Fig. 1.23), the pore structure is unique, consisting of intersecting channels. Shape-selective catalysis by zeolites depends on the pore structure. The smallest pore size in zeolites is about 2.6 Å and the largest is about 7.4 Å. The pore opening is determined by the free aperture of the oxygen ring that limits the pore. Maximum values of the pore openings are 2.6, 3.6, 4.2, 6.3 and 7.4 Å for 4-, 6-, 8-, 10- and 12-membered rings respectively.

Hydroxyl groups associated with the aluminium atoms are the active centres for zeolite catalysis. They are readily generated by deammoniation of ammonium-exchanged zeolites on heating around 670 K. When the zeolites are heated at higher temperatures (> 800 K), the Brönsted acid sites are converted to Lewis acid sites by loss of H_2O. Since zeolites can be readily synthesized over a wide range of Si/Al ratios, it is possible to vary the concentration of active sites in a controlled manner. In a study of active sites in acidic catalysts based on ZSM-5, Hagg et al. (1984) have shown that the activity towards cracking of n-hexane is linearly related to aluminium content and extrapolates to zero activity at zero aluminium content. The quantitative correlation can be explained only if all the Al atoms are at tetrahedral sites. This conclusion is consistent with a high-resolution MASNMR study (Fyfe et al., 1982), which has shown that all the aluminium incorporated during the synthesis of ZSM-5, even at the lowest concentration level, goes to tetrahedral sites. Such correlations between structural detail and catalytic activity are of great significance.

Diffusion of reactant molecules through the intracrystalline pores is an essential prerequisite for adsorption and shape-selective catalysis (Rees, 1984). As distinct from classical diffusion and Knudsen diffusion, we come across a new type of diffusion phenomenon in zeolites called *configurational diffusion*. Here, diffusion is controlled by matching of the size, shape and configuration of the diffusing species to the corresponding parameters in the host lattice. Configurational diffusion forms the basis of shape-selective catalysis of zeolites. The relation between diffusion and shape-selectivity is illustrated in Fig. 8.21, where diffusion coefficients of C_3–C_{14} n-alkanes at 610 K through potassium T-zeolite are compared with cracking-product distribution obtained from n-tricosane over H-erionite at the same temperature. The two zeolites are structurally related. Molecules that are too small (up to n–C_6) as well as those that are too large (n–C_9–n–C_{12}) diffuse faster through the zeolite, while intermediate ones (n-heptane and n-octane), whose lengths (11.56 Å and 12.82 Å) compare closely with the length (13 Å) of the erionite cage, get trapped and hence have low diffusivity. The close parallel between product distribution for the cracking of n-tricosane and diffusion data shows the interrelationship between diffusion and shape-selectivity (Derouane, 1982).

Shape-selective catalysis in zeolite requires that the reactants diffuse inwards to the active sites located at the intracrystalline volume (pores), and that products counterdiffuse after the reaction. At the active sites, presence of a high local electric field may direct the reaction according to steric requirements to yield specific products. Thus, shape-selectivity may be achieved by virtue of geometric factors, Coulombic field at the active sites and/or difference in diffusion rates. Accordingly, three different kinds of shape-selectivity are distinguished (Dwyer, 1984). If the geometric factors are such that only certain reactant molecules can diffuse through the pores, a *reactant selectivity* is achieved. If the geometric restrictions apply to the diffusion of product molecules, the selectivity is called *product selectivity*. If the spatial configuration around the active site is such that it allows the formation of one type of intermediate to the exclusion of other possible ones, a *transition state selectivity* is achieved. Thus o-xylene preferentially undergoes isomerization rather than disproportionation in zeolites such as ZSM-5 and mordenite, where the cavities are too small to accommodate the relatively large diaryl intermediate involved in the disproportionation reaction:

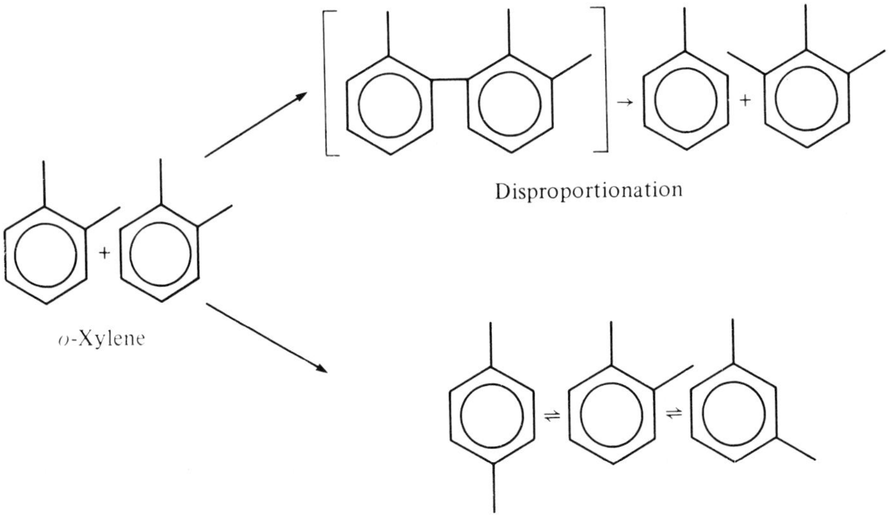

Disproportionation

o-Xylene

Isomerization

The shape-selectivity of H^+ ZSM-5 is particularly remarkable. Active centres at the inner walls of the catalyst readily release protons to organic reactant molecules forming carbonium ions, which in turn, through loss of water and a succession of C—C forming steps, yield a mixture of hydrocarbons that is similar to gasoline. The feedstock can be methanol, ethanol, corn oil or jojoba oil. The shape-selectivity of this catalyst is particularly striking, as can be seen from the product distribution obtained for the dehydration of three different alcohols (Table 8.2). The product distribution can be understood in terms of the intermediate pore size of ZSM-5, which can accommodate linear alkanes and isoalkanes as well as monocyclic aromatic hydrocarbons smaller than 1, 3, 5-trimethyl benzene. In Table 8.3, we list some of the recent innovations in catalysis, to highlight the important place occupied by molecular-sieve catalysts.

Oxides of transition metals can act as acid–base or redox catalysts. Oxides of non-transition metals (Al_2O_3, SiO_2) are, however, good acid–base catalysts. There is a large family of aluminosilicate zeolitic acids (e.g. H^+-ZSM-5, H-mordenite). Micropor-ous aluminium phosphates (ALPOs) can be modified to yield acidic SAPOs (Si replaces

Table 8.2. *Product distribution percentages for conversion of alcohols[a] over ZSM-5 zeolite at 644 K*

Products	Reactant		
	Methanol	*t*-Butanol	l-Heptanol
Methane	1.0	0.1	0.0
Ethane	0.6	0.7	0.3
Ethylene	0.5	0.5	<0.1
Propane	16.2	18.8	16.4
Propylene	1.0	1.1	0.2
i-Butane	18.7	18.4	19.3
n-Butane	5.6	8.7	11.0
Butenes	1.3	0.7	<0.1
i-Pentane	7.8	6.2	8.7
n-Pentane	1.3	1.4	1.5
Pentenes	0.5	0.2	0.1
C_{6+} aliphatics	4.3	7.6	3.0
Benzene	1.7	3.3	3.4
Toluene	10.5	11.6	14.3
Ethylbenzene	0.8	1.3	1.2
Xylenes	17.2	12.4	11.6
C_9 aromatics	7.5	6.1	5.3
C_{10} aromatics	3.3	0.4	2.9
C_{11+} aromatics	0.2	0.6	0.6

[a]After Chang & Silvestri (1977).

Table 8.3. *Some recent catalytic processes*

Catalyst	Process
Ga-ZSM-5	Dehydrocyclization of alkanes and conversion of alkanes to aromatics
Pt or Ni/Zeolite H^+ ferrierite	Hydrotreating of hydrocarbons
Phase-transfer catalyst	Polymerization of tetrahydrofuran
Mo-V-heteropolyacid	Oxidation of methacrolein
CuCl	Production of dimethylcarbonate from acetone
SAPO-11	Cyclohexanone oxime to C-caprolactam
Immobilized nitrile hydratase	Acrylamide from vinylcyanide
Mixed metal oxides	H_2S oxidation to S
Zeolites, SAPO	Benzene to phenol (*via* cyclohexene)
Zeolite	Oligomerization of olefins
Mordenite (zeolite)	2,6-diisopropylnaphthalene using propene as alkylating agent
NiO	Hypochlorite decomposition

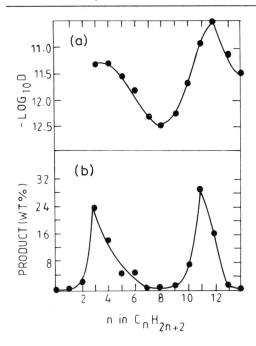

Figure 8.21   Relation between diffusion and shape-selective catalysis by zeolites: (a) diffusion coefficients of *n*-alkanes in potassium T-zeolite and (b) product distribution for the cracking of *n*-tricosane over H-erionite. (Following Gorring, 1973.)

some of the phosphorus) and MAPOs (a divalent metal, M = Mg, Zn, Co etc, replaces some of the Al). A variety of such highly acidic derivatives have been prepared recently, recent examples being DAF-1 and MAPO-18 (Wright *et al.*, 1993; Chen & Thomas, 1994). Such acids could indeed replace environmentally hazardous acids such as H_2SO_4.

References
Addadi, L., Ariel, S., Lahav, M., Leiserowitz, L., Papovitz-Biro, R. & Tang, C. P. (1980) in *Chemical Physics of Solids and their Surfaces* (Senior Reporters Roberts, M. W. & Thomas, J. M.) The Royal Society of Chemistry, London.

Addadi, L., Berkovitch,-Yellin, Z., Weissbuch, I., Mil, J. V., Shimon, L. J. W., Lahav, M. & Leiserowitz, L. (1985) *Angew. Chem. Int. Ed. (Engl.)* **24**, 466.

Alberti, G. & Constantino, U. (1982) in *Intercalation Chemistry* (eds Whittingham, M. S. & Jacobson, A. J.) Academic Press, New York.

Arunarkavalli, T., Kulkarni, G. U. & Rao, C. N. R. (1993) *Catal. Lett.* **20**, 259.

Bahl, M. K., Tsai, S. C. & Chung, Y. W. (1980) *Phys. Rev.* **B21**, 1344.

Balandin, A. A. (1969) *Advances in Catalysis* **19**, 1.

Baughman, R. H. (1978) *J. Chem. Phys.* **68**, 3110.

Benard, J., Gronlund, F., Oudar, J. & Duret, M. (1959) *Z. Electrochem.* **63**, 799.

Bhuvanesh, N. S. P., Uma, S., Subbanna, G. N. & Gopalakrishnan, J. (1995) *J. Mater. Chem.* **5**, 927.

Boldyrev, V. V. (1979) *Ann. Rev. Mater. Sci.* **9**, 455.

Boldyrev, V. V. (ed) (1996) *Reactivity of Solids*, Blackwell, Oxford.

Boudart, M. (1969) *Adv. Catal.* **19**, 153.

Bouwens, S. M. A. M., Koningsberger, D. C., de Beer, V. H. J., Louwers, S. P. A. & Princes, R. (1990) *Catal. Lett.* **5**, 273.

Burdett, J. K. (1980) *J. Amer. Chem. Soc.* **102**, 5458.

Burwell, Jr, R. L. (1973) in *Catalysis-Progress in Research* (eds Basolo, F. & Burwell, Jr, R. L.) Plenum Press, New York.

Byrn, S. R., Pfeiffer, R. R., Stephenson, G., Grant, D. J. W. & Gleason, W. B. (1994) *Chem. Mater.* **6**, 1148.

Cabrera, N. & Mott, N. F. (1948) *Repts. Prog. Phys.* **12**, 163.

Carter, R. E. (1961) *J. Chem. Phys.* **34**, 2010; **35**, 1137.

Casey, W. H., Westrich, H. R., Banfield, J. F., Ferruzzi, G. & Arnold, G. W. (1993) *Nature* **366**, 253.

Centi, G., Trifiro, F., Ebner, J. R. & Franchetti, V. M. (1988) *Chem. Rev.* **88**, 55.

Chang, C. D. & Silvestri, A. J. (1977) *J. Catal.* **47**, 249.

Chen, J. & Thomas, J. M. (1994) *J. Chem. Soc. Chem. Commun.* 603.

Clearfield, A. (1981) in *Preparation and Characterization of Materials* (eds Honig, J. M. & Rao, C. N. R.) Academic Press, New York.

Clearfield, A. (1988) *Chem. Rev.* **88**, 125.

Cohen, M. D. & Green, B. S. (1973) *Chem. Britain* **9**, 490.

Cohen, M. D. (1975) *Angew. Chem. Int. Ed. (Engl.)* **14**, 386.

Cohn, H. & Prater, C. D. (1966) *Disc. Faraday Soc.* **41**, 380.

Colpaert, M. N., Clauws, P., Fiermans, L. & Vernik, J. (1973) *Surface Science* **36**, 513.

Cox, S. D., Gier, T. E., Stucky, G. D. & Bierlein, J. (1988) *J. Amer. Chem. Soc.* **110**, 2986.

Curtin, D. Y. & Paul, I. C. (1981) *Chem. Rev.* **81**, 525.

Daumas, N. & Herold, A. (1969) *C. R. Acad. Sci. (Paris)* **C268**, 373.

Derouane, E. G. (1982) in *Intercalation Chemistry* (eds Whittingham, M. S. & Jacobson, A. J.) Academic Press, New York.

Desiraju, G. R. (1984a) *Proc. Indian Acad. Sci. (Chem. Sci.)* **93**, 407.

Desiraju, G. R. (1984b) *Endeavour* **8**, 201.

Desiraju, G. R. (1989) *Crystal Engineering: The Design of Organic Solids*, Elsevier, Amsterdam.

Dines, M. B. (1975) *Science* **188**, 210.

Dowden, D. A. (1972) *Catal. Rev.* **5**, 1.

Dwyer, J. (1984) *Chemistry & Industry*, 2 April Issue, p. 258.

Dyer, A. (1984) *Chemistry & Industry*, 2 April Issue, p. 241.

Enzel, P. & Bein, T. (1989) *J. Phys. Chem.* **93**, 6270.

Etter, M. C. (1990) *Acc. Chem. Res.* **23**, 120.

Fehlner, F. P. & Mott, N. F. (1970) *Oxidation of Metals* **2**, 59.

Fister, L., Johnson, D. C. & Brown, R. (1994) *J. Amer. Chem. Soc.* **116**, 629.

Fyfe, C. A., Gobbi, G. C., Klinowski, J., Thomas, J. M. & Ramdas, S. (1982) *Nature* **296**, 530.

Galiotis, C., Read, R. T., Yeung, P. H. J., Young, R. J., Chalmers, I. F. & Bloor, D. (1984) *J. Polymer Sci.* **22**, 1589.

Ganapathi, L., Ramanan, A., Gopalakrishnan, J. & Rao, C. N. R. (1986) *J. Chem. Soc. Chem. Comm.* 62.

Garner, W. E. (ed.) (1955) *Chemistry of the Solid State*, Butterworths, London.

Ginsburg, D. (ed.) (1976) *Solid State Photochemistry*, Verlag Chemie, Weinheim.

Gnanaguru, K., Ramasubbu, N., Venkatesan, K. & Ramamurthy, V. (1985) *J. Org. Chem.* **50**, 2337.

Gomes, W. P. & Dekeyser, W. (1976) in *Treatise on Solid State Chemistry* (ed. Hannay, N. B.) Vol. 4, Plenum Press, New York.

Goodman, D. W. (1984) *Acc. Chem. Res.* **17**, 194.

Gorring, R. L. (1973) *J. Catal.* **31**, 13.

Grasselli, R. K. & Burrington, J. D. (1981) *Advances in Catalysis* **30**, 133.

Grasselli, R. K., Burrington, J. D. & Brazdil, J. F. (1981) *Faraday Discussions of the Chemical Society* **72**, 203.

Grasselli, R. K. & Brazdil, J. F. (1985) *Solid State Chemistry in Catalysis*, American Chemical Society, Washington D.C.

Green, B. S., Lahav, M. & Rabinovich, D. (1979) *Acc. Chem. Res.* **12**, 191.

Gross, H. & Sixl, H. (1982) *Chem. Phys. Lett.* **91**, 262.

Günter, J. R. (1972) *J. Solid State Chem.* **5**, 354.

Guru Row, T. N., Swamy, H. R., Acharya, K. R., Ramamurthy, V., Venkatesan, K. & Rao, C. N. R. (1983) *Tetrahedron Lett.* **24**, 3263.

Gwathmey, A. T. & Lawless, K. R. (1960) in *The Surface Chemistry of Metals and Semiconductors* (ed. Gatos, H. C.) Wiley, New York.

Hagg, W. O., Lago, R. M. & Weisz, P. B. (1984) *Nature* **309**, 589.

Haggin, J. (1982) *Chem. Engg. News* **60**(6), 13.

Halbert, T. R. (1982) in *Intercalation Chemistry* (eds Whittingham, M. S. & Jacobson, A. J.) Academic Press, New York.

Hannay, N. B. (ed.) (1976) *Treatise on Solid State Chemistry*, Vol. 4, Plenum Press, New York.

Hasegawa, M., Iguchi, M. & Nakanishi, H. (1968) *J. Polymer Sci.* **A-1**, 6, 1054.

Hauffe, K. (1966) *Reactionen in und an Festenstoffen*, (2nd Edn) Springer-Verlag, Berlin.

Hauffe, K. (1976) in *Treatise on Solid State Chemistry*, (ed. Hannay, N. B.) Vol. 4, Plenum Press, New York.

Herron, N., Wang, Y., Eddy, M. M., Stucky, G. D., Cox, D. E., Moller, K. & Bein, T. (1989) *J. Amer. Chem. Soc.* **111**, 530.

Horsley, J. A. (1979) *J. Amer. Chem. Soc.* **101**, 2870.

Iguchi, E. & Tilley, R. J. D. (1978) *J. Solid State Chem.* **24**, 121.

Jacobson, A. J. (1982) in *Intercalation Chemistry* (eds Whittingham, M. S. & Jacobson, A. J.) Academic Press, New York.

Jander, W. (1927) *Z. Anorg. Allgem. Chem* **163**, 1.

Jiang, X. Z., Hayden, T. F. & Dumesic, J. A. (1983) *J. Catal.* **83**, 168.

Johnson, J. W. (1982) in *Intercalation Chemistry* (eds. Whittingham, M. S. & Jacobson, A. J.) Academic Press, New York.

Johnson, J. W., Johnston, D. C., Jacobson, A. J. & Brody, J. F. (1984) *J. Amer. Chem. Soc.* **106**, 8123.

Jones, W. & Thomas, J. M. (1979) *Prog. Solid State Chem.* **12**, 101.

Jones, W., Nakanishi, H., Theocharis, C. R. & Thomas, J. M. (1980) *J. Chem. Soc. Chem. Comm.* 610.

Jones, W., Ramdas, S., Theocharis, C. R., Thomas, J. M. & Thomas, N. W. (1981) *J. Phys. Chem.* **85**, 2594.

Jost, W. (1937) *Diffusion und Chemische Reaction in Festenstoffen, Steinkopf-Verlag*, Dresden.

Kearsley, S. K. & Desiraju, G. R. (1985) *Proc. Roy. Soc. London* **A397**, 157.

Ko, C. S. & Gorte, R. J. (1984) *J. Catal.* **90**, 59.

Krylov, O. V. (1970) *Catalysis by Nonmetals*, Academic Press, New York.

Kulkarni, G. U. & Rao, C. N. R. (1991) *Catal. Lett.* **9**, 427.

Lehn, J. M., Mascal, M., DeCian, A. & Fischer, J. (1990) *J. Chem. Soc. Chem. Comm.* 479.

Lehn, J. M., Mascal, M., DeCian, A. & Fischer, J. (1992) *J. Chem. Soc. Perkin Trans.* **2**, 461.

Leiserowitz, L. & Hagler, A. T. (1983) *Proc. Roy. Soc. London* **A388**, 133.

Lem, G., Kaprinidis, N. A., Shuster, D. I., Ghatlia, N. D. & Turro, N. J. (1993) *J. Amer. Chem. Soc.* **115**, 7009.

Levy, F. (ed.) (1979) *Physics and Chemistry of Materials with Layered Structures*, Vol. 6, D. Reidel, Dordrecht.

Lewis, J. & Green, M. L. H. (eds) (1982) *Phil. Trans. Roy. Soc. London* **A308**, 1.

Mady, T. E., Yates, Jr, J. T., Sandstrom, D. R. & Voorhoeve, R. J. H. (1976) in *Treatise on Solid State Chemistry* (ed. Hannay, N. B.) Vol. 6B, Plenum Press, New York.

Mallouk, E., Hawkins, B. L., Conrad, M. P., Zilm, K., Maciel, G. E. & Bartlett, N. (1985) *Phil. Trans. Roy. Soc. London* **A314**, 179.

Manohar, H. (1974) in *Solid States Chemistry* (ed. Rao, C. N. R.) Marcel Dekker, New York.

Marder, S. R., Perry, J. W. & Yakymyshyn, C. P. (1994) *Chem. Mater.* **6**, 1137.

McBridge, J. M. (1983) *Acc. Chem. Res.* **16**, 304.

Mizuno, T., Hattori, S. & Tawata, M. (1977) *J. Opt. Soc. Amer.* **67**, 1651.

Moran, H. F., Ohstuki, M., Hibino, A. & Hough, C. (1971) *Science* **174**, 498.

Morawetz, H. (1977) in *Reactivity of Solids* (eds Woods, J., Lindquist, O., Helgesson, C. & Vannerberg, N-G) Plenum Press, New York.

Muetterties, E. L., Rhodin, T. N., Band, E., Brucker, C. F. & Pretzer, W. R. (1979) *Chem. Rev.* **79**, 91.

Nakanishi, F., Tasai, T. & Hasegawa, M. (1975) *Polymer* **16**, 218.

Nakanishi, H., Hasegawa, M. & Sasada, Y. (1972) *J. Polymer Sci.* **10**, 1537.

Nakanishi, H., Nakanishi, F., Suzuki, Y. & Hasegawa, M. (1973) *J. Polymer Sci.* **11**, 2501.

Nakanishi, H., Jones, W., Thomas, J. M., Hasegawa, M. & Rees, W. L. (1980) *Proc. Roy. Soc. London* **A369**, 307.

Niederwald, H., Richter, K.-H., Güttler, W. & Schwoerer, M. (1983) *Mol. Cryst. Liq. Cryst.* **93**, 247.

Ozin, G. A., Özkar, S. & Prokopowicz, P. A. (1992) *Acc. Chem. Res.* **25**, 553.

Parasnis, A. S. (1970) in *Modern Aspects of Solid State Chemistry* (ed. Rao, C. N. R.) Plenum Press, New York.

Pauling, L. (1949) *Proc. Roy. Soc. London* **A196**, 343.

Pilling, N. B. & Bedworth, R. E. (1923) *J. Inst. Metals* **29**, 529.

Poeppelmeier, K. R. & Kipp, D. O. (1988) *Inorg. Chem.* **27**, 766.

Prabhakaran, K. & Rao, C. N. R. (1985) *Surf. Science*, **163**, L 771.

Rabinovich, D. & Hope, H. (1975) *Acta Crystallogr.* **A31**, S128.

Rabo, J. A. (ed.) (1976) *Zeolite Chemistry and Catalysis*, ACS Monograph No. 171, Amer. Chem. Soc. Washington.

Ramamurthy, V., Eaton, D. F. & Caspar, J. V. (1992) *Acc. Chem. Res.* **25**, 299.

Rangarao, M. & Rao, C. N. R. (1990) *J. Phys. Chem.* **94**, 7986.

Rao, C. N. R. (1984) *Solid State Chemistry: A Perspective*, Golden Jubilee Publication, Indian National Science Academy.

Rao, C. N. R. (1985) *Proc. Ind. Nat. Sci. Acad.* **A51**, 1.

Rao, C. N. R., Kulkarni, G. U., Kannan, K. R. & Chaturvedi, S. (1992) *J. Phys. Chem.* **96**, 7379.

Rees, L. V. C. (1984) *Chemistry and Industry*, 2 April Issue, p. 252.

Sachtler, W. M. H. (1981) *Faraday Discussions of the Chemical Society* **72**, 7.

Sachtler, W. M. H. & Fahrenfort, J. (1960) in *Actes du Deuxieme Congres International de Catalyse*, 831.

Salmeron, M., Gale, R. J. & Somorjai, G. A. (1979) *J. Chem. Phys.* **70**, 2807.

Santos, J., Phillips, J. & Dumesic, J. A. (1983) *J. Catal.* **81**, 147.

Schafer-Stahl, H. & Abele, R. (1980) *Angew. Chem. Int. Edn. (Engl.)* **19**, 477.

Schlögl, R. (1985) in *Proc. 5th Int. Conf. on Rapidly Quenched Metals*, North-Holland, Amsterdam.

Schmalzried, H. (1971) *Solid State Reactions*, Verlag Chemie, Weinheim.

Schmalzried, H. (1976) in *Treatise on Solid State Chemistry* (ed. Hannay, N. B.) Vol. 4, Plenum Press, New York.

Schmidt, G. M. J. (1971) *Pure and Appl. Chem.* **27**, 647.

Schöllhorn, R. (1980) *Angew. Chem. Int. Edn. (Engl.)* **19**, 983.

Schöllhorn, R. (1984) *Pure and Appl. Chem.* **56**, 1739.

Sermon, P. A. & Bond, G. C. (1974) *Catal. Rev.* **8**, 211.

Sinfelt, J. A. (1975) *Prog. Solid State Chem.* **10**, 55.

Sinfelt, J. A. (1977) *Science* **195**, 641.

Singh, N. B., Singh, R. J. & Singh, N. P. (1994) *Tetrahedron* **50**, 6441.

Sleight, A. W. (1977) in *Advanced Materials in Catalysis* (eds. Burton, J. J. & Garten, R. L.) Academic Press, New York.

Sleight, A. W. (1980) *Science* **208**, 895.

Smeltzer, W. W. & Young, D. J. (1975) *Prog. Solid State Chem.* **10**, 17.

Somorjai, G. A. (1981) *Chemistry in Two Dimensions: Surfaces*, Cornell University Press, Ithaca.

Stinson, S. C. (1983) *Chem. Engg. News* **61**(49), 19.

Stucky, G. D. (1992) *Prog. Inorg. Chem.* **40**, 99.

Stucky, G. D. & MacDougall, J. E. (1990) *Science* **247**, 669.

Subba, Rao, G. V. & Shafer, M. W. (1979) in *Physics and Chemistry of Materials with Layered Structures* (ed. Levy, F.) Vol. 6, D. Reidel, Dordrecht.

Sudo, M., Ichikawa, M., Soma, M., Ovisha, T. & Tamaru, K. (1969) *J. Phys. Chem.* **73**, 1174.

Tachez, M., Theobald, F., Bernard, J. & Hewat, A. W. (1982) *Rev. Chim. Miner.* **19**, 291.

Takatani, S. & Chung, Y. W. (1984) *J. Catalysis* **90**, 75.

Tammann, G. (1920) *Z. Anorg. Allgem. Chem.* **111**, 78.

Tauster, S. J., Fung, S. C. & Garten, R. L. (1978) *J. Amer. Chem. Soc.* **100**, 170.

Tauster, S. J., Fung, S. C., Barker, R. T. K. & Horsley, J. A. (1981) *Science* **211**, 1121.

Taylor, H. S. (1952) *Proc. Roy. Soc. London* **A108**, 105.

Theocharis, C. R., Desiraju, G. R. & Jones, W. (1984) *J. Amer. Chem. Soc.* **106**, 3606.

Thomas, J. M. (1967) *J. Chem. Soc. Chem. Comm.* 432.

Thomas, J. M. (1974) *Phil. Trans. Roy. Soc. London* **A277**, 251.

Thomas, J. M. (1979) *Pure and Appl. Chem.* **51**, 1065.

Thomas, J. M. (1982) in *Intercalation Chemistry* (eds Whittingham, M. S. and Jacobson, A. J.) Academic Press, New York.

Thomas, J. M. (1983) in *Proceedings of the NATO DAVY ASI*, Cambridge, UK. *Physics and Chemistry of Electrons and Ions in Condensed Matter.*

Thomas, J. M. (1994) *Angew. Chem. Int. Ed. (Engl.)* **33**, 913.

Thomas, J. M., Evans, E. L. & Clarke, T. A. (1971) *J. Chem. Soc.* A 2338.

Thomas, J. M. & Williams, J. O. (1971) *Prog. Solid State Chem.* **6**, 119.

Thomas, J. M. & Lambert, R. M. (eds) (1980) *Characterization of Catalysts*, John Wiley, Chichester.

Thomas, J. M., Millward, G. R., Schlögl, R. F. & Boechm, H. P. (1980) *Mater. Res. Bull.* **15**, 671.

Thomas, J. M. & Zamaraev, K. I. (eds) (1992) *Perspectives in Catalysis*, Blackwell, Oxford.

Thomas, N. W., Ramdas, S. & Thomas, J. M. (1985) *Proc. Roy. Soc. London* **A400**, 219.

Thompson, A. H. (1981) *Physica (B + C)* (Amsterdam) **105B**, 461.

Toda, F., Hyoda, S., Okada, K. & Hirotsu, K. (1995) *JCS Chem. Commun.* 1531.

Venkataraman, D., Lee, S., Zhang, J. & Moore, J. S. (1994) *Nature* **371**, 591.

Wagner, C. (1936) *Z. Phys. Chem.* **32**, 447.

Wagner, C. (1938) *Z. Anorg. Allgem. Chem.* **236**, 320.

Wagner, C. (1975) *Prog. Solid State Chem.* **10**, 3.

Wagner, C. & Grünewald, K. (1938) *Z. Phys. Chem.* **40**, 455.

Wegner, G. (1971) *Makromol. Chem.* **154**, 35.

Wegner, G. (1979) in *Molecular Metals* (ed. Hatfield, W. E.) Plenum Press, New York.

Whittingham, M. S. (1978) *Prog. Solid State Chem.* **12**, 41.

Whittingham, M. S. & Chianelli, R. R. (1980) *J. Chem. Educ.* **57**, 569.

Whittingham, M. S. & Jacobson, A. J. (eds) (1982) *Intercalation Chemistry*, Academic Press, New York.

Wolkenstein, Th. (1960) *Advances in Catalysis* **12**, 189.

Wright, P. A., Jones, R. H., Natarajan, S., Bell, R. G., Chen, J., Hursthouse, M. B. & Thomas, J. M. (1993) *J. Chem. Soc. Chem. Comm.* 633.

Wu, C. G. & Bein, T. (1994a) *Science* **264**, 1757.

Wu, C. G. & Bein, T. (1994b) *Science* **266**, 1013.

Zerkowski, J. A., MacDonald, J. C., Seto, C. T., Wierda, D. A. & Whitesides, G. M. (1994) *J. Amer. Chem. Soc.* **116**, 2382.

Index

Numbers in **bold face** refer to the first page number of a sub-section or a section dealing with the subject. The symbol Ln refers to lanthanide.